Multiagent Systems

Multiagent systems combine multiple autonomous entities, each having diverging interests or different information. This comprehensive overview of the field offers a computer science perspective but also draws on ideas from game theory, economics, operations research, logic, philosophy, and linguistics. It will serve as a reference for researchers in each of these fields and be used as a text for advanced undergraduate or graduate courses.

The authors emphasize foundations to create a broad and rigorous treatment of their subject, with thorough presentations of distributed problem solving, noncooperative game theory, multiagent communication and learning, social choice, mechanism design, auctions, cooperative game theory, and modal logics of knowledge and belief. For each topic, basic concepts are introduced, examples are given, proofs of key results are offered, and algorithmic considerations are examined. An appendix covers background material in probability theory, classical logic, Markov decision processes, and mathematical programming.

Yoav Shoham is a professor of computer science at Stanford University.

Kevin Leyton-Brown is an associate professor of computer science at the University of British Columbia.

T0332638

Multiagent Systems

Algorithmic, Game-Theoretic, and Logical Foundations

YOAV SHOHAM

Stanford University

KEVIN LEYTON-BROWN

University of British Columbia

CAMBRIDGE
UNIVERSITY PRESS

32 Avenue of the Americas, New York NY 10013-2473, USA

Cambridge University Press is part of the University of Cambridge.

It furthers the University's mission by disseminating knowledge in the pursuit of education, learning and research at the highest international levels of excellence.

www.cambridge.org
Information on this title: www.cambridge.org/9780521899437

© Yoav Shoham and Kevin Leyton-Brown 2009

First published 2009

A catalogue record for this publication is available from the British Library

Library of Congress Cataloguing in Publication data

Shoham, Yoav.
Multiagent systems : algorithmic, game-theoretic, and logical
foundations / Yoav Shoham, Kevin Leyton-Brown.
p.   cm.
Includes index.
ISBN 978-0-521-89943-7 (hardback)
1. Intelligent agents (Computer software)   2. Electronic data processing – Distributed
processing.   I. Leyton-Brown, Kevin, 1975–   II. Title.
QA76.76.I58S75   2008
006.3 – dc22   2008012063

ISBN 978-0-521-89943-7 Hardback

To my wife, Noa, and my daughters, Maia, Talia, and Ella

—YS

To Jude

—KLB

Contents

Credits and Acknowledgments

We should start off by explaining the order of authorship. Yoav conceived of the project and started it, in late 2001, working on it alone and with several colleagues (see below). Sometime in 2004 Yoav realized he needed help if the project were ever to come to conclusion, and he enlisted the help of Kevin. The result was a true partnership and a complete overhaul of the material. The current book is vastly different from the draft that existed when the partnership was formed—in depth, breadth, and form. Yoav and Kevin have made equal contributions to the book; the order of authorship reflects the history of the book, but nothing else.

In six years of book-writing we accumulated many debts. The following is our best effort to acknowledge those. If we omit any names, it is due solely to our poor memories and record keeping, and we apologize in advance.

When the book started out, Teg Grenager served as a prolific ghost writer. While little of the original writing remains (though some does, for example, in Section 8.3.1 on speech acts), the project would not have gotten off the ground without him.

Several past and present graduate students made substantial contributions. Chapter 12 (coalitional games) is based entirely on writing by Sam Ieong, who was also closely involved in the editing. Section 3.3.4 (the existence of Nash equilibria) and parts of Section 6.5 (compact game representations) are based entirely on writing by Albert Xin Jiang, who also worked extensively with us to refine the material. Albert also contributed to the proof of Theorem 3.4.4 (the minmax theorem). Some of the material in Chapter 4 on computing solution concepts is based on writing by Ryan Porter, who also contributed much of the material in Section 6.1.3 (bounded rationality). The material in Chapter 7 (multiagent learning) is based in part on joint work with Rob Powers, who also contributed text. Section 10.6.4 (mechanisms for matching) is based entirely on text by Baharak Rastegari, and David R. M. Thompson contributed material to Sections 10.6.3 (mechanisms for multicast routing) and 6.3.4 (*ex post* equilibria). Finally, all of the past and present students listed here offered invaluable comments on drafts. Other students also offered valuable comments. Samantha Leung deserves special mention; we also received useful feedback from Michael Cheung, Matthew Chudek, Farhad Ghassemi, Ryan Golbeck, James Wright, and Erik Zawadzki. We apologize in advance to any others whose names we have missed.

Several of our colleagues generously contributed material to the book, in addition to lending their insight. They include Geoff Gordon (Matlab code to generate Figure 3.13, showing the saddle point for zero-sum games), Carlos Guestrin (material on action selection in distributed MDPs in Section 2.2, and Figure 1.1, showing a deployed sensor network), Michael Littman (Section 5.1.4 on computing all subgame-perfect equilibria), Amnon Meisels (much of the material on heuristic distributed constraint satisfaction in Chapter 1), Marc Pauly (material on coalition logic in Section 14.3), Christian Shelton (material on computing Nash equilibria for n-player games in Section 4.3), and Moshe Tennenholtz (material on restricted mechanism design in Section 10.7). We thank Éva Tardos and Tim Roughgarden for making available notes that we drew on for our proofs of Lemma 3.3.14 (Sperner's lemma) and Theorem 3.3.21 (Brouwer's fixed-point theorem for simplotopes), respectively.

Many colleagues around the world generously gave us comments on drafts, or provided counsel otherwise. Felix Brandt and Vince Conitzer deserve special mention for their particularly detailed and insightful comments. Other colleagues to whom we are indebted include Alon Altman, Krzysztof Apt, Navin A. R. Bhat, Ronen Brafman, Yiling Chen, Yossi Feinberg, Jeff Fletcher, Nando de Freitas, Raul Hakli, Joe Halpern, Jason Hartline, Jean-Jacques Herings, Ramesh Johari, Bobby Kleinberg, Daphne Koller, Fangzhen Lin, David Parkes, David Poole, Maurice Queyranne, Tim Roughgarden, Tuomas Sandholm, Peter Stone, Nikos Vlasis, Mike Wellman, Bob Wilson, Mike Wooldridge, and Dongmo Zhang.

Many others pointed out errors in the first printing of the book through our errata wiki: B. J. Buter, Nicolas Dudebout, Marco Guazzone, Joel Kammet, Nicolas Lambert, Nimalan Mahendran, Mike Rogers, Ivomar Brito Soares, Michael Styer, Sean Sutherland, Grigorios Tsoumakas, Steve Wolfman, and James Wright.

Several people provided critical editorial and production assistance of various kinds. Most notably, David R. M. Thompson overhauled our figures, code formatting, bibliography, and index. Chris Manning was kind enough to let us use the LaTeX macros from his own book, and Ben Galin added a few miracles of his own. Ben also composed several of the examples, found some bugs, drew many figures, and more generally for two years served as an intelligent jack-of-all-trades on this project. Erik Zawadzki helped with the bibliography and with some figures. Maia Shoham helped with some historical notes and bibliography entries, as well as with some copy-editing.

We thank all these friends and colleagues. Their input has contributed to a better book, but of course they are not to be held accountable for any remaining shortcomings. We claim sole credit for those.

We also thank Cambridge University Press for publishing the book, and for their enlightened online-publishing policy, which has enabled us to provide the broadest possible access to it. Specific thanks to Lauren Cowles, an editor of unusual intelligence, good judgment, and sense of humor.

Last, and certainly not the least, we thank our families, for supporting us through this time-consuming project. We dedicate this book to them, with love.

Introduction

Imagine a personal software agent engaging in electronic commerce on your behalf. Say the task of this agent is to track goods available for sale in various online venues over time, and to purchase some of them on your behalf for an attractive price. In order to be successful, your agent will need to embody your preferences for products, your budget, and in general your knowledge about the environment in which it will operate. Moreover, the agent will need to embody your knowledge of other similar agents with which it will interact (e.g., agents who might compete with it in an auction or agents representing store owners)—including their own preferences and knowledge. A collection of such agents forms a multiagent system. The goal of this book is to bring under one roof a variety of ideas and techniques that provide foundations for modeling, reasoning about, and building multiagent systems.

Somewhat strangely for a book that purports to be rigorous, we will not give a precise definition of a multiagent system. The reason is that many competing, mutually inconsistent answers have been offered in the past. Indeed, even the seemingly simpler question—What is a (single) agent?—has resisted a definitive answer. For our purposes, the following loose definition will suffice: Multiagent systems are those systems that include multiple autonomous entities with either diverging information or diverging interests, or both.

Scope of the book

The motivation for studying multiagent systems often stems from interest in artificial (software or hardware) agents, for example software agents living on the Internet. Indeed, the Internet can be viewed as the ultimate platform for interaction among self-interested, distributed computational entities. Such agents can be trading agents of the sort discussed above, "interface agents" that facilitate the interaction between the user and various computational resources (including other interface agents), game-playing agents that assist (or replace) human players in a multiplayer game, or autonomous robots in a multi-robot environment. However, although the material is written by computer scientists with computational sensibilities, it is quite interdisciplinary, and the material is in general fairly abstract. Many of the ideas apply to—and indeed are often taken from—inquiries about human individuals and institutions.

The material spans disciplines as diverse as computer science (including artificial intelligence, theory, and distributed systems), economics (chiefly microeconomic theory), operations research, analytic philosophy, and linguistics. The technical material includes logic, probability theory, game theory, and optimization. Each of the topics covered easily supports multiple independent books and courses, and this book does not aim to replace them. Rather, the goal has been to gather the most important elements from each discipline and weave them together into a balanced and accurate introduction to this broad field. The intended reader is a graduate student or an advanced undergraduate, prototypically, but not necessarily, in computer science.

Because the umbrella of multiagent systems is so broad, the questions of what to include in any book on the topic and how to organize the selected material are crucial. To begin with, this book concentrates on foundational topics rather than surface applications. Although we will occasionally make reference to real-world applications, we will do so primarily to clarify the concepts involved; this is despite the practical motivations professed earlier. And so this is the wrong text for the reader interested in a practical guide to building this or that sort of software. The emphasis is rather on important concepts and the essential mathematics behind them. The intention is to delve in enough detail into each topic to be able to tackle some technical material, and then to point the reader in the right directions for further education on particular topics.

Our decision was thus to include predominantly established, rigorous material that is likely to withstand the test of time, and to emphasize computational perspectives where appropriate. This still left us with vast material from which to choose. In understanding the selection made here, it is useful to keep in mind the following keywords: *coordination*, *competition*, *algorithms*, *game theory*, and *logic*. These terms will help frame the chapter overview that follows.

Overview of the chapters

Starting with issues of coordination, we begin in **Chapter 1** and **Chapter 2** with distributed problem solving. In these multiagent settings there is no question of agents' individual preferences; there is some global problem to be solved, but for one reason or another it is either necessary or advantageous to distribute the task among multiple agents, whose actions may require coordination. These chapters are thus strongly algorithmic. The first one looks at distributed constraint-satisfaction problems. The latter addresses distributed optimization and specifically examines four algorithmic methods: distributed dynamic programming, action selection in distributed MDPs, auction-like optimization procedures for linear and integer programming, and social laws.

We then begin to embrace issues of competition as well as coordination. Whereas the area of multiagent systems is not synonymous with game theory, there is no question that game theory is a key tool to master within the field, and so we devote several chapters to it. **Chapters 3, 5**, and **6** constitute a crash course in noncooperative game theory. They cover, respectively, the normal form, the extensive form, and a host of other game representations. In these chapters, as in others that draw on game theory, we culled the material that in our judgment is needed in order to be a knowledgeable consumer of modern-day game theory. Unlike traditional game theory texts, we also include discussion of algorithmic considerations. In the context of the normal-form representation, that material is sufficiently substantial to warrant its own chapter, **Chapter 4**.

We then switch to two specialized topics in multiagent systems. In **Chapter 7** we cover multiagent learning. The topic is interesting for several reasons. First, it is a key facet of multiagent systems. Second, the very problems addressed in the area are diverse and sometimes ill understood. Finally, the techniques used, which draw equally on computer science and game theory (as well as some other disciplines), are not straightforward extensions of learning in the single-agent case.

In **Chapter 8** we cover another element unique to multiagent systems: communication. We cover communication in a game-theoretic setting, as well as in cooperative settings traditionally considered by linguists and philosophers (except that we see that there too a game-theoretic perspective can creep in).

Next is a three-chapter sequence that might be called "protocols for groups." **Chapters 9** covers social-choice theory, including voting methods. This is a nonstrategic theory, in that it assumes that the preferences of agents are known, and the only question is how to aggregate them properly. **Chapter 10** covers mechanism design, which looks at how such preferences can be aggregated by a central designer even when agents *are* strategic. Finally, **Chapter 11** looks at the special case of auctions.

Chapter 12 covers coalitional game theory, in recent times somewhat neglected within game theory and certainly underappreciated in computer science.

The material in Chapters 1–12 is mostly Bayesian and/or algorithmic in nature. And thus the tools used in them include probability theory, utility theory, algorithms, Markov decision problems (MDPs), and linear/integer programming. We conclude with two chapters on logical theories in multiagent systems. In **Chapter 13** we cover modal logic of knowledge and belief. This material hails from philosophy and computer science, but it turns out to dovetail very nicely with the discussion of Bayesian games in Chapter 6. Finally, in **Chapter 14** we extend the discussion in several directions—we discuss how beliefs change over time, logical models of games, and how one might begin to use logic to model motivational attitudes (such as "intention") in addition to the informational ones (knowledge, belief).

Required background

The book is rigorous and requires mathematical thinking, but only basic background knowledge. In much of the book we assume knowledge of basic computer science (algorithms, complexity) and basic probability theory. In more technical parts we assume familiarity with Markov decision problems (MDPs), mathematical programming (specifically, linear and integer programming), and classical logic. All of these (except basic computer science) are covered briefly in **appendices**, but those are meant as refreshers and to establish notation, not as a substitute for background in those subjects. This is true in particular of probability theory. However, above all, a prerequisite is a capacity for clear thinking.

How to teach (and learn) from this book

There are partial dependencies among the 13 chapters. To understand them, it is useful to think of the book as consisting of the following "blocks."

- **Block 1**, Chapters 1–2: Distributed problem solving
- **Block 2**, Chapters 3–6: Noncooperative game theory
- **Block 3**, Chapter 7: Learning
- **Block 4**, Chapter 8: Communication
- **Block 5**, Chapters 9–11: Protocols for groups
- **Block 6**, Chapter 12: Coalitional game theory
- **Block 7**, Chapters 13–14: Logical theories

Within every block there is a sequential dependence (except within Block 1, in which the sections are largely independent of each other). Among the blocks, however, there is only one strong dependence: Blocks 3, 4, and 5 each depend on some elements of noncooperative game theory and thus on block 2 (though none requires the entire block). Otherwise there are some interesting local pairwise connections between blocks, but none that require that both blocks be covered, whether sequentially or in parallel.

Given this weak dependence among the chapters, there are many ways to craft a course out of the material, depending on the background of the students, their interests, and the time available. The book's Web site

```
http://www.masfoundations.org
```

contains several specific syllabi that have been used by us and other colleagues, as well as additional resources for both students and instructors.

On pronouns and gender

We use male pronouns to refer to agents throughout the book. We debated this between us, not being happy with any of the alternatives. In the end we reluctantly settled on the "standard" male convention rather than the reverse female convention or the grammatically dubious "they." We urge the reader not to read patriarchal intentions into our choice.

1

Distributed Constraint Satisfaction

In this chapter and the next we discuss cooperative situations in which agents collaborate to achieve a common goal. This goal can be viewed as shared between the agents or, alternatively, as the goal of a central designer who is designing the various agents. Of course, if such a designer exists, a natural question is why it matters that there are multiple agents; they can be viewed merely as end sensors and effectors for executing the plan devised by the designer. However, there exist situations in which a problem needs to be solved in a distributed fashion, either because a central controller is not feasible or because one wants to make good use of the distributed resources. A good example is provided by
sensor network *sensor networks*. Such networks consist of multiple processing units, each with local sensor capabilities, limited processing power, limited power supply, and limited communication bandwidth. Despite these limitations, these networks aim to provide some global service. Figure 1.1 shows an example of a fielded sensor network used for monitoring environmental quantities like humidity, temperature and pressure in an office environment. Each sensor can monitor only its local area and, similarly, can communicate only with other sensors in its local vicinity. The question is what algorithm the individual sensors should run so that the center can still piece together a reliable global picture.

Distributed algorithms have been widely studied in computer science. We concentrate on distributed problem-solving algorithms of the sort studied in artificial intelligence. We divide the discussion into two parts. In this chapter we cover distributed constraint satisfaction, where agents attempt in a distributed fashion to find a feasible solution to a problem with global constraints. In the next chapter we look at agents who try not only to satisfy constraints, but also to optimize some objective function subject to these constraints.

Later in this book we will encounter additional examples of distributed problem solving. Each of them requires specific background, however, which is why they are not discussed here. Two of them stand out in particular.

- In Chapter 7 we encounter a family of techniques that involve learning, some of them targeted at purely cooperative situations. In these situations the agents learn through repeated interactions how to coordinate a choice of action. This material requires some discussion of noncooperative game theory (discussed in Chapter 3) as well as general discussion of multiagent learning (discussed in Chapter 7).

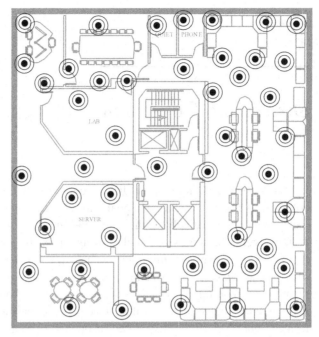

Figure 1.1 Part of a real sensor network used for indoor environmental monitoring.

- In Chapter 13 we discuss the use of logics of knowledge (introduced in that chapter) to establish the knowledge conditions required for coordination, including an application to distributed control of multiple robots.

1.1 Defining distributed constraint satisfaction problems

constraint
satisfaction
problem (CSP)

A *constraint satisfaction problem (CSP)* is defined by a set of variables, domains for each of the variables, and constraints on the values that the variables might take on simultaneously. The role of constraint satisfaction algorithms is to assign values to the variables in a way that is consistent with all the constraints, or to determine that no such assignment exists.

Constraint satisfaction techniques have been applied in diverse domains, including machine vision, natural language processing, theorem proving, and planning and scheduling, to name but a few. Here is a simple example taken from the domain of sensor networks. Figure 1.2 depicts a three-sensor snippet from the scenario illustrated in Figure 1.1. Each of the sensors has a certain radius that, in combination with the obstacles in the environment, gives rise to a particular coverage area. These coverage areas are shown as ellipses in Figure 1.2. As you can see, some of the coverage areas overlap. We consider a specific problem in this setting. Suppose that each sensor can choose one of three possible radio frequencies. All the frequencies work equally well so long as no two sensors with overlapping coverage areas use the same frequency. The question is which

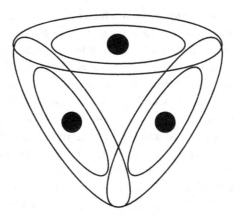

Figure 1.2 A simple sensor net problem.

algorithms the sensors should employ to select their frequencies, assuming that this decision cannot be made centrally.

The essence of this problem can be captured as a graph-coloring problem. Figure 1.3 shows such a graph, corresponding to the sensor network CSP above. The nodes represent the individual units; the different frequencies are represented by colors; and two nodes are connected by an undirected edge if and only if the coverage areas of the corresponding sensors overlap. The goal of graph coloring is to choose one color for each node so that no two adjacent nodes have the same color.

Formally speaking, a CSP consists of a finite set of variables $X = \{X_1, \ldots, X_n\}$, a domain D_i for each variable X_i, and a set of constraints $\{C_1, \ldots, C_m\}$. Although in general CSPs allow infinite domains, we assume here that all the domains are finite. In the graph-coloring example above there were three variables, and they each had the same domain, {*red, green, blue*}. Each constraint is a predicate on some subset of the variables, say, $X_{i_1}, \ldots, X_{i_j}$; the predicate defines a relation that is a subset of the Cartesian product $D_{i_1} \times \cdots \times D_{i_j}$. Each such constraint restricts the values that may be simultaneously assigned to the variables participating in the constraint. In this chapter we restrict the discussion to

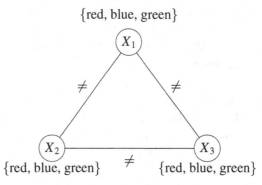

Figure 1.3 A graph-coloring problem equivalent to the sensor net problem of Figure 1.2.

binary constraints, each of which constrains exactly two variables. For example, in the map-coloring case, each "not-equal" constraint applied to two nodes.

Given a subset S of the variables, an *instantiation of S* is an assignment of a unique domain value for each variable in S; it is *legal* if it does not violate any constraint that mentions only variables in S. A *solution* to a network is a legal instantiation of all variables. Typical tasks associated with constraint networks are to determine whether a solution exists, to find one or all solutions, to determine whether a legal instantiation of some of the variables can be extended to a solution, and so on. We will concentrate on the most common task, which is to find one solution to a CSP, or to prove that none exists.

In a *distributed* CSP, each variable is owned by a different agent. The goal is still to find a global variable assignment that meets the constraints, but each agent decides on the value of his own variable with relative autonomy. While he does not have a global view, each agent can communicate with his neighbors in the constraint graph. A distributed algorithm for solving a CSP has each agent engage in some protocol that combines local computation with communication with his neighbors. A good algorithm ensures that such a process terminates with a legal solution (or with a realization that no legal solution exists) and does so quickly.

We discuss two types of algorithms. Algorithms of the first kind embody a least-commitment approach and attempt to rule out impossible variable values without losing any possible solutions. Algorithms of the second kind embody a more adventurous spirit and select tentative variable values, backtracking when those choices prove unsuccessful. In both cases we assume that the communication between neighboring nodes is perfect, but nothing about its timing; messages can take more or less time without rhyme or reason. We do assume, however, that if node i sends multiple messages to node j, those messages arrive in the order in which they were sent.

1.2   Domain-pruning algorithms

Under domain-pruning algorithms, nodes communicate with their neighbors in order to eliminate values from their domains. We consider two such algorithms. In the first, the *filtering algorithm*, each node communicates its domain to its neighbors, eliminates from its domain the values that are not consistent with the values received from the neighbors, and the process repeats. Specifically, each node x_i with domain D_i repeatedly executes the procedure **Revise**(x_i, x_j) for each neighbor x_j.

filtering algorithm

> **procedure** Revise(x_i, x_j)
> **forall** $v_i \in D_i$ **do**
> > **if** *there is no value $v_j \in D_j$ such that v_i is consistent with v_j* **then**
> > > delete v_i from D_i

arc consistency

The process, known also under the general term *arc consistency*, terminates when no further elimination takes place, or when one of the domains becomes

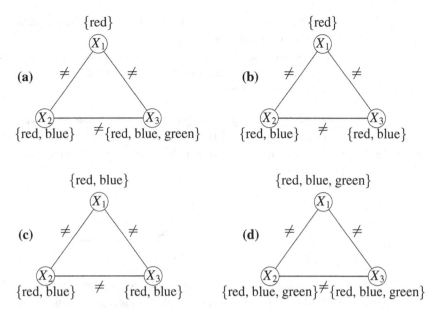

Figure 1.4 A family of graph coloring problems

empty (in which case the problem has no solution). If the process terminates with one value in each domain, that set of values constitutes a solution. If it terminates with multiple values in each domain, the result is inconclusive; the problem might or might not have a solution.

Clearly, the algorithm is guaranteed to terminate, and furthermore it is sound (in that if it announces a solution, or announces that no solution exists, it is correct), but it is not complete (i.e., it may fail to pronounce a verdict). Consider, for example, the family of very simple graph-coloring problems shown in Figure 1.4. (Note that problem (d) is identical to the problem in Figure 1.3.)

In this family of CSPs the three variables (i.e., nodes) are fixed, as are the "not-equal" constraints between them. What are not fixed are the domains of the variables. Consider the four instances of Figure 1.4.

(a) Initially, as the nodes communicate with one another, only x_1's messages result in any change. Specifically, when either x_2 or x_3 receive x_1's message they remove *red* from their domains, ending up with $D_2 = \{blue\}$ and $D_3 = \{blue, green\}$. Then, when x_2 communicates his new domain to x_3, x_3 further reduces his domain to $\{green\}$. At this point no further changes take place and the algorithm terminates with a correct solution.

(b) The algorithm starts as before, but once x_2 and x_3 receive x_1's message they each reduce their domains to $\{blue\}$. Now, when they update each other on their new domains, they each reduce their domains to $\{\}$, the empty set. At this point the algorithm terminates and correctly announces that no solution exists.

(c) In this case the initial set of messages yields no reduction in any domain. The algorithm terminates, but all the nodes have multiple values remaining.

And so the algorithm is not able to show that the problem is overconstrained and has no solution.

(d) Filtering can also fail when a solution exists. For similar reasons as in instance (c), the algorithm is unable to show that in this case the problem *does* have a solution.

In general, filtering is a very weak method and, at best, is used as a preprocessing step for more sophisticated methods. The algorithm is directly based on the notion of *unit resolution* from propositional logic. Unit resolution is the following inference rule:

$$A_1$$
$$\neg(A_1 \wedge A_2 \wedge \cdots \wedge A_n)$$
$$\overline{\hspace{3cm}}$$
$$\neg(A_2 \wedge \cdots \wedge A_n)$$

To see how the filtering algorithm corresponds to unit resolution, we must first write the constraints as forbidden value combinations, called *Nogoods*. For example, the constraint that x_1 and x_2 cannot both take the value "red" would give rise to the propositional sentence $\neg(x_1 = red \wedge x_2 = red)$, which we write as the Nogood $\{x_1, x_2\}$. In instance (b) of Figure 1.4, agent X_2 updated his domain based on agent X_1's announcement that $x_1 = red$ and the Nogood $\{x_1 = red, x_2 = red\}$.

$$x_1 = red$$
$$\neg(x_1 = red \wedge x_2 = red)$$
$$\overline{\hspace{3cm}}$$
$$\neg(x_2 = red)$$

Unit resolution is a weak inference rule, and so it is not surprising that the filtering algorithm is weak as well. *Hyper-resolution* is a generalization of unit resolution and has the following form:

$$A_1 \vee A_2 \vee \cdots \vee A_m$$
$$\neg(A_1 \wedge A_{1,1} \wedge A_{1,2} \wedge \cdots)$$
$$\neg(A_2 \wedge A_{2,1} \wedge A_{2,2} \wedge \cdots)$$
$$\vdots$$
$$\neg(A_m \wedge A_{m,1} \wedge A_{m,2} \wedge \cdots)$$
$$\overline{\hspace{4cm}}$$
$$\neg(A_{1,1} \wedge \cdots \wedge A_{2,1} \wedge \cdots \wedge A_{m,1} \wedge \cdots)$$

Hyper-resolution is both sound and complete for propositional logic, and indeed it gives rise to a complete distributed CSP algorithm. In this algorithm, each agent repeatedly generates new constraints for his neighbors, notifies them of these new constraints, and prunes his own domain based on new constraints passed to him by his neighbors. Specifically, he executes the following algorithm, where NG_i is the set of all Nogoods of which agent i is aware and NG_j^* is a set of new Nogoods communicated from agent j to agent i.

[margin notes:] unit resolution · Nogood · hyper-resolution

procedure ReviseHR(NG_i, NG_j^*)

repeat

$\quad NG_i \leftarrow NG_i \bigcup NG_j^*$

$\quad$ let NG_i^* denote the set of new Nogoods that i can derive from NG_i and his domain using hyper-resolution

$\quad$ **if** NG_i^* *is nonempty* **then**

$\quad\quad NG_i \leftarrow NG_i \bigcup NG_i^*$

$\quad\quad$ send the Nogoods NG_i^* to all neighbors of i

$\quad\quad$ **if** $\{\} \in NG_i^*$ **then**

$\quad\quad\quad$ stop

until *there is no change in i's set of Nogoods NG_i*

The algorithm is guaranteed to converge in the sense that after sending and receiving a finite number of messages, each agent will stop sending messages and generating Nogoods. Furthermore, the algorithm is complete. The problem has a solution iff, on completion, no agent has generated the empty Nogood. (Obviously, every superset of a Nogood is also forbidden, and thus if a single node ever generates an empty Nogood then the problem has no solution.)

Consider again instance (c) of the CSP problem in Figure 1.4. In contrast to the filtering algorithm, the hyper-resolution-based algorithm proceeds as follows. Initially, x_1 maintains four Nogoods—$\{x_1 = red, x_2 = red\}$, $\{x_1 = red, x_3 = red\}$, $\{x_1 = blue, x_2 = blue\}$, $\{x_1 = blue, x_3 = blue\}$—which are derived directly from the constraints involving x_1. Furthermore, x_1 must adopt one of the values in his domain, so $x_1 = red \lor x_1 = blue$. Using hyper-resolution, x_1 can reason:

$$x_1 = red \lor x_1 = blue$$
$$\neg(x_1 = red \land x_2 = red)$$
$$\neg(x_1 = blue \land x_3 = blue)$$
$$\overline{}$$
$$\neg(x_2 = red \land x_3 = blue)$$

Thus, x_1 constructs the new Nogood $\{x_2 = red, x_3 = blue\}$; in a similar way he can also construct the Nogood $\{x_2 = blue, x_3 = red\}$. x_1 then sends both Nogoods to his neighbors x_2 and x_3. Using his domain, an existing Nogood and one of these new Nogoods, x_2 can reason:

$$x_2 = red \lor x_2 = blue$$
$$\neg(x_2 = red \land x_3 = blue)$$
$$\neg(x_2 = blue \land x_3 = blue)$$
$$\overline{}$$
$$\neg(x_3 = blue)$$

Using the other new Nogood from x_1, x_2 can also construct the Nogood $\{x_3 = red\}$. These two singleton Nogoods are communicated to x_3 and allow him to generate the empty Nogood. This proves that the problem does not have a solution.

This example, while demonstrating the greater power of the hyper-resolution-based algorithm relative to the filtering algorithm, also exposes its weakness; the number of Nogoods generated can grow to be unmanageably large. (Indeed, we only described the minimal number of Nogoods needed to derive the empty Nogood; many others would be created as all the agents processed each other's messages in parallel. Can you find an example?) Thus, the situation in which we find ourselves is that we have one algorithm that is too weak and another that is impractical. The problem lies in the least-commitment nature of these algorithms; they are restricted to removing only provably impossible value combinations. The alternative to such "safe" procedures is to explore a subset of the space, making tentative value selections for variables, and backtracking when necessary. This is the topic of the next section. However, the algorithms we have just described are not irrelevant; the filtering algorithm is an effective preprocessing step, and the algorithm we discuss next is based on the hyper-resolution-based algorithm.

1.3 Heuristic search algorithms

A straightforward *centralized* trial-and-error solution to a CSP is to first order the variables (e.g., alphabetically). Then, given the ordering $x_1, x_2, \ldots, x_n$, invoke the procedure ChooseValue(x_1, {}). The procedure ChooseValue is defined recursively as follows, where $\{v_1, v_2, \ldots, v_{i-1}\}$ is the set of values assigned to variables $x_1, \ldots, x_{i-1}$.

> **procedure** ChooseValue(x_i, $\{v_1, v_2, \ldots, v_{i-1}\}$)
> $v_i \leftarrow$ value from the domain of x_i that is consistent with $\{v_1, v_2, \ldots, v_{i-1}\}$
> **if** *no such value exists* **then**
> | backtrack[1]
> **else if** $i = n$ **then**
> | stop
> **else**
> | ChooseValue(x_{i+1}, $\{v_1, v_2, \ldots, v_i\}$)

chronological backtracking

This exhaustive search of the space of assignments has the advantage of completeness. But it is "distributed" only in the uninteresting sense that the different agents execute sequentially, mimicking the execution of a centralized algorithm.

The following attempt at a distributed algorithm has the opposite properties; it allows the agents to execute in parallel and asynchronously, is sound, but is not complete. Consider the following naive procedure, executed by all agents in parallel and asynchronously.

---

1. There are various ways to implement the backtracking in this procedure. The most straightforward way is to undo the choices made thus far in reverse chronological order, a procedure known as *chronological backtracking*. It is well known that more sophisticated backtracking procedures can be more efficient, but that does not concern us here.

select a value from your domain
repeat
| **if** *your current value is consistent with the current values of your*
| *neighbors, or if none of the values in your domain are consistent with*
| *them* **then**
| | do nothing
| **else**
| | select a value in your domain that is consistent with those of your
| | neighbors and notify your neighbors of your new value
until *there is no change in your value*

Clearly, when the algorithm terminates because no constraint violations have occurred, a solution has been found. But in all other cases, all bets are off. If the algorithm terminates because no agent can find a value consistent with those of his neighbors, there might still be a consistent global assignment. And the algorithm may never terminate even if there is a solution. For example, consider example (d) of Figure 1.4: if every agent cycles sequentially between red, green, and blue, the algorithm will never terminate.

ABT algorithm   We have given these two straw-man algorithms for two reasons. Our first reason is to show that reconciling true parallelism and asynchrony with soundness and completeness is likely to require somewhat complex algorithms. And second, the fundamental heuristic algorithm for distributed CSPs—the asynchronous backtracking (or ABT) algorithm—shares much with the two algorithms. From the first algorithm it borrows the notion of a global total ordering on the agents. From the second it borrows a message-passing protocol, albeit a more complex one, which relies on the global ordering. We will describe the ABT in its simplest form. After demonstrating it on an extended example, we will point to ways in which it can be improved upon.

1.3.1 *The asynchronous backtracking algorithm*

As we said, the asynchronous backtracking (ABT) algorithm assumes a total ordering (the "priority order") on the agents. Each binary constraint is known to both the constrained agents and is checked in the algorithm by the agent with the lower priority between the two. A link in the constraint network is always directed from an agent with higher priority to an agent with lower priority.

Agents instantiate their variables concurrently and send their assigned values to the agents that are connected to them by outgoing links. All agents wait for and respond to messages. After each update of his assignment, an agent sends his new assignment along all outgoing links. An agent who receives an assignment (from the higher-priority agent of the link), tries to find an assignment for his variable that does not violate a constraint with the assignment he received.

ok? messages are messages carrying an agent's variable assignment. When an agent A_i receives an **ok?** message from agent A_j, A_i places the received assignment in a data structure called *agent_view*, which holds the last assignment A_i

received from higher-priority neighbors such as A_j. Next, A_i checks if his current assignment is still consistent with his *agent_view*. If it is consistent, A_i does nothing. If not, then A_i searches his domain for a new consistent value. If he finds one, he assigns his variable that value and sends **ok?** messages to all lower-priority agents linked to him informing them of this value. Otherwise, A_i backtracks.

The *backtrack* operation is executed by sending a Nogood message. Recall that a Nogood is simply an inconsistent partial assignment, that is, assignments of specific values to some of the variables that together violate the constraints on those variables. In this case, the Nogood consists of A_i's *agent_view*.[2] The Nogood is sent to the agent with the lowest priority among the agents whose assignments are included in the inconsistent tuple in the Nogood. Agent A_i who sends a Nogood message to agent A_j assumes that A_j will change his assignment. Therefore, A_i removes from his *agent_view* the assignment of A_j and makes an attempt to find an assignment for A_j's variable that is consistent with the updated *agent_view*.

Because of its reliance on building up a set of Nogoods, the ABT algorithm can be seen as a greedy version of the hyper-resolution algorithm of the previous section. In the latter, all possible Nogoods are generated by each agent and communicated to all neighbors, even though the vast majority of these messages are not useful. Here, agents make tentative choices of a value for their variables, only generate Nogoods that incorporate values already generated by the agents above them in the order, and—importantly—communicate new values only to some agents and new Nogoods to only one agent.

Below is the pseudocode of the ABT algorithm, specifying the protocol for agent A_i.

> **when** *received (Ok?, (A_j, d_j))* **do**
>   add (A_j, d_j) to *agent_view*
>   **check_agent_view**
> **when** *received (Nogood, nogood)* **do**
>   add *nogood* to Nogood list
>   **forall** *(A_k, d_k) ∈* nogood, *if A_k is not a neighbor of A_i* **do**
>   add (A_k, d_k) to *agent_view*
>   request A_k to add A_i as a neighbor
>   **check_agent_view**

> **procedure** check_agent_view
> **when** agent_view *and* current_value *are inconsistent* **do**
>   **if** *no value in D_i is consistent with* agent_view **then**
>   **backtrack**
>   **else**
>   select $d ∈ D_i$ consistent with *agent_view*
>   *current_value* ← d
>   send (**ok?**, (A_i, d)) to lower-priority neighbors

---

2. We later discuss schemes that achieve better performance by avoiding always sending this entire set.

procedure backtrack
nogood ← some inconsistent set, using hyper-resolution or similar
procedure
if nogood *is the empty set* **then**
  | broadcast to other agents that there is no solution
  | terminate this algorithm
else
  | select $(A_j, d_j) \in$ *nogood* where A_j has the lowest priority in *nogood*
  | send (**Nogood**, *nogood*) to A_j
  | remove (A_j, d_j) from *agent_view*
  | **check_agent_view**

Notice a certain wrinkle in the pseudocode, having to do with the addition
of edges. Since the Nogood can include assignments of some agent A_j, which
A_i was not previously constrained with, after adding A_j's assignment to its
agent_view A_i sends a message to A_j asking it to add A_i to its list of outgoing
links. Furthermore, after adding the link, A_j sends an **ok?** message to A_i each
time it reassigns its variable. After storing the Nogood, A_i checks if its assignment
is still consistent. If it is, a message is sent to the agent the Nogood was received
from. This resending of the assignment is crucial since, as mentioned earlier, the
agent sending a Nogood assumes that the receiver of the Nogood replaces its
assignment. Therefore it needs to know that the assignment is still valid. If the
old assignment that was forbidden by the Nogood is inconsistent, A_i tries to find
a new assignment similarly to the case when an **ok?** message is received.

1.3.2 *A simple example*

In Section 1.3.3 we give a more elaborate example, but here is a brief illustration of
the operation of the ABT algorithm on one of the simple problems encountered
earlier. Consider again the instance (c) of the CSP in Figure 1.4, and assume
the agents are ordered alphabetically: x_1, x_2, x_3. They initially select values at
random; suppose they all select *blue*. x_1 notifies x_2 and x_3 of his choice, and
x_2 notifies x_3. x_2's local view is thus $\{x_1 = blue\}$, and x_3's local view is $\{x_1 =$
$blue, x_2 = blue\}$. x_2 and x_3 must check for consistency of their local views with
their own values. x_2 detects the conflict, changes his own value to *red*, and notifies
x_3. In the meantime, x_3 also checks for consistency and similarly changes his
value to *red*; he, however, notifies no one. Then x_3 receives a second message from
x_2, and updates his local view to $\{x_1 = blue, x_2 = red\}$. At this point he cannot
find a value from his domain consistent with his local view, and, using hyper
resolution, generates the Nogood $\{x_1 = blue, x_2 = red\}$. He communicates this
Nogood to x_2, the lowest ranked agent participating in the Nogood. x_2 now cannot
find a value consistent with his local view, generates the Nogood $\{x_1 = blue\}$, and
communicates it to x_1. x_1 detects the inconsistency with his current value, changes
his value to *red*, and communicates the new value to x_2 and x_3. The process now
continues as before; x_2 changes his value back to *blue*, x_3 finds no consistent

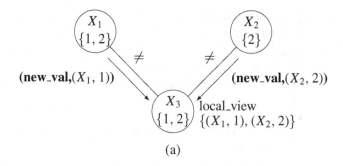

(a)

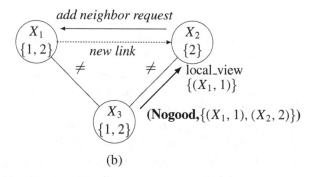

(b)

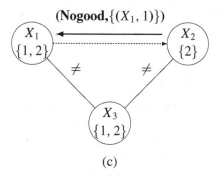

(c)

Figure 1.5 Asynchronous backtracking with dynamic link addition.

value and generates the Nogood $\{x_1 = red, x_2 = blue\}$, and then x_2 generates the Nogood $\{x_1 = red\}$. At this point x_1 has the Nogood $\{x_1 = blue\}$ as well as the Nogood $\{x_1 = red\}$, and using hyper-resolution he generates the Nogood $\{\}$, and the algorithm terminates having determined that the problem has no solution.

The need for the addition of new edges is seen in a slightly modified example, shown in Figure 1.5.

As in the previous example, here too x_3 generates the Nogood $\{x_1 = blue, x_2 = red\}$ and notifies x_2. x_2 is not able to regain consistency by changing his own value. However, x_1 is not a neighbor of x_2, and so x_2 does not have the value $x_1 = blue$ in his local view and is not able to send the Nogood $\{x_1 = blue\}$ to x_1. So x_2 sends

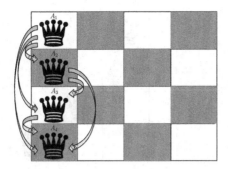
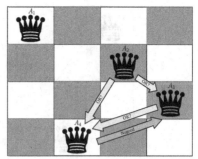

Figure 1.6 Cycle 1 of ABT for four queens. All agents are active.

Figure 1.7 Cycle 2 of ABT for four queens. A_2, A_3, and A_4 are active. The Nogood message is $A_1 = 1 \land A_2 = 1 \to A_3 \neq 1$.

a request to x_1 to add x_2 to its list of neighbors and to send x_2 his current value. From there onward the algorithm proceeds as before.

1.3.3   An extended example: the four queens problem

In order to gain additional feeling for the ABT algorithm beyond the didactic example in the previous section, let us look at one of the canonical CSP problems: the n-queens problem. More specifically, we will consider the four queens problem, which asks how four queens can be placed on a 4×4 chessboard so that no queen can (immediately) attack any other. We will describe ABT's behavior in terms of cycles of computation, which we somewhat artificially define to be the receiving of messages, the computations triggered by received messages, and the sending of messages due to these computations.

In the first cycle (Figure 1.6) all agents select values for their variables, which represent the positions of their queens along their respective rows. Arbitrarily, we assume that each begins by positioning his queen at the first square of his row. Each agent 1, 2, and 3 sends **ok?** messages to the agents ordered after him: A_1 sends three messages, A_2 sends two, and agent A_3 sends a single message. Agent A_4 does not have any agent after him, so he sends no messages. All agents are active in this first cycle of the algorithm's run.

In the second cycle (Figure 1.7) agents A_2, A_3, and A_4 receive the **ok?** messages sent to them and proceed to assign consistent values to their variables. Agent A_3 assigns the value 4 that is consistent with the assignments of A_1 and A_2 that he receives. Agent A_4 has no value consistent with the assignments of A_1, A_2, and A_3, and so he sends a *Nogood* containing these three assignments to A_3 and removes the assignment of A_3 from his *Agent_view*. Then, he assigns the value 2 which is consistent with the assignments that he received from A_1 and A_2 (having erased the assignment of A_3, assuming that it will be replaced because of the Nogood message). The active agents in this cycle are A_2, A_3, and A_4. Agent A_2 acts according to his information about A_1's position and moves to square 3, sending two **ok?** messages to inform his successors about his value. As can be seen in Figure 1.7, A_3 has moved to square 4 after receiving the **ok?**

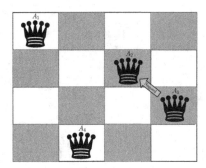
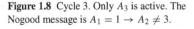

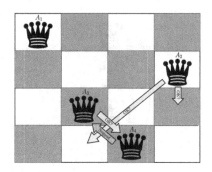

Figure 1.8 Cycle 3. Only A_3 is active. The Nogood message is $A_1 = 1 \rightarrow A_2 \neq 3$.

Figure 1.9 Cycles 4 and 5. A_2, A_3, and A_4 are active. The Nogood message is $A_1 = 1 \wedge A_2 = 4 \rightarrow A_3 \neq 4$.

messages of agents A_1 and A_2. Note that agent A_3 thinks that these agents are still in the first column of their respective rows. This is a manifestation of concurrency that causes each agent to act at all times in a form that is based only on his *Agent_View*. The *Agent_view* of agent A_3 includes the **ok?** messages he received.

The third cycle is described in Figure 1.8; only A_3 is active. After receiving the assignment of agent A_2, A_3 sends back a Nogood message to agent A_2. He then erases the assignment of agent A_2 from his *Agent_view* and validates that his current assignment (the value 4) is consistent with the assignment of agent A_1. Agents A_1 and A_2 continue to be idle, having received no messages that were sent in cycle 2. The same is true for agent A_4. Agent A_3 also receives the Nogood sent by A_4 in cycle 2 but ignores it since it includes an invalid assignment for A_2 (i.e., (2, 1) and not the currently correct (2, 4)).

Cycles 4 and 5 are depicted in Figure 1.9. In cycle 4 agent A_2 moves to square 4 because of the Nogood message he received. His former value was ruled out and the new value is the next valid one. He informs his successors A_3 and A_4 of his new position by sending two **ok?** messages. In cycle 5 agent A_3 receives agent A_2's new position and selects the only value that is compatible with the positions of his two predecessors, square 2. He sends a message to his successor informing him about this new value. Agent A_4 is now left with no valid value to assign and sends a Nogood message to A_3 that includes all his conflicts. The Nogood message appears at the bottom of Figure 1.9. Note that the Nogood message is no longer valid. Agent A_4, however, assumes that A_3 will change his position and moves to his only valid position (given A_3's anticipated move)—column 3.

Consider now cycle 6. Agent A_4 receives the new assignment of agent A_3 and sends him a Nogood message. Having erased the assignment of A_3 after sending the Nogood message, he then decides to stay at his current assignment (column 3), since it is compatible with the positions of agents A_1 and A_2. Agent A_3 is idle in cycle 6, since he receives no messages from either agent A_1 or agent A_2 (who are idle too). So, A_4 is the only active agent at cycle 6 (see Figure 1.10).

In each of cycles 7 and 8, one Nogood is sent. Both are depicted in Figure 1.11. First, agent A_3, after receiving the Nogood message from A_4, finds that he has

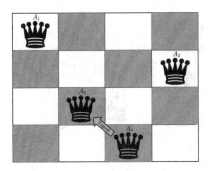

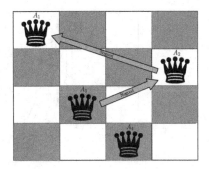

Figure 1.10 Cycle 6. Only A_4 is active. The Nogood message is $A_1 = 1 \wedge A_2 = 4 \rightarrow A_3 \neq 2$.

Figure 1.11 Cycles 7 and 8. A_3 is active in the first cycle and A_2 is active in the second. The Nogood messages are $A_1 = 1 \rightarrow A_2 \neq 4$ and $A_1 \neq 1$.

no valid values left and sends a Nogood to A_2. Next, in cycle 8, agent A_2 also discovers that his domain of values is exhausted and sends a Nogood message to A_1. Both sending agents erase the values of their successors (to whom the Nogood messages were sent) from their *agent_views* and therefore remain in their positions, which are now conflict free.

Cycle 9 involves only agent A_1, who receives the Nogood message from A_2 and so moves to his next value—square 2. Next, he sends **ok?** messages to his three successors.

The final cycle is cycle 10. Agent A_3 receives the **ok?** message of A_1 and so moves to a consistent value—square 1 of his row. Agents A_2 and A_4 check their *Agent_views* after receiving the same **ok?** messages from agent A_1 and find that their current values are consistent with the new position of A_1. Agent A_3 sends an **ok?** message to his successor A_4, informing of his move, but A_4 finds no reason to move. His value is consistent with all value assignments of all his predecessors. After cycle 10 all agents remain idle, having no constraint violations with assignments on their *agent_views*. Thus, this is a final state of the ABT algorithm in which it finds a solution.

1.3.4   *Beyond the ABT algorithm*

The ABT algorithm is the backbone of modern approaches to distributed constraint satisfaction, but it admits many extensions and modifications.

A major modification has to do with which inconsistent partial assignment (i.e., Nogood) is sent in the backtrack message. In the version presented earlier, which is the early version of ABT, the full *agent_view* is sent. However, the full *agent_view* is in many cases not a minimal Nogood; a strict subset of it may also be inconsistent. In general, shorter Nogoods can lead to a more efficient search process, since they permit backjumping further up the search tree.

Here is an example. Consider an agent A_6 holding an inconsistent *agent_view* with the assignments of agents A_1, A_2, A_3, A_4 and A_5. If we assume that A_6 is

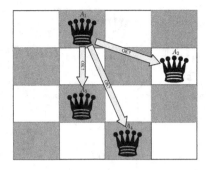

Figure 1.12 Cycle 9. Only A_1 is active.

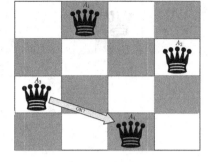

Figure 1.13 Cycle 10. Only A_3 is active.

only constrained by the current assignments of A_1 and A_3, sending a Nogood message to A_5 that contains all the assignments in the *agent_view* seems to be a waste. After sending the Nogood to A_5, A_6 will remove his assignment from the *agent_view* and make another attempt to assign his variable, which will be followed by an additional Nogood sent to A_4 and the removal of A_4's assignment from the *agent_view*. These attempts will continue until a minimal subset is sent as a Nogood. In this example, it is the Nogood sent to A_3. The assignment with the lower priority in the minimal inconsistent subset is removed from the *agent_view* and a consistent assignment can now be found. In this example the computation ended by sending a Nogood to the culprit agent, which would have been the outcome if the agent computed a minimal subset.

The solution to this inefficiency, however, is not straightforward, since finding a minimal Nogood is in general intractable (specifically, NP-hard). And so various heuristics are needed to cut down on the size of the Nogood, without sacrificing correctness.

A related issue is the number of Nogoods stored by each agent. In the preceding ABT version, each Nogood is recorded by the receiving agent. Since the number of inconsistent subsets can be exponential, constraint lists with exponential size will be created, and a search through such lists requires exponential time in the worst case. Various proposals have been made to cut down on this number while preserving correctness. One proposal is that agents keep only Nogoods consistent with their *agent_view*. While this prunes some of the Nogoods, in the worst case it still leaves a number of Nogoods that is exponential in the size of the *agent_view*. A further improvement is to store only Nogoods that are consistent with both the agent's *agent_view* and his current assignment. This approach, which is considered by some the best implementation of the ABT algorithm, ensures that the number of Nogoods stored by any single agent is no larger than the size of the domain.

asynchronous   Finally, there are approaches to distributed constraint satisfaction that do
forward   not follow the ABT scheme, including *asynchronous forward checking* and
checking   *concurrent dynamic backtracking*. Discussion of them is beyond the scope of

concurrent   this book, but the references point to further reading on the topic.
dynamic
backtracking

1.4   History and references

Distributed constraint satisfaction is discussed in detail in Yokoo [2001], and reviewed in Yokoo and Hirayama [2000]. The ABT algorithm was initially introduced in Yokoo [1994]. More comprehensive treatments, including the latest insights into distributed CSPs, appear in Meisels [2008] and Faltings [2006]. The sensor net figure is due to Carlos Guestrin.

2

Distributed Optimization

In the previous chapter we looked at distributed ways of meeting global constraints. Here we up the ante; we ask how agents can, in a distributed fashion, optimize a global objective function. Specifically, we consider four families of techniques and associated sample problems. They are, in order:

- distributed dynamic programming (as applied to path-planning problems);
- distributed solutions to Markov Decision Problems (MDPs);
- optimization algorithms with an economic flavor (as applied to matching and scheduling problems); and
- coordination via social laws and conventions, and the example of traffic rules.

2.1 Distributed dynamic programming for path planning

Like graph coloring, path planning constitutes another common abstract problem-solving framework. A path-planning problem consists of a weighted directed graph with a set of n nodes N, directed links L, a weight function $w : L \mapsto \mathbb{R}^+$, and two nodes $s, t \in N$. The goal is to find a directed path from s to t having minimal total weight. More generally, we consider a set of goal nodes $T \subset N$, and are interested in the shortest path from s to any of the goal nodes $t \in T$.

This abstract framework applies in many domains. Certainly it applies when there is some concrete network at hand (e.g., a transportation or telecommunication network). But it also applies in more roundabout ways. For example, in a planning problem the nodes can be states of the world, the arcs actions available to the agent, and the weights the cost (or, alternatively, time) of each action.

2.1.1 *Asynchronous dynamic programming*

principle of optimality

Path planning is a well-studied problem in computer science and operations research. We are interested in distributed solutions, in which each node performs a local computation, with access only to the state of its neighbors. Underlying our solutions will be the *principle of optimality*: if node x lies on a shortest path from s to t, then the portion of the path from s to x (or, respectively, from x to t) must also be the shortest paths between s and x (resp., x and t). This dynamic
programming allows an incremental divide-and-conquer procedure, also known as *dynamic programming*.

Let us represent the shortest distance from any node i to the goal t as $h^*(i)$. asynchronous
dynamic
programming Thus the shortest distance from i to t via a node j neighboring i is given by $f^*(i, j) = w(i, j) + h^*(j)$, and $h^*(i) = \min_j f^*(i, j)$. Based on these facts, the ASYNCHDP algorithm has each node repeatedly perform the following procedure. In this procedure, given in Figure 2.1, each node i maintains a variable $h(i)$, which is an estimate of $h^*(i)$.

Figure 2.2 shows this algorithm in action. The h values are initialized to ∞, and incrementally decrease to their correct values. The figure shows three iterations; note that after the first iteration, not all finite h values are correct; in particular, the value 3 in node d still overestimates the true distance, which is corrected in the next iteration.

procedure ASYNCHDP (node i)
if *i is a goal node* **then**
  | $h(i) \leftarrow 0$
else
  └   initialize $h(i)$ arbitrarily (e.g., to ∞ or 0)
repeat
  | **forall** *neighbors j* **do**
  | └ $f(j) \leftarrow w(i, j) + h(j)$
  └ $h(i) \leftarrow \min_j f(j)$

Figure 2.1 The asynchronous dynamic programming algorithm.

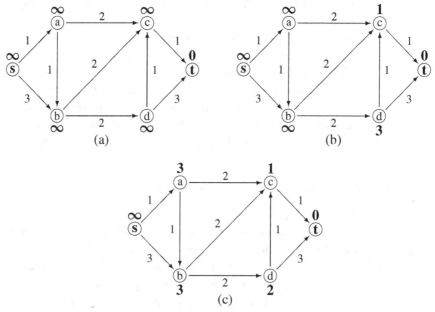

Figure 2.2 Asynchronous dynamic programming in action.

procedure LRTA*

$i \leftarrow s$ // the start node

while *i is not a goal node* **do**

> **foreach** *neighbor j* **do**
>> $\lfloor \ f(j) \leftarrow w(i, j) + h(j)$
>
> $i' \leftarrow \arg\min_j f(j)$ // breaking ties at random
>
> $h(i) \leftarrow \max(h(i), f(i'))$
>
> $\lfloor \ i \leftarrow i'$

Figure 2.3 The learning real-time A* algorithm.

One can prove that the ASYNCHDP procedure is guaranteed to converge to the true values, that is, h will converge to h^*. Specifically, convergence will require one step for each node in the shortest path, meaning that in the worst case convergence will require n iterations. However, for realistic problems this is of little comfort. Not only can convergence be slow, but this procedure assumes a process (or agent) for each node. In typical search spaces one cannot effectively enumerate all nodes, let alone allocate them each a process. (For example, chess has approximately 10^{120} board positions, whereas there are fewer than 10^{81} atoms in the universe and there have only been 10^{26} nanoseconds since the Big Bang.) So to be practical we turn to heuristic versions of the procedure, which require a smaller number of agents. Let us start by considering the opposite extreme in which we have only one agent.

2.1.2   *Learning real-time A**

learning
real-time A*
(LRTA*)

In the *learning real-time A**, or LRTA*, algorithm, the agent starts at a given node, performs an operation similar to that of asynchronous dynamic programming, and then moves to the neighboring node with the shortest estimated distance to the goal, and repeats. The procedure is given in Figure 2.3.

As earlier, we assume that the set of nodes is finite and that all weights $w(i, j)$ are positive and finite. Note that this procedure uses a given heuristic function $h(\cdot)$ that serves as the initial value for each newly encountered node. For our purposes it is not important what the precise function is. However, to

admissible
heuristic

guarantee certain properties of LRTA*, we must assume that h is *admissible*. This means that h never overestimates the distance to the goal, that is, $h(i) \leq h^*(i)$. Because weights are nonnegative we can ensure admissibility by setting $h(i) = 0$ for all i, although less conservative admissible heuristic functions (built using knowledge of the problem domain) can speed up the convergence to the optimal solution. Finally, we must assume that there exists some path from every node in the graph to a goal node. With these assumptions, LRTA* has the following properties:

- The h-values never decrease, and remain admissible.
- LRTA* terminates; the complete execution from the start node to termination at the goal node is called a *trial*.

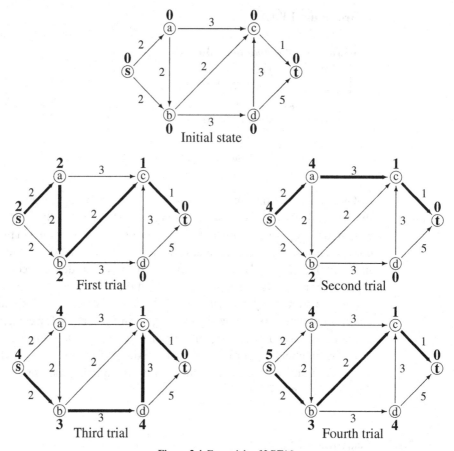

Figure 2.4 Four trials of LRTA*.

- If LRTA* is repeated while maintaining the h-values from one trial to the next, it eventually discovers the shortest path from the start to a goal node.
- If LRTA* find the same path on two sequential trials, this is shortest path. (However, this path may also be found in one or more previous trials before it is found twice in a row. Do you see why?)

Figure 2.4 shows four trials of LRTA*. Do you see why admissibility of the heuristic is necessary?

LRTA* is a centralized procedure. However, we note that rather than have a single agent execute this procedure, one can have multiple agents execute it. The properties of the algorithm (call it LRTA*(n), with n agents) are not altered, but the convergence to the shortest path can be sped up dramatically. First, if the agents each break ties differently, some will reach the goal much faster than others. Furthermore, if they all have access to a shared h-value table, the learning of one agent can teach the others. Specifically, after every round and for every i, $h(i) = \max_j h_j(i)$, where $h_j(i)$ is agent j's updated value for $h(i)$. Figure 2.5 shows an execution of LRTA*(2)—that is, LRTA* with two agents—starting from

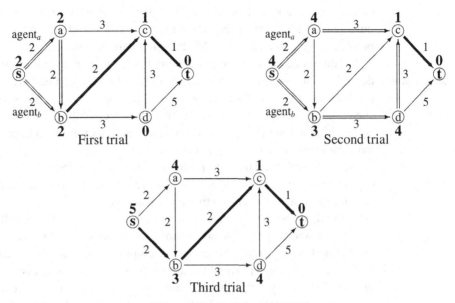

Figure 2.5 Three trials of LRTA*(2).

the same initial state as in Figure 2.4. (The hollow arrows show paths traversed by a single agent, while the dark arrows show paths traversed by both agents.)

2.2 Action selection in multiagent MDPs

In this section we discuss the problem of optimal action selection in multiagent MDPs.[1] Recall that in a single-agent MDP the optimal policy π^* is characterized by the mutually-recursive Bellman equations:

$$Q^{\pi^*}(s, a) = r(s, a) + \beta \sum_{\hat{s}} p(s, a, \hat{s}) V^{\pi^*}(\hat{s})$$

$$V^{\pi^*}(s) = \max_a Q^{\pi^*}(s, a)$$

value iteration   Furthermore, these equations turn into an algorithm—specifically, the dynamic-programming-style *value iteration algorithm*—by replacing the equality signs "=" with assignment operators "←" and iterating repeatedly through those assignments.

However, in real-world applications the situation is not that simple. For example, the MDP may not be known by the planning agent and thus may have to be learned. This case is discussed in Chapter 7. But more basically, the MDP may simply be too large to iterate over all instances of the equations. In this case, one approach is to exploit independence properties of the MDP. One case where this arises is when the states can be described by feature vectors; each feature can take on many values, and thus the number of states is exponential in the number

---

1. The basics of single-agent MDPs are covered in Appendix C.

of features. One would ideally like to solve the MDP in time polynomial in the number of features rather than the number of states, and indeed techniques have been developed to tackle such MDPs with factored state spaces.

multiagent MDP

We do not address that problem here, but instead on a similar one that has to do with the modularity of actions rather than of states. In a *multiagent MDP* any (global) action a is really a vector of local actions $(a_1, \ldots, a_n)$, one by each of n agents. The assumption here is that the reward is common, so there is no issue of competition among the agents. There is not even a problem of coordination; we have the luxury of a central planner (but see discussion at the end of this section of parallelizability). The only problem is that the number of global actions is exponential in the number of agents. Can we somehow solve the MDP other than by enumerating all possible action combinations?

We will not address this problem, which is quite involved, in full generality. Instead we will focus on an easier subproblem. Assuming that the Q values for the optimal policy have already been computed. How hard is it to decide on which action each agent should take? Since we are assuming away the problem of coordination by positing a central planner, on the face of it the problem is straightforward. In Appendix C we state that once the optimal (or close to optimal) Q values are computed, the optimal policy is "easily" recovered; the optimal action in state s is $\arg \max_a Q^{\pi^*}(s, a)$. But of course if a ranges over an exponential number of choices by all agents, "easy" becomes "hard." Can we do better than naively enumerating over all action combinations by the agents?

In general the answer is no, but in practice, the interaction among the agents' actions can be quite limited, which can be exploited both in the representation of the Q function and in the maximization process. Specifically, in some cases we can associate an individual Q_i function with each agent i, and express the Q function (either precisely or approximately) as a linear sum of the individual Q_is:

$$Q(s, a) = \sum_{i=1}^{n} Q_i(s, a).$$

The maximization problem now becomes

$$\arg \max_a \sum_{i=1}^{n} Q_i(s, a).$$

This in and of itself is not very useful, as one still needs to look at the set of all global actions a, which is exponential in n, the number of agents. However, it is often also the case that each individual Q_i depends only on a small subset of the variables.

For example, imagine a metal-reprocessing plant with four locations, each with a distinct function: one for loading contaminated material and unloading reprocessed material; one for cleaning the incoming material; one for reprocessing the cleaned material; and one for eliminating the waste. The material flow among them is depicted in Figure 2.6.

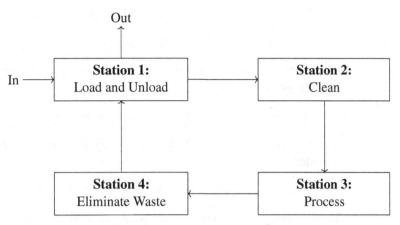

Figure 2.6 A metal-reprocessing plant.

Each station can be in one of several states, depending on the load at that time. The operator of the station has two actions available: "pass material to next station in process," and "suspend flow." The state of the plant is a function of the state of each of the stations; the higher the utilization of existing capacity the better, but exceeding full capacity is detrimental. Clearly, in any given global state of the system, the optimal action of each local station depends only on the action of the station directly "upstream" from it. Thus in our example the global Q function becomes

$$Q(a_1, a_2, a_3, a_4) = Q_1(a_1, a_2) + Q_2(a_2, a_4) + Q_3(a_1, a_3) + Q_4(a_3, a_4)$$

and we wish to compute

$$\arg\max_{(a_1, a_2, a_3, a_4)} Q_1(a_1, a_2) + Q_2(a_2, a_4) + Q_3(a_1, a_3) + Q_4(a_3, a_4).$$

variable elimination

Note that in the preceding expressions we omit the state argument, since that is being held fixed; we are looking at optimal action selection at a given state.

In this case we can employ a *variable elimination* algorithm, which optimizes the choice for the agents one at a time. We explain the operation of the algorithm via our example. Let us begin our optimization with agent 4. To optimize a_4, functions Q_1 and Q_3 are irrelevant. Hence, we obtain

$$\max_{a_1, a_2, a_3} Q_1(a_1, a_2) + Q_3(a_1, a_3) + \max_{a_4}[Q_2(a_2, a_4) + Q_4(a_3, a_4)].$$

conditional strategy

We see that to make the optimal choice over a_4, the values of a_2 and a_3 must be known. Thus, what must be computed for agent 4 is a *conditional strategy*, with a (possibly) different action choice for each action choice of agents 2 and 3. The value that agent 4 brings to the system in the different circumstances can be summarized using a new function $e_4(A_2, A_3)$ whose value at the point a_2, a_3 is the value of the internal max expression

$$e_4(a_2, a_3) = \max_{a_4}[Q_2(a_2, a_4) + Q_4(a_3, a_4)].$$

Agent 4 has now been "eliminated," and our problem now reduces to computing

$$\max_{a_1,a_2,a_3} Q_1(a_1, a_2) + Q_3(a_1, a_3) + e_4(a_2, a_3),$$

having one fewer agent involved in the maximization. Next, the choice for agent 3 is made, giving

$$\max_{a_1,a_2} Q_1(a_1, a_2) + e_3(a_1, a_2).$$

where $e_3(a_1, a_2) = \max_{a_3}[Q_3(a_1, a_3) + e_4(a_2, a_3)]$ Next, the choice for agent 2 is made:

$$e_2(a_1) = \max_{a_2}[Q_1(a_1, a_2) + e_3(a_1, a_2)].$$

The remaining decision for agent 1 is now the following maximization:

$$e_1 = \max_{a_1} e_2(a_1).$$

The result e_1 is simply a number, the required maximization over $a_1, \ldots, a_4$. Note that although this expression is short, there is no free lunch; in order to perform this optimization, one needs to iterate not only over all actions a_1 of the first agent, but also over the action of the other agents as needed to unwind the internal maximizations. However, in general the total number of combinations will be smaller than the full exponential combination of agent actions.[2]

We can recover the maximizing set of actions by performing the process in reverse. The maximizing choice for e_1 defines the action a_1^* for agent 1:

$$a_1^* = \arg\max_{a_1} e_2(a_1).$$

To fulfill its commitment to agent 1, agent 2 must choose the value a_2^* that yielded $e_2(a_1^*)$,

$$a_2^* = \arg\max_{a_2}[Q_1(a_1^*, a_2) + e_3(a_1^*, a_2)].$$

This, in turn, forces agent 3 and then agent 4 to select their actions appropriately:

$$a_3^* = \arg\max_{a_3}[Q_3(a_1^*, a_3) + e_4(a_2^*, a_3)];$$

$$a_4^* = \arg\max_{a_4}[Q_2(a_2^*, a_4) + Q_4(a_3^*, a_4)].$$

The actual implementation of this procedure allows several versions. Here are a few of them:

A quick-down, slow-up two-pass sequential implementation: This follows the example in that variables are eliminated symbolically one at a time

---

2. Full discussion of this point is beyond the scope of this book, but for the record, the complexity of the algorithm is exponential in the tree width of the *coordination graph*; this is the graph whose nodes are the agents and whose edges connect agents whose Q values share one or more arguments. The tree width is also the maximum clique size minus one in the triangulation of the graph; each triangulation essentially corresponds to one of the variable elimination orders. Unfortunately, it is NP-hard to compute the optimal ordering. The notes at the end of the chapter provide additional references on the topic.

starting with a_n. This is done in $O(n)$ time. When up to a_1 the actual maximization starts; all values of a_1 are tried, alongside all values of the variables appearing in the unwinding of the expression. This phase requires $O(k^n)$ time in the worst case, where k is the bound on domain sizes.

A slow-down, quick-up two-phase sequential implementation: A similar procedure, except here the actual best-response table is built as variables are eliminated. This requires $O(k^n)$ time in the worst case. The payoff is in the second phase, where the optimization requires a simple table-lookup for each value of the variable, resulting in a complexity of $O(kn)$.

Asynchronous versions: The full linear pass in both directions is not necessary, given only partial dependence among variables. Thus in the down phase variables need await a signal from the higher-indexed variables with which they interact (as opposed to all higher-indexed variables) before computing their best-response functions, and similarly in the pass up they need await the signal from only the lower-indexed variables with which they interact.

One final comment. We have discussed variable elimination in the particular context of multiagent MDPs, but it is relevant in any context in which multiple agents wish to perform a distributed optimization of an factorable objective function.

2.3   Negotiation, auctions, and optimization

In this section we consider distributed problem solving that has a certain economic flavor. In the first section below we will informally give the general philosophy and background; in the following two sections we will be more precise.

2.3.1   *From contract nets to auction-like optimization*

contract net   *Contract nets* were one of the earliest proposals for such an economic approach. Contract nets are not a specific algorithm, but a framework, a protocol for implementing specific algorithms. In a contract net the global problem is broken down into subtasks, and these are distributed among a set of agents. Each agent has different capabilities; for each agent i there is a function c_i such that for any set of tasks T, $c_i(T)$ is the cost for agent i to achieve all the tasks in T. Each agent starts out with some initial set of tasks, but in general this assignment is not optimal, in the sense that the sum of all agents' costs is not minimal. The agents then enter into a negotiation process which improves on the assignment and, hopefully, culminates in an optimal assignment, that is, one with minimal cost. Furthermore, the process can have a so-called *anytime property*; even if it is anytime property   interrupted prior to achieving optimality, it can achieve significant improvements over the initial allocation.

The negotiation consists of agents repeatedly contracting out assignments among themselves, each contract involving the exchange of tasks as well as money. The question is how the bidding process takes place and what contracts hold based on this bidding. The general contract-net protocol is open on these issues. One particular approach has each agent bid for each set of tasks the agent's marginal cost for the task, that is, the agent's additional cost for adding that task to its current set. The tasks are allocated to the lowest bidders, and the process repeats. It can be shown that there always exists a sequence of contracts that result in the optimal allocation. If one is restricted to basic contract types in which one agent contracts a single task to another agent, and receives from him some money in return, then in general achieving optimality requires that agents enter into "money-losing" contracts in the process. However, there exist more complex contracts—which involve contracting for a bundle of tasks

cluster contracts

swap contracts

multiagent
contracts

("*cluster contracts*"), or a swap of tasks among two agents ("*swap contracts*"), or simultaneous transfers among many agents ("*multiagent contracts*")—whose combination allows for a sequence of contracts that are not money losing and which culminate in the optimal solution.

At this point several questions may naturally occur to the reader.

- We start with some global problem to be solved, but then speak about minimizing the total cost to the agents. What is the connection between the two?
- When exactly do agents make offers, and what is the precise method by which the contracts are decided on?
- Since we are in a cooperative setting, why does it matter whether agents "lose money" or not on a given contract?

We will provide an answer to the first question in the next section. We will see that, in certain settings (specifically, those of linear programming and integer programming), finding an optimal solution is closely related to the individual utilities of the agents.

Regarding the second question, indeed one can provide several instantiations of even the specific, marginal-cost version of the contract-net protocol. In the next two sections we will be much more specific. We will look at a particular class of negotiation schemes, namely (specific kinds of) auctions. Every negotiation

bidding rule

market clearing
rule

information
dissemination
rule

scheme consists of three elements: (1) permissible ways of making offers (*bidding rules*), (2) definition of the outcome based on the offers (*market clearing rules*), and (3) the information made available to the agents throughout the process (*information dissemination rules*). Auctions are a structured way of settling each of these dimensions, and we will look at auctions that do so in specific ways. It should be mentioned, however, that this specificity is not without a price. While convergence to optimality in contract nets depends on particular sequences of contracts taking place, and thus on some coordinating hand, the process is inherently distributed. The auction algorithms we will study include an auctioneer, an explicit centralized component.

The last of our questions deserves particular attention. As we said, we start with some problem to be solved. We then proceed to define an auction-like process for solving it in a distributed fashion. However it is no accident that this section precedes our (rather detailed) discussion of auctions in Chapter 11. As we see there, auctions are a way to allocate scarce resources among *self-interested* agents.

Auction theory thus belongs to the realm of game theory. In this chapter we also speak about auctions, but the discussion has little to do with game theory. In the spirit of the contract-net paradigm, in our auctions agents will engage in a series of bids for resources, and at the end of the auction the assignment of the resources to the "winners" of the auction will constitute an optimal (or near optimal, in some cases) solution. However, in the standard treatment of auctions (and thus in Chapter 11) the bidders are assumed to bid in a way that maximizes their personal payoff. Here there is no question of the agents deviating from the prescribed bidding protocol for personal gain. For this reason, despite the surface similarity, the discussion of these auction-like methods makes no reference to game theory or mechanism design. In particular, while these methods have some nice properties—for example, they are intuitive, provably correct, naturally parallelizable, appropriate for deployment in distributed systems settings, and tend to be robust to slight perturbations of the problem specification—no claim is made about their usefulness in adversarial situations. For this reason it is indeed something of a red herring, in this cooperative setting, to focus on questions such as whether a given contract is profitable for a given agent. In noncooperative settings, where contract nets are also sometimes pressed into service, the situation is of course different.

In the next two sections we will be looking at two classical optimization problems, one representable as a linear program (LP) and one only as an integer program (IP) (for a brief review of LPs and IPs, see Appendix B). There exists a vast literature on how to solve LPs and IPs, and it is not our aim in this chapter (or in the appendix) to capture this broad literature. Our more limited aim here is to look at the auction-style solutions for them. First we will look at an LP problem—the problem of *weighted matching in a bipartite graph*, also known as the *assignment problem*. We will then look at a more complex, IP problem—that of *scheduling*. As we shall see, since the LP problem is relatively easy (specifically, solvable in polynomial time), it admits an auction-like procedure with tight guarantees. The IP problem is NP-complete, and so it is not surprising that the auction-like procedure does not come with such guarantees.

2.3.2 *The assignment problem and linear programming*

The problem and its LP formulation

weighted
matching

The problem of *weighted matching in a bipartite graph*, otherwise known as the *assignment problem*, is defined as follows.

assignment
problem

Definition 2.3.1 (Assignment problem)
A (symmetric) assignment problem *consists of:*

- *a set N of n agents;*
- *a set X of n objects;*
- *a set $M \subseteq N \times X$ of possible assignment pairs; and*
- *a function $v : M \mapsto \mathbb{R}$ giving the value of each assignment pair.*

feasible
assignment

An assignment is a set of pairs $S \subseteq M$ such that each agent $i \in N$ and each object $j \in X$ is in at most one pair in S. A feasible assignment *is one in which all agents are assigned an object. A feasible assignment S is* optimal *if it maximizes* $\sum_{(i,j) \in S} v(i, j)$.

An example of an assignment problem is the following (in this example, $X = \{x_1, x_2, x_3\}$ and $N = \{1, 2, 3\}$).

i	$v(i, x_1)$	$v(i, x_2)$	$v(i, x_3)$
1	2	4	0
2	1	5	0
3	1	3	2

In this small example it is not hard to see that $(1, x_1), (2, x_2), (3, x_3)$ is an optimal assignment. In larger problems, however, the solution is not obvious, and the question is how to compute it algorithmically.

We first note that an assignment problem can be encoded as a linear program. Given a general assignment problem as defined earlier, we introduce the indicator matrix $\mathbf{x}$; $x_{i,j} = 1$ indicates that the pair (i, j) is selected, and $x_{i,j} = 0$ otherwise. Then we express the linear program as follows.

$$\text{maximize} \quad \sum_{(i,j) \in M} v(i, j) x_{i,j}$$

$$\text{subject to} \quad \sum_{j \mid (i,j) \in M} x_{i,j} \leq 1 \qquad \forall i \in N$$

$$\sum_{i \mid (i,j) \in M} x_{i,j} \leq 1 \qquad \forall j \in X$$

On the face of it the LP formulation is inappropriate since it allows for fractional matches (i.e., for $0 < x_{i,j} < 1$). But as it turns out this LP has integral solutions.

Lemma 2.3.2 *The LP encoding of the assignment problem has a solution such that for every i, j it is the case that $x_{i,j} = 0$ or $x_{i,j} = 1$. Furthermore, any optimal fractional solution can be converted in polynomial time to an optimal integral solution.*

Since any LP can be solved in polynomial time, we have the following.

Corollary 2.3.3 *The assignment problem can be solved in polynomial time.*

This corollary might suggest that we are done. However, there are a number of reasons to not stop there. First, the polynomial-time solution to the LP problem is of complexity roughly $O(n^3)$, which may be too high in some cases. Furthermore, the solution is not obviously parallelizable, and is not particularly robust to changes in the problem specification (if one of the input parameters changes, the program must essentially be solved from scratch). One solution that suffers less from these shortcomings is based on the economic notion of competitive equilibrium, which we explore next.

The assignment problem and competitive equilibrium

competitive equilibrium

Imagine that each of the objects in X has an associated price; the price vector is $p = (p_1, \ldots, p_n)$, where p_j is the price of object j. Given an assignment $S \subseteq M$ and a price vector p, define the "utility" from an assignment j to agent i as $u(i, j) = v(i, j) - p_j$. An assignment and a set of prices are in *competitive equilibrium* when each agent is assigned the object that maximizes his utility given the current prices. More formally, we have the following.

Definition 2.3.4 (Competitive equilibrium) *A feasible assignment S and a price vector p are in* competitive equilibrium *when for every pairing $(i, j) \in S$ it is the case that $\forall k, \ u(i, j) \geq u(i, k)$.*

It might seem strange to drag an economic notion into a discussion of combinatorial optimization, but as the following theorem shows there are good reasons for doing so.

Theorem 2.3.5 *If a feasible assignment S and a price vector p satisfy the competitive equilibrium condition then S is an optimal assignment. Furthermore, for any optimal solution S, there exists a price vector p such that p and S satisfy the competitive equilibrium condition.*

For example, in the previous example, it is not hard to see that the optimal assignment $(1, x_1), (2, x_2), (3, x_3)$ is a competitive equilibrium given the price vector $(2, 4, 1)$; the "utilities" of the agents are 0, 1, and 1, respectively, and none of them can increase their profit by bidding for one of the other objects at the current prices. We outline the proof of a more general form of the theorem in the next section.

This last theorem means that one way to search for solutions of the LP is to search the space of competitive equilibria. And a natural way to search that space involves auction-like procedures, in which the individual agents "bid" for the different resources in a prespecified way. We will look at open outcry, ascending auction-like procedures, resembling the English auction discussed in Chapter 11. Before that, however, we take a slightly closer look at the connection between optimization problems and competitive equilibrium.

Competitive equilibrium and primal-dual problems

Theorem 2.3.5 may seem at first almost magical; why would an economic notion prove relevant to an optimization problem? However, a slightly closer look

removes some of the mystery. Rather than looking at the specific LP corresponding to the assignment problem, consider the general ("primal") form of an LP.

$$\text{maximize} \quad \sum_{i=1}^{n} c_i x_i$$

$$\text{subject to} \quad \sum_{i=1}^{n} a_{ij} x_i \leq b_j \qquad \forall j \in \{1, \ldots, m\}$$

$$x_i \geq 0 \qquad \forall i \in \{1, \ldots, n\}$$

Note that this formulation makes reverse use the $\leq$ and $\geq$ signs as compared
to the formulation in Appendix B. As we remark there, this is simply a matter of
the signs of the constants used.

production economy
  The primal problem has a natural economic interpretation, regardless of its
actual origin. Imagine a *production economy*, in which you have a set of resources
and a set of products. Each product consumes a certain amount of each resource,
and each product is sold at a certain price. Interpret x_i as the amount of product
i produced, and c_i as the price of product i. Then the optimization problem can
be interpreted as profit maximization. Of course, this must be done within the
constraints of available resources. If we interpret b_j as the available amount of
resource j and a_{ij} as the amount of resource j needed to produce a unit of
product i, then the constraint $\sum_i a_{ij} x_i \leq b_j$ appropriately captures the limitation
on resource j.

Now consider the dual problem.

$$\text{minimize} \quad \sum_{i=1}^{m} b_i y_i$$

$$\text{subject to} \quad \sum_{i=1}^{m} a_{ij} y_i \geq c_j \qquad \forall j \in \{1, \ldots, n\}$$

$$y_i \geq 0 \qquad \forall i \in \{1, \ldots, m\}$$

shadow price
  It turns out that y_i can also be given a meaningful economic interpretation,
namely, as the *marginal value* of resource i, also known as its *shadow price*. The
shadow price captures the sensitivity of the optimal solution to a small change
in the availability of that particular resource, holding everything else constant.
A high shadow price means that increasing its availability would have a large
impact on the optimal solution, and vice versa.[3]

This helps explain why the economic perspective on optimization, at least in
the context of linear programming, is not that odd. Indeed, armed with these
intuitions, one can look at traditional algorithms such as the Simplex method and
give them an economic interpretation. In the next section we look at a specific
auction-like algorithm, which is overtly economic in nature.

---

3. To be precise, the shadow price is the value of the Lagrange multiplier at the optimal solution.

A naive auction algorithm

We start with a naive auction-like procedure which is "almost" right; it contains the main ideas, but has a major flaw. In the next section we will fix that flaw. The naive procedure begins with no objects allocated, and terminates once it has found a feasible solution. We define the naive auction algorithm formally as follows.

> **Naive Auction Algorithm**
> // Initialization:
> $S \leftarrow \emptyset$
> **forall** $j \in X$ **do**
> $\lfloor\ p_j \leftarrow 0$
> **repeat**
>   // Bidding Step:
>   let $i \in N$ be an unassigned agent
>   // Find an object $j \in X$ that offers i maximal value at current prices:
> $j \in \arg\max_{k|(i,k)\in M}(v(i,k) - p_k)$
>   // Compute i's bid increment for j:
> $b_i \leftarrow (v(i,j) - p_j) - \max_{k|(i,k)\in M; k \neq j}(v(i,k) - p_k)$
>   // which is the difference between the value to i of the best and second-best objects
>   at current prices (note that i's bid will be the current price plus this bid increment).
>   // Assignment Step:
>   add the pair (i, j) to the assignment S
>   **if** *there is another pair* (i', j) **then**
> $\lfloor$ remove it from the assignment S
>   increase the price p_j by the increment b_i
> **until** *S is feasible*   // that is, it contains an assignment for all $i \in N$

It is not hard to verify that the following is true of the algorithm.

Theorem 2.3.6 *The naive algorithm terminates only at a competitive equilibrium.*

Here, for example, is a possible execution of the algorithm on our current example. The following table shows each round of bidding. In this execution we pick the unassigned agents in order, round-robin style.

round	p_1	p_2	p_3	bidder	preferred object	bid incr.	current assignment
0	0	0	0	1	x_2	2	$(1, x_2)$
1	0	2	0	2	x_2	2	$(2, x_2)$
2	0	4	0	3	x_3	1	$(2, x_2), (3, x_3)$
3	0	4	1	1	x_1	2	$(2, x_2), (3, x_3), (1, x_1)$

At first agents 1 and 2 compete for x_2, but quickly x_2 becomes too expensive for agent 1, who opts for x_1. By the time agent 3 gets to bid he is priced out of his preferred item, x_2, and settles for x_3.

Thus when the procedure terminates we have our solution. The problem, though, is that it may not terminate. This can occur when more than one object offers maximal value for a given agent; in this case the agent's bid increment will be zero. If these two items also happen to be the best items for another agent, they will enter into an infinite bidding war in which the price never rises. Consider a modification of our previous example, in which the value function is given by the following table.

i	$v(i, x_1)$	$v(i, x_2)$	$v(i, x_3)$
1	1	1	0
2	1	1	0
3	1	1	0

The naive auction protocol would proceed as follows.

round	p_1	p_2	p_3	bidder	preferred object	bid incr.	current assignment
0	0	0	0	1	x_1	0	$(1, x_1)$
1	0	0	0	2	x_2	0	$(1, x_1), (2, x_2)$
2	0	0	0	3	x_1	0	$(3, x_1), (2, x_2)$
3	0	0	0	1	x_2	0	$(3, x_1), (1, x_2)$
4	0	0	0	2	x_1	0	$(2, x_1), (1, x_2)$
⋮	⋮	⋮	⋮	⋮	⋮	⋮	⋮

Clearly, in this example the naive algorithm will have the three agents forever fight over the two desired objects.

A terminating auction algorithm

To remedy the flaw exposed previously, we must ensure that prices continue to increase when objects are contested by a group of agents. The extension is quite straightforward: we add a small amount to the bidding increment. Thus we calculate the bid increment of agent $i \in N$ as follows.

$$b_i = u(i, j) - \max_{k \mid (i,k) \in M; k \neq j} u(i, k) + \epsilon$$

Otherwise, the algorithm is as stated earlier.

Consider again the problematic assignment problem on which the naive algorithm did not terminate. The terminating auction protocol would proceed as follows.

round	p_1	p_2	p_3	bidder	preferred object	bid incr.	current assignment
0	ϵ	0	0	1	x_1	ϵ	$(1, x_1)$
1	ϵ	2ϵ	0	2	x_2	2ϵ	$(1, x_1), (2, x_2)$
2	3ϵ	2ϵ	0	3	x_1	2ϵ	$(3, x_1), (2, x_2)$
3	3ϵ	4ϵ	0	1	x_2	2ϵ	$(3, x_1), (1, x_2)$
4	5ϵ	4ϵ	0	2	x_1	2ϵ	$(2, x_1), (1, x_2)$

Note that at each iteration, the price for the preferred item increases by at least ϵ. This gives us some hope that we will avoid nontermination. We must first though make sure that, if we terminate, we terminate with the "right" results.

First, because the prices must increase by at least ϵ at every round, the competitive equilibrium property is no longer preserved over the iteration. Agents may "overbid" on some objects. For this reason we will need to define a notion of ϵ-competitive equilibrium.

Definition 2.3.7 (ϵ-competitive equilibrium) *S and p satisfy ϵ-competitive equilibrium when for each $i \in N$, if there exists a pair $(i, j) \in S$ then $\forall k,\ u(i, j) + \epsilon \geq u(i, k)$.*

In other words, in an ϵ-equilibrium no agent can profit more than ϵ by bidding for an object other than his assigned one, given current prices.

Theorem 2.3.8 *A feasible assignment S with n goods that forms an ϵ-competitive equilibrium with some price vector is within $n\epsilon$ of optimal.*

Corollary 2.3.9 *Consider a feasible assignment problem with an integer valuation function $v : M \mapsto \mathbb{Z}$. If $\epsilon < \frac{1}{n}$ then any feasible assignment found by the terminating auction algorithm will be optimal.*

This leaves the question of whether the algorithm indeed terminates, and if so, how quickly. To see why the algorithm must terminate, note that if an object receives a bid in k iterations, its price must exceed its initial price by at least $k\epsilon$. Thus, for sufficiently large k, the object will become expensive enough to be judged inferior to some object that has not received a bid so far. The total number of iterations in which an object receives a bid must be no more than

$$\frac{\max_{(i, j)} v(i, j) - \min_{(i, j)} v(i, j)}{\epsilon}.$$

Once all objects receive at least one bid, the auction terminates (do you see why?). If each iteration involves a bid by a single agent, the total number of iterations is no more than n times the preceding quantity. Thus, since each bid requires $O(n)$ operations, the running time of the algorithm is $O(n^2 \max_{(i, j)} \frac{|v(i, j)|}{\epsilon})$. Observe that if $\epsilon = O(1/n)$ (as discussed in Corollary 2.3.9), the algorithm's running time is $O(n^3 k)$, where k is a constant that does not depend on n, yielding worst-case performance similar to linear programming.

2.3.3 *The scheduling problem and integer programming*

The problem and its integer program

scheduling
problem

The *scheduling problem* involves a set of time slots and a set of agents. Each agent requires some number of time slots and has a deadline. Intuitively, the agents each have a task that requires the use of a shared resource, and that task lasts a certain number of hours and has a certain deadline. Each agent also has some value for completing the task by the deadline. Formally, we have the following definition.

Definition 2.3.10 (Scheduling problem) *A scheduling problem consists of a tuple $C = (N, X, q, v)$, where:*

- *N is a set of n agents;*
- *X is a set of m discrete and consecutive time slots;*
- *$q = (q_1, \ldots, q_m)$ is a reserve price vector, where q_j is a reserve value for time slot x_j; q can be thought of as the value for the slot of the owner of the resource, the value he could get for it by allocating it other than to one of the n agents; and*
- *$v = (v_1, \ldots, v_n)$, where v_i, the valuation function of agent i, is a function over possible allocations of time slots that is parameterized by two arguments: d_i, the deadlines of agent i, and λ_i, the required number of time slots required by agent i. Thus for an allocation $F_i \subseteq 2^X$, we have that:*

$$v_i(F_i) = \begin{cases} w_i & \text{if } F_i \text{ includes } \lambda_i \text{ hours before } d_i; \\ 0 & \text{otherwise.} \end{cases}$$

A solution to a scheduling problem is a vector $F = (F_\emptyset, F_1, \ldots, F_n)$, where F_i is the set of time slots assigned to agent i, and $F_\emptyset$ is the time slots that are not assigned. The value of a solution is defined as

$$V(F) = \sum_{j | x_j \in F_\emptyset} q_j + \sum_{i \in N} v_i(F_i).$$

A solution is optimal if no other solution has a higher value.

Here is an example, involving scheduling jobs on a busy processor. The processor has several discrete time slots for the day—specifically, eight one-hour time slots from 9:00 A.M. to 5:00 P.M. Its operating costs force it to have a reserve price of \$3 per hour. There are four jobs, each with its own length, deadline, and worth. They are shown in the following table.

job	length (λ)	deadline (d)	worth (w)
1	2 hours	1:00 P.M.	\$10.00
2	2 hours	12:00 P.M.	\$16.00
3	1 hours	12:00 P.M.	\$6.00
4	4 hours	5:00 P.M.	\$14.50

Even in this small example it takes a moment to see that an optimal solution is to allocate the machines as follows.

time slot	agent
9:00 A.M.	2
10:00 A.M.	2
11:00 A.M.	1
12:00 P.M.	1
13:00 P.M.	4
14:00 P.M.	4
15:00 P.M.	4
16:00 P.M.	4

The question is again how to find the optimal schedule algorithmically. The scheduling problem is inherently more complex than the assignment problem. The reason is that the dependence of agents' valuation functions on the job length and deadline exhibits both *complementarity* and *substitutability*. For example, for agent 1 any two blocks of two hours prior to 1:00 are perfect substitutes. On the other hand, any two single time slots before the deadline are strongly complementary; alone they are worth nothing, but together they are worth the full $10. This makes for a more complex search space than in the case of the assignment problem, and whereas the assignment problem is polynomial, the scheduling problem is NP-complete. Indeed, the scheduling application is merely an instance of the general *set packing problem*.[4]

complementarity

substitutes

set packing problem

The complex nature of the scheduling problem has many ramifications. Among other things, this means that we cannot hope to find a polynomial LP encoding of the problem (since linear programming has a polynomial-time solution). We can, however, encode it as an integer program. In the following, for every subset $S \subseteq X$, the boolean variable $x_{i,S}$ will represent the fact that agent i was allocated the bundle S, and $v_i(S)$ his valuation for that bundle.

$$\text{maximize} \quad \sum_{S \subseteq X, i \in N} v_i(S) x_{i,S}$$

$$\text{subject to} \quad \sum_{S \subseteq X} x_{i,S} \leq 1 \qquad \forall i \in N$$

$$\sum_{S \subseteq X : j \in S, i \in N} x_{i,S} \leq 1 \qquad \forall j \in X$$

$$x_{i,S} \in \{0, 1\} \qquad \forall S \subseteq X, i \in N$$

In general, the length of the optimized quantity is exponential in the size of X. In practice, many of the terms can be assumed to be zero, and thus dropped. However, even when the IP is small, our problems are not over. IPs are not in general solvable in polynomial time, so we cannot hope for easy answers. However, it turns out that a generalization of the auction-like procedure can be

---

4. Even the scheduling problem can be defined much more broadly. It could involve earliest start times as well as deadlines, could require contiguous blocks of time for a given agent (this turns out that this requirement does not matter in our current formulation), could involve more than one resource, and so on. But the current problem formulation is rich enough for our purposes.

applied in this case too. The price we will pay for the higher complexity of the problem is that the generalized algorithm will not come with the same guarantees that we had in the case of the assignment problem.

A more general form of competitive equilibrium

competitive equilibrium

We start by revisiting the notion of *competitive equilibrium*. The definition really does not change, but rather is generalized to apply to assignments of bundles of time slots rather than single objects.

Definition 2.3.11 (Competitive equilibrium, generalized form) *Given a scheduling problem, a solution F is in* competitive equilibrium *at prices p if and only if:*

- *For all $i \in N$, it is the case that $F_i = \arg\max_{T \subseteq X}(v_i(T) - \sum_{j|x_j \in T} p_j)$ (the set of time slots allocated to agent i maximizes his surplus at prices p);*
- *For all j such that $x_j \in F_\emptyset$, it is the case that $p_j = q_j$ (the price of all unallocated time slots is the reserve price); and*
- *For all j such that $x_j \notin F_\emptyset$ it is the case that $p_j \geq q_j$ (the price of all allocated time slots is greater than the reserve price).*

As was the case in the assignment problem, a solution that is in competitive equilibrium is guaranteed to be optimal.

Theorem 2.3.12 *If a solution F to a scheduling problem C is in equilibrium at prices p, then F is also optimal for C.*

We give an informal proof to facilitate understanding of the theorem. Assume that F is in equilibrium at prices p; we would like to show that the total value of F is higher than the total value of any other solution F'. Starting with the definition of the total value of the solution F, the following equations show this inequality for an arbitrary F'.

$$
\begin{aligned}
V(F) &= \sum_{j|x_j \in F_\emptyset} q_j + \sum_{i \in N} v_i(F_i) \\
&= \sum_{j|x_j \in F_\emptyset} p_j + \sum_{i \in N} v_i(F_i) \\
&= \sum_{j|x_j \in X} p_j + \sum_{i \in N} \left[v_i(F_i) - \sum_{j|x_j \in F_i} p_j \right] \\
&\geq \sum_{j|x_j \in X} p_j + \sum_{i \in N} \left[v_i(F_i') - \sum_{j|x_j \in F_i'} p_j \right] = V(F')
\end{aligned}
$$

The last line comes from the definition of a competitive equilibrium, for each agent i, there does not exist another allocation F_i' that would yield a larger profit at the current prices (formally, $\forall i, F_i' \; v_i(F_i) - \sum_{j|x_j \in F_i} p_j \geq v_i(F_i') - \sum_{j|x_j \in F_i'} p_j$). Applying this condition to all agents, it follows that there exists no alternative allocation F' with a higher total value.

Consider our sample scheduling problem. A competitive equilibrium for that problem is shown in the following table.

time slot	agent	price
9:00 A.M.	2	$6.25
10:00 A.M.	2	$6.25
11:00 A.M.	1	$6.25
12:00 P.M.	1	$3.25
13:00 P.M.	4	$3.25
14:00 P.M.	4	$3.25
15:00 P.M.	4	$3.25
16:00 P.M.	4	$3.25

Note that the price of all allocated time slots is higher than the reserve prices of $3.00. Also note that the allocation of time slots to each agent maximizes his surplus at the prices p. Finally, also notice that the solution is stable, in that no agent can profit by making an offer for an alternative bundle at the current prices.

Even before we ask how we might find such a competitive equilibrium, we should note that one does not always exist. Consider a modified version of our scheduling example, in which the processor has two one-hour time slots, at 9:00 A.M. and at 10:00 A.M., and there are two jobs as in Table 2.1. The reserve price is $3 per hour. We show that no competitive equilibrium exists by case analysis. Clearly, if agent 1 is allocated a slot he must be allocated both slots. But then their combined price cannot exceed $10, and thus for at least one of those hours the price must not exceed $5. However, agent 2 is willing to pay as much as $6 for that hour, and thus we are out of equilibrium. Similarly, if agent 2 is allocated at least one of the two slots, their combined price cannot exceed $6, his value. But then agent 1 would happily pay more and get both slots. Finally, we cannot have both slots unallocated, since in this case their combined price would be $6, the sum of the reserve prices, in which case both agents would have the incentive to buy.

This instability arises from the fact that the agents' utility functions are superadditive (or, equivalently, that there are complementary goods). This suggest some restrictive conditions under which we are guaranteed the existence of a competitive equilibrium solution. The first theorem captures the essential connection to linear programming.

Theorem 2.3.13 *A scheduling problem has a competitive equilibrium solution if and only if the LP relaxation of the associated integer program has a integer solution.*

job	length (λ)	deadline (d)	worth (w)
1	2 hours	11:00 A.M.	$10.00
2	1 hour	11:00 A.M.	$6.00

Table 2.1 A problematic scheduling example.

The following theorem captures weaker sufficient conditions for the existence of a competitive equilibrium solution.

Theorem 2.3.14 *A scheduling problem has a competitive equilibrium solution if any one of the following conditions hold:*

- *For all agents $i \in N$, there exists a time slot $x \in X$ such that for all $T \subseteq X$, $v_i(T) = v_i(\{x\})$. (Each agent desires only a single time slot, which must be the first one in the current formulation.)*
- *For all agents $i \in N$, and for all $R, T \subseteq X$, such that $R \cap T = \emptyset$, $v_i(R \cup T) = v_i(R) + v_i(T)$. (The utility functions are additive.)*
- *Time slots are gross substitutes; demand for one time slot does not decrease if the price of another time slot increases.*

An auction algorithm

ascending-
auction
algorithm Perhaps the best-known distributed protocol for finding a competitive equilibrium is the so-called *ascending-auction algorithm*. In this protocol, the center advertises an *ask price*, and the agents bid the ask price for bundles of time slots that maximize their surplus at the given ask prices. This process repeats until there is no change.

Let $b = (b_1, \ldots, b_m)$ be the bid price vector, where b_j is the highest bid so far for time slot $x_j \in X$. Let $F = (F_1, \ldots, F_n)$ be the set of allocated slots for each agent. Finally, let ϵ be the price increment. The ascending-auction algorithm is given in Figure 2.7.

The ascending-auction algorithm is very similar to the assignment problem auction presented in the previous section, with one notable difference. Instead of calculating a bid increment from the difference between the surplus gained from the best and second-best objects, the bid increment here is always constant.

Let us consider a possible execution of the algorithm to the sample scheduling problem discussed earlier. We use an increment of $0.25 for this execution of the algorithm.

round	bidder	slots bid on	$F = (F_1, F_2, F_3, F_4)$	b
0	1	(9,10)	({9, 10}, {∅}, {∅}, {∅})	(3.25,3.25,3,3,3,3,3,3)
1	2	(10,11)	({9}, {10, 11}, {∅}, {∅})	(3.25,3.5,3.25,3,3,3,3,3)
2	3	(9)	({∅}, {10, 11}, {9}, {∅})	(3.5,3.5,3.25,3,3,3,3,3)
⋮	⋮	⋮	⋮	⋮
24	1	∅	({11, 12}, {9, 10}, {∅}, {12, 13, 14, 15})	(6.25,6.25,6.25,3.25, 3.25,3.25,3.25,3.25)

At this point, no agent has a profitable bid, and the algorithm terminates. However, this convergence depended on our choice of the increment. Let us consider what happens if we select an increment of $1.

foreach *slot* x_j **do**
$\quad b_j \leftarrow q_j$
$\quad$ // Set the initial bids to be the reserve price
foreach *agent i* **do**
$\quad F_i \leftarrow \emptyset$
repeat
$\quad$ **foreach** *agent i* $= 1$ *to n* **do**
$\qquad$ **foreach** *slot* x_j **do**
$\qquad\quad$ **if** $x_j \in F_i$ **then**
$\qquad\qquad p_j \leftarrow b_j$
$\qquad\quad$ **else**
$\qquad\qquad p_j \leftarrow b_j + \epsilon$
$\qquad\quad$ // Agents assume that they will get slots they are currently the high bidder
$\qquad\qquad\quad$ on at that price, while they must increment the bid by ϵ to get any other
$\qquad\qquad\quad$ slot.
$\qquad S^* \leftarrow \arg\max_{S \subseteq X | S \supseteq F_i}(v_i(S) - \sum_{j \in S} p_j)$
$\qquad$ // Find the best subset of slots, given your current outstanding bids
$\qquad$ // Agent *i* becomes the high bidder for all slots in $S^* \setminus F_i$.
$\qquad$ **foreach** *slot* $x_j \in S^* \setminus F_i$ **do**
$\qquad\quad b_j \leftarrow b_j + \epsilon$
$\qquad\quad$ **if** *there exists an agent* $k \neq i$ *such that* $x_j \in F_k$ **then**
$\qquad\qquad$ set $F_k \leftarrow F_k \setminus \{x_j\}$
$\qquad\quad$ // Update the bidding price and current allocations of the other bidders.
$\qquad F_i \leftarrow S^*$
until *F does not change*

Figure 2.7 The ascending-auction algorithm.

round	bidder	slots bid on	F = (F₁, F₂, F₃, F₄)	b
0	1	(9,10)	({9, 10}, {∅}, {∅}, {∅})	(4,4,3,3,3,3,3,3)
1	2	(10,11)	({9}, {10, 11}, {∅}, {∅})	(4,5,4,3,3,3,3,3)
2	3	(9)	({∅}, {10, 11}, {9}, {∅})	(5,5,4,3,3,3,3,3)
3	4	(12,13,14,15)	({∅}, {10, 11}, {9}, {12, 13, 14, 15})	(5,5,4,4,4,4,4,3)
4	1	(11,12)	({11, 12}, {10}, {9}, {13, 14, 15})	(5,5,5,5,4,4,4,3)
5	2	(9,10)	({11, 12}, {9, 10}, {∅}, {13, 14, 15})	(6,6,5,5,4,4,4,3)
6	3	(11)	({12}, {9, 10}, {11}, {13, 14, 15})	(6,6,6,5,4,4,4,3)
7	4	∅	({12}, {9, 10}, {11}, {13, 14, 15})	(6,6,6,5,4,4,4,3)
8	1	∅	({12}, {9, 10}, {11}, {13, 14, 15})	(6,6,6,5,4,4,4,3)

Unfortunately, this bidding process does not reach the competitive equilibrium because the bidding increment is not small enough.

It is also possible for the ascending-auction algorithm algorithm to not converge to an equilibrium independently of how small the increment is. Consider another problem of scheduling jobs on a busy processor. The processor has three one-hour time slots, at 9:00 A.M., 10:00 A.M., and 11:00 A.M., and there are three jobs as shown in the following table. The reserve price is $0 per hour.

job	length (λ)	deadline (d)	worth (w)
1	1 hour	11:00 A.M.	$2.00
2	2 hours	12:00 P.M.	$20.00
3	2 hours	12:00 P.M.	$8.00

Here an equilibrium exists, but the ascending auction can miss it, if agent 2 bids up the 11:00 A.M. slot.

Despite a lack of a guarantee of convergence, we might still like to be able to claim that if we do converge then we converge to an optimal solution. Unfortunately, not only can we not do that, we cannot even bound how far the solution is from optimal. Consider the following problem. The processor has two one-hour time slots, at 9:00 A.M. and 10:00 A.M. (with reserve prices of $1 and $9, respectively), and there are two jobs as shown in the following table.

job	length (λ)	deadline (d)	worth (w)
1	1 hour	10:00 A.M.	$3.00
2	2 hours	11:00 A.M.	$11.00

The ascending-auction algorithm will stop with the first slot allocated to agent 1 and the second to agent 2. By adjusting the value to agent 2 and the reserve price of the 11:00 A.M. time slot, we can create examples in which the allocation is arbitrarily far from optimal.

One property we can guarantee, however, is termination. We show this by contradiction. Assume that the algorithm does not converge. It must be the case that at each round at least one agent bids on at least one time slot, causing the price of that slot to increase. After some finite number of bids on bundles that include a particular time slot, it must be the case that the price on this slot is so high that every agent prefers the empty bundle to all bundles that include this slot. Eventually, this condition will hold for all time slots, and thus no agent will bid on a nonempty bundle, contradicting the assumption that the algorithm does not converge. In the worst case, in each iteration only one of the n agents bids, and this bid is on a single slot. Once the sum of the prices exceeds the maximum total value for the agents, the algorithm must terminate, giving us the worst-case running time $O(n \max_{F_i} \frac{\sum_{i \in N} v_i(F_i)}{\epsilon})$.

2.4 Social laws and conventions

social law

Consider the task of a city transportation official who wishes to optimize traffic flow in the city. While he cannot redesign cars or create new roads, he can impose *traffic rules*. A traffic rule is a form of a *social law*: a restriction on the given strategies of the agents. A typical traffic rule prohibits people from driving on the left side of the road or through red lights. For a given agent, a social law presents a tradeoff; he suffers from loss of freedom, but can benefit from the fact that others lose some freedom. A good social law is designed to benefit all agents.

One natural formal view of social laws is from the perspective of game theory. We discuss game theory in detail starting in Chapter 3, but here we need very little of that material. For our purposes here, suffice it to say that in a game each agent has a number of possible strategies (in our traffic example, driving plans), and depending on the strategies selected by each agent, each agent receives a certain payoff. In general, agents are free to choose their own strategies, which they will do based on their guesses about the strategies of other agents. Sometimes the interests of the agents are at odds with each other, but sometimes they are not. In the extreme case the interests are perfectly aligned, and the only problem is that of coordination among the agents. Again, traffic presents the perfect example; agents are equally happy driving on the left or on the right, provided everyone does the same.

social convention

A social law simply eliminates from a given game certain strategies for each of the agents, and thus induces a subgame. When the subgame consists of a single strategy for each agent, we call it a *social convention*. In many cases the setting is naturally symmetric (the game is symmetric, as are the restrictions), but it need not be that way. A social law is good if the induced subgame is "preferred" to the original one. There can be different notions of preference here; we will discuss this further after we discuss the notion of *solution concepts* in Chapter 3. For now we leave the notion of preference at the intuitive level; intuitively, a world where everyone (say) drives on the right and stops at red lights is preferable to one in which drivers cannot rely on such laws and must constantly coordinate with each other.

This leaves the question of how one might find such a good social law or social convention. In Chapter 7 we adopt a democratic perspective; we look at how conventions can emerge dynamically as a result of a learning process within the population. Here we adopt a more autocratic perspective, and imagine a social planner imposing a good social law (or even a single convention). The question is how such a benign dictator arrives at such a good social law. In general the problem is hard; specifically, when formally defined, the general problem of finding a good social law (under an appropriate notion of "good") can be shown to be NP-hard. However, the news is not all bad. First, there exist restrictions that render the problem polynomial. Furthermore, in specific situations, one can simply hand craft good social laws.

Indeed, traffic rules provide an excellent example. Consider a set of k robots $\{0, 1, \ldots, k - 1\}$ belonging to Deliverobot, Inc., who must navigate a road system connecting seven locations as depicted in Figure 2.8.

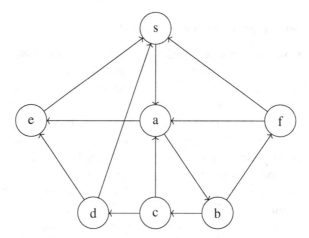

Figure 2.8 Seven locations in a transportation domain.

Assume these k robots are the only vehicles using the road, and their main challenge is to avoid collisions among themselves. Assume further that they all start at point s, the company's depot, at the start of the day. We assume a discrete model of time, and that each robot requires one unit of time to traverse any given edge, though the robots can also travel more slowly if they wish. At each of the first k time steps one robot is assigned initial tasks and sent on its way, with robot i sent at time i $(i = 0, 1, \ldots, k - 1)$. Thereafter they are in continuous motion; as soon as they arrive at their current destination they are assigned a new task, and off they go. A collision is defined as two robots occupying the same location at the same time. How can collisions be avoided without the company constantly planning routes for the robots, and without the robots constantly having to negotiate with each other? The tools they have at their disposal are the speed with which they traverse each edge and the common clock they implicitly share with the other robots.

Here is one simple solution: Each robot drives so that traversing each link takes exactly k time units. In this case, at any time t the only robot who will arrive at a node—any node—is $i \equiv t \mod k$. This is an example of a simple social convention that is useful, but that comes at a price. Each robot is free to travel along the shortest path, but will traverse this path k times more slowly than he would without this particular social law.

Here is a more efficient convention. Assign each vertex an arbitrary label between 0 and $k - 1$, and define the time to traverse an edge between vertices labeled x and y to be $(y - x) \mod k$ if $(y - x) \mod k > 0$, and k otherwise. Observe that the difference in this expression will sometimes be negative; this is not a problem because the modulo nevertheless returns a nonnegative value. To consider an example, if agent i follows the sequence of nodes labeled s, x_1, x_2, x_3 then its travel times are $(x_1 - s) \mod k$, $(x_2 - x_1) \mod k$, $(x_3 - x_2) \mod k$, presuming that none of these expressions evaluated to zero. Adding these travel times to the start time we see that i reaches node x_3 at time $t \equiv i + x_1 + (x_2 - x_1) + (x_3 - x_2) \equiv x_3 + i \mod k$. In general, we have that at time t agent i will always either be on an edge or waiting at a node labeled $(t - i) \mod k$, and thus there will be no collisions.

A final comment is in order. In the discussion so far we have assumed that once a social law is imposed (or agreed upon) it is adhered to. This is of course a tenuous assumption when applied to fallible and self-interested agents. In Chapter 10 (and specifically in Section 10.7) we return to this topic.

2.5 History and references

Distributed dynamic programming is discussed in detail in Bertsekas [1982]. LRTA* is introduced in Korf [1990], and our section follows that material, as well as Yokoo and Ishida [1999].

Distributed solutions to Markov Decision Problems are discussed in detail in Guestrin [2003]; the discussion there goes far beyond the specific problem of joint action selection covered here. Additional discussion specifically on the issue of problem selection in distributed MDPs can be found in Vlassis et al. [2004].

Contract nets were introduced in Smith [1980], and Davis and Smith [1983] is perhaps the most influential publication on the topic. The marginal-cost interpretation of contract nets was introduced in Sandholm [1993], and the discussion of the capabilities and limitations of the various contract types (O, C, S, and M) followed in Sandholm [1998]. Auction algorithms for linear programming are discussed broadly in Bertsekas [1991]. The specific algorithm for the matching problem is taken from Bertsekas [1992]. Its extension to the combinatorial setting is discussed in Parkes and Ungar [2000]. Auction algorithms for combinatorial problems in general are introduced in Wellman [1993], and the specific auction algorithms for the scheduling problem appear in Wellman et al. [2001].

Social laws and conventions, and the example of traffic laws, were introduced in Shoham and Tennenholtz [1995]. The treatment there includes many additional tweaks on the basic traffic grid discussed here, as well as an algorithmic analysis of the problem in general.

3

Introduction to Noncooperative Game Theory: Games in Normal Form

Game theory is the mathematical study of interaction among independent, self-interested agents. It has been applied to disciplines as diverse as economics (historically, its main area of application), political science, biology, psychology, linguistics—and computer science. In this chapter we will concentrate on what has become the dominant branch of game theory, called *noncooperative* game theory, and specifically on normal-form games, a canonical representation in this discipline.

As an aside, the name "noncooperative game theory" could be misleading, since it may suggest that the theory applies exclusively to situations in which the interests of different agents conflict. This is not the case, although it is fair to say that the theory is most interesting in such situations. By the same token, in Chapter 12 we will see that *coalitional game theory* (also known as *cooperative game theory*) does not apply only in situations in which the interests of the agents align with each other. The essential difference between the two branches is that in noncooperative game theory the basic modeling unit is the individual (including his beliefs, preferences, and possible actions) while in coalitional game theory the basic modeling unit is the group. We will return to that later in Chapter 12, but for now let us proceed with the individualistic approach.

coalitional game theory

3.1 Self-interested agents

What does it mean to say that agents are self-interested? It does not necessarily mean that they want to cause harm to each other, or even that they care only about themselves. Instead, it means that each agent has his own description of which states of the world he likes—which can include good things happening to other agents—and that he acts in an attempt to bring about these states of the world. In this section we will consider how to model such interests.

utility theory

The dominant approach to modeling an agent's interests is *utility theory*. This theoretical approach aims to quantify an agent's degree of preference across a set of available alternatives. The theory also aims to understand how these preferences change when an agent faces uncertainty about which alternative he will receive. When we refer to an agent's *utility function*, as we will do throughout much of this book, we will be making an implicit assumption that the agent has

utility function

desires about how to act that are consistent with utility-theoretic assumptions. Thus, before we discuss game theory (and thus interactions between *multiple* utility-theoretic agents), we should examine some key properties of utility functions and explain why they are believed to form a solid basis for a theory of preference and rational action.

A utility function is a mapping from states of the world to real numbers. These numbers are interpreted as measures of an agent's level of happiness in the given states. When the agent is uncertain about which state of the world he faces, his utility is defined as the expected value of his utility function with respect to the appropriate probability distribution over states.

3.1.1   *Example: friends and enemies*

We begin with a simple example of how utility functions can be used as a basis for making decisions. Consider an agent Alice, who has three options: going to the club (c), going to a movie (m), or watching a video at home (h). If she is on her own, Alice has a utility of 100 for c, 50 for m, and 50 for h. However, Alice is also interested in the activities of two other agents, Bob and Carol, who frequent both the club and the movie theater. Bob is Alice's nemesis; he is downright painful to be around. If Alice runs into Bob at the movies, she can try to ignore him and only suffers a disutility of 40; however, if she sees him at the club he will pester her endlessly, yielding her a disutility of 90. Unfortunately, Bob prefers the club: he is there 60% of the time, spending the rest of his time at the movie theater. Carol, on the other hand, is Alice's friend. She makes everything more fun. Specifically, Carol increases Alice's utility for either activity by a factor of 1.5 (after taking into account the possible disutility of running into Bob). Carol can be found at the club 25% of the time, and the movie theater 75% of the time.

It will be easier to determine Alice's best course of action if we list Alice's utility for each possible state of the world. There are 12 outcomes that can occur: Bob and Carol can each be in either the club or the movie theater, and Alice can be in the club, the movie theater, or at home. Alice has a baseline level of utility for each of her three actions, and this baseline is adjusted if either Bob, Carol, or both are present. Following the description of our example, we see that Alice's utility is always 50 when she stays home, and for her other two activities it is given by Figure 3.1.

So how should Alice choose among her three activities? To answer this question we need to combine her utility function with her knowledge of Bob and Carol's randomized entertainment habits. Alice's expected utility for going to the club can be calculated as $0.25(0.6 \cdot 15 + 0.4 \cdot 150) + 0.75(0.6 \cdot 10 + 0.4 \cdot 100) = 51.75$. In the same way, we can calculate her expected utility for going to the movies as $0.25(0.6 \cdot 50 + 0.4 \cdot 10) + 0.75(0.6(75) + 0.4(15)) = 46.75$. Of course, Alice gets an expected utility of 50 for staying home. Thus, Alice prefers to go to the club (even though Bob is often there and Carol rarely is) and prefers staying home to going to the movies (even though Bob is usually not at the movies and Carol almost always is).

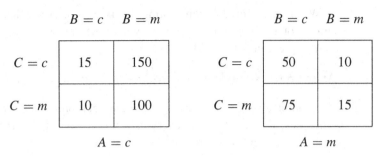

Figure 3.1 Alice's utility for the actions c and m.

3.1.2   *Preferences and utility*

Because the idea of utility is so pervasive, it may be hard to see why anyone would argue with the claim that it provides a sensible formal model for reasoning about an agent's happiness in different situations. However, when considered more carefully this claim turns out to be substantive, and hence requires justification. For example, why should a single-dimensional function be enough to explain preferences over an arbitrarily complicated set of alternatives (rather than, say, a function that maps to a point in a three-dimensional space, or to a point in a space whose dimensionality depends on the number of alternatives being considered)? And why should an agent's response to uncertainty be captured purely by the expected value of his utility function, rather than also depending on other properties of the distribution such as its standard deviation or number of modes?

preferences   Utility theorists respond to such questions by showing that the idea of utility can be grounded in a more basic concept of *preferences*. The most influential such theory is due to von Neumann and Morgenstern, and thus the utility functions are sometimes called von Neumann–Morgenstern utility functions to distinguish them from other varieties. We present that theory here.

Let O denote a finite set of outcomes. For any pair $o_1, o_2 \in O$, let $o_1 \succeq o_2$ denote the proposition that the agent weakly prefers o_1 to o_2. Let $o_1 \sim o_2$ denote the proposition that the agent is indifferent between o_1 and o_2. Finally, by $o_1 \succ o_2$, denote the proposition that the agent strictly prefers o_1 to o_2. Note that while the second two relations are notationally convenient, the first relation $\succeq$ is the only one we actually need. This is because we can define $o_1 \succ o_2$ as "$o_1 \succeq o_2$ and not $o_2 \succeq o_1$," and $o_1 \sim o_2$ as "$o_1 \succeq o_2$ and $o_2 \succeq o_1$."

We need a way to talk about how preferences interact with uncertainty about which outcome will be selected. In utility theory this is achieved through the lottery   concept of *lotteries*. A lottery is the random selection of one of a set of outcomes according to specified probabilities. Formally, a lottery is a probability distribution over outcomes written $[p_1 : o_1, \ldots, p_k : o_k]$, where each $o_i \in O$, each $p_i \geq 0$ and $\sum_{i=1}^{k} p_i = 1$. Let $\mathcal{L}$ denote the set of all lotteries. We will extend the $\succeq$ relation to apply to the elements of $\mathcal{L}$ as well as to the elements of O, effectively considering lotteries over outcomes to be outcomes themselves.

We are now able to begin stating the axioms of utility theory. These are constraints on the $\succeq$ relation which, we will argue, make it consistent with our ideas of how preferences should behave.

Axiom 3.1.1 (Completeness) $\forall o_1, o_2, \; o_1 \succ o_2 \; or \; o_2 \succ o_1 \; or \; o_1 \sim o_2.$

The completeness axiom states that the $\succeq$ relation induces an ordering over the outcomes, allowing ties. For every pair of outcomes, either the agent prefers one to the other or he is indifferent between them.

Axiom 3.1.2 (Transitivity) *If* $o_1 \succeq o_2$ *and* $o_2 \succeq o_3$, *then* $o_1 \succeq o_3$.

There is good reason to feel that every agent should have transitive preferences. If an agent's preferences were nontransitive, then there would exist some triple of outcomes o_1, o_2, and o_3 for which $o_1 \succeq o_2$, $o_2 \succeq o_3$, and $o_3 \succ o_1$. We can show that such an agent would be willing to engage in behavior that is hard to call rational. Consider a world in which o_1, o_2, and o_3 correspond to owning three different items, and an agent who currently owns the item o_3. Since $o_2 \succeq o_3$, there must be some nonnegative amount of money that the agent would be willing to pay in order to exchange o_3 for o_2. (If $o_2 \succ o_3$ then this amount would be strictly positive; if $o_2 \sim o_3$, then it would be zero.) Similarly, the agent would pay a nonnegative amount of money to exchange o_2 for o_1. However, from non-transitivity ($o_3 \succ o_1$) the agent would *also* pay a strictly positive amount of money to exchange o_1 for o_3. The agent would thus be willing to pay a strictly positive sum to exchange o_3 for o_3 in three steps. Such an agent could quickly be separated from *any* amount of money, which is why such a scheme is known as
money pump   a *money pump*.

Axiom 3.1.3 (Substitutability) *If* $o_1 \sim o_2$, *then for all sequences of one or more outcomes* $o_3, \ldots, o_k$ *and sets of probabilities* $p, p_3, \ldots, p_k$ *for which* $p + \sum_{i=3}^{k} p_i = 1$, $[p : o_1, p_3 : o_3, \ldots, p_k : o_k] \sim [p : o_2, p_3 : o_3, \ldots, p_k : o_k]$.

Let $P_\ell(o_i)$ denote the probability that outcome o_i is selected by lottery ℓ. For example, if $\ell = [0.3 : o_1; 0.7 : [0.8 : o_2; 0.2 : o_1]]$, then $P_\ell(o_1) = 0.44$ and $P_\ell(o_3) = 0$.

Axiom 3.1.4 (Decomposability) *If* $\forall o_i \in O$, $P_{\ell_1}(o_i) = P_{\ell_2}(o_i)$ *then* $\ell_1 \sim \ell_2$.

These axioms describe the way preferences change when lotteries are introduced. Substitutability states that if an agent is indifferent between two outcomes, he is also indifferent between two lotteries that differ only in which of these outcomes is offered. Decomposability states that an agent is always indifferent between lotteries that induce the same probabilities over outcomes, no matter whether these probabilities are expressed through a single lottery or nested in a lottery over lotteries. For example, $[p : o_1, 1 - p : [q : o_2, 1 - q : o_3]] \sim [p : o_1, (1 - p)q : o_2, (1 - p)(1 - q) : o_3]$. Decomposability is sometimes called the "no fun in gambling" axiom because it implies that, all else being equal, the number of times an agent "rolls dice" has no effect on his preferences.

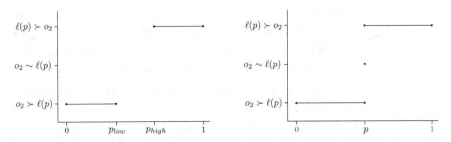

Figure 3.2 Relationship between o_2 and $\ell(p)$.

Axiom 3.1.5 (Monotonicity) *If $o_1 \succ o_2$ and $p > q$ then $[p : o_1, 1 - p : o_2] \succ [q : o_1, 1 - q : o_2]$.*

The monotonicity axiom says that agents prefer more of a good thing. When an agent prefers o_1 to o_2 and considers two lotteries over these outcomes, he prefers the lottery that assigns the larger probability to o_1. This property is called monotonicity because it does not depend on the numerical values of the probabilities—the more weight o_1 receives, the happier the agent will be.

Lemma 3.1.6 *If a preference relation $\succeq$ satisfies the axioms completeness, transitivity, decomposability, and monotonicity, and if $o_1 \succ o_2$ and $o_2 \succ o_3$, then there exists some probability p such that for all $p' < p$, $o_2 \succ [p' : o_1; (1 - p') : o_3]$, and for all $p'' > p$, $[p'' : o_1; (1 - p'') : o_3] \succ o_2$.*

Proof. Denote the lottery $[p : o_1; (1 - p) : o_3]$ as $\ell(p)$. Consider some p_{low} for which $o_2 \succ \ell(p_{low})$. Such a p_{low} must exist since $o_2 \succ o_3$; for example, by decomposability $p_{low} = 0$ satisfies this condition. By monotonicity, $\ell(p_{low}) \succ \ell(p')$ for any $0 \le p' < p_{low}$, and so by transitivity $\forall p' \le p_{low}$, $o_2 \succ \ell(p')$. Consider some p_{high} for which $\ell(p_{high}) \succ o_2$. By monotonicity, $\ell(p') \succ \ell(p_{high})$ for any $1 \ge p' > p_{high}$, and so by transitivity $\forall p' \ge p_{high}$, $\ell(p') \succ o_2$. We thus know the relationship between $\ell(p)$ and o_2 for all values of p except those on the interval (p_{low}, p_{high}). This is illustrated in Figure 3.2 (left).

Consider $p^* = (p_{low} + p_{high})/2$, the midpoint of our interval. By completeness, $o_2 \succ \ell(p^*)$ or $\ell(p^*) \succ o_2$ or $o_2 \sim \ell(p^*)$. First consider the case $o_2 \sim \ell(p^*)$. It cannot be that there is also another point $p' \ne p^*$ for which $o_2 \sim \ell(p')$: this would entail $\ell(p^*) \sim \ell(p')$ by transitivity, and since $o_1 \succ o_3$, this would violate monotonicity. For all $p' \ne p^*$, then, it must be that either $o_2 \succ \ell(p')$ or $\ell(p') \succ o_2$. By the arguments earlier, if there was a point $p' > p^*$ for which $o_2 \succ \ell(p')$, then $\forall p'' < p'$, $o_2 \succ \ell(p'')$, contradicting $o_2 \sim \ell(p^*)$. Similarly there cannot be a point $p' < p^*$ for which $\ell(p') \succ o_2$. The relationship that must therefore hold between o_2 and $\ell(p)$ is illustrated in Figure 3.2 (right). Thus, in the case $o_2 \sim \ell(p^*)$, we have our result.

Otherwise, if $o_2 \succ \ell(p^*)$, then by the argument given earlier $o_2 \succ \ell(p')$ for all $p' \le p^*$. Thus we can redefine p_{low}—the lower bound of the interval of values for which we do not know the relationship between o_2 and $\ell(p)$—to be p^*. Likewise, if $\ell(p^*) \succ o_2$ then we can redefine $p_{high} = p^*$. Either way, our interval (p_{low}, p_{high}) is halved. We can continue to iterate the above argument, examining the midpoint of the updated interval (p_{low}, p_{high}). Either we will encounter a p^* for which $o_2 \sim \ell(p^*)$, or in the limit p_{low} will approach some p from below, and p_{high} will approach that p from above.   ■

Something our axioms do not tell us is what preference relation holds between o_2 and the lottery $[p : o_1; (1 - p) : o_3]$. It could be that the agent strictly prefers o_2 in this case, that the agent prefers the lottery, or that the agent is indifferent. Our final axiom says that the third alternative—depicted in Figure 3.2 (right)—always holds.

Axiom 3.1.7 (Continuity) *If $o_1 \succ o_2$ and $o_2 \succ o_3$, then $\exists p \in [0, 1]$ such that $o_2 \sim [p : o_1, 1 - p : o_3]$.*

utility function   If we accept Axioms 3.1.1, 3.1.2, 3.1.4, 3.1.5, and 3.1.7, it turns out that we have no choice but to accept the existence of single-dimensional *utility functions* whose expected values agents want to maximize. (And if we do *not* want to reach this conclusion, we must therefore give up at least one of the axioms.) This fact is stated as the following theorem.

Theorem 3.1.8 (von Neumann and Morgenstern, 1944) *If a preference relation $\succeq$ satisfies the axioms completeness, transitivity, substitutability, decomposability, monotonicity, and continuity, then there exists a function $u : O \mapsto [0, 1]$ with the properties that:*

1. *$u(o_1) \geq u(o_2)$ iff $o_1 \succeq o_2$; and*
2. *$u([p_1 : o_1, \ldots, p_k : o_k]) = \sum_{i=1}^{k} p_i u(o_i)$.*

Proof. If the agent is indifferent among all outcomes, then for all $o_i \in O$ set $u(o_i) = 0$ and for all $\ell \in \mathcal{L}$ set $\ell = 0$. In this case Part 1 follows trivially (both sides of the implication are always true), and Part 2 is immediate from decomposability.

Otherwise, there must be a set of one or more most-preferred outcomes and a disjoint set of one or more least-preferred outcomes. (There may of course be other outcomes belonging to neither set.) Label one of the most-preferred outcomes as $\overline{o}$ and one of the least-preferred outcomes as $\underline{o}$. For any outcome o_i, define $u(o_i)$ to be the number p_i such that $o_i \sim [p_i : \overline{o}, (1 - p_i) : \underline{o}]$. By continuity such a number exists; by Lemma 3.1.6 it is unique.

Part 1: $u(o_1) \geq u(o_2)$ iff $o_1 \succeq o_2$.

We know that $o_1 \sim [u(o_1) : \overline{o}; 1 - u(o_1) : \underline{o}]$; denote this lottery ℓ_1. Likewise, $o_2 \sim [u(o_2) : \overline{o}; 1 - u(o_2) : \underline{o}]$; denote this lottery ℓ_2. First, we show that $u(o_1) \geq u(o_2) \Rightarrow o_1 \succeq o_2$. If $u(o_1) > u(o_2)$ then, since $\overline{o} \succ \underline{o}$ we can conclude that $\ell_1 \succ \ell_2$ by monotonicity. Thus, we have $o_1 \sim \ell_1 \succ \ell_2 \sim o_2$; by transitivity and completeness, this gives $o_1 \succ o_2$. If $u(o_1) = u(o_2)$, the ℓ_1 and ℓ_2 are identical lotteries; thus, $o_1 \sim \ell_1 \equiv \ell_2 \sim o_2$, and transitivity gives $o_1 \sim o_2$.

Now we must show that $o_1 \succeq o_2 \Rightarrow u(o_1) \geq u(o_2)$. It suffices to prove the contrapositive of this statement, $u(o_1) \not\geq u(o_2) \Rightarrow o_1 \not\succeq o_2$, which can be rewritten as $u(o_2) > u(o_1) \Rightarrow o_2 \succ o_1$ by completeness. This statement was already proved earlier (with the labels o_1 and o_2 swapped).

Part 2: $u([p_1 : o_1, \ldots, p_k : o_k]) = \sum_{i=1}^{k} p_i u(o_i)$.

Let $u^* = u([p_1 : o_1, \ldots, p_k : o_k])$. From the construction of u we know that $o_i \sim [u(o_i) : \overline{o}, (1 - u(o_i)) : \underline{o}]$. By substitutability, we can replace each o_i in the definition of u^* by the lottery $[u(o_i) : \overline{o}, (1 - u(o_i)) : \underline{o}]$, giving us $u^* = u([p_1 : [u(o_1) : \overline{o}, (1 - u(o_1)) : \underline{o}], \ldots, p_k : [u(o_k) : \overline{o}, (1 - u(o_k)) : \underline{o}]])$. This nested lottery only selects between the two outcomes $\overline{o}$ and $\underline{o}$. This means that we can use decomposability to conclude $u^* = u\left(\left[\left(\sum_{i=1}^{k} p_i u(o_i)\right) : \overline{o}, 1 - \left(\sum_{i=1}^{k} p_i u(o_i)\right) : \underline{o}\right]\right)$. By our definition of u, $u^* = \sum_{i=1}^{k} p_i u(o_i)$. ∎

One might wonder why we do not use money to express the real-valued quantity that rational agents want to maximize, rather than inventing the new concept of utility. The reason is that while it is reasonable to assume that all agents get happier the more money they have, it is often not reasonable to assume that agents care only about the *expected values* of their bank balances. For example, consider a situation in which an agent is offered a gamble between a payoff of two million and a payoff of zero, with even odds. When the outcomes are measured in units of utility ("utils") then Theorem 3.1.8 tells us that the agent would prefer this gamble to a sure payoff of 999,999 utils. However, if the outcomes were measured in money, few of us would prefer to gamble—most people would prefer a guaranteed payment of nearly a million dollars to a double-or-nothing bet. This is not to say that utility-theoretic reasoning goes out the window when money is involved. It simply points out that utility and money are often not linearly related. This issue is discussed in more detail in Section 10.3.1.

What if we want a utility function that is not confined to the range [0, 1], such as the one we had in our friends and enemies example? Luckily, Theorem 3.1.8 does not *require* that every utility function maps to this range; it simply shows that one such utility function must exist for every set of preferences that satisfy the required axioms. Indeed, von Neumann and Morgenstern also showed that the absolute magnitudes of the utility function evaluated at different outcomes are unimportant. Instead, every positive affine transformation of a utility function yields another utility function for the same agent (in the sense that it will also satisfy both properties of Theorem 3.1.8). In other words, if $u(o)$ is a utility function for a given agent then $u'(o) = au(o) + b$ is also a utility function for the same agent, as long as a and b are constants and a is positive.

3.2   Games in normal form

We have seen that under reasonable assumptions about preferences, agents will always have utility functions whose expected values they want to maximize. This suggests that acting optimally in an uncertain environment is conceptually straightforward—at least as long as the outcomes and their probabilities are known to the agent and can be succinctly represented. Agents simply need to choose the course of action that maximizes expected utility. However, things can

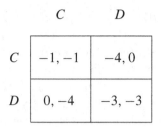

Figure 3.3 The TCP user's (aka the Prisoner's) Dilemma.

get considerably more complicated when the world contains *two or more* utility-maximizing agents whose actions can affect each other's utilities. (To augment our example from Section 3.1.1, what if Bob hates Alice and wants to avoid her too, while Carol is indifferent to seeing Alice and has a crush on Bob? In this case, we might want to revisit our previous assumption that Bob and Carol will act randomly without caring about what the other two agents do.) To study such settings, we turn to game theory.

3.2.1 *Example: the TCP user's game*

TCP user's game   Let us begin with a simpler example to provide some intuition about the type of phenomena we would like to study. Imagine that you and another colleague are the only people using the internet. Internet traffic is governed by the TCP protocol. One feature of TCP is the *backoff* mechanism; if the rates at which you and your colleague send information packets into the network causes congestion, you each back off and reduce the rate for a while until the congestion subsides. This is how a correct implementation works. A defective one, however, will not back off when congestion occurs. You have two possible strategies: C (for using a correct implementation) and D (for using a defective one). If both you and your colleague adopt C then your average packet delay is 1 ms. If you both adopt D the delay is 3 ms, because of additional overhead at the network router. Finally, if one of you adopts D and the other adopts C then the D adopter will experience no delay at all, but the C adopter will experience a delay of 4 ms.

These consequences are shown in Figure 3.3. Your options are the two rows, and your colleague's options are the columns. In each cell, the first number represents your payoff (or, the negative of your delay) and the second number Prisoner's   represents your colleague's payoff.[1]
Dilemma game
Given these options what should you adopt, C or D? Does it depend on what you think your colleague will do? Furthermore, from the perspective of the network operator, what kind of behavior can he expect from the two users? Will any two users behave the same when presented with this scenario? Will the behavior change if the network operator allows the users to communicate with each other before making a decision? Under what changes to the delays would the users' decisions still be the same? How would the users behave if they have

---

1. A more standard name for this game is the Prisoner's Dilemma; we return to this in Section 3.2.3.

the opportunity to face this same decision with the same counterpart multiple times? Do answers to these questions depend on how rational the agents are and how they view each other's rationality?

Game theory gives answers to many of these questions. It tells us that any rational user, when presented with this scenario once, will adopt *D*—regardless of what the other user does. It tells us that allowing the users to communicate beforehand will not change the outcome. It tells us that for perfectly rational agents, the decision will remain the same even if they play multiple times; however, if the number of times that the agents will play is infinite, or even uncertain, we may see them adopt *C*.

3.2.2 *Definition of games in normal form*

The normal form, also known as the strategic form, is the most familiar representation of strategic interactions in game theory. A game written in this way amounts to a representation of every player's utility for every state of the world, in the special case where states of the world depend only on the players' combined actions. Consideration of this special case may seem uninteresting. However, it turns out that settings in which the state of the world also depends on randomness in the environment—called Bayesian games and introduced in Section 6.3—can be reduced to (much larger) normal-form games. Indeed, there also exist normal-form reductions for other game representations, such as games that involve an element of time (extensive-form games, introduced in Chapter 5). Because most other representations of interest can be reduced to it, the normal-form representation is arguably the most fundamental in game theory.

Definition 3.2.1 (Normal-form game) *A (finite, n-person) normal-form game is a tuple* (N, A, u), *where:*

action
action profile

- *N is a finite set of n players, indexed by i;*
- $A = A_1 \times \cdots \times A_n$, *where* A_i *is a finite set of* actions *available to player i. Each vector* $a = (a_1, \ldots, a_n) \in A$ *is called an* action profile;
- $u = (u_1, \ldots, u_n)$, *where* $u_i : A \mapsto \mathbb{R}$ *is a real-valued* utility *(or* payoff*) function for player i.*

Note that we previously argued that utility functions should map from the set of *outcomes*, not the set of *actions*. Here we make the implicit assumption that $O = A$.

A natural way to represent games is via an *n*-dimensional matrix. We already saw a two-dimensional example in Figure 3.3. In general, each row corresponds to a possible action for player 1, each column corresponds to a possible action for player 2, and each cell corresponds to one possible outcome. Each player's utility for an outcome is written in the cell corresponding to that outcome, with player 1's utility listed first.

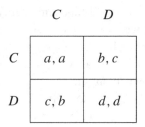

$$
\begin{array}{c|c|c|}
 & C & D \\
\hline
C & a, a & b, c \\
\hline
D & c, b & d, d \\
\hline
\end{array}
$$

Figure 3.4 Any $c > a > d > b$ define an instance of Prisoner's Dilemma.

3.2.3 *More examples of normal-form games*

Prisoner's Dilemma

Previously, we saw an example of a game in normal form, namely, the Prisoner's (or the TCP user's) Dilemma. However, as discussed in Section 3.1.2, the precise payoff numbers play a limited role. The essence of the Prisoner's Dilemma example would not change if the −4 was replaced by −5, or if 100 was added to each of the numbers. In its most general form, the Prisoner's Dilemma is any normal-form game shown in Figure 3.4, in which $c > a > d > b$.[2]

Incidentally, the name "Prisoner's Dilemma" for this famous game-theoretic situation derives from the original story accompanying the numbers. The players of the game are two prisoners suspected of a crime rather than two network users. The prisoners are taken to separate interrogation rooms, and each can either "confess" to the crime or "deny" it (or, alternatively, "cooperate" or "defect"). If the payoff are all nonpositive, their absolute values can be interpreted as the length of jail term each of prisoner gets in each scenario.

Common-payoff games

There are some restricted classes of normal-form games that deserve special mention. The first is the class of *common-payoff games*. These are games in which, for every action profile, all players have the same payoff.

common-payoff game

Definition 3.2.2 (Common-payoff game) *A common-payoff game is a game in which for all action profiles $a \in A_1 \times \cdots \times A_n$ and any pair of agents i, j, it is the case that $u_i(a) = u_j(a)$.*

pure coordination game

Common-payoff games are also called *pure coordination games* or *team games*. In such games the agents have no conflicting interests; their sole challenge is to coordinate on an action that is maximally beneficial to all.

team game

As an example, imagine two drivers driving towards each other in a country having no traffic rules, and who must independently decide whether to drive on

---

2. Under some definitions, there is the further requirement that $a > \frac{b+c}{2}$, which guarantees that the outcome (C, C) maximizes the sum of the agents' utilities.

Left   Right

	Left	Right
Left	1, 1	0, 0
Right	0, 0	1, 1

Figure 3.5 Coordination game.

the left or on the right. If the drivers choose the same side (left or right) they have some high utility, and otherwise they have a low utility. The game matrix is shown in Figure 3.5.

Zero-sum games

zero-sum game   At the other end of the spectrum from pure coordination games lie *zero-sum games*, which (bearing in mind the comment we made earlier about positive affine transformations) are more properly called *constant-sum games*. Unlike common-payoff games, constant-sum games are meaningful primarily in the context of two-player (though not necessarily two-strategy) games.

constant-sum game

Definition 3.2.3 (Constant-sum game) *A two-player normal-form game is constant-sum if there exists a constant c such that for each strategy profile $a \in A_1 \times A_2$ it is the case that $u_1(a) + u_2(a) = c$.*

For convenience, when we talk of constant-sum games going forward we will always assume that $c = 0$, that is, that we have a zero-sum game. If common-payoff games represent situations of pure coordination, zero-sum games represent situations of pure competition; one player's gain must come at the expense of the other player. This property requires that there be exactly two agents. Indeed, if you allow more agents, any game can be turned into a zero-sum game by adding a dummy player whose actions do not impact the payoffs to the other agents, and whose own payoffs are chosen to make the payoffs in each outcome sum to zero.

Matching   A classical example of a zero-sum game is the game of *Matching Pennies*.
Pennies game   In this game, each of the two players has a penny and independently chooses to display either heads or tails. The two players then compare their pennies. If they are the same then player 1 pockets both, and otherwise player 2 pockets them. The payoff matrix is shown in Figure 3.6.

The popular children's game of Rock, Paper, Scissors, also known as Rochambeau, provides a three-strategy generalization of the matching-pennies game. The payoff matrix of this zero-sum game is shown in Figure 3.7. In this game, each of the two players can choose either rock, paper, or scissors. If both players choose the same action, there is no winner and the utilities are zero. Otherwise,

	Heads	Tails
Heads	1, −1	−1, 1
Tails	−1, 1	1, −1

Figure 3.6 Matching Pennies game.

	Rock	Paper	Scissors
Rock	0, 0	−1, 1	1, −1
Paper	1, −1	0, 0	−1, 1
Scissors	−1, 1	1, −1	0, 0

Figure 3.7 Rock, Paper, Scissors game.

each of the actions wins over one of the other actions and loses to the other remaining action.

Battle of the Sexes

Battle of the
Sexes game

In general, games can include elements of both coordination and competition. Prisoner's Dilemma does, although in a rather paradoxical way. Here is another well-known game that includes both elements. In this game, called *Battle of the Sexes*, a husband and wife wish to go to the movies, and they can select among two movies: "Lethal Weapon (LW)" and "Wondrous Love (WL)." They much prefer to go together rather than to separate movies, but while the wife (player 1) prefers LW, the husband (player 2) prefers WL. The payoff matrix is shown in Figure 3.8. We will return to this game shortly.

3.2.4 *Strategies in normal-form games*

pure strategy
pure-strategy
profile

We have so far defined the actions available to each player in a game, but not yet his set of *strategies* or his available choices. Certainly one kind of strategy is to select a single action and play it. We call such a strategy a *pure strategy*, and we will use the notation we have already developed for actions to represent it. We call a choice of pure strategy for each agent a *pure-strategy profile*.

Husband

LW   WL

	LW	WL
LW	2, 1	0, 0
WL	0, 0	1, 2

Wife

Figure 3.8 Battle of the Sexes game.

Players could also follow another, less obvious type of strategy: randomizing over the set of available actions according to some probability distribution. Such a strategy is called a mixed strategy. Although it may not be immediately obvious why a player should introduce randomness into his choice of action, in fact in a multiagent setting the role of mixed strategies is critical. We define a mixed strategy for a normal-form game as follows.

mixed strategy

Definition 3.2.4 (Mixed strategy) *Let* (N, A, u) *be a normal-form game, and for any set X let $\Pi(X)$ be the set of all probability distributions over X. Then the set of* mixed strategies *for player i is $S_i = \Pi(A_i)$.*

mixed-strategy
profile

Definition 3.2.5 (Mixed-strategy profile) *The set of* mixed-strategy profiles *is simply the Cartesian product of the individual mixed-strategy sets, $S_1 \times \cdots \times S_n$.*

By $s_i(a_i)$ we denote the probability that an action a_i will be played under mixed strategy s_i. The subset of actions that are assigned positive probability by the mixed strategy s_i is called the *support* of s_i.

support of a
mixed strategy

Definition 3.2.6 (Support) *The* support *of a mixed strategy s_i for a player i is the set of pure strategies $\{a_i | s_i(a_i) > 0\}$.*

fully mixed
strategy

Note that a pure strategy is a special case of a mixed strategy, in which the support is a single action. At the other end of the spectrum we have *fully mixed strategies*. A strategy is fully mixed if it has full support (i.e., if it assigns every action a nonzero probability).

expected utility

We have not yet defined the payoffs of players given a particular strategy profile, since the payoff matrix defines those directly only for the special case of pure-strategy profiles. But the generalization to mixed strategies is straightforward, and relies on the basic notion of decision theory—*expected utility*. Intuitively, we first calculate the probability of reaching each outcome given the strategy profile, and then we calculate the average of the payoffs of the outcomes, weighted by the probabilities of each outcome. Formally, we define the expected utility as follows (overloading notation, we use u_i for both utility and expected utility).

Definition 3.2.7 (Expected utility of a mixed strategy) *Given a normal-form game* (N, A, u)*, the expected utility* u_i *for player* i *of the mixed-strategy profile* $s = (s_1, \ldots, s_n)$ *is defined as*

$$u_i(s) = \sum_{a \in A} u_i(a) \prod_{j=1}^{n} s_j(a_j).$$

3.3   Analyzing games: from optimality to equilibrium

Now that we have defined what games in normal form are and what strategies are available to players in them, the question is how to reason about such games.

optimal strategy   In single-agent decision theory the key notion is that of an *optimal strategy*, that is, a strategy that maximizes the agent's expected payoff for a given environment in which the agent operates. The situation in the single-agent case can be fraught with uncertainty, since the environment might be stochastic, partially observable, and spring all kinds of surprises on the agent. However, the situation is even more complex in a multiagent setting. In this case the environment includes—or, in many cases we discuss, consists entirely of—other agents, all of whom are also hoping to maximize their payoffs. Thus the notion of an optimal strategy for a given agent is not meaningful; the best strategy depends on the choices of others.

Game theorists deal with this problem by identifying certain subsets of outcomes, called *solution concepts*, that are interesting in one sense or another. In this section we describe two of the most fundamental solution concepts: Pareto optimality and Nash equilibrium.

3.3.1   *Pareto optimality*

First, let us investigate the extent to which a notion of optimality can be meaningful in games. From the point of view of an outside observer, can some outcomes of a game be said to be better than others?

This question is complicated because we have no way of saying that one agent's interests are more important than another's. For example, it might be tempting to say that we should prefer outcomes in which the sum of agents' utilities is higher. However, recall from Section 3.1.2 that we can apply any positive affine transformation to an agent's utility function and obtain another valid utility function. For example, we could multiply all of player 1's payoffs by 1,000, which could clearly change which outcome maximized the sum of agents' utilities.

Thus, our problem is to find a way of saying that some outcomes are better than others, even when we only know agents' utility functions up to a positive affine transformation. Imagine that each agent's utility is a monetary payment that you will receive, but that each payment comes in a different currency, and you do not know anything about the exchange rates. Which outcomes should you prefer? Observe that, while it is not usually possible to identify the best outcome, there *are* situations in which you can be sure that one outcome is better than another. For example, it is better to get 10 units of currency A and 3 units of currency B than to get 9 units of currency A and 3 units of currency B, regardless of the exchange rate. We formalize this intuition in the following definition.

Pareto domination

Definition 3.3.1 (Pareto domination) *Strategy profile s* Pareto dominates *strategy profile s' if for all $i \in N$, $u_i(s) \geq u_i(s')$, and there exists some $j \in N$ for which $u_j(s) > u_j(s')$.*

In other words, in a Pareto-dominated strategy profile some player can be made better off without making any other player worse off. Observe that we define Pareto domination over strategy profiles, not just action profiles. Thus, here we treat strategy profiles as outcomes, just as we treated lotteries as outcomes in Section 3.1.2.

Pareto domination gives us a partial ordering over strategy profiles. Thus, in answer to our question before, we cannot generally identify a single "best" outcome; instead, we may have a set of noncomparable optima.

Pareto optimality

strict Pareto efficiency

Definition 3.3.2 (Pareto optimality) *Strategy profile s is* Pareto optimal, *or* strictly Pareto efficient, *if there does not exist another strategy profile $s' \in S$ that Pareto dominates s.*

We can easily draw several conclusions about Pareto optimal strategy profiles. First, every game must have at least one such optimum, and there must always exist at least one such optimum in which all players adopt pure strategies. Second, some games will have multiple optima. For example, in zero-sum games, *all* strategy profiles are strictly Pareto efficient. Finally, in common-payoff games, all Pareto optimal strategy profiles have the same payoffs.

3.3.2 *Defining best response and Nash equilibrium*

Now we will look at games from an individual agent's point of view, rather than from the vantage point of an outside observer. This will lead us to the most influential solution concept in game theory, the *Nash equilibrium.*

Our first observation is that if an agent knew how the others were going to play, his strategic problem would become simple. Specifically, he would be left with the single-agent problem of choosing a utility-maximizing action that we discussed in Section 3.1. Formally, define $s_{-i} = (s_1, \ldots, s_{i-1}, s_{i+1}, \ldots, s_n)$, a strategy profile s without agent i's strategy. Thus we can write $s = (s_i, s_{-i})$. If the agents other than i (whom we denote $-i$) were to commit to play s_{-i}, a utility-maximizing agent i would face the problem of determining his best response.

best response

Definition 3.3.3 (Best response) *Player i's* best response *to the strategy profile s_{-i} is a mixed strategy $s_i^* \in S_i$ such that $u_i(s_i^*, s_{-i}) \geq u_i(s_i, s_{-i})$ for all strategies $s_i \in S_i$.*

The best response is not necessarily unique. Indeed, except in the extreme case in which there is a unique best response that is a pure strategy, the number of best responses is always infinite. When the support of a best response s^* includes two or more actions, the agent must be indifferent among them—otherwise, the agent

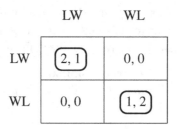

Figure 3.9 Pure-strategy Nash equilibria in the Battle of the Sexes game.

would prefer to reduce the probability of playing at least one of the actions to zero. But thus *any* mixture of these actions must also be a best response, not only the particular mixture in s^*. Similarly, if there are two pure strategies that are individually best responses, any mixture of the two is necessarily also a best response.

Of course, in general an agent will not know what strategies the other players plan to adopt. Thus, the notion of best response is not a solution concept—it does not identify an interesting set of outcomes in this general case. However, we can leverage the idea of best response to define what is arguably the most central notion in noncooperative game theory, the Nash equilibrium.

Nash equilibrium **Definition 3.3.4 (Nash equilibrium)** *A strategy profile $s = (s_1, \ldots, s_n)$ is a Nash equilibrium if, for all agents i, s_i is a best response to s_{-i}.*

Intuitively, a Nash equilibrium is a *stable* strategy profile: no agent would want to change his strategy if he knew what strategies the other agents were following.

We can divide Nash equilibria into two categories, strict and weak, depending on whether or not every agent's strategy constitutes a *unique* best response to the other agents' strategies.

strict Nash equilibrium **Definition 3.3.5 (Strict Nash)** *A strategy profile $s = (s_1, \ldots, s_n)$ is a strict Nash equilibrium if, for all agents i and for all strategies $s_i' \neq s_i$, $u_i(s_i, s_{-i}) > u_i(s_i', s_{-i})$.*

weak Nash equilibrium **Definition 3.3.6 (Weak Nash)** *A strategy profile $s = (s_1, \ldots, s_n)$ is a weak Nash equilibrium if, for all agents i and for all strategies $s_i' \neq s_i$, $u_i(s_i, s_{-i}) \geq u_i(s_i', s_{-i})$, and s is not a strict Nash equilibrium.*

Intuitively, weak Nash equilibria are less stable than strict Nash equilibria, because in the former case at least one player has a best response to the other players' strategies that is not his equilibrium strategy. Mixed-strategy Nash equilibria are necessarily weak, while pure-strategy Nash equilibria can be either strict or weak, depending on the game.

3.3.3 *Finding Nash equilibria*

Consider again the Battle of the Sexes game. We immediately see that it has two pure-strategy Nash equilibria, depicted in Figure 3.9.

We can check that these are Nash equilibria by confirming that whenever one of the players plays the given (pure) strategy, the other player would only lose by deviating.

	Heads	Tails
Heads	$1, -1$	$-1, 1$
Tails	$-1, 1$	$1, -1$

Figure 3.10 The Matching Pennies game.

Are these the only Nash equilibria? The answer is no; although they are indeed the only pure-strategy equilibria, there is also another mixed-strategy equilibrium. In general, it is tricky to compute a game's mixed-strategy equilibria; we consider this problem in detail in Chapter 4. However, we will show here that this computational problem is easy when we know (or can guess) the *support* of the equilibrium strategies, particularly so in this small game. Let us now guess that both players randomize, and let us assume that husband's strategy is to play LW with probability p and WL with probability $1 - p$. Then if the wife, the row player, also mixes between her two actions, she must be indifferent between them, given the husband's strategy. (Otherwise, she would be better off switching to a pure strategy according to which she only played the better of her actions.) Then we can write the following equations.

$$U_{\text{wife}}(\text{LW}) = U_{\text{wife}}(\text{WL})$$
$$2 * p + 0 * (1 - p) = 0 * p + 1 * (1 - p)$$
$$p = \frac{1}{3}$$

We get the result that in order to make the wife indifferent between her actions, the husband must choose LW with probability $1/3$ and WL with probability $2/3$. Of course, since the husband plays a mixed strategy he must also be indifferent between his actions. By a similar calculation it can be shown that to make the husband indifferent, the wife must choose LW with probability $2/3$ and WL with probability $1/3$. Now we can confirm that we have indeed found an equilibrium: since both players play in a way that makes the other indifferent, they are both best responding to each other. Like all mixed-strategy equilibria, this is a weak Nash equilibrium. The expected payoff of both agents is $2/3$ in this equilibrium, which means that each of the pure-strategy equilibria Pareto-dominates the mixed-strategy equilibrium.

Earlier, we mentioned briefly that mixed strategies play an important role. The previous example may not make it obvious, but now consider again the Matching Pennies game, reproduced in Figure 3.10. It is not hard to see that no pure strategy could be part of an equilibrium in this game of pure competition. Therefore, likewise there can be no strict Nash equilibrium in this game. But using the aforementioned procedure, the reader can verify that again there exists a mixed-strategy equilibrium; in this case, each player chooses one of the two available actions with probability $1/2$.

What does it mean to say that an agent plays a mixed-strategy Nash equilibrium? Do players really sample probability distributions in their heads? Some people have argued that they really do. One well-known motivating example for mixed strategies involves soccer: specifically, a kicker and a goalie getting ready for a penalty kick. The kicker can kick to the left or the right, and the goalie can jump to the left or the right. The kicker scores if and only if he kicks to one side and the goalie jumps to the other; this is thus best modeled as Matching Pennies. Any pure strategy on the part of either player invites a winning best response on the part of the other player. It is only by kicking or jumping in either direction with equal probability, goes the argument, that the opponent cannot exploit your strategy.

Of course, this argument is not uncontroversial. In particular, it can be argued that the strategies of each player are deterministic, but each player has uncertainty regarding the other player's strategy. This is indeed a second possible interpretation of mixed strategies: the mixed strategy of player i is everyone else's assessment of how likely i is to play each pure strategy. In equilibrium, i's mixed strategy has the further property that every action in its support is a best response to player i's beliefs about the other agents' strategies.

Finally, there are two interpretations that are related to learning in multiagent systems. In one interpretation, the game is actually played many times repeatedly, and the probability of a pure strategy is the fraction of the time it is played in the limit (its so-called *empirical frequency*). In the other interpretation, not only is the game played repeatedly, but each time it involves two different agents selected at random from a large population. In this interpretation, each agent in the population plays a pure strategy, and the probability of a pure strategy represents the fraction of agents playing that strategy. We return to these learning interpretations in Chapter 7.

empirical
frequency

3.3.4   *Nash's theorem: proving the existence of Nash equilibria*

We have now seen two examples in which we managed to find Nash equilibria (three equilibria for Battle of the Sexes, one equilibrium for Matching Pennies). Did we just luck out? Here there is some good news—it was not just luck. In this section we prove that every game has at least one Nash equilibrium.

First, a disclaimer: this section is more technical than the rest of the chapter. A reader who is prepared to take the existence of Nash equilibria on faith can safely skip to the beginning of Section 3.4 on p. 71. For the bold of heart who remain, we begin with some preliminary definitions.

convexity   **Definition 3.3.7 (Convexity)** *A set $C \subset \mathbb{R}^m$ is* convex *if for every $x, y \in C$ and $\lambda \in [0, 1]$, $\lambda x + (1 - \lambda)y \in C$. For vectors $x^0, \ldots, x^n$ and nonnegative scalars $\lambda_0, \ldots, \lambda_n$ satisfying $\sum_{i=0}^{n} \lambda_i = 1$, the vector $\sum_{i=0}^{n} \lambda_i x^i$ is called a* convex combination *of $x^0, \ldots, x^n$.*

convex
combination

For example, a cube is a convex set in $\mathbb{R}^3$; a bowl is not.

Definition 3.3.8 (Affine independence) *A finite set of vectors $\{x^0, \ldots, x^n\}$ in a Euclidean space is* affinely independent *if $\sum_{i=0}^{n} \lambda_i x^i = 0$ and $\sum_{i=0}^{n} \lambda_i = 0$ imply that $\lambda_0 = \cdots = \lambda_n = 0$.*

affine independence

An equivalent condition is that $\{x^1 - x^0, x^2 - x^0, \ldots, x^n - x^0\}$ are linearly independent. Intuitively, a set of points is affinely independent if no three points from the set lie on the same line, no four points from the set lie on the same plane, and so on. For example, the set consisting of the origin 0 and the unit vectors $e^1, \ldots, e^n$ is affinely independent.

Next we define a simplex, which is an n-dimensional generalization of a triangle.

n-simplex

Definition 3.3.9 (n-simplex) *An n-simplex, denoted $x^0 \cdots x^n$, is the set of all convex combinations of the affinely independent set of vectors $\{x^0, \ldots, x^n\}$, that is,*

$$x^0 \cdots x^n = \left\{ \sum_{i=0}^{n} \lambda_i x^i : \forall i \in \{0, \ldots, n\}, \ \lambda_i \geq 0; \ and \sum_{i=0}^{n} \lambda_i = 1 \right\}.$$

vertex

k-face

Each x^i is called a *vertex* of the simplex $x^0 \cdots x^n$ and each k-simplex $x^{i_0} \cdots x^{i_k}$ is called a *k-face* of $x^0 \cdots x^n$, where $i_0, \ldots, i_k \in \{0, \ldots, n\}$. For example, a triangle (i.e., a 2-simplex) has one 2-face (itself), three 1-faces (its sides) and three 0-faces (its vertices).

Definition 3.3.10 (Standard n-simplex) *The* standard n-simplex $\triangle_n$ *is*

$$\left\{ y \in \mathbb{R}^{n+1} : \sum_{i=0}^{n} y_i = 1, \forall i = 0, \ldots, n, \ y_i \geq 0 \right\}.$$

In other words, the standard n-simplex is the set of all convex combinations of the $n + 1$ unit vectors $e^0, \ldots, e^n$.

simplicial subdivision

Definition 3.3.11 (Simplicial subdivision) *A* simplicial subdivision *of an n-simplex T is a finite set of simplexes $\{T_i\}$ for which $\bigcup_{T_i \in T} T_i = T$, and for any $T_i, T_j \in T$, $T_i \cap T_j$ is either empty or equal to a common face.*

Intuitively, this means that a simplex is divided up into a set of smaller simplexes that together occupy exactly the same region of space and that overlap only on their boundaries. Furthermore, when two of them overlap, the intersection must be an entire face of both subsimplexes. Figure 3.11 (left) shows a 2-simplex subdivided into 16 subsimplexes.

Let $y \in x^0 \cdots x^n$ denote an arbitrary point in a simplex. This point can be written as a convex combination of the vertices: $y = \sum_i \lambda_i x^i$. Now define a function that gives the set of vertices "involved" in this point: $\chi(y) = \{i : \lambda_i > 0\}$. We use this function to define a proper labeling.

Definition 3.3.12 (Proper labeling) *Let $T = x^0 \cdots x^n$ be simplicially subdivided, and let V denote the set of all distinct vertices of all the subsimplexes. A*

proper labeling

function $\mathcal{L} : V \mapsto \{0, \ldots, n\}$ is a proper labeling *of a subdivision if $\mathcal{L}(v) \in \chi(v)$.*

One consequence of this definition is that the vertices of a simplex must all receive different labels. (Do you see why?) As an example, the subdivided simplex in Figure 3.11 (left) is properly labeled.

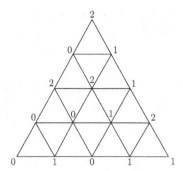 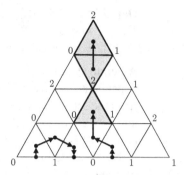

Figure 3.11 A properly labeled simplex (left), and the same simplex with completely labeled subsimplexes shaded and three walks indicated (right).

completely
labeled
subsimplex

Definition 3.3.13 (Complete labeling) *A subsimplex is* completely labeled *if $\mathcal{L}$ assumes all the values $0, \ldots, n$ on its set of vertices.*

For example in the subdivided triangle in Figure 3.11 (left), the subtriangle at the very top is completely labeled.

Sperner's lemma

Lemma 3.3.14 (Sperner's lemma) *Let $T_n = x^0 \cdots x^n$ be simplicially subdivided and let $\mathcal{L}$ be a proper labeling of the subdivision. Then there are an odd number of completely labeled subsimplexes in the subdivision.*

> **Proof.** We prove this by induction on n. The case $n = 0$ is trivial. The simplex consists of a single point x^0. The only possible simplicial subdivision is $\{x^0\}$. There is only one possible labeling function, $\mathcal{L}(x^0) = 0$. Note that this is a proper labeling. So there is one completely labeled subsimplex, x^0 itself.
>
> We now assume the statement to be true for $n - 1$ and prove it for n. The simplicial subdivision of T_n induces a simplicial subdivision on its face $x^0 \cdots x^{n-1}$. This face is an $(n - 1)$-simplex; denote it as T_{n-1}. The labeling function $\mathcal{L}$ restricted to T_{n-1} is a proper labeling of T_{n-1}. Therefore by the induction hypothesis there exist an odd number of $(n - 1)$-subsimplexes in T_{n-1} that bear the labels $(0, \ldots, n - 1)$. (To provide graphical intuition, we will illustrate the induction argument on a subdivided 2-simplex. In Figure 3.11 (left), observe that the bottom face $x^0 x^1$ is a subdivided 1-simplex—a line segment—containing four subsimplexes, three of which are completely labeled.)
>
> We now define rules for "walking" across our subdivided, labeled simplex T_n. The walk begins at an $(n - 1)$-subsimplex with labels $(0, \ldots, n - 1)$ on the face T_{n-1}; call this subsimplex b. There exists a unique n-subsimplex d that has b as a face; d's vertices consist of the vertices of b and another vertex z. If z is labeled n, then we have a completely labeled subsimplex and the walk ends. Otherwise, d has the labels $(0, \ldots, n - 1)$, where one of the labels (say j) is repeated, and the label n is missing. In this case there exists exactly one other $(n - 1)$-subsimplex that is a face of d and bears the labels $(0, \ldots, n - 1)$. This is because each $(n - 1)$-face of d is defined by all but one of d's vertices; since only the label j is repeated, an $(n - 1)$-face of d has

labels $(0, \ldots, n - 1)$ if and only if one of the two vertices with label j is left out. We know b is one such face, so there is exactly one other, which we call e. (For example, you can confirm in Figure 3.11 (left) that if a subtriangle has an edge with labels $(0, 1)$, then it is either completely labeled, or it has exactly one other edge with labels $(0, 1)$.) We continue the walk from e. We make use of the following property: an $(n - 1)$-face of an n-subsimplex in a simplicially subdivided simplex T_n is either on an $(n - 1)$-face of T_n, or the intersection of two n-subsimplexes. If e is on an $(n - 1)$-face of T_n we stop the walk. Otherwise we walk into the unique other n-subsimplex having e as a face. This subsimplex is either completely labeled or has one repeated label, and we continue the walk in the same way we did with subsimplex d earlier.

Note that the walk is completely determined by the starting $(n - 1)$-subsimplex. The walk ends either at a completely labeled n-subsimplex, or at a $(n - 1)$-subsimplex with labels $(0, \ldots, n - 1)$ on the face T_{n-1}. (It cannot end on any other face because $\mathcal{L}$ is a proper labeling.) Note also that every walk can be followed backward: beginning from the end of the walk and following the same rule as earlier, we end up at the starting point. This implies that if a walk starts at t on T_{n-1} and ends at t' on T_{n-1}, t and t' must be different, because otherwise we could reverse the walk and get a different path with the same starting point, contradicting the uniqueness of the walk. (Figure 3.11 (right) illustrates one walk of each of the kinds we have discussed so far: one that starts and ends at different subsimplexes on the face $x^0 x^1$, and one that starts on the face $x^0 x^1$ and ends at a completely labeled subtriangle.) Since by the induction hypothesis there are an odd number of $(n - 1)$-subsimplexes with labels $(0, \ldots, n - 1)$ at the face T_{n-1}, there must be at least one walk that does not end on this face. Since walks that start and end on the face "pair up," there are thus an odd number of walks starting from the face that end at completely labeled subsimplexes. All such walks end at *different* completely labeled subsimplexes, because there is exactly one $(n - 1)$-simplex face labeled $(0, \ldots, n - 1)$ for a walk to enter from in a completely labeled subsimplex.

Not all completely labeled subsimplexes are led to by such walks. To see why, consider reverse walks starting from completely labeled subsimplexes. Some of these reverse walks end at $(n - 1)$-simplexes on T_{n-1}, but some end at other completely labeled n-subsimplexes. (Figure 3.11 (right) illustrates one walk of this kind.) However, these walks just pair up completely labeled subsimplexes. There are thus an even number of completely labeled subsimplexes that pair up with each other, and an odd number of completely labeled subsimplexes that are led to by walks from the face T_{n-1}. Therefore the total number of completely labeled subsimplexes is odd. ∎

compactness **Definition 3.3.15 (Compactness)** *A subset of* $\mathbb{R}^n$ *is* compact *if the set is closed and bounded.*

It is straightforward to verify that $\triangle_m$ is compact. A compact set has the property that every sequence in the set has a convergent subsequence.

centroid   **Definition 3.3.16 (Centroid)** *The* centroid *of a simplex $x^0 \cdots x^m$ is the "average" of its vertices,* $\frac{1}{m+1} \sum_{i=0}^{m} x^i$.

We are now ready to use Sperner's lemma to prove Brouwer's fixed-point theorem.

Brouwer's
fixed-point
theorem   **Theorem 3.3.17 (Brouwer's fixed-point theorem)** *Let $f : \Delta_m \mapsto \Delta_m$ be continuous. Then f has a fixed point—that is, there exists some $z \in \Delta_m$ such that $f(z) = z$.*

> **Proof.** We prove this by first constructing a proper labeling of Δ_m, then showing that as we make finer and finer subdivisions, there exists a subsequence of completely labeled subsimplexes that converges to a fixed point of f.
>
> **Part 1: $\mathcal{L}$ is a proper labeling.** Let $\epsilon > 0$. We simplicially subdivide[3] Δ_m such that the Euclidean distance between any two points in the same m-subsimplex is at most ϵ. We define a labeling function $\mathcal{L} : V \mapsto \{0, \ldots, m\}$ as follows. For each v we choose a label satisfying
>
> $$\mathcal{L}(v) \in \chi(v) \cap \{i : f_i(v) \leq v_i\}, \tag{3.1}$$
>
> where v_i is the ith component of v and $f_i(v)$ is the ith component of $f(v)$. In other words, $\mathcal{L}(v)$ can be any label i such that $v_i > 0$ and f weakly decreases the ith component of v. To ensure that $\mathcal{L}$ is well defined, we must show that the intersection on the right side of Equation (3.1) is always nonempty. (Intuitively, since v and $f(v)$ are both on the standard simplex Δ_m, and on Δ_m each point's components sum to 1, there must exist a component of v that is weakly decreased by f. This intuition holds even though we restrict to the components in $\chi(v)$ because these are exactly all the positive components of v.) We now show this formally. For contradiction, assume otherwise. This assumption implies that $f_i(v) > v_i$ for all $i \in \chi(v)$. Recall from the definition of a standard simplex that $\sum_{i=0}^{m} v_i = 1$. Since by the definition of χ, $v_j > 0$ if and only if $j \in \chi(v)$, we have
>
> $$\sum_{j \in \chi(v)} v_j = \sum_{i=0}^{m} v_i = 1. \tag{3.2}$$
>
> Since $f_j(v) > v_j$ for all $j \in \chi(v)$,
>
> $$\sum_{j \in \chi(v)} f_i(v) > \sum_{j \in \chi(v)} v_j = 1. \tag{3.3}$$
>
> But since $f(v)$ is also on the standard simplex Δ_m,
>
> $$\sum_{j \in \chi(v)} f_i(v) \leq \sum_{i=0}^{m} f_i(v) = 1. \tag{3.4}$$
>
> Equations (3.3) and (3.4) lead to a contradiction. Therefore, $\mathcal{L}$ is well defined; it is a proper labeling by construction.

---

3. Here, we implicitly assume that simplices can always be subdivided regardless of dimension. This is true, but surprisingly difficult to show.

Part 2: As $\epsilon \to 0$, completely labeled subsimplexes converge to fixed points of f. Since $\mathcal{L}$ is a proper labeling, by Sperner's lemma (3.3.14) there is at least one completely labeled subsimplex $p^0 \cdots p^m$ such that $f_i(p^i) \leq p^i$ for each i. Let $\epsilon \to 0$ and consider the sequence of centroids of completely labeled subsimplexes. Since $\triangle_m$ is compact, there is a convergent subsequence. Let z be its limit; then for all $i = 0, \ldots, m$, $p^i \to z$ as $\epsilon \to 0$. Since f is continuous we must have $f_i(z) \leq z_i$ for all i. This implies $f(z) = z$, because otherwise (by an argument similar to the one in Part 1) we would have $1 = \sum_i f_i(z) < \sum_i z_i = 1$, a contradiction. ∎

Theorem 3.3.17 cannot be used directly to prove the existence of Nash equilibria. This is because a Nash equilibrium is a point in the set of mixed-strategy profiles S. This set is not a simplex but rather a *simplotope*: a Cartesian product of simplexes. (Observe that each individual agent's mixed strategy *can* be understood as a point in a simplex.) However, it turns out that Brouwer's theorem can be extended beyond simplexes to simplotopes.[4] In essence, this is because every simplotope is topologically the same as a simplex (formally, they are *homeomorphic*).

simplotope

Definition 3.3.18 (Bijective function) *A function f is* injective *(or one-to-one) if $f(a) = f(b)$ implies $a = b$. A function $f : X \mapsto Y$ is* onto *if for every $y \in Y$ there exists $x \in X$ such that $f(x) = y$. A function is* bijective *if it is both injective and onto.*

bijective

Definition 3.3.19 (Homeomorphism) *A set A is* homeomorphic *to a set B if there exists a continuous, bijective function $h : A \mapsto B$ such that h^{-1} is also continuous. Such a function h is called a* homeomorphism.

homeomorphism

Definition 3.3.20 (Interior) *A point x is an* interior *point of a set $A \subset \mathbb{R}^m$ if there is an open m-dimensional ball $B \subset \mathbb{R}^m$ centered at x such that $B \subset A$. The* interior *of a set A is the set of all its interior points.*

interior

Corollary 3.3.21 (Brouwer's fixed-point theorem, simplotopes) *Let $K = \prod_{j=1}^{k} \triangle_{m_j}$ be a simplotope and let $f : K \mapsto K$ be continuous. Then f has a fixed point.*

Proof. Let $m = \sum_{j=1}^{k} m_j$. First we show that if K is homeomorphic to $\triangle_m$, then a continuous function $f : K \mapsto K$ has a fixed point. Let $h : \triangle_m \mapsto K$ be a homeomorphism. Then $h^{-1} \circ f \circ h : \triangle_m \mapsto \triangle_m$ is continuous, where $\circ$ denotes function composition. By Theorem 3.3.17 there exists a z' such that $h^{-1} \circ f \circ h(z') = z'$. Let $z = h(z')$, then $h^{-1} \circ f(z) = z' = h^{-1}(z)$. Since h^{-1} is injective, $f(z) = z$.

We must still show that $K = \prod_{j=1}^{k} \triangle_{m_j}$ is homeomorphic to $\triangle_m$. K is convex and compact because each $\triangle_{m_j}$ is convex and compact, and a product of convex and compact sets is also convex and compact. Let the *dimension* of a subset of an Euclidean space be the number of independent parameters

---

4. An argument similar to our proof below can be used to prove a generalization of Theorem 3.3.17 to arbitrary convex and compact sets.

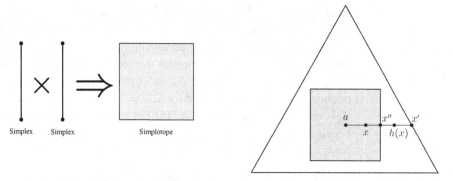

Figure 3.12 A product of two standard 1-simplexes is a square (a simplotope; left). The square is scaled and put inside a triangle (a 2-simplex), and an example of radial projection h is shown (right).

required to describe each point in the set. For example, an n-simplex has dimension n. Since each $\triangle_{m_j}$ has dimension m_j, K has dimension m. Since $K \subset \mathbb{R}^{m+k}$ and $\triangle_m \subset \mathbb{R}^{m+1}$ both have dimension m, they can be embedded in $\mathbb{R}^m$ as K' and $\triangle'_m$ respectively. Furthermore, whereas $K \subset \mathbb{R}^{m+k}$ and $\triangle_m \subset \mathbb{R}^{m+1}$ have no interior points, both K' and $\triangle'_m$ have nonempty interior. For example, a standard 2-simplex is defined in $\mathbb{R}^3$, but we can embed the triangle in $\mathbb{R}^2$. As illustrated in Figure 3.12 (left), the product of two standard 1-simplexes is a square, which can also be embedded in $\mathbb{R}^2$. We scale and translate K' into K'' such that K'' is strictly inside $\triangle'_m$. Since scaling and translation are homeomorphisms, and a chain of homeomorphisms is still a homeomorphism, we just need to find a homeomorphism $h : K'' \mapsto \triangle'_m$. Fix a point a in the interior of K''. Define h to be the "radial projection" with respect to a, where $h(a) = a$ and for $x \in K'' \setminus \{a\}$,

$$h(x) = a + \frac{||x' - a||}{||x'' - a||}(x - a),$$

where x' is the intersection point of the boundary of $\triangle'_m$ with the ray that starts at a and passes through x, and x'' is the intersection point of the boundary of K'' with the same ray. Because K'' and $\triangle'_m$ are convex and compact, x'' and x' exist and are unique. Since a is an interior point of K'' and $\triangle_m$, $||x' - a||$ and $||x'' - a||$ are both positive. Intuitively, h scales x along the ray by a factor of $\frac{||x'-a||}{||x''-a||}$. Figure 3.12 (right) illustrates an example of this radial projection from a square simplotope to a triangle.

Finally, it remains to show that h is a homeomorphism. It is relatively straightforward to verify that h is continuous. Since we know that $h(x)$ lies on the ray that starts at a and passes through x, given $h(x)$ we can reconstruct the same ray by drawing a ray from a that passes through $h(x)$. We can then recover x' and x'', and find x by scaling $h(x)$ along the ray by a factor of $\frac{||x''-a||}{||x'-a||}$. Thus h is injective. h is onto because given any point $y \in \triangle'_m$, we can construct the ray and find x such that $h(x) = y$. So, h^{-1} has the same form as h except that the scaling factor is inverted, thus h^{-1} is also continuous. Therefore, h is a homeomorphism.   ∎

We are now ready to prove the existence of Nash equilibrium. Indeed, now that we have Corollary 3.3.21 and notation for discussing mixed strategies (Section 3.2.4), it is surprisingly easy. The proof proceeds by constructing a continuous $f : S \mapsto S$ such that each fixed point of f is a Nash equilibrium. Then we use Corollary 3.3.21 to argue that f has at least one fixed point, and thus that Nash equilibria always exist.

Theorem 3.3.22 (Nash, 1951) *Every game with a finite number of players and action profiles has at least one Nash equilibrium.*

Proof. Given a strategy profile $s \in S$, for all $i \in N$ and $a_i \in A_i$ we define

$$\varphi_{i,a_i}(s) = \max\{0, u_i(a_i, s_{-i}) - u_i(s)\}.$$

We then define the function $f : S \mapsto S$ by $f(s) = s'$, where

$$
\begin{aligned}
s_i'(a_i) &= \frac{s_i(a_i) + \varphi_{i,a_i}(s)}{\sum_{b_i \in A_i} s_i(b_i) + \varphi_{i,b_i}(s)} \\
&= \frac{s_i(a_i) + \varphi_{i,a_i}(s)}{1 + \sum_{b_i \in A_i} \varphi_{i,b_i}(s)}.
\end{aligned}
\tag{3.5}
$$

Intuitively, this function maps a strategy profile s to a new strategy profile s' in which each agent's actions that are better responses to s receive increased probability mass.

The function f is continuous since each φ_{i,a_i} is continuous. Since S is convex and compact and $f : S \mapsto S$, by Corollary 3.3.21 f must have at least one fixed point. We must now show that the fixed points of f are the Nash equilibria.

First, if s is a Nash equilibrium then all φ's are 0, making s a fixed point of f.

Conversely, consider an arbitrary fixed point of f, s. By the linearity of expectation there must exist at least one action in the support of s, say a_i', for which $u_{i,a_i'}(s) \leq u_i(s)$. From the definition of φ, $\varphi_{i,a_i'}(s) = 0$. Since s is a fixed point of f, $s_i'(a_i') = s_i(a_i')$. Consider Equation (3.5), the expression defining $s_i'(a_i')$. The numerator simplifies to $s_i(a_i')$, and is positive since a_i' is in i's support. Hence the denominator must be 1. Thus for any i and $b_i \in A_i$, $\varphi_{i,b_i}(s)$ must equal 0. From the definition of φ, this can occur only when no player can improve his expected payoff by moving to a pure strategy. Therefore, s is a Nash equilibrium.   ∎

3.4   Further solution concepts for normal-form games

solution concept

As described earlier at the beginning of Section 3.3, we reason about multiplayer games using *solution concepts*, principles according to which we identify interesting subsets of the outcomes of a game. While the most important solution concept is the Nash equilibrium, there are also a large number of others, only some of which we will discuss here. Some of these concepts are more restrictive than the Nash equilibrium, some less so, and some noncomparable. In Chapters 5 and 6 we will introduce some additional solution concepts that are only applicable to game representations other than the normal form.

3.4.1   *Maxmin and minmax strategies*

The *maxmin strategy* of player i in an n-player, general-sum game is a (not necessarily unique, and in general mixed) strategy that maximizes i's worst-case payoff, in the situation where all the other players happen to play the strategies

security level   which cause the greatest harm to i. The *maxmin value* (or *security level*) of the game for player i is that minimum amount of payoff guaranteed by a maxmin strategy.

maxmin strategy   **Definition 3.4.1 (Maxmin)** *The* maxmin strategy *for player i is* $\arg\max_{s_i}$
maxmin value $\min_{s_{-i}} u_i(s_i, s_{-i})$, *and the* maxmin value *for player i is* $\max_{s_i} \min_{s_{-i}} u_i(s_i, s_{-i})$.

Although the maxmin strategy is a concept that makes sense in simultaneous-move games, it can be understood through the following temporal intuition. The maxmin strategy is i's best choice when first i must commit to a (possibly mixed) strategy, and then the remaining agents $-i$ observe this strategy (but not i's action choice) and choose their own strategies to minimize i's expected payoff. In the Battle of the Sexes game (Figure 3.8), the maxmin value for either player is 2/3, and requires the maximizing agent to play a mixed strategy. (Do you see why?)

While it may not seem reasonable to assume that the other agents would be solely interested in minimizing i's utility, it is the case that if i plays a maxmin strategy and the other agents play arbitrarily, i will still receive an expected payoff of at least his maxmin value. This means that the maxmin strategy is a sensible choice for a conservative agent who wants to maximize his expected utility without having to make any assumptions about the other agents, such as that they will act rationally according to their own interests, or that they will draw their action choices from known distributions.

The *minmax strategy* and *minmax value* play a dual role to their maxmin counterparts. In two-player games the minmax strategy for player i against player $-i$ is a strategy that keeps the maximum payoff of $-i$ at a minimum, and the minmax value of player $-i$ is that minimum. This is useful when we want to consider the amount that one player can punish another without regard for his own payoff. Such punishment can arise in repeated games, as we will see in Section 6.1. The formal definitions follow.

minmax strategy   **Definition 3.4.2 (Minmax, two-player)** *In a two-player game, the* minmax
                  strategy *for player i against player $-i$ is* $\arg\min_{s_i} \max_{s_{-i}} u_{-i}(s_i, s_{-i})$, *and player*
minmax value   *$-i$'s* minmax value *is* $\min_{s_i} \max_{s_{-i}} u_{-i}(s_i, s_{-i})$.

In n-player games with $n > 2$, defining player i's minmax strategy against player j is a bit more complicated. This is because i will not usually be able to guarantee that j achieves minimal payoff by acting unilaterally. However, if we assume that all the players other than j choose to "gang up" on j—and that they are able to coordinate appropriately when there is more than one strategy profile that would yield the same minimal payoff for j—then we can define minmax strategies for the n-player case.

minmax strategy　**Definition 3.4.3 (Minmax, n-player)** *In an n-player game, the* minmax strategy *for player i against player $j \neq i$ is i's component of the mixed-strategy profile s_{-j} in the expression* $\arg\min_{s_{-j}} \max_{s_j} u_j(s_j, s_{-j})$, *where $-j$ denotes the set of players other than j. As before, the* minmax value *for player j is* $\min_{s_{-j}} \max_{s_j} u_j(s_j, s_{-j})$.

As with the maxmin value, we can give temporal intuition for the minmax value. Imagine that the agents $-i$ must commit to a (possibly mixed) strategy profile, to which i can then play a best response. Player i receives his minmax value if players $-i$ choose their strategies in order to minimize i's expected utility after he plays his best response.

In two-player games, a player's minmax value is always equal to his maxmin value. For games with more than two players a weaker condition holds: a player's maxmin value is always less than or equal to his minmax value. (Can you explain why this is?)

Since neither an agent's maxmin strategy nor his minmax strategy depend on the strategies that the other agents actually choose, the maxmin and minmax strategies give rise to solution concepts in a straightforward way. We will call a mixed-strategy profile $s = (s_1, s_2, \ldots)$ a *maxmin strategy profile* of a given game if s_1 is a maxmin strategy for player 1, s_2 is a maxmin strategy for player 2 and so on. In two-player games, we can also define *minmax strategy profiles* analogously. In two-player, zero-sum games, there is a very tight connection between minmax and maxmin strategy profiles. Furthermore, these solution concepts are also linked to the Nash equilibrium.

Theorem 3.4.4 (Minimax theorem (von Neumann, 1928)) *In any finite, two-player, zero-sum game, in any Nash equilibrium*[5] *each player receives a payoff that is equal to both his maxmin value and his minmax value.*

> **Proof.** At least one Nash equilibrium must exist by Theorem 3.3.22. Let (s_i', s_{-i}') be an arbitrary Nash equilibrium, and denote i's equilibrium payoff as v_i. Denote i's maxmin value as $\bar{v}_i$ and i's minmax value as $\underline{v}_i$.
>
> First, show that $\bar{v}_i = v_i$. Clearly we cannot have $\bar{v}_i > v_i$, as if this were true then i would profit by deviating from s_i' to his maxmin strategy, and hence (s_i', s_{-i}') would not be a Nash equilibrium. Thus it remains to show that $\bar{v}_i$ cannot be less than v_i.
>
> Assume that $\bar{v}_i < v_i$. By definition, in equilibrium each player plays a best response to the other. Thus
>
> $$v_{-i} = \max_{s_{-i}} u_{-i}(s_i', s_{-i}).$$

---

5. The attentive reader might wonder how a theorem from 1928 can use the term "Nash equilibrium," when Nash's work was published in 1950. Von Neumann used different terminology and proved the theorem in a different way; however, the given presentation is probably clearer in the context of modern game theory.

Equivalently, we can write that $-i$ minimizes the negative of his payoff, given i's strategy,

$$-v_{-i} = \min_{s_{-i}} -u_{-i}(s'_i, s_{-i}).$$

Since the game is zero sum, $v_i = -v_{-i}$ and $u_i = -u_{-i}$. Thus,

$$v_i = \min_{s_{-i}} u_i(s'_i, s_{-i}).$$

We defined $\bar{v}_i$ as $\max_{s_i} \min_{s_{-i}} u_i(s_i, s_{-i})$. By the definition of max, we must have

$$\max_{s_i} \min_{s_{-i}} u_i(s_i, s_{-i}) \geq \min_{s_{-i}} u_i(s'_i, s_{-i}).$$

Thus $\bar{v}_i \geq v_i$, contradicting our assumption.

We have shown that $\bar{v}_i = v_i$. The proof that $\underline{v}_i = v_i$ is similar, and is left as an exercise.   ■

Why is the minmax theorem important? It demonstrates that maxmin strategies, minmax strategies and Nash equilibria coincide in two-player, zero-sum games. In particular, Theorem 3.4.4 allows us to conclude that in two-player, zero-sum games:

value of a zero-sum game

1. Each player's maxmin value is equal to his minmax value. By convention, the maxmin value for player 1 is called the *value of the game*;
2. For both players, the set of maxmin strategies coincides with the set of minmax strategies; and
3. Any maxmin strategy profile (or, equivalently, minmax strategy profile) is a Nash equilibrium. Furthermore, these are all the Nash equilibria. Consequently, all Nash equilibria have the same payoff vector (namely, those in which player 1 gets the value of the game).

For example, in the Matching Pennies game in Figure 3.6, the value of the game is 0. The unique Nash equilibrium consists of both players randomizing between heads and tails with equal probability, which is both the maxmin strategy and the minmax strategy for each player.

Nash equilibria in zero-sum games can be viewed graphically as a "saddle" in a high-dimensional space. At a saddle point, any deviation of the agent lowers his utility and increases the utility of the other agent. It is easy to visualize in the simple case in which each agent has two pure strategies. In this case the space of mixed strategy profiles can be viewed as the points on the square between $(0,0)$ and $(1,1)$. Adding a third dimension representing player 1's expected utility, the payoff to player 1 under these mixed strategy profiles (and thus the negative of the payoff to player 2) is a saddle-shaped surface. Figure 3.13 (left) gives a pictorial example, illustrating player 1's expected utility in Matching Pennies as a function of both players' probabilities of playing heads. Figure 3.13 (right) adds a plane at $z = 0$ to make it easier to see that it is an equilibrium for both players to play heads 50% of the time and that zero is both the maxmin value and the minmax value for both players.

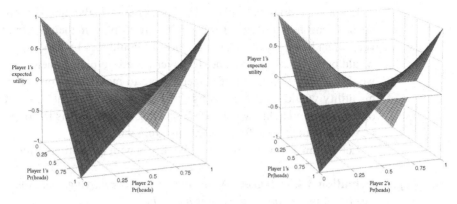

Figure 3.13 The saddle point in Matching Pennies, with and without a plane at $z = 0$.

	L	R
T	$100, a$	$1 - \epsilon, b$
B	$2, c$	$1, d$

Figure 3.14 A game for contrasting maxmin with minimax regret. The numbers refer only to player 1's payoffs; ϵ is an arbitrarily small positive constant. Player 2's payoffs are the arbitrary (and possibly unknown) constants a, b, c, and d.

3.4.2 *Minimax regret*

We argued earlier that agents might play maxmin strategies in order to achieve good payoffs in the worst case, even in a game that is not zero sum. However, consider a setting in which the other agent is not believed to be malicious, but is instead entirely unpredictable. (Crucially, in this section we do not approach the problem as Bayesians, saying that agent i's beliefs can be described by a probability distribution; instead, we use a "pre-Bayesian" model in which i does not know such a distribution and indeed has no beliefs about it.) In such a setting, it can make sense for agents to care about minimizing their worst-case *losses*, rather than maximizing their worst-case payoffs.

Consider the game in Figure 3.14. Let ϵ be an arbitrarily small positive constant. For this example it does not matter what agent 2's payoffs a, b, c, and d are, and we can even imagine that agent 1 does not know these values. Indeed, this could be one reason why player 1 would be unable to form beliefs about how player 2 would play, even if he were to believe that player 2 was rational. Let us imagine that agent 1 wants to determine a strategy to follow that makes sense despite his uncertainty about player 2. First, agent 1 might play his maxmin, or "safety level" strategy. In this game it is easy to see that player 1's maxmin strategy is to play B; this is because player 2's minmax strategy is to play R, and B is a best response to R.

If player 1 does not believe that player 2 is malicious, however, he might instead reason as follows. If player 2 were to play R then it would not matter very much how player 1 plays: the most he could lose by playing the wrong way would be ϵ. On the other hand, if player 2 were to play L then player 1's action would be very significant: if player 1 were to make the wrong choice here then his utility would be decreased by 98. Thus player 1 might choose to play T in order to minimize his worst-case loss. Observe that this is the opposite of what he would choose if he followed his maxmin strategy.

Let us now formalize this idea. We begin with the notion of regret.

regret   **Definition 3.4.5 (Regret)** *An agent i's regret for playing an action a_i if the other agents adopt action profile a_{-i} is defined as*

$$\left[\max_{a_i' \in A_i} u_i(a_i', a_{-i})\right] - u_i(a_i, a_{-i}).$$

In words, this is the amount that i loses by playing a_i, rather than playing his best response to a_{-i}. Of course, i does not know what actions the other players will take; however, he can consider those actions that would give him the highest regret for playing a_i.

maximum regret   **Definition 3.4.6 (Max regret)** *An agent i's maximum regret for playing an action a_i is defined as*

$$\max_{a_{-i} \in A_{-i}} \left(\left[\max_{a_i' \in A_i} u_i(a_i', a_{-i})\right] - u_i(a_i, a_{-i})\right).$$

This is the amount that i loses by playing a_i rather than playing his best response to a_{-i}, if the other agents chose the a_{-i} that makes this loss as large as possible. Finally, i can choose his action in order to minimize this worst-case regret.

Definition 3.4.7 (Minimax regret) *Minimax regret actions for agent i are defined as*

$$\arg\min_{a_i \in A_i} \left[\max_{a_{-i} \in A_{-i}} \left(\left[\max_{a_i' \in A_i} u_i(a_i', a_{-i})\right] - u_i(a_i, a_{-i})\right)\right].$$

Thus, an agent's minimax regret action is an action that yields the smallest maximum regret. Minimax regret can be extended to a solution concept in the natural way, by identifying action profiles that consist of minimax regret actions for each player. Note that we can safely restrict ourselves to actions rather than mixed strategies in the definitions above (i.e., maximizing over the sets A_i and A_{-i} instead of S_i and S_{-i}), because of the linearity of expectation. We leave the proof of this fact as an exercise.

3.4.3 *Removal of dominated strategies*

We first define what it means for one strategy to dominate another. Intuitively, one strategy dominates another for a player i if the first strategy yields i a greater payoff than the second strategy, for *any* strategy profile of the remaining players.[6] There are, however, three gradations of dominance, which are captured in the following definition.

Definition 3.4.8 (Domination) *Let s_i and s_i' be two strategies of player i, and S_{-i} the set of all strategy profiles of the remaining players. Then:*

strict domination

1. *s_i strictly dominates s_i' if for all $s_{-i} \in S_{-i}$, it is the case that $u_i(s_i, s_{-i}) > u_i(s_i', s_{-i})$.*

weak domination

2. *s_i weakly dominates s_i' if for all $s_{-i} \in S_{-i}$, it is the case that $u_i(s_i, s_{-i}) \geq u_i(s_i', s_{-i})$, and for at least one $s_{-i} \in S_{-i}$, it is the case that $u_i(s_i, s_{-i}) > u_i(s_i', s_{-i})$.*

very weak domination

3. *s_i very weakly dominates s_i' if for all $s_{-i} \in S_{-i}$, it is the case that $u_i(s_i, s_{-i}) \geq u_i(s_i', s_{-i})$.*

If one strategy dominates all others, we say that it is (strongly, weakly or very weakly) *dominant*.

Definition 3.4.9 (Dominant strategy) *A strategy is* strictly *(resp., weakly; very weakly)* dominant *for an agent if it strictly (weakly; very weakly) dominates any other strategy for that agent.*

It is obvious that a strategy profile $(s_1, \ldots, s_n)$ in which every s_i is dominant

equilibrium in dominant strategies

for player i (whether strictly, weakly, or very weakly) is a Nash equilibrium. Such a strategy profile forms what is called an *equilibrium in dominant strategies* with the appropriate modifier (*strictly*, etc). An equilibrium in strictly dominant strategies is necessarily the unique Nash equilibrium. For example, consider again the Prisoner's Dilemma game. For each player, the strategy D is strictly dominant, and indeed (D, D) is the unique Nash equilibrium. Indeed, we can now explain the "dilemma" which is particularly troubling about the Prisoner's Dilemma game: the outcome reached in the unique equilibrium, which is an equilibrium in strictly dominant strategies, is also the only outcome that is *not* Pareto optimal.

Games with dominant strategies play an important role in game theory, espe-

mechanism design

cially in games handcrafted by experts. This is true in particular in *mechanism design*, as we will see in Chapter 10. However, dominant strategies are rare in naturally-occurring games. More common are dominated strategies.

dominated strategy

Definition 3.4.10 (Dominated strategy) *A strategy s_i is* strictly *(weakly; very weakly)* dominated *for an agent i if some other strategy s_i' strictly (weakly; very weakly) dominates s_i.*

---

6. Note that here we consider strategy domination from one individual player's point of view; thus, this notion is unrelated to the concept of Pareto domination discussed earlier.

Let us focus for the moment on strictly dominated strategies. Intuitively, all strictly dominated pure strategies can be ignored, since they can never be best responses to any moves by the other players. There are several subtleties, however. First, once a pure strategy is eliminated, another strategy that was not dominated can become dominated. And so this process of elimination can be continued. Second, a pure strategy may be dominated by a mixture of other pure strategies without being dominated by any of them independently. To see this, consider the game in Figure 3.15.

	L	C	R
U	3, 1	0, 1	0, 0
M	1, 1	1, 1	5, 0
D	0, 1	4, 1	0, 0

Figure 3.15 A game with dominated strategies.

Column R can be eliminated, since it is dominated by, for example, column L. We are left with the reduced game in Figure 3.16.

	L	C
U	3, 1	0, 1
M	1, 1	1, 1
D	0, 1	4, 1

Figure 3.16 The game from Figure 3.15 after removing the dominated strategy R.

In this game M is dominated by neither U nor D, but it is dominated by the mixed strategy that selects either U or D with equal probability. (Note, however, that it was not dominated before the elimination of the R column.) And so we are left with the maximally reduced game in Figure 3.17.

This yields us a solution concept: the set of all strategy profiles that assign zero probability to playing any action that would be removed through iterated removal of strictly dominated strategies. Note that this is a much weaker solution concept than Nash equilibrium—the set of strategy profiles will include all the Nash equilibria, but it will include many other mixed strategies as well. In some games, it will be equal to S, the set of all possible mixed strategies.

$$\begin{array}{cc} & L \qquad\quad C \end{array}$$

	L	C
U	3, 1	0, 1
D	0, 1	4, 1

Figure 3.17 The game from Figure 3.16 after removing the dominated strategy M.

Since iterated removal of strictly dominated strategies preserves Nash equilibria, we can use this technique to computational advantage. In the previous example, rather than computing the Nash equilibria of the original 3×3 game, we can now compute them for this 2×2 game, applying the technique described earlier. In some cases, the procedure ends with a single cell; this is the case, for example, with the Prisoner's Dilemma game. In this case we say that the game is *solvable* by iterated elimination.

Clearly, in any finite game, iterated elimination ends after a finite number of iterations. One might worry that, in general, the order of elimination might affect the final outcome. It turns out that this elimination order does not matter when we

Church–Rosser property

remove *strictly* dominated strategies. (This is called a *Church–Rosser* property.) However, the elimination order can make a difference to the final reduced game if we remove weakly or very weakly dominated strategies.

Which flavor of domination should we concern ourselves with? In fact, each flavor has advantages and disadvantages, which is why we present all of them here. Strict domination leads to better-behaved iterated elimination: it yields a reduced game that is independent of the elimination order, and iterated elimination is more computationally manageable. (This and other computational issues regarding domination are discussed in Section 4.5.3.) There is also a further related advantage that we will defer to Section 3.4.4. Weak domination can yield smaller reduced games, but under iterated elimination the reduced game can depend on the elimination order. Very weak domination can yield even smaller reduced games, but again these reduced games depend on elimination order. Furthermore, very weak domination does not impose a strict order on strategies: when two strategies are equivalent, each very weakly dominates the other. For this reason, this last form of domination is generally considered the least important.

3.4.4 *Rationalizability*

rationalizable strategy

A strategy is *rationalizable* if a perfectly rational player could justifiably play it against one or more perfectly rational opponents. Informally, a strategy profile for player i is rationalizable if it is a best response to some beliefs that i could have about the strategies that the other players will take. The wrinkle, however, is that i cannot have arbitrary beliefs about the other players' actions—his beliefs must take into account his knowledge of *their* rationality, which incorporates their knowledge of *his* rationality, their knowledge of his knowledge of their rationality, and so on in an infinite regress. A rationalizable strategy profile is a strategy profile that consists only of rationalizable strategies.

For example, in the Matching Pennies game given in Figure 3.6, the pure strategy *heads* is rationalizable for the row player. First, the strategy *heads* is a best response to the pure strategy *heads* by the column player. Second, believing that the column player would also play *heads* is consistent with the column player's rationality: the column player could believe that the row player would play *tails*, to which the column player's best response is *heads*. It would be rational for the column player to believe that the row player would play *tails* because the column player could believe that the row player believed that the column player would play *tails*, to which *tails* is a best response. Arguing in the same way, we can make our way up the chain of beliefs.

However, not every strategy can be justified in this way. For example, considering the Prisoner's Dilemma game given in Figure 3.3, the strategy *C* is not rationalizable for the row player, because *C* is not a best response to any strategy that the column player could play. Similarly, consider the game from Figure 3.15. *M* is not a rationalizable strategy for the row player: although it *is* a best response to a strategy of the column player's (*R*), there do not exist any beliefs that the column player could hold about the row player's strategy to which *R* would be a best response.

Because of the infinite regress, the formal definition of rationalizability is somewhat involved; however, it turns out that there are some intuitive things that we can say about rationalizable strategies. First, Nash equilibrium strategies are always rationalizable: thus, the set of rationalizable strategies (and strategy profiles) is always nonempty. Second, in two-player games rationalizable strategies have a simple characterization: they are those strategies that survive the iterated elimination of strictly dominated strategies. In n-player games there exist strategies that survive iterated removal of dominated strategies but are not rationalizable. In this more general case, rationalizable strategies are those strategies that survive iterative removal of strategies that are never a best response to any strategy profile by the other players.

We now define rationalizability more formally. First we will define an infinite sequence of (possibly mixed) strategies $S_i^0, S_i^1, S_i^2, \ldots$ for each player i. Let $S_i^0 = S_i$; thus, for each agent i, the first element in the sequence is the set of all i's mixed strategies. Let $CH(S)$ denote the convex hull of a set S: the smallest convex set containing all the elements of S. Now we define S_i^k as the set of all strategies $s_i \in S_i^{k-1}$ for which there exists some $s_{-i} \in \prod_{j \neq i} CH(S_j^{k-1})$ such that for all $s_i' \in S_i^{k-1}$, $u_i(s_i, s_{-i}) \geq u_i(s_i', s_{-i})$. That is, a strategy belongs to S_i^k if there is some strategy s_{-i} for the other players in response to which s_i is at least as good as any other strategy from S_i^{k-1}. The convex hull operation allows i to best respond to uncertain beliefs about which strategies from S_j^{k-1} player j will adopt. $CH(S_j^{k-1})$ is used instead of $\Pi(S_j^{k-1})$, the set of all probability distributions over S_j^{k-1}, because the latter would allow consideration of mixed strategies that are dominated by some pure strategies for j. Player i

	LW	WL
LW	2, 1	0, 0
WL	0, 0	1, 2

Figure 3.18 Battle of the Sexes game.

could not believe that j would play such a strategy because such a belief would be inconsistent with i's knowledge of j's rationality.

Now we define the set of rationalizable strategies for player i as the intersection of the sets $S_i^0, S_i^1, S_i^2, \ldots$.

rationalizable strategy **Definition 3.4.11 (Rationalizable strategies)** *The* rationalizable strategies *for player i are* $\bigcap_{k=0}^{\infty} S_i^k$.

3.4.5 *Correlated equilibrium*

The correlated equilibrium is a solution concept that generalizes the Nash equilibrium. Some people feel that this is the most fundamental solution concept of all.[7]

In a standard game, each player mixes his pure strategies independently. For example, consider again the Battle of the Sexes game (reproduced here as Figure 3.18) and its mixed-strategy equilibrium.

As we saw in Section 3.3.3, this game's unique mixed-strategy equilibrium yields each player an expected payoff of $2/3$. But now imagine that the two players can observe the result of a fair coin flip and can condition their strategies based on that outcome. They can now adopt strategies from a richer set; for example, they could choose "WL if heads, LW if tails." Indeed, this pair forms an equilibrium in this richer strategy space; given that one player plays the strategy, the other player only loses by adopting another. Furthermore, the expected payoff to each player in this so-called correlated equilibrium is $.5 * 2 + .5 * 1 = 1.5$. Thus both agents receive higher utility than they do under the mixed-strategy equilibrium in the uncorrelated case (which had expected payoff of $2/3$ for both agents), and the outcome is fairer than either of the pure-strategy equilibria in the sense that the worst-off player achieves higher expected utility. Correlating devices can thus be quite useful.

The aforementioned example had both players observe the exact outcome of the coin flip, but the general setting does not require this. Generally, the setting includes some random variable (the "external event") with a commonly-known probability distribution, and a private signal to each player about the instantiation

---

7. A Nobel-prize-winning game theorist, R. Myerson, has gone so far as to say that "if there is intelligent life on other planets, in a majority of them, they would have discovered correlated equilibrium before Nash equilibrium."

of the random variable. A player's signal can be correlated with the random variable's value and with the signals received by other players, without uniquely identifying any of them. Standard games can be viewed as the degenerate case in which the signals of the different agents are probabilistically independent.

To model this formally, consider n random variables, with a joint distribution over these variables. Imagine that nature chooses according to this distribution, but reveals to each agent only the realized value of his variable, and that the agent can condition his action on this value.[8]

Definition 3.4.12 (Correlated equilibrium) *Given an n-agent game $G = (N, A, u)$, a correlated equilibrium is a tuple (v, π, σ), where v is a tuple of random variables $v = (v_1, \ldots, v_n)$ with respective domains $D = (D_1, \ldots, D_n)$, π is a joint distribution over v, $\sigma = (\sigma_1, \ldots, \sigma_n)$ is a vector of mappings $\sigma_i : D_i \mapsto A_i$, and for each agent i and every mapping $\sigma_i' : D_i \mapsto A_i$ it is the case that*

$$\sum_{d \in D} \pi(d) u_i \left(\sigma_1(d_1), \ldots, \sigma_i(d_i), \ldots, \sigma_n(d_n) \right)$$

$$\geq \sum_{d \in D} \pi(d) u_i \left(\sigma_1(d_1), \ldots, \sigma_i'(d_i), \ldots, \sigma_n(d_n) \right).$$

Note that the mapping is to an action—that is, to a pure strategy rather than a mixed one. One could allow a mapping to mixed strategies, but that would add no greater generality. (Do you see why?)

For every Nash equilibrium, we can construct an equivalent correlated equilibrium, in the sense that they induce the same distribution on outcomes.

Theorem 3.4.13 *For every Nash equilibrium σ^* there exists a corresponding correlated equilibrium σ.*

The proof is straightforward. Roughly, we can construct a correlated equilibrium from a given Nash equilibrium by letting each $D_i = A_i$ and letting the joint probability distribution be $\pi(d) = \prod_{i \in N} \sigma_i^*(d_i)$. Then we choose σ_i as the mapping from each d_i to the corresponding a_i. When the agents play the strategy profile σ, the distribution over outcomes is identical to that under σ^*. Because the v_i's are uncorrelated and no agent can benefit by deviating from σ^*, σ is a correlated equilibrium.

On the other hand, not every correlated equilibrium is equivalent to a Nash equilibrium; the Battle-of-the-Sexes example given earlier provides a counter-example. Thus, correlated equilibrium is a strictly weaker notion than Nash equilibrium.

Finally, we note that correlated equilibria can be combined together to form new correlated equilibria. Thus, if the set of correlated equilibria of a game G does not contain a single element, it is infinite. Indeed, any convex combination of correlated equilibrium payoffs can itself be realized as the payoff profile of some correlated equilibrium. The easiest way to understand this claim is to imagine

---

8. This construction is closely related to two other constructions later in the book, one in connection with Bayesian Games in Chapter 6, and one in connection with knowledge and probability (KP) structures in Chapter 13.

a public random device that selects which of the correlated equilibria will be played; next, another random number is chosen in order to allow the chosen equilibrium to be played. Overall, each agent's expected payoff is the weighted sum of the payoffs from the correlated equilibria that were combined. Since no agent has an incentive to deviate regardless of the probabilities governing the first random device, we can achieve any convex combination of correlated equilibrium payoffs. Finally, observe that having two stages of random number generation is not necessary: we can simply derive new domains D and a new joint probability distribution π from the D's and π's of the original correlated equilibria, and so perform the random number generation in one step.

3.4.6 *Trembling-hand perfect equilibrium*

Another important solution concept is the *trembling-hand perfect equilibrium*, or simply *perfect equilibrium*. While rationalizability is a weaker notion than that of a Nash equilibrium, perfection is a stronger one. Several equivalent definitions of the concept exist. In the following definition, recall that a fully mixed strategy is one that assigns every action a strictly positive probability.

trembling-hand
perfect
equilibrium

Definition 3.4.14 (Trembling-hand perfect equilibrium) *A mixed-strategy profile s is a* (trembling-hand) perfect equilibrium *of a normal-form game G if there exists a sequence $s^0, s^1, \ldots$ of fully mixed-strategy profiles such that $\lim_{n \to \infty} s^n = s$, and such that for each s^k in the sequence and each player i, the strategy s_i is a best response to the strategies s_{-i}^k.*

proper
equilibrium

Perfect equilibria are relevant to one aspect of multiagent learning (see Chapter 7), which is why we mention them here. However, we do not discuss them in any detail; they are an involved topic, and relate to other subtle refinements of the Nash equilibrium such as the *proper equilibrium*. The notes at the end of the chapter point the reader to further readings on this topic. We should, however, at least explain the term "trembling hand." One way to think about the concept is as requiring that the equilibrium be robust against slight errors—"trembles"—on the part of players. In other words, one's action ought to be the best response not only against the opponents' equilibrium strategies, but also against small perturbation of those. However, since the mathematical definition speaks about arbitrarily small perturbations, whether these trembles in fact model player fallibility or are merely a mathematical device is open to debate.

3.4.7 *ε-Nash equilibrium*

Our final solution concept reflects the idea that players might not care about changing their strategies to a best response when the amount of utility that they could gain by doing so is very small. This leads us to the idea of an ϵ-Nash equilibrium.

Definition 3.4.15 (ϵ-Nash) *Fix $\epsilon > 0$. A strategy profile $s = (s_1, \ldots, s_n)$ is an ϵ-Nash equilibrium if, for all agents i and for all strategies $s_i' \neq s_i$, $u_i(s_i, s_{-i}) \geq u_i(s_i', s_{-i}) - \epsilon$.*

This concept has various attractive properties. ϵ-Nash equilibria always exist; indeed, every Nash equilibrium is surrounded by a region of ϵ-Nash equilibria for any $\epsilon > 0$. The argument that agents are indifferent to sufficiently small gains is convincing to many. Further, the concept can be computationally useful: algorithms that aim to identify ϵ-Nash equilibria need to consider only a finite set of mixed-strategy profiles rather than the whole continuous space. (Of course, the size of this finite set depends on both ϵ and on the game's payoffs.) Since computers generally represent real numbers using a floating-point approximation, it is usually the case that even methods for the "exact" computation of Nash equilibria (see e.g., Section 4.2) actually find only ϵ-equilibria where ϵ is roughly the "machine precision" (on the order of 10^{-16} or less for most modern computers). ϵ-Nash equilibria are also important to multiagent learning algorithms; we discuss them in that context in Section 7.3.

However, ϵ-Nash equilibria also have several drawbacks. First, although Nash equilibria are always surrounded by ϵ-Nash equilibria, the reverse is not true. Thus, a given ϵ-Nash equilibrium is not necessarily close to any Nash equilibrium. This undermines the sense in which ϵ-Nash equilibria can be understood as approximations of Nash equilibria. Consider the game in Figure 3.19.

	L	R
U	1, 1	0, 0
D	$1 + \frac{\epsilon}{2}$, 1	500, 500

Figure 3.19 A game with an interesting ϵ-Nash equilibrium.

This game has a unique Nash equilibrium of (D, R), which can be identified through the iterated removal of dominated strategies. (D dominates U for player 1; on the removal of U, R dominates L for player 2.) (D, R) is also an ϵ-Nash equilibrium, of course. However, there is also another ϵ-Nash equilibrium: (U, L). This game illustrates two things.

First, neither player's payoff under the ϵ-Nash equilibrium is within ϵ of his payoff in a Nash equilibrium; indeed, in general both players' payoffs under an ϵ-Nash equilibrium can be arbitrarily less than in any Nash equilibrium. The problem is that the requirement that player 1 cannot gain more than ϵ by deviating from the ϵ-Nash equilibrium strategy profile of (U, L) does not imply that *player 2* would not be able to gain more than ϵ by best responding to player 1's deviation.

Second, some ϵ-Nash equilibria might be very unlikely to arise in play. Although player 1 might not care about a gain of $\frac{\epsilon}{2}$, he might reason that the fact that D dominates U would lead player 2 to expect him to play D, and that player 2 would thus play R in response. Player 1 might thus play D because it is his best response to R. Overall, the idea of ϵ-approximation is much messier when applied to the identification of a fixed point than when it is applied to a (single-objective) optimization problem.

3.5 History and references

There exist several excellent technical introductory textbooks for game theory, including Osborne and Rubinstein [1994], Fudenberg and Tirole [1991], and Myerson [1991]. The reader interested in gaining deeper insight into game theory should consult not only these, but also the most relevant strands of the the vast literature on game theory which has evolved over the years.

The origins of the material covered in the chapter are as follows. In 1928, John von Neumann derived the "maximin" solution concept to solve zero-sum normal-form games [von Neumann, 1928]. Our proof of his minimax theorem is similar to the one in Luce and Raiffa [1957b]. In 1944, von Neumann together with Oskar Morgenstern authored what was to become the founding document of game theory [von Neumann and Morgenstern, 1944]; a second edition quickly followed in 1947. Among the many contributions of this work are the axiomatic foundations for "objective probabilities" and what became known as von Neumann–Morgenstern utility theory. The classical foundation of "subjective probabilities" is Savage [1954], but we do not cover those since they do not play a role in the book. A comprehensive overview of these foundational topics is provided by Kreps [1988], among others. Our own treatment of utility theory draws on Poole et al. [1997]; see also Russell and Norvig [2003].

But von Neumann and Morgenstern [1944] did much more; they introduced the normal-form game, the extensive form (to be discussed in Chapter 5), the concepts of pure and mixed strategies, as well as other notions central to game theory. Schelling [1960] was one of the first to show that interesting social interactions could usefully be modeled using game theory, for which he was recognized in 2005 with a Nobel Prize.

Shortly afterward John Nash introduced the concept of what would become known as the "Nash equilibrium" [Nash, 1950, 1951], without a doubt the most influential concept in game theory to this date. Indeed, Nash received a Nobel Prize in 1994 because of this work.[9] The proof in Nash [1950] uses Kakutani's fixed-point theorem; our proof of Theorem 3.3.22 follows Nash [1951]. Lemma 3.3.14 is due to Sperner [1928] and Theorem 3.3.17 is due to Brouwer [1912]; our proof of the latter follows Border [1985].

This work opened the floodgates to a series of refinements and alternative solution concepts which continues to this day. We covered several of these solution concepts. The literature on Pareto optimality and social optimization dates back to the early twentieth century, including seminal work by Pareto and Pigou, but perhaps was best established by Arrow in his seminal work on social choice [Arrow, 1970]. The minimax regret decision criterion was first proposed by Savage [1954], and further developed in Loomes and Sugden [1982] and Bell [1982]. Recent work from a computer science perspective includes Hyafil and Boutilier [2004], which also applies this criterion to the Bayesian games setting we introduce in Section 6.3. Iterated removal of dominated strategies, and the closely

---

9. John Nash was also the topic of the Oscar-winning 2001 movie *A Beautiful Mind*; however, the movie had little to do with his scientific contributions and indeed got the definition of Nash equilibrium wrong.

related rationalizability, enjoy a long history, though modern discussion of them is most firmly anchored in two independent and concurrent publications: Pearce [1984] and Bernheim [1984]. Correlated equilibria were introduced in Aumann [1974]; Myerson's quote is taken from Solan and Vohra [2002]. Trembling-hand perfection was introduced in Selten [1975]. An even stronger notion than (trembling-hand) perfect equilibrium is that of proper equilibrium [Myerson, 1978]. In Chapter 7 we discuss the concept of evolutionarily stable strategies [Maynard Smith and Price, 1973] and their connection to Nash equilibria. In
stable
equilibrium addition to such single-equilibrium concepts, there are concepts that apply to sets of equilibria, not single ones. Of note are the notions of *stable equilibria* as originally defined in Kohlberg and Mertens [1986], and various later refine-
hyperstable set ments such as *hyperstable sets* defined in Govindan and Wilson [2005a]. Good surveys of many of these concepts can be found in Hillas and Kohlberg [2002] and Govindan and Wilson [2005b].

4

Computing Solution Concepts of Normal-Form Games

The discussion of strategies and solution concepts in Chapter 3 largely ignored issues of computation. We start by asking the most basic question: How hard is it to compute the Nash equilibria of a game? The answer turns out to be quite subtle, and to depend on the class of games being considered.

We have already seen how to compute the Nash equilibria of simple games. These calculations were deceptively easy, partly because there were only two players and partly because each player had only two actions. In this chapter we discuss several different classes of games, starting with the simple two-player, zero-sum normal-form game. Dropping only the zero-sum restriction yields a problem of different complexity—while it is generally believed that any algorithm that guarantees a solution must have an exponential worst case complexity, it is also believed that a proof to this effect may not emerge for some time. We also consider procedures for n-player games. In each case, we describe how to formulate the problem, the algorithm (or algorithms) commonly used to solve them, and the complexity of the problem. While we focus on the problem of finding a sample Nash equilibrium, we will briefly discuss the problem of finding all Nash equilibria and finding equilibria with specific properties. Along the way we also discuss the computation of other game-theoretic solution concepts: maxmin and minmax strategies, strategies that survive iterated removal of dominated strategies, and correlated equilibria.

4.1 Computing Nash equilibria of two-player, zero-sum games

The class of two-player, zero-sum games is the easiest to solve. The Nash equilibrium problem for such games can be expressed as a *linear program (LP)*, which means that equilibria can be computed in polynomial time.[1] Consider a two-player, zero-sum game $G = (\{1, 2\}, A_1 \times A_2, (u_1, u_2))$. Let U_i^* be the expected utility for player i in equilibrium (the value of the game); since the game is zero-sum, $U_1^* = -U_2^*$. The minmax theorem (see Section 3.4.1 and Theorem 3.4.4)

---

1. Appendix B reviews the basics of linear programming.

tells us that U_1^* holds constant in all equilibria and that it is the same as the value that player 1 achieves under a minmax strategy by player 2. Using this result, we can construct the linear program that follows.

$$\text{minimize} \quad U_1^* \tag{4.1}$$

$$\text{subject to} \quad \sum_{k \in A_2} u_1\left(a_1^j, a_2^k\right) \cdot s_2^k \leq U_1^* \qquad \forall j \in A_1 \tag{4.2}$$

$$\sum_{k \in A_2} s_2^k = 1 \tag{4.3}$$

$$s_2^k \geq 0 \qquad \forall k \in A_2 \tag{4.4}$$

Note first of all that the utility terms $u_1(\cdot)$ are constants in the linear program, while the mixed strategy terms $s_2^\cdot$ and U_1^* are variables. Let us start by looking at constraint (4.2). This states that for every pure strategy j of player 1, his expected utility for playing any action $j \in A_1$ given player 2's mixed strategy s_2 is at most U_1^*. Those pure strategies for which the expected utility is exactly U_1^* will be in player 1's best response set, while those pure strategies leading to lower expected utility will not. Of course, as mentioned earlier U_1^* is a variable; the linear program will choose player 2's mixed strategy in order to minimize U_1^* subject to the constraint just discussed. Thus, lines (4.1) and (4.2) state that player 2 plays the mixed strategy that minimizes the utility player 1 can gain by playing his best response. This is almost exactly what we want. All that is left is to ensure that the values of the variables s_2^k are consistent with their interpretation as probabilities. Thus, the linear program also expresses the constraints that these variables must sum to one (4.3) and must each be nonnegative (4.4).

This linear program gives us player 2's mixed strategy in equilibrium. In the same fashion, we can construct a linear program to give us player 1's mixed strategies. This program reverses the roles of player 1 and player 2 in the constraints; the objective is to *maximize* U_1^*, as player 1 wants to maximize his own payoffs. This corresponds to the *dual* of player 2's program.

$$\text{maximize} \quad U_1^* \tag{4.5}$$

$$\text{subject to} \quad \sum_{j \in A_1} u_1(a_1^j, a_2^k) \cdot s_1^j \geq U_1^* \qquad \forall k \in A_2 \tag{4.6}$$

$$\sum_{j \in A_1} s_1^j = 1 \tag{4.7}$$

$$s_1^j \geq 0 \qquad \forall j \in A_1 \tag{4.8}$$

Finally, we give a formulation equivalent to our first linear program from Equations (4.1)–(4.4), which will be useful in the next section. This program

slack variable   works by introducing *slack variables* r_1^j for every $j \in A_1$ and then replacing the inequality constraints with equality constraints. This LP formulation follows.

$$\text{minimize} \quad U_1^* \tag{4.9}$$

$$\text{subject to} \quad \sum_{k \in A_2} u_1(a_1^j, a_2^k) \cdot s_2^k + r_1^j = U_1^* \qquad \forall j \in A_1 \tag{4.10}$$

$$\sum_{k \in A_2} s_2^k = 1 \tag{4.11}$$

$$s_2^k \geq 0 \qquad \forall k \in A_2 \tag{4.12}$$

$$r_1^j \geq 0 \qquad \forall j \in A_1 \tag{4.13}$$

Comparing the LP formulation given in Equations (4.9)–(4.12) with our first formulation given in Equations (4.1)–(4.4), observe that constraint (4.2) changed to constraint (4.10) and that a new constraint (4.13) was introduced. To see why the two formulations are equivalent, note that since constraint (4.13) requires only that each slack variable must be positive, the requirement of equality in constraint (4.10) is equivalent to the inequality in constraint (4.2).

4.2 Computing Nash equilibria of two-player, general-sum games

Unfortunately, the problem of finding a Nash equilibrium of a two-player, general-sum game cannot be formulated as a linear program. Essentially, this is because the two players' interests are no longer diametrically opposed. Thus, we cannot state our problem as an optimization problem: one player is not trying to minimize the other's utility.

4.2.1 *Complexity of computing a sample Nash equilibrium*

The issue of characterizing the complexity of computing a sample Nash equilibrium is tricky. No known reduction exists from our problem to a decision problem that is NP-complete, nor has our problem been shown to be easier. An intuitive stumbling block is that *every* game has at least one Nash equilibrium, whereas known NP-complete problems are expressible in terms of decision problems that do not always have solutions.

Current knowledge about the complexity of computing a sample Nash equilibrium thus relies on another, less familiar complexity class that describes the PPAD problem of finding a solution which always exists. This class is called PPAD, which stands for "polynomial parity argument, directed version." To describe this class we must first define a family of directed graphs which we will denote $\mathcal{G}(n)$. Let each graph in this family be defined on a set N of 2^n nodes. Although each graph in $\mathcal{G}(n)$ thus contains a number of nodes that is exponential in n, we want to restrict our attention to graphs that can be described in polynomial space. There is no need to encode the set of nodes explicitly; we encode the set of edges in a given graph as follows. Let *Parent* : $N \mapsto N$ and *Child* : $N \mapsto N$

be two functions that can be encoded as arithmetic circuits with sizes polynomial in n.[2] Let there be one graph $G \in \mathcal{G}(n)$ for every such pair of *Parent* and *Child* functions, as long as G satisfies one additional restriction that is described later. Given such a graph G, an edge exists from a node j to a node k iff $Parent(k) = j$ and $Child(j) = k$. Thus, each node has either zero parents or one parent and either zero children or one child. The additional restriction is that there must exist one distinguished node $0 \in N$ with exactly zero parents.

The aforementioned constraints on the in- and out-degrees of the nodes in graphs $G \in \mathcal{G}(n)$ imply that every node is either part of a cycle or part of a path from a source (a parentless node) to a sink (a childless node). The computational task of problems in the class PPAD is finding either a sink or a source other than 0 for a given graph $G \in \mathcal{G}(n)$. Such a solution always exists: because the node 0 is a source, there must be some sink which is either a descendent of 0 or 0 itself.

We can now state the main complexity result.[3]

Theorem 4.2.1 *The problem of finding a sample Nash equilibrium of a general-sum finite game with two or more players is PPAD-complete.*

Of course, this proof is achieved by showing that the problem is in PPAD and that any other problem in PPAD can be reduced to it in polynomial time. To show that the problem is in PPAD, a reduction is given, which expresses the problem of finding a Nash equilibrium as the problem of finding source or sink nodes in a graph as described earlier. This reduction proceeds quite directly from the proof that every game has a Nash equilibrium that appeals to Sperner's lemma. The harder part is the other half of the puzzle: showing that Nash equilibrium computation is PPAD-hard, or in other words that every problem in PPAD can be reduced to finding a Nash equilibrium of some game with size polynomial in the size of the original problem. This result, obtained in 2005, is a culmination of a series of intermediate results obtained over more than a decade. The initial

graphical game   results relied in part on the concept of *graphical games* (see Section 6.5.2) which, in equilibrium, simulate the behavior of the arithmetic circuits *Parent* and *Child* used in the definition of PPAD. More details are given in the notes at the end of the chapter. What are the practical implications of the result that the problem of finding a sample Nash equilibrium is PPAD-complete? As is the case with other complexity classes such as NP, it is not known whether or not P = PPAD. However, it is generally believed (e.g., due to oracle arguments) that the two classes are not equivalent. Thus, the common belief is that in the worst case, computing a sample Nash equilibrium will take time that is exponential in the size of the game. We do know for sure that finding a Nash equilibrium of a two-player game is no easier than finding an equilibrium of an n-player game—a result that may be surprising, given that in practice different algorithms are used for the two-player case than for the n-player case—and that finding a Nash equilibrium is no easier than finding an arbitrary Brouwer fixed point.

---

2. We warn the reader that some technical details are glossed over here.

3. This theorem describes the problem of approximating a Nash equilibrium to an arbitrary, specified degree of precision (i.e., computing an ϵ-equilibrium for a given ϵ). The equilibrium computation problem is defined in this way partly because games with three or more players can have equilibria involving irrational-valued probabilities.

4.2.2 *An LCP formulation and the Lemke–Howson algorithm*

Lemke–Howson algorithm

We now turn to algorithms for computing sample Nash equilibria, notwithstanding the discouraging computational complexity of this problem. We start with the *Lemke–Howson algorithm*, for two reasons. First, it is the best known algorithm for the two-player, general-sum case (however, it must be said, not the fastest algorithm, experimentally speaking). Second, it provides insight into the structure of Nash equilibria, and indeed constitutes an independent, constructive proof of Nash's theorem (Theorem 3.3.22).

The LCP formulation

linear complementarity problem (LCP)

Unlike in the special zero-sum case, the problem of finding a sample Nash equilibrium cannot be formulated as a linear program. However, the problem of finding a Nash equilibrium of a two-player, general-sum game can be formulated as a *linear complementarity problem* (LCP). In this section we show how to construct this formulation by starting with the slack variable formulation given in Equations (4.9)–(4.12). After giving the formulation, we present the Lemke–Howson algorithm, which can be used to solve this LCP.

feasibility program

As it turns out, our LCP will have no objective function at all, and is thus a constraint satisfaction problem, or a *feasibility program*, rather than an optimization problem. Also, we can no longer determine one player's equilibrium strategy by only considering the other player's payoff; instead, we will need to discuss both players explicitly. The LCP for computing the Nash equilibrium of a general-sum two-player game follows.

$$\sum_{k \in A_2} u_1(a_1^j, a_2^k) \cdot s_2^k + r_1^j = U_1^* \qquad \forall j \in A_1 \qquad (4.14)$$

$$\sum_{j \in A_1} u_2(a_1^j, a_2^k) \cdot s_1^j + r_2^k = U_2^* \qquad \forall k \in A_2 \qquad (4.15)$$

$$\sum_{j \in A_1} s_1^j = 1, \quad \sum_{k \in A_2} s_2^k = 1 \qquad (4.16)$$

$$s_1^j \geq 0, \quad s_2^k \geq 0 \qquad \forall j \in A_1, \forall k \in A_2 \qquad (4.17)$$

$$r_1^j \geq 0, \quad r_2^k \geq 0 \qquad \forall j \in A_1, \forall k \in A_2 \qquad (4.18)$$

$$r_1^j \cdot s_1^j = 0, \quad r_2^k \cdot s_2^k = 0 \qquad \forall j \in A_1, \forall k \in A_2 \qquad (4.19)$$

Observe that this formulation bears a strong resemblance to the LP formulation with slack variables given earlier in Equations (4.9)–(4.12). Let us go through the differences. First, as discussed earlier the LCP has no objective function. Second, constraint (4.14) is the same as constraint (4.10) in our LP formulation; however, here we also include constraint (4.15) which constrains player 2's actions in the same way. We also give the standard constraints that probabilities sum to one (4.16), that probabilities are nonnegative (4.17) and that slack variables are nonnegative (4.18), but now state these constraints for both players rather than only for player 1.

0, 1	6, 0
2, 0	5, 2
3, 4	3, 3

Figure 4.1 A game for the exposition of the Lemke–Howson algorithm.

If we included only constraints (4.14)–(4.18)), we would still have a linear program. However, we would also have a flaw in our formulation: the variables U_1^* and U_2^* would be insufficiently constrained. We want these values to express the expected utility that each player would achieve by playing his best response to the other player's chosen mixed strategy. However, with the constraints we have described so far, U_1^* and U_2^* would be allowed to take unboundedly large values, because all of these constraints remain satisfied when both U_i^* and r_i^j are increased by the same constant, for any given i and j. We solve this problem by adding the nonlinear constraint (4.19), called the *complementarity condition*. The addition of this constraint means that we no longer have a linear program; instead, we have a linear complementarity problem.

complementarity
condition

Why does the complementarity condition fix our problem formulation? This constraint requires that whenever an action is played by a given player with positive probability (i.e., whenever an action is in the support of a given player's mixed strategy) then the corresponding slack variable must be zero. Under this requirement, each slack variable can be viewed as the player's incentive to deviate from the corresponding action. Thus, the complementarity condition captures the fact that, in equilibrium, all strategies that are played with positive probability must yield the same expected payoff, while all strategies that lead to lower expected payoffs are not played. Taking all of our constraints together, we are left with the requirement that each player plays a best response to the other player's mixed strategy: the definition of a Nash equilibrium.

The Lemke–Howson algorithm: a graphical exposition

Lemke–Howson
algorithm

The best-known algorithm designed to solve this LCP formulation is the *Lemke–Howson algorithm*. We will explain it initially through a graphical exposition. Consider the game in Figure 4.1. Figure 4.2 shows a graphical representation of the two players' mixed-strategy spaces in this game. Each player's strategy space is shown in a separate graph. Within a graph, each axis corresponds to one of the corresponding player's pure strategies and the region spanned by these axes represents all the mixed strategies (as discussed in Section 3.3.4, with $k + 1$ axes, the region forms a k-dimensional simplex). For example, in the right-hand side of the figure, the two dots show player 2's two pure strategies and the line connecting them (a one-dimensional simplex) represents all his possible mixed strategies.

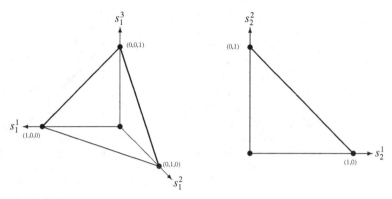

Figure 4.2 Strategy spaces for player 1 (left) and player 2 (right) in the game from Figure 4.1.

Similarly, player 1's three pure strategies are represented by the points $(0, 0, 1)$, $(0, 1, 0)$, and $(1, 0, 0)$, while the set of his mixed strategies (a two-dimensional simplex) is represented by the region bounded by the triangle having these three points as its vertices. (Can you identify the point corresponding to the strategy that randomizes equally among the three pure strategies?)

Our next step in defining the Lemke–Howson algorithm is to define a labeling on the strategies. Every possible mixed strategy s_i is given a set of labels $L(s_i^j) \subseteq A_1 \cup A_2$ drawn from the set of available actions for both players. Denoting a given player as i and the other player as $-i$, mixed strategy s_i for player i is labeled as follows:

- with each of player i's actions a_i^j that is *not* in the support of s_i; and
- with each of player $-i$'s actions a_{-i}^j that *is* a best response by player $-i$ to s_i.

This labeling is useful because a pair of strategies (s_1, s_2) is a Nash equilibrium if and only if it is completely labeled (i.e., $L(s_1) \cup L(s_2) = A_1 \cup A_2$). For a pair to be completely labeled, each action a_i^j must either played by player i with zero probability, or be a best response by player i to the mixed strategy of player $-i$.[4,5]

The requirement that a pair of mixed strategies must be completely labeled can be understood as a restatement of the complementarity condition given in constraint (4.19) in the LCP for computing the Nash equilibrium of a general-sum two-player game, because the slack variable r_i^j is zero exactly when its corresponding action a_i^j is a best response to the mixed strategy s_{-i}.

---

4. We must introduce a certain caveat here. In general, it is possible that some actions will satisfy both of these conditions and thus belong to both $L(s_1)$ and $L(s_2)$; however, this will not occur when a game is nondegenerate. Full discussion of degenericity lies beyond the scope of the book, but for the record, one definition is as follows: A two-player game is *degenerate* if there exists some mixed strategy for either player such that the number of pure strategy best responses of the other player is greater than the size of the support of the mixed strategy. Here we will assume that the game is nondegenerate.
5. Some readers may be reminded of the labeling of simplex vertices in the proof of Sperner's Lemma in Section 3.3.4. These readers should note that these are rather different kinds of labeling, which should not be confused with each other.

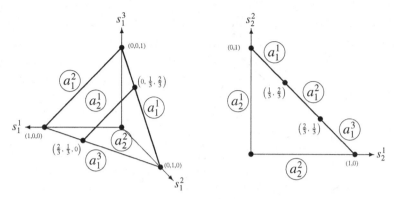

Figure 4.3 Labeled strategy spaces for player 1 (left) and player 2 (right) in the game from Figure 4.1.

It turns out that it is convenient to add one fictitious point in the strategy space of each agent, the origin; that is, $(0, 0, 0)$ for player 1 and $(0, 0)$ for player 2. Thus, we want to be able to consider these points as belonging to the players' strategy spaces. While discussing this algorithm, therefore, we redefine the players' strategy spaces to be the convex hull of their true strategy spaces and the origin of the graph. (This can be understood as replacing the constraint that $\sum_j s_i^j = 1$ with the constraint that $\sum_j s_i^j \leq 1$.) Thus, player 2's strategy space is a triangle with vertices $(0, 0)$, $(1, 0)$, and $(0, 1)$, while player 1's strategy space is a pyramid with vertices $(0, 0, 0)$, $(1, 0, 0)$, $(0, 1, 0)$, and $(0, 0, 1)$.

Returning to our running example, the labeled version of the strategy spaces is given in Figure 4.3. Consider first the right side of Figure 4.3, which describes player 2's strategy space, and examine the two regions labeled with player 2's actions. The line from $(0, 0)$ to $(0, 1)$ is labeled with a_2^1, because none of these mixed strategies assign any probability to playing action a_2^1. In the same way, the line from $(0, 0)$ to $(1, 0)$ is labeled with a_2^2. Now consider the three regions labeled with player 1's actions. Examining the payoff matrix in Figure 4.1, you can verify that, for example, the action a_1^1 is a best response by player 1 to any of the mixed strategies represented by the line from $(0, 1)$ to $(\frac{1}{3}, \frac{2}{3})$. Notice that the point $(\frac{1}{3}, \frac{2}{3})$ is labeled by both a_1^1 and a_1^2, because both of these actions are best responses by player 1 to the mixed strategy $(\frac{1}{3}, \frac{2}{3})$ by player 2.[6]

Similarly, consider now the left side of Figure 4.3, representing player 1's strategy space. There is a region labeled with each action a_1^j of player 1, which is the triangle having a vertex at the origin and running orthogonal to the axis s_1^j. (Can you see why these are the only mixed strategies for player 1 that do not involve the action a_1^j?) The two regions for the labels corresponding to actions of player 2 (a_2^1 and a_2^2) divide the outer triangle. As earlier, note that some mixed strategies are multiply labeled: for example, the point $\left(0, \frac{1}{3}, \frac{2}{3}\right)$ is labeled with a_2^1, a_2^2, and a_1^1.

---

6. The reader may note a subtlety here. Since we added the point $(0, 0)$ and are considering the entire triangle and not just the line $(1, 0) - (0, 1)$, it might be expected that we would attach best-response labels also to interior points within the triangle. However, it turns out that the Lemke–Howson algorithm traverses only the edges of the polygon containing the simplexes and has no use for interior points, and so we ignore them.

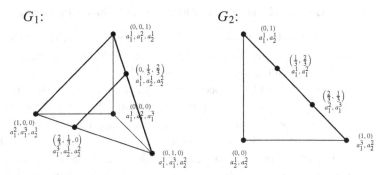

Figure 4.4 Graph of triply labeled strategies for player 1 (left) and doubly labeled strategies for player 2 (right) derived from the game in Figure 4.1.

The Lemke–Howson algorithm can be understood as searching these pairs of labeled spaces for a completely labeled pair of points. Define G_1 and G_2 to be graphs, for players 1 and 2 respectively. The nodes in the graph are fully labeled points in the labeled space, that is, triply labeled points in G_1 and doubly labeled points in G_2. An edge exists between pairs of points that differ in exactly one label. These graphs for our example are shown in Figure 4.4; each node is annotated with the mixed strategy to which it corresponds as well as the actions with which it is labeled.

When the game is nondegenerate, there are no points with more labels than the given player has actions, which implies that a completely labeled pair of strategies must consist of two points that have no labels in common. In our example it is easy to find the three Nash equilibria of the game by inspection: $((0,0,1),(1,0))$, $\left(\left(0, \frac{1}{3}, \frac{2}{3}\right), \left(\frac{2}{3}, \frac{1}{3}\right)\right)$, and $\left(\left(\frac{2}{3}, \frac{1}{3}, 0\right), \left(\frac{1}{3}, \frac{2}{3}\right)\right)$.

The Lemke–Howson algorithm finds an equilibrium by following a path through pairs $(s_1, s_2) \in G_1 \times G_2$ in the cross product of the two graphs. Alternating between the two graphs, each iteration changes one of the two points to a new point that is connected by an edge to the original point. Starting from $(\mathbf{0}, \mathbf{0})$, which is completely labeled, the algorithm picks one of the two graphs and moves from $\mathbf{0}$ in that graph to some adjacent node x. The node x, together with the $\mathbf{0}$ from the other graph, together form an almost completely labeled pair, in that between them they miss exactly one label. The algorithm then moves from the remaining $\mathbf{0}$ to a neighboring node that picks up that missing label, but in the process loses a different label. The process thus proceeds, alternating between the two graphs, until an equilibrium (i.e., a totally labeled pair) is reached.

In our running example, a possible execution of the algorithm starts at $(\mathbf{0}, \mathbf{0})$ and then changes s_1 to $(0, 1, 0)$. Now, our pair $\left((0, 1, 0), (0, 0)\right)$ is a_2^1-almost completely labeled, and the duplicate label is a_2^2. For its next step in G_2 the algorithm moves to $(0, 1)$ because the other possible choice, $(1, 0)$, has the label a_2^2. Returning to G_1 for the next iteration, we move to $\left(\frac{2}{3}, \frac{1}{3}, 0\right)$ because it is the point adjacent to $(0, 1, 0)$ that does not have the duplicate label a_1^1. The final step is to change s_2 to $\left(\frac{1}{3}, \frac{2}{3}\right)$ in order to move away from the label a_2^1. We have now reached the completely labeled pair $\left(\left(\frac{2}{3}, \frac{1}{3}, 0\right), \left(\frac{1}{3}, \frac{2}{3}\right)\right)$, and the algorithm terminates. This execution trace can be summarized by the path $((0, 0, 0), (0, 0)) \rightarrow ((0, 1, 0), (0, 0)) \rightarrow ((0, 1, 0), (0, 1)) \rightarrow ((\frac{2}{3}, \frac{1}{3}, 0), (0, 1)) \rightarrow ((\frac{2}{3}, \frac{1}{3}, 0), (\frac{1}{3}, \frac{2}{3}))$.

The Lemke–Howson algorithm: A deeper look at pivoting

The graphical description of the Lemke–Howson algorithm in the previous section provides good intuition but glosses over elements that only a close look at the algebraic formulation reveals. Specifically, in abstracting away to the graphical exposition we did not specify how to compute the graph nodes from the game description. This is the role of this section. The two sections complement each other: This one provides a clear recipe for implementing the algorithm, but on its own would provide little intuition. The previous section did the opposite.

In fact, we do not compute the nodes in advance at all. Instead, we compute them incrementally along the path being explored. At each step, we find the missing label to be added (called the *entering variable*), add it, find out which label has been lost (it is called the *leaving variable*), and the process repeats until no variable is lost in which case a solution has been obtained. This procedure is called *pivoting*, and also underlies the *simplex algorithm* for solving linear programming problems. The high-level description of the Lemke–Howson algorithm is given in Figure 4.5.

pivot algorithms

simplex
algorithm

initialize *the two systems of equations at the origin*
arbitrarily pick *one dependent variable from one of the two systems. This variable* enters *the basis.*
repeat
    identify *one of the previous basis variables which must* leave, *according to the minimum ratio test. The result is a new basis.*
    if *this basis is completely labeled* **then**
       | **return** *the basis*   // we have found an equilibrium.
    else
       | *the variable dual to the variable that last left* enters *the basis.*

Figure 4.5 Pseudocode for the Lemke–Howson algorithm.

As can be seen from the pseudocode, identifying the entering variable follows immediately from the current labeling (except in the first step, in which the choice is arbitrary). The only nontrivial step is identifying the leaving variable. We explain it by tracing the operation of the algorithm on our example.

We start with a reformulation of the first two constraints (4.14) and (4.15) from our LCP formulation.[7]

$$\begin{aligned} r_1 &= 1 & & & -6y_5' \\ r_2 &= 1 & -2y_4' & & -5y_5' \\ r_3 &= 1 & -3y_4' & & -3y_5' \end{aligned} \tag{4.20}$$

$$\begin{aligned} s_4 &= 1 & -x_1' & & -4x_3' \\ s_5 &= 1 & & -2x_2' & -3x_3' \end{aligned} \tag{4.21}$$

---

7. Beside the minor rearrangement of terms and slight notational change, the reader will note that we have lost the different U values and replaced them by the unit values 1; this turns out to be convenient computationally and does not alter the solutions.

This system admits the trivial solution of assigning 0 to all variables on the right-hand side, which is our fictitious starting point. At this point, r_1, r_2, r_3, s_4, s_5 form the basis of our system of equations, and the other variables (the y's and the x's) are the dependent variables.[8] Note that each basis variable has a dual dependent one; the dual pairs are (r_1, x_1'), (r_2, x_2'), (r_3, x_3'), (s_4, y_4'), and (s_5, y_5'). We will now iteratively remove some of the variables from the basis and replace them with what were previously dependent variables to get a new basis. The rule for which variable enters is simple; initially the choice is arbitrary, and thereafter it is the dual to the variable that previously left. The rule for which variable leaves is more complicated and is called the *minimum ratio test*. When a variable enters, the candidates to leave are all the "clashing variables"; these are all the current basis variables in whose equation the entering variable appears. If there is only one such equation we are done, but otherwise we choose as follows. Each such equation has the form $v = c + qu + T$, where v is the clashing variable, c is a constant (initially they are all 1), u is the entering variable, q is a constant coefficient, and T is a linear combination of variables other than v or u. The clashing variable to leave is the one in whose equation the q/c ratio is smallest.

minimum ratio test

We illustrate the procedure on our example. Let us arbitrarily pick x_2' as the first entering variable. In this case we see immediately that s_5 must leave, since it is the only clashing variable. (x_2' does not appear in the equation of any other basis variable.) With x_2' in the basis the equations much be updated to remove any occurrence of x_2' on the right-hand side, which in this case is achieved simply by rearranging the terms of the second equation in (4.21). This gives us the following.

$$\begin{aligned} s_4 &= 1 &-x_1' &-4x_3' \\ x_2' &= \tfrac{1}{2} & &-\tfrac{3}{2}x_3' &-\tfrac{1}{2}s_5 \end{aligned} \qquad (4.22)$$

The next variable that must enter the basis y_5', s_5's dual. Now the choice for which variable should leave the basis is less obvious; all three variables r_1, r_2, r_3 clash with y_5'. The variable we choose is r_1, since it has the lowest ratio: $\tfrac{1}{6}$, versus $\tfrac{1}{5}$ for r_2 and $\tfrac{1}{3}$ for r_3. Equation (4.20) is now replaced by the following.

$$\begin{aligned} y_5' &= \tfrac{1}{6} & & &-\tfrac{1}{6}r_1 \\ r_2 &= \tfrac{1}{6} &-2y_4' & &+\tfrac{5}{6}r_1 \\ r_3 &= \tfrac{1}{2} &-3y_4' & &+\tfrac{1}{2}r_1 \end{aligned} \qquad (4.23)$$

In this case the first equation is rearranged as above, and then, in the second two equations, the occurrences of y_5' are replaced by $\tfrac{1}{6} - \tfrac{1}{6}r_1$.

With r_1 having left x_1' must enter. This entails that s_4 must leave (in this case again, the only clashing variable). Equation (4.22) now changes as follows.

$$\begin{aligned} x_1' &= 1 &-4x_3' &-s_4 \\ x_2' &= \tfrac{1}{2} &-\tfrac{3}{2}x_3' & &-\tfrac{1}{2}s_5 \end{aligned} \qquad (4.24)$$

---

8. From the definitions of matrix theory, in our particular system the basis variables are independent of each other (i.e., their values can be chosen independently), but together they determine the values of all other variables.

With y_4' entering, either r_2 or r_3 must leave, and it is r_2 that leaves since its ratio of $\frac{\frac{1}{6}}{2} = \frac{1}{12}$ is lower than r_3's ratio of $\frac{\frac{1}{3}}{2} = \frac{1}{6}$. Equation (4.23) changes as follows.

$$
\begin{aligned}
y_5' &= \tfrac{1}{6} &-\tfrac{1}{6}r_1 & \\
y_4' &= \tfrac{1}{12} &+\tfrac{5}{12}r_1 &-\tfrac{1}{2}r_2 \\
r_3 &= \tfrac{1}{4} &-\tfrac{3}{4}r_1 &+\tfrac{3}{2}r_2
\end{aligned}
\tag{4.25}
$$

At this point the algorithm terminates since, between them, Equations (4.25) and (4.24) contain all the labels. Renormalizing the vectors x' and y' to be proper probabilities, one gets the solution $((\frac{2}{3}, \frac{1}{3}, 0), (\frac{1}{3}, \frac{2}{3}))$ with payoffs 4 and $\frac{2}{3}$ to the row and column players, respectively.

Properties of the Lemke–Howson algorithm

The Lemke–Howson algorithm has some good properties. First, it is guaranteed to find a sample Nash equilibrium. Indeed, its constructive nature constitutes an alternative proof of the existence of a Nash equilibrium (Theorem 3.3.22). Also, note the following interesting fact: Since the algorithm repeatedly seeks to cover a missing label, after choosing the initial move away from $(\mathbf{0}, \mathbf{0})$, the path through almost completely labeled pairs to an equilibrium is unique. So while the algorithm is nondeterministic, all the nondeterminism is concentrated in its first move. Finally, it can be used to find more than one Nash equilibrium. The reason the algorithm is initialized to start at the origin is that this is the only pair that is known *a priori* to be completely labeled. However, once we have found another completely labeled pair, we can use *it* as the starting point, allowing us to reach additional equilibria. For example, starting at the equilibrium we just found and making an appropriate first choice, we can quickly find another equilibrium by the path $\left((\frac{2}{3}, \frac{1}{3}, 0), (\frac{1}{3}, \frac{2}{3})\right) \to \left((0, \frac{1}{3}, \frac{2}{3}), (\frac{1}{3}, \frac{2}{3})\right) \to \left((0, \frac{1}{3}, \frac{2}{3}), (\frac{2}{3}, \frac{1}{3})\right)$. The remaining equilibrium can be found using the following path from the origin: $((0, 0, 0), (0, 0)) \to ((0, 0, 1), (0, 0)) \to ((0, 0, 1), (1, 0))$.

However, the algorithm is not without its limitations. While we were able to use the algorithm to find all equilibria in our running example, in general we are not guaranteed to be able to do so. As we have seen, the Lemke–Howson algorithm can be thought of as exploring a graph of all completely and almost completely labeled pairs. The bad news is that this graph can be disconnected, and the algorithm is only able to find the equilibria in the connected component that contains the origin (although luckily, there is guaranteed to be at least one such equilibrium). Not only are we unable to guarantee that we will find all equilibria—there is not even an efficient way to determine whether or not all equilibria have been found.

Even with respect to finding a single equilibrium we are not trouble free. First, there is still indeterminacy in the first move, and the algorithm provides no guidance on how to make a good first choice, one that will lead to a relatively short path to the equilibrium, if one exists. And one may not exist—there are cases in which *all* paths are of exponential length (and thus the time complexity of the Lemke–Howson algorithm is provably exponential). Finally, even if one gives up on worst-case guarantees and hopes for good heuristics, the fact that the algorithm has no objective function means that it provides no obvious guideline to assess how close it is to a solution before actually finding one.

Nevertheless, despite all these limitations, the Lemke–Howson algorithm remains a key element in understanding the algorithmic structure of Nash equilibria in general two-person games.

4.2.3 *Searching the space of supports*

One can identify a spectrum of approaches to the design of algorithms. At one end of the spectrum one can develop deep insight into the structure of the problem, and craft a highly specialized algorithm based on this insight. The Lemke–Howson algorithm lies close to this end of the spectrum. At the other end of the spectrum, one identifies relatively shallow heuristics and hopes that these, coupled with ever-increasing computing power, will do the job. Of course, in order to be effective, even these heuristics must embody some insight into the problem. However, this insight tends to be limited and local, yielding rules of thumb that aid in guiding the search through the space of possible solutions, but that do not directly yield a solution. One of the lessons from computer science is that sometimes heuristic approaches can outperform more sophisticated algorithms in practice. In this section we discuss such a heuristic algorithm.

The basic idea behind the algorithm is straightforward. We first note that while the general problem of computing a Nash equilibrium (NE) is a complementarity problem, computing whether there exists a NE with a *particular support*[9] for each player is a relatively simple feasibility program. So the problem is reduced to searching the space of supports. Of course the size of this space is exponential in the number of actions, and this is where the heuristics come in.

We start with the feasibility program. Given a support profile $\sigma = (\sigma_1, \sigma_2)$ as input (where each $\sigma_i \subseteq A_i$), feasibility program TGS (for "test given supports") finds a NE p consistent with σ or proves that no such strategy profile exists. In this program, v_i corresponds to the expected utility of player i in an equilibrium, and the subscript $-i$ indicates the player other than i as usual. The complete program follows.

$$\sum_{a_{-i} \in \sigma_{-i}} p(a_{-i}) u_i(a_i, a_{-i}) = v_i \qquad \forall i \in \{1, 2\}, a_i \in \sigma_i \qquad (4.26)$$

$$\sum_{a_{-i} \in \sigma_{-i}} p(a_{-i}) u_i(a_i, a_{-i}) \leq v_i \qquad \forall i \in \{1, 2\}, a_i \notin \sigma_i \qquad (4.27)$$

$$p_i(a_i) \geq 0 \qquad \forall i \in \{1, 2\}, a_i \in \sigma_i \qquad (4.28)$$

$$p_i(a_i) = 0 \qquad \forall i \in \{1, 2\}, a_i \notin \sigma_i \qquad (4.29)$$

$$\sum_{a_i \in \sigma_i} p_i(a_i) = 1 \qquad \forall i \in \{1, 2\} \qquad (4.30)$$

---

9. Recall that the support specifies the pure strategies played with nonzero probability (see Definition 3.2.6).

Constraints (4.26) and (4.27) require that each player must be indifferent between all actions within his support and must not strictly prefer an action outside of his support. These imply that neither player can deviate to a pure strategy that improves his expected utility, which is exactly the condition for the strategy profile to be a NE. Constraints (4.28) and (4.29) ensure that each S_i can be interpreted as the support of player i's mixed strategy: the pure strategies in S_i must be played with zero or positive probability, and the pure strategies not in S_i must be played with zero probability.[10] Finally, constraint (4.30) ensures that each p_i can be interpreted as a probability distribution. A solution will be returned only when there exists an equilibrium with support S (subject to the caveat in footnote 10).

With this feasibility program in our arsenal, we can proceed to search the space of supports. There are three keys to the efficiency of the following algorithm, called SEM (for *support-enumeration method*). The first two are the factors used to order the search space. Specifically, SEM considers every possible support size profile separately, favoring support sizes that are balanced and small. The third key to SEM is that it separately instantiates each player's support, making use of what we will call *conditional strict dominance* to prune the search space.

support-enumeration method

conditional strict dominance

Definition 4.2.2 (Conditionally strictly dominated action) *An action $a_i \in A_i$ is conditionally strictly dominated, given a profile of sets of available actions $R_{-i} \subseteq A_{-i}$ for the remaining agents, if the following condition holds: $\exists a_i' \in A_i \ \forall a_{-i} \in R_{-i}: \ u_i(a_i, a_{-i}) < u_i(a_i', a_{-i})$.*

Observe that this definition is strict because, in a Nash equilibrium, no action that is played with positive probability can be conditionally dominated given the actions in the support of the opponents' strategies. The problem of checking whether an action is conditionally strictly dominated is equivalent to the problem of checking whether the action is strictly dominated by a pure strategy in a reduced version of the original game. As we show in Section 4.5.1, this problem can be solved in time linear in the size of the game.

The preference for small support sizes amplifies the advantages of checking for conditional dominance. For example, after instantiating a support of size two for the first player, it will often be the case that many of the second player's actions are pruned, because only two inequalities must hold for one action to conditionally dominate another.

Pseudocode for SEM is given in Figure 4.6.

Note that SEM is complete, because it considers all support size profiles and because it prunes only those actions that are *strictly* dominated. As mentioned earlier, the number of supports is exponential in the number of actions and hence this algorithm has an exponential worst-case running time.

---

10. Note that constraint (4.28) allows an action $a_i \in S_i$ to be played with zero probability, and so the feasibility program may sometimes find a solution even when some S_i includes actions that are not in the support. However, player i must still be indifferent between action a_i and each other action $a_i' \in S_i$. Thus, simply substituting in $S_i = A_i$ would not necessarily yield a Nash equilibrium as a solution.

forall *support size profiles* $x = (x_1, x_2)$, *sorted in increasing order of, first,* $|x_1 - x_2|$ *and, second,* $(x_1 + x_2)$ **do**

    forall $\sigma_1 \subseteq A_1$ *s.t.* $|\sigma_1| = x_1$ **do**

        $A_2' \leftarrow \{a_2 \in A_2$ not conditionally dominated, given $\sigma_1 \}$

        if $\nexists a_1 \in \sigma_1$ *conditionally dominated, given* A_2' **then**

            forall $\sigma_2 \subseteq A_2'$ *s.t.* $|\sigma_2| = x_2$ **do**

                if $\nexists a_1 \in \sigma_1$ *conditionally dominated, given* σ_2 **and** *TGS is satisfiable for* $\sigma = (\sigma_1, \sigma_2)$ **then**

                    return *the solution found; it is a NE*

Figure 4.6 The SEM algorithm

Of course, any enumeration order would yield a solution; the particular ordering here has simply been shown to yield solutions quickly in practice. In fact, extensive testing on a wide variety of games encountered throughout the literature has shown SEM to perform *better* than the more sophisticated algorithms. Of course, this result tells us as much about the games in the literature (e.g., they tend to have small-support equilibria) as it tells us about the algorithms.

4.2.4 *Beyond sample equilibrium computation*

In this section we consider two problems related to the computation of Nash equilibria in two-player, general-sum games that go beyond simply identifying a sample equilibrium.

First, instead of just searching for a sample equilibrium, we might want to find an equilibrium with a specific property. Listed below are several different questions we could ask about the existence of such an equilibrium.

1. **(Uniqueness)** Given a game G, does there exist a unique equilibrium in G?
2. **(Pareto optimality)** Given a game G, does there exist a strictly Pareto efficient equilibrium in G?
3. **(Guaranteed payoff)** Given a game G and a value v, does there exist an equilibrium in G in which some player i obtains an expected payoff of at least v?
4. **(Guaranteed social welfare)** Given a game G, does there exist an equilibrium in which the sum of agents' utilities is at least k?
5. **(Action inclusion)** Given a game G and an action $a_i \in A_i$ for some player $i \in N$, does there exist an equilibrium of G in which player i plays action a_i with strictly positive probability?
6. **(Action exclusion)** Given a game G and an action $a_i \in A_i$ for some player $i \in N$, does there exist an equilibrium of G in which player i plays action a_i with zero probability?

The answers to these questions are more useful that they might appear at first glance. For example, the ability to answer the *guaranteed payoff* question in polynomial time could be used to find, in polynomial time, the maximum expected payoff that can be guaranteed in a Nash equilibrium. Unfortunately, all of these questions are hard in the worst case.

Theorem 4.2.3 *The following problems are NP-hard when applied to Nash equilibria:* uniqueness, Pareto optimality, guaranteed payoff, guaranteed social welfare, action inclusion, *and* action exclusion.

This result holds even for two-player games. Further, it is possible to show that the *guaranteed payoff* and *guaranteed social welfare* properties cannot even be approximated to any constant factor by a polynomial-time algorithm.

A second problem is to determine *all* equilibria of a game.

Theorem 4.2.4 *Computing all of the equilibria of a two-player, general-sum game requires worst-case time that is exponential in the number of actions for each player.*

This result follows straightforwardly from the observation that a game with k actions can have $2^k - 1$ Nash equilibria, even if the game is nondegenerate (when the game is degenerate, it can have an infinite number of equilibria). Consider a two-player Coordination game in which both players have k actions and a utility function given by the identity matrix possesses $2^k - 1$ Nash equilibria: one for each nonempty subset of the k actions. The equilibrium for each subset is for both players to randomize uniformly over each action in the subset. Any algorithm that finds all of these equilibria must have a running time that is at least exponential in k.

4.3   Computing Nash equilibria of n-player, general-sum games

For n-player games where $n \geq 3$, the problem of finding a Nash equilibrium can no longer be represented even as an LCP. While it does allow a formulation

nonlinear com-
plementarity
problem

as a *nonlinear complementarity problem*, such problems are often hopelessly impractical to solve exactly. Unlike the two-player case, therefore, it is unclear how to best formulate the problem as input to an algorithm. In this section we discuss three possibilities.

Instead of solving the nonlinear complementarity problem exactly, there has been some success approximating the solution using a *sequence of linear complementarity problems (SLCP)*. Each LCP is an approximation of the problem, and its solution is used to create the next approximation in the sequence. This method can be thought of as a generalization to *Newton's method* of approximating the local maximum of a quadratic equation. Although this method is not globally convergent, in practice it is often possible to try a number of different starting points because of its relative speed.

Another approach is to formulate the problem as a minimum of a function. First, we need to define some more notation. Starting from a strategy profile s, let $c_i^j(s)$ be the change in utility to player i if he switches to playing action a_i^j as a pure strategy. Then, define $d_i^j(s)$ as $c_i^j(s)$ bounded from below by zero.

$$c_i^j(s) = u_i(a_i^j, s_{-i}) - u_i(s)$$

$$d_i^j(s) = \max(c_i^j(s), 0)$$

Note that $d_i^j(s)$ is positive if and only if player i has an incentive to deviate to action a_i^j. Thus, strategy profile s is a Nash equilibrium if and only if $d_i^j(s) = 0$ for all players i, and all actions j for each player.

We capture this property in the objective function given in Equation (4.31); we will refer to this function as $f(s)$.

$$\text{minimize} \quad f(s) = \sum_{i \in N} \sum_{j \in A_i} \left(d_i^j(s) \right)^2 \tag{4.31}$$

$$\text{subject to} \quad \sum_{j \in A_i} s_i^j = 1 \qquad\qquad \forall i \in N \tag{4.32}$$

$$s_i^j \geq 0 \qquad\qquad \forall i \in N, \forall j \in A_i \tag{4.33}$$

This function has one or more global minima at 0, and the set of all s such that $f(s) = 0$ is exactly the set of Nash equilibria. Of course, this property holds even if we did not square each $d_i^j(s)$, but doing so makes the function differentiable everywhere. The constraints on the function are the obvious ones: each player's distribution over actions must sum to one, and all probabilities must be nonnegative. The advantage of this method is its flexibility. We can now apply any method for constrained optimization.

If we instead want to use an unconstrained optimization method, we can roll the constraints into the objective function (which we now call $g(s)$) in such a way that we still have a differentiable function that is zero if and only if s is a Nash equilibrium. This optimization problem follows.

$$\text{minimize} \quad \sum_{i \in N} \sum_{j \in A_i} \left(d_i^j(s) \right)^2 + \sum_{i \in N} \left(1 - \sum_{j \in A_i} s_i^j \right)^2 + \sum_{i \in N} \sum_{j \in A_i} \left(\min(s_i^j, 0) \right)^2$$

Observe that the first term in $g(s)$ is just $f(s)$ from Equation (4.31). The second and third terms in $g(s)$ enforce the constraints given in Equations (4.32) and (4.33) respectively.

A disadvantage in the formulations given in both Equations (4.31)–(4.33) and Equation (4.3) is that both optimization problems have local minima which do not correspond to Nash equilibria. Thus, global convergence is an issue. For example, considering the commonly-used optimization methods hill-climbing and simulated annealing, the former get stuck in local minima while the latter often converge globally only for parameter settings that yield an impractically long running time.

simplicial subdivision When global convergence is required, a common choice is to turn to the class of *simplicial subdivision algorithms*. Before describing these algorithms we will revisit some properties of the Nash equilibrium. Recall from the Nash existence theorem (Theorem 3.3.22) that Nash equilibria are fixed points of the best response function, f. (As defined previously, given a strategy profile $s = (s_1, s_2, \ldots, s_n)$, $f(s)$ consists of all strategy profiles $(s_1', s_2', \ldots, s_n')$ such that s_i' is a best response by player i to s_{-i}.) Since the space of mixed-strategy profiles

simplotope can be viewed as a product of simplexes—a so-called *simplotope*—f is a function mapping from a simplotope to a set of simplotopes.

Scarf's algorithm is a simplicial subdivision method for finding the fixed point of any function on a simplex or simplotope. It divides the simplotope into small regions and then searches over the regions. Unfortunately, such a search is approximate, since a continuous space is approximated by a mesh of small regions. The quality of the approximation can be controlled by refining the meshes into smaller and smaller subdivisions. One way to do this is by restarting the algorithm with a finer mesh after an initial solution has been found. Alternately,

homotopy a *homotopy method* can be used. In this approach, a new variable is added that
method represents the fidelity of the approximation, and the variable's value is gradually adjusted until the algorithm converges.

An alternative approach, due to Govindan and Wilson, uses a homotopy method in a different way. (This homotopy method actually turns out to be an n-player extension of the Lemke–Howson algorithm, although this correspondence is not obvious.) Instead of varying between coarse and fine approximations, the new added variable interpolates between the given game and an easy-to-solve game. That is, we define a set of games indexed by a scalar $\lambda \in [0, 1]$ such that when $\lambda = 0$, we have our original game, and when $\lambda = 1$, we have a very simple game. (One way to do this is to change the original game by adding a "bonus" λk to each player's payoff in one outcome $a = (a_1, \ldots, a_n)$. Consider a choice of k big enough that for each player i, playing a_i is a strictly dominant strategy. Then, when $\lambda = 1$, a will be a (unique) Nash equilibrium, and when $\lambda = 0$, we will have our original game.) We begin with an equilibrium to the simple game and $\lambda = 1$ and let both the equilibrium to the game and the index vary in a continuous fashion to trace the path of game-equilibrium pairs. Along this path λ may both decrease and increase; however, if the path is followed correctly, it will necessarily pass through a point where $\lambda = 0$. This point's corresponding equilibrium is a sample Nash equilibrium of the original game.

Finally, it is possible to generalize the SEM algorithm to the n-player case. Unfortunately, the feasibility program becomes nonlinear, as follows. We call this feasibility program TGS-n.

$$\sum_{a_{-i} \in \sigma_{-i}} \left(\prod_{j \neq i} p_j(a_j) \right) u_i(a_i, a_{-i}) = v_i \qquad \forall i \in N, a_i \in \sigma_i \qquad (4.34)$$

$$\sum_{a_{-i} \in \sigma_{-i}} \left(\prod_{j \neq i} p_j(a_j) \right) u_i(a_i, a_{-i}) \leq v_i \qquad \forall i \in N, a_i \notin \sigma_i \qquad (4.35)$$

$$p_i(a_i) \geq 0 \qquad \forall i \in N, a_i \in \sigma_i \qquad (4.36)$$

$$p_i(a_i) = 0 \qquad \forall i \in N, a_i \notin \sigma_i \qquad (4.37)$$

$$\sum_{a_i \in \sigma_i} p_i(a_i) = 1 \qquad \forall i \in N \qquad (4.38)$$

The expression $p(a_{-i})$ from constraints (4.26) and (4.27) is no longer a single variable, but must now be written as $\prod_{j \neq i} p_j(a_j)$ in constraints (4.34) and (4.35). The resulting feasibility problem can be solved using standard numerical techniques for nonlinear optimization. As with two-player games, in principle any enumeration method would work; the question is which search heuristic works the fastest. It turns out that a minor modification of the SEM heuristic described in Figure 4.6 is effective for the general case as well: one simply reverses the lexicographic ordering between size and balance of supports (SEM first sorts them by size, and then by a measure of balance; in the n-player case we reverse the ordering). The resulting heuristic algorithm performs very well in practice, and better than the algorithms discussed earlier. We should note that while the ordering between balance and size becomes extremely important to the efficiency of the algorithm as n increases, this reverse ordering does not perform substantially worse than SEM in the two-player case, because the smallest of the balanced support size profiles still appears very early in the ordering.

4.4 Computing maxmin and minmax strategies for two-player, general-sum games

Recall from Section 3.4.1 that in a two-player, general-sum game a maxmin strategy for player i is a strategy that maximizes his worst-case payoff, presuming that the other player j follows the strategy that will cause the greatest harm to i. A minmax strategy for j against i is such a maximum-harm strategy. Maxmin and minmax strategies can be computed in polynomial time because they correspond to Nash equilibrium strategies in related zero-sum games.

Let G be an arbitrary two-player game $G = (\{1, 2\}, A_1 \times A_2, (u_1, u_2))$. Let us consider how to compute a maxmin strategy for player 1. It will be useful to define the zero-sum game $G' = (\{1, 2\}, A_1 \times A_2, (u_1, -u_1))$, in which player 1's utility function is unchanged and player 2's utility is the negative of player 1's. By the minmax theorem (Theorem 3.4.4), since G' is zero sum every strategy for player 1 which is part of a Nash equilibrium strategy profile for G' is a maxmin strategy for player 1 in G'. Notice that by definition, player 1's maxmin strategy is independent of player 2's utility function. Thus, player 1's maxmin strategy is the same in G and in G'. Our problem of finding a maxmin strategy in G thus reduces to finding a Nash equilibrium of G', a two-player, zero-sum game. We can thus solve the problem by applying the techniques given earlier in Section 4.1.

The computation of minmax strategies follows the same pattern. We can again use the minmax theorem to argue that player 2's Nash equilibrium strategy in G' is a minmax strategy for him against player 1 in G. (If we wanted to compute player 1's minmax strategy, we would have to construct another game G'' where player 1's payoff is $-u_2$, the negative of player 2's payoff in G.) Thus, both maxmin and minmax strategies can be computed efficiently for two-player games.

4.5   Identifying dominated strategies

Recall that one strategy dominates another when the first strategy is always at least as good as the second, regardless of the other players' actions. (Section 3.4.3 gave the formal definitions.) In this section we discuss some computational uses for identifying dominated strategies, and consider the computational complexity of this process.

As discussed earlier, iterated removal of strictly dominated strategies is conceptually straightforward: the same set of strategies will be identified regardless of the elimination order, and all Nash equilibria of the original game will be contained in this set. Thus, this method can be used to narrow down the set of strategies to consider before attempting to identify a sample Nash equilibrium. In the worst case this procedure will have no effect—many games have *no* dominated strategies. In practice, however, it can make a big difference to iteratively remove dominated strategies before attempting to compute an equilibrium.

Things are a bit trickier with the iterated removal of *weakly* or *very weakly* dominated strategies. In this case the elimination order does make a difference: the set of strategies that survive iterated removal can differ depending on the order in which dominated strategies are removed. As a consequence, removing weakly or very weakly dominated strategies *can* eliminate some equilibria of the original game. There is still a computational benefit to this technique, however. Since no new equilibria are ever created by this elimination (and since every game has at least one equilibrium), at least one of the original equilibria always survives. This is enough if all we want to do is to identify a sample Nash equilibrium. Furthermore, iterative removal of weakly or very weakly dominated strategies can eliminate a larger set of strategies than iterative removal of strictly dominated strategies and so will often produce a smaller game.

What is the complexity of determining whether a given strategy can be removed? This depends on whether we are interested in checking the strategy for domination by a pure or mixed strategies, whether we are interested in strict, weak or very weak domination, and whether we are interested only in domination or in survival under iterated removal of dominated strategies.

4.5.1   *Domination by a pure strategy*

The simplest case is checking whether a (not necessarily pure) strategy s_i for player i is (strictly; weakly; very weakly) dominated by any pure strategy for i. For concreteness, let us consider the case of strict dominance. To solve the problem we must check every pure strategy a_i for player i and every pure-strategy profile for the other players to determine whether there exists some a_i for which it is never weakly better for i to play s_i instead of a_i. If so, s_i is strictly dominated. An algorithm for this case is given in Figure 4.7.

Observe that this algorithm works because we do not need to check every *mixed*-strategy profile of the other players, even though the definition of dominance refers to such strategies. Why can we get away with this? If it is the case (as the inner loop of our algorithm attempts to prove) that for every

forall *pure strategies* $a_i \in A_i$ *for player i where* $a_i \neq s_i$ **do**
  $dom \leftarrow true$
  forall *pure-strategy profiles* $a_{-i} \in A_{-i}$ *for the players other than i* **do**
    if $u_i(s_i, a_{-i}) \geq u_i(a_i, a_{-i})$ **then**
      $dom \leftarrow false$
      break
  if $dom = true$ **then**
    return *true*
return *false*

Figure 4.7 Algorithm for determining whether s_i is strictly dominated by any pure strategy.

pure-strategy profile $a_{-i} \in A_{-i}$, $u_i(s_i, a_{-i}) < u_i(a_i, a_{-i})$, then there cannot exist any mixed-strategy profile $s_{-i} \in S_{-i}$ for which $u_i(s_i, s_{-i}) \geq u_i(a_i, s_{-i})$. This holds because of the linearity of expectation.

The case of very weak dominance can be tested using essentially the same algorithm as in Figure 4.7, except that we must test the condition $u_i(s_i, s_{-i}) > u_i(s_i', s_{-i})$. For weak dominance we need to do a bit more book-keeping: we can test the same condition as for very weak dominance, but we must also set $dom \leftarrow false$ if there is not at least one s_{-i} for which $u_i(s_i, s_{-i}) < u_i(s_i', s_{-i})$. For all of the definitions of domination, the complexity of the procedure is $O(|A|)$, linear in the size of the normal-form game.

4.5.2 Domination by a mixed strategy

Recall that sometimes a strategy is not dominated by any pure strategy, but *is* dominated by some mixed strategy. (We saw an example of this in Figure 3.16.) We cannot use a simple algorithm like the one in Figure 4.7 to test whether a given strategy s_i is dominated by a mixed strategy because these strategies cannot be enumerated. However, it turns out that we can still answer the question in polynomial time by solving a linear program. In this section, we will assume that player i's utilities are strictly positive. This assumption is without loss of generality because if any player i's utilities were negative, we could add a constant to all of i's payoffs without changing the game (see Section 3.1.2).

Each flavor of domination requires a somewhat different linear program. First, let us consider strict domination by a mixed strategy. This would seem to have the following straightforward LP formulation (indeed, a mere feasibility program).

$$\sum_{j \in A_i} p_j u_i(a_j, a_{-i}) > u_i(s_i, a_{-i}) \qquad \forall a_{-i} \in A_{-i} \qquad (4.39)$$

$$p_j \geq 0 \qquad \forall j \in A_i \qquad (4.40)$$

$$\sum_{j \in A_i} p_j = 1 \qquad (4.41)$$

While constraints (4.39)–(4.41) do indeed describe strict domination by a mixed strategy, they do not constitute a linear program. The problem is that the constraints in linear programs must be *weak* inequalities (see Appendix B), and thus we cannot write constraint (4.39) as we have done here. Instead, we must use the LP that follows.

$$\text{minimize} \quad \sum_{j \in A_i} p_j \tag{4.42}$$

$$\text{subject to} \quad \sum_{j \in A_i} p_j u_i(a_j, a_{-i}) \geq u_i(s_i, a_{-i}) \qquad \forall a_{-i} \in A_{-i} \tag{4.43}$$

$$p_j \geq 0 \qquad \forall j \in A_i \tag{4.44}$$

This linear program simulates the strict inequality of constraint (4.39) through the objective function, as we will describe in a moment. Because no constraints restrict the p_j's from above, this LP will always be feasible. However, in the optimal solution the p_j's may not sum to 1; indeed, their sum can be greater than 1 or less than 1. In the optimal solution the p_j's will be set so that their sum cannot be reduced any further without violating constraint (4.43). Thus for at least some $a_{-i} \in A_{-i}$ we will have $\sum_{j \in A_i} p_j u_i(a_j, a_{-i}) = u_i(s_i, a_{-i})$. A strictly dominating mixed strategy therefore exists if and only if the optimal solution to the LP has objective function value strictly less than 1. In this case, we can add a positive amount to each p_j in order to cause constraint (4.43) to hold in its strict version everywhere while achieving the condition $\sum_j p_j = 1$.

Next, let us consider very weak domination. This flavor of domination does not require any strict inequalities, so things are easy here. Here we *can* construct a feasibility program—nearly identical to our earlier failed attempt from Equations (4.39)–(4.41)—which follows.

$$\sum_{j \in A_i} p_j u_i(a_j, a_{-i}) \geq u_i(s_i, a_{-i}) \qquad \forall a_{-i} \in A_{-i} \tag{4.45}$$

$$p_j \geq 0 \qquad \forall j \in A_i \tag{4.46}$$

$$\sum_{j \in A_i} p_j = 1 \tag{4.47}$$

Finally, let us consider weak domination by a mixed strategy. Again our inability to write a strict inequality will make things more complicated. However, we can derive an LP by adding an objective function to the feasibility program given in Equations (4.45)–(4.47).

$$\text{maximize} \quad \sum_{a_{-i} \in A_{-i}} \left[\left(\sum_{j \in A_i} p_j \cdot u_i(a_j, a_{-i}) \right) - u_i(s_i, a_{-i}) \right] \tag{4.48}$$

$$\text{subject to} \quad \sum_{j \in A_i} p_j u_i(a_j, a_{-i}) \geq u_i(s_i, a_{-i}) \qquad \forall a_{-i} \in A_{-i} \tag{4.49}$$

$$p_j \geq 0 \qquad \forall j \in A_i \tag{4.50}$$

$$\sum_{j \in A_i} p_j = 1 \tag{4.51}$$

Because of constraint (4.49), any feasible solution will have a nonnegative objective value. If the optimal solution has a strictly positive objective, the mixed strategy given by the p_j's achieves strictly positive expected utility for at least one $a_{-i} \in A_{-i}$, meaning that s_i is weakly dominated by this mixed strategy.

As a closing remark, observe that all of our linear programs can be modified to check whether a strategy s_i is strictly dominated by any mixed strategy that only places positive probability on some subset of i's actions $T \subset A_i$. This can be achieved simply by replacing all occurrences of A_i by T in the linear programs given earlier.

4.5.3 *Iterated dominance*

Finally, we consider the iterated removal of dominated strategies. We only consider pure strategies as candidates for removal; indeed, as it turns out, it never helps to remove dominated mixed strategies when performing iterated removal. It *is* important, however, that we consider the possibility that pure strategies may be dominated *by* mixed strategies, as we saw in Section 3.4.3.

For all three flavors of domination, it requires only polynomial time to iteratively remove dominated strategies until the game has been maximally reduced (i.e., no strategy is dominated for any player). A single step of this process consists of checking whether every pure strategy of every player is dominated by any other mixed strategy, which requires us to solve at worst $\sum_{i \in N} |A_i|$ linear programs. Each step removes one pure strategy for one player, so there can be at most $\sum_{i \in N} (|A_i| - 1)$ steps.

However, recall that some forms of dominance can produce different reduced games depending on the order in which dominated strategies are removed. We might therefore want to ask other computational questions, regarding which strategies remain in reduced games. Listed below are some such questions.

1. **(Strategy elimination)** Does there exist some elimination path under which the strategy s_i is eliminated?
2. **(Reduction identity)** Given action subsets $A_i' \subseteq A_i$ for each player i, does there exist a maximally reduced game where each player i has the actions A_i'?
3. **(Reduction size)** Given constants k_i for each player i, does there exist a maximally reduced game where each player i has exactly k_i actions?

It turns out that the complexity of answering these questions depends on the form of domination under consideration.

Theorem 4.5.1 *For iterated strict dominance, the* strategy elimination, reduction identity, uniqueness *and* reduction size *problems are in P. For iterated weak dominance, these problems are NP-complete.*

The first part of this result, considering iterated strict dominance, is straightforward: it follows from the fact that iterated strict dominance always arrives at the same set of strategies regardless of elimination order. The second part is trickier; indeed, our statement of this theorem sweeps under the carpet some subtleties about whether domination by mixed strategies is considered (it is in some cases, and is not in others) and the minimum number of utility values permitted for each player. For all the details, the reader should consult the papers cited at the end of the chapter.

4.6 Computing correlated equilibria

The final solution concept that we will consider is correlated equilibrium. It turns out that correlated equilibria are (probably) easier to compute than Nash equilibria: a sample correlated equilibrium can be found in polynomial time using a linear programming formulation. It is not hard to see (e.g., from the proof of Theorem 3.4.13) that every game has at least one correlated equilibrium in which the value of the random variable can be interpreted as a recommendation to each agent of what action to play, and in equilibrium the agents all follow these recommendations. Thus, we can find a sample correlated equilibrium if we can find a probability distribution over pure action profiles with the property that each agent would prefer to play the action corresponding to a chosen outcome when told to do so, given that the other agents are doing the same.

As in Section 3.2, let $a \in A$ denote a pure-strategy profile, and let $a_i \in A_i$ denote a pure strategy for player i. The variables in our linear program are $p(a)$, the probability of realizing a given pure-strategy profile a; since there is a variable for every pure-strategy profile there are thus $|A|$ variables. Observe that as above the values $u_i(a)$ are constants. The linear program follows.

$$\sum_{a \in A | a_i \in a} p(a)u_i(a) \geq \sum_{a \in A | a_i \in a} p(a)u_i(a'_i, a_{-i}) \quad \forall i \in N, \forall a_i, a'_i \in A_i \quad (4.52)$$

$$p(a) \geq 0 \quad\quad\quad\quad\quad\quad\quad\quad\quad\quad \forall a \in A \quad (4.53)$$

$$\sum_{a \in A} p(a) = 1 \quad\quad\quad\quad\quad\quad\quad\quad\quad\quad (4.54)$$

Constraints (4.53) and (4.54) ensure that p is a valid probability distribution. The interesting constraint is (4.52), which expresses the requirement that player i must be (weakly) better off playing action a when he is told to do so than playing any other action a'_i, given that other agents play their prescribed actions. This constraint effectively restates the definition of a correlated equilibrium given in Definition 3.4.12. Note that it can be rewritten as $\sum_{a \in A | a_i \in a} [u_i(a) - u_i(a'_i, a_{-i})] p(a) \geq 0$; in other words, whenever agent i is "recommended" to play action a_i with positive probability, he must get at least as much utility from doing so as he would from playing any other action a'_i.

We can select a desired correlated equilibrium by adding an objective function to the linear program. For example, we can find a correlated equilibrium that maximizes the sum of the agents' expected utilities by adding the objective function

$$\text{maximize:} \sum_{a \in A} p(a) \sum_{i \in N} u_i(a). \quad (4.55)$$

Furthermore, all of the questions discussed in Section 4.2.4 can be answered about correlated equilibria in polynomial time, making them (most likely) fundamentally easier problems.

Theorem 4.6.1 *The following problems are in the complexity class P when applied to correlated equilibria:* uniqueness, *Pareto optimal, guaranteed payoff,* subset inclusion, *and* subset containment.

Finally, it is worthwhile to consider the reason for the computational difference between correlated equilibria and Nash equilibria. Why can we express the definition of a correlated equilibrium as a linear constraint (4.52), while we cannot do the same with the definition of a Nash equilibrium, even though both definitions are quite similar? The difference is that a correlated equilibrium involves a single randomization over action profiles, while in a Nash equilibrium agents randomize separately. Thus, the (nonlinear) version of constraint (4.52) which would instruct a feasibility program to find a Nash equilibrium would be

$$\sum_{a \in A} u_i(a) \prod_{j \in N} p_j(a_j) \geq \sum_{a \in A} u_i(a_i', a_{-i}) \prod_{j \in N \setminus \{i\}} p_j(a_j) \quad \forall i \in N, \forall a_i' \in A_i.$$

This constraint now mimics constraint (4.52), directly expressing the definition of Nash equilibrium. It states that each player i attains at least as much expected utility from following his mixed strategy p_i as from any pure strategy deviation a_i', given the mixed strategies of the other players. However, the constraint is nonlinear because of the product $\prod_{j \in N} p_j(a_j)$.

4.7 History and references

The complexity of finding a sample Nash equilibrium is explored in a series of articles. First came the original definition of the class TFNP [Megiddo and Papadimitriou, 1991], a super-class of PPAD followed by the definition of PPAD by Papadimitriou [1994]. Next, Goldberg and Papadimitriou [2006] showed that finding an equilibrium of a game with any constant number of players is no harder than finding the equilibrium of a four-player game, and Daskalakis et al. [2006b] showed that these computational problems are PPAD-complete. The result was almost immediately tightened to encompass two-player games by Chen and Deng [2006]. The NP-completeness results for Nash equilibria with specific properties are due to Gilboa and Zemel [1989] and Conitzer and Sandholm [2003b]; the inapproximability result appeared in Conitzer [2006].

A general survey of the classical algorithms for computing Nash equilibria in 2-person games is provided in von Stengel [2002]. Another good survey is McKelvey and McLennan [1996]. Some specific references, both to these classical algorithms and to the newer ones discussed in the chapter, are as follows. The Lemke–Howson algorithm [Lemke and Howson, 1964] can be understood as a a specialization of Lemke's pivoting procedure for solving linear complementarity problems [Lemke, 1978]. The graphical exposition of the Lemke–Howson algorithm appeared first in Shapley [1974], and then in a modified version in von Stengel [2002]. Our description of the Lemke–Howson algorithm is based on the latter. An example of games for which *all* Lemke–Howson paths are of exponential length appears in Savani and von Stengel [2004]. Scarf's simplicial-subdivision-based algorithm is described in Scarf [1967]. Homotopy-based approximation methods are covered, for example, in García and Zangwill [1981]. Govindan and Wilson's homotopy method was presented in Govindan and Wilson [2003]; its path-following procedure depends on topological results due to Kohlberg and Mertens [1986]. The support-enumeration method for finding a sample Nash equilibrium is described in Porter et al. [2004a]. The complexity of iteratedly eliminating dominated strategies is described in Gilboa et al. [1989] and Conitzer and Sandholm [2005].

GAMBIT   Two online resources are of particular note. *GAMBIT* [McKelvey et al., 2006] (`http://econweb.tamu.edu/gambit`) is a library of game-theoretic algorithms for finite normal-form and extensive-form games. It includes many different algorithms for finding Nash equilibria. In addition to several algorithms that can be used on general sum, n-player games, it includes implementations of algorithms designed for special cases, including two-player games, zero-sum
GAMUT   games, and finding all equilibria. Finally, *GAMUT* [Nudelman et al., 2004] (`http://gamut.stanford.edu`) is a suite of game generators designed for testing game-theoretic algorithms.

5

Games with Sequential Actions: Reasoning and Computing with the Extensive Form

In Chapter 3 we assumed that a game is represented in normal form: effectively, as a big table. In some sense, this is reasonable. The normal form is conceptually straightforward, and most see it as fundamental. While many other representations exist to describe finite games, we will see in this chapter and in Chapter 6 that each of them has an "induced normal form": a corresponding normal-form representation that preserves game-theoretic properties such as Nash equilibria. Thus the results given in Chapter 3 hold for all finite games, no matter how they are represented; in that sense the normal-form representation is universal.

In this chapter we will look at extensive-form games, a finite representation that does not always assume that players act simultaneously. This representation is in general exponentially smaller than its induced normal form, and furthermore can be much more natural to reason about. While the Nash equilibria of an extensive-form game can be found through its induced normal form, computational benefit can be had by working with the extensive form directly. Furthermore, there are other solution concepts, such as subgame-perfect equilibrium (see Section 5.1.3), which explicitly refer to the sequence in which players act and which are therefore not meaningful when applied to normal-form games.

5.1 Perfect-information extensive-form games

The normal-form game representation does not incorporate any notion of sequence, or time, of the actions of the players. The *extensive (or tree) form* is an alternative representation that makes the temporal structure explicit. We start by discussing the special case of *perfect information* extensive-form games, and then move on to discuss the more general class of *imperfect-information* extensive-form games in Section 5.2. In both cases we will restrict the discussion to finite games, that is, to games represented as finite trees.

5.1.1 *Definition*

Informally speaking, a perfect-information game in extensive form (or, more simply, a perfect-information game) is a tree in the sense of graph theory, in which each node represents the choice of one of the players, each edge represents a possible action, and the leaves represent final outcomes over which each player has a utility function. Indeed, in certain circles (in particular, in artificial intelligence), these are known simply as game trees. Formally, we define them as follows.

Perfect-
information
game

Definition 5.1.1 (Perfect-information game) *A   (finite)   perfect-information game (in extensive form) is a tuple $G = (N, A, H, Z, \chi, \rho, \sigma, u)$, where:*

- *N is a set of n players;*
- *A is a (single) set of actions;*
- *H is a set of nonterminal choice nodes;*
- *Z is a set of terminal nodes, disjoint from H;*
- *$\chi : H \mapsto 2^A$ is the action function, which assigns to each choice node a set of possible actions;*
- *$\rho : H \mapsto N$ is the player function, which assigns to each nonterminal node a player $i \in N$ who chooses an action at that node;*
- *$\sigma : H \times A \mapsto H \cup Z$ is the successor function, which maps a choice node and an action to a new choice node or terminal node such that for all $h_1, h_2 \in H$ and $a_1, a_2 \in A$, if $\sigma(h_1, a_1) = \sigma(h_2, a_2)$ then $h_1 = h_2$ and $a_1 = a_2$; and*
- *$u = (u_1, \ldots, u_n)$, where $u_i : Z \mapsto \mathbb{R}$ is a real-valued utility function for player i on the terminal nodes Z.*

Since the choice nodes form a tree, we can unambiguously identify a node with its *history*, that is, the sequence of choices leading from the root node to it. We can also define the *descendants* of a node h, namely all the choice and terminal nodes in the subtree rooted at h.

An example of such a game is the *Sharing game*. Imagine a brother and sister following the following protocol for sharing two indivisible and identical presents from their parents. First the brother suggests a split, which can be one of three—he keeps both, she keeps both, or they each keep one. Then the sister chooses whether to accept or reject the split. If she accepts they each get their allocated present(s), and otherwise neither gets any gift. Assuming both siblings value the two presents equally and additively, the tree representation of this game is shown in Figure 5.1.

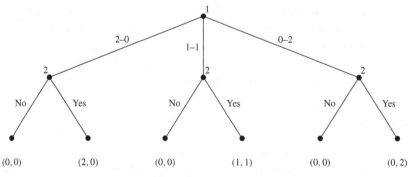

Figure 5.1 The Sharing game.

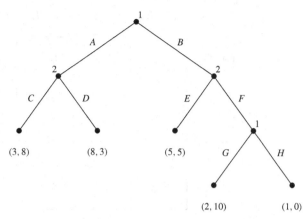

Figure 5.2 A perfect-information game in extensive form.

5.1.2   Strategies and equilibria

A pure strategy for a player in a perfect-information game is a complete specification of which deterministic action to take at every node belonging to that player. A more formal definition follows.

Definition 5.1.2 (Pure strategies) *Let $G = (N, A, H, Z, \chi, \rho, \sigma, u)$ be a perfect-information extensive-form game. Then the pure strategies of player i consist of the Cartesian product $\prod_{h \in H, \rho(h)=i} \chi(h)$.*

Notice that the definition contains a subtlety. An agent's strategy requires a decision at each choice node, regardless of whether or not it is possible to reach that node given the other choice nodes. In the Sharing game above the situation is straightforward—player 1 has three pure strategies, and player 2 has eight, as follows.

$$S_1 = \{2\text{--}0,\ 1\text{--}1,\ 0\text{--}2\}$$

$$S_2 = \{(yes, yes, yes),\ (yes, yes, no),\ (yes, no, yes),\ (yes, no, no),\ (no, yes, yes),\ (no, yes, no),\ (no, no, yes),\ (no, no, no)\}$$

But now consider the game shown in Figure 5.2.

In order to define a complete strategy for this game, each of the players must choose an action at each of his two choice nodes. Thus we can enumerate the pure strategies of the players as follows.

$$S_1 = \{(A, G), (A, H), (B, G), (B, H)\}$$

$$S_2 = \{(C, E), (C, F), (D, E), (D, F)\}$$

It is important to note that we have to include the strategies (A, G) and (A, H), even though once player 1 has chosen A then his own G-versus-H choice is moot.

	(C, E)	(C, F)	(D, E)	(D, F)
(A, G)	3, 8	3, 8	8, 3	8, 3
(A, H)	3, 8	3, 8	8, 3	8, 3
(B, G)	5, 5	2, 10	5, 5	2, 10
(B, H)	5, 5	1, 0	5, 5	1, 0

Figure 5.3 The game from Figure 5.2 in normal form.

The definition of best response and Nash equilibria in this game are exactly as they are for normal-form games. Indeed, this example illustrates how every perfect-information game can be converted to an equivalent normal-form game. For example, the perfect-information game of Figure 5.2 can be converted into the normal-form image of the game, shown in Figure 5.3. Clearly, the strategy spaces of the two games are the same, as are the pure-strategy Nash equilibria. (Indeed, both the mixed strategies and the mixed-strategy Nash equilibria of the two games are also the same; however, we defer further discussion of mixed strategies until we consider imperfect-information games in Section 5.2.)

In this way, for every perfect-information game there exists a corresponding normal-form game. Note, however, that the temporal structure of the extensive-form representation can result in a certain redundancy within the normal form. For example, in Figure 5.3 there are 16 different outcomes, while in Figure 5.2 there are only 5 (the payoff (3, 8) occurs only once in Figure 5.2 but four times in Figure 5.3 etc.). One general lesson is that while this transformation can always be performed, it can result in an exponential blowup of the game representation. This is an important lesson, since the didactic examples of normal-form games are very small, wrongly suggesting that this form is more compact.

The normal form gets its revenge, however, since the reverse transformation—from the normal form to the perfect-information extensive form—does not always exist. Consider, for example, the Prisoner's Dilemma game from Figure 3.3. A little experimentation will convince the reader that there does not exist a perfect-information game that is equivalent in the sense of having the same strategy profiles and the same payoffs. Intuitively, the problem is that perfect-information extensive-form games cannot model simultaneity. The general characterization of the class of normal-form games for which there exist corresponding perfect-information games in extensive form is somewhat complex.

The reader will have noticed that we have so far concentrated on pure strategies and pure Nash equilibria in extensive-form games. There are two reasons for this, or perhaps one reason and one excuse. The reason is that mixed strategies introduce a new subtlety, and it is convenient to postpone discussion of

	(C, E)	(C, F)	(D, E)	(D, F)
(A, G)	3, 8	(3, 8)	8, 3	8, 3
(A, H)	3, 8	(3, 8)	8, 3	8, 3
(B, G)	5, 5	2, 10	5, 5	2, 10
(B, H)	(5, 5)	1, 0	5, 5	1, 0

Figure 5.4 Equilibria of the game from Figure 5.2.

it. The excuse (which also allows the postponement, though not for long) is the following theorem.

Theorem 5.1.3 *Every (finite) perfect-information game in extensive form has a pure-strategy Nash equilibrium.*

This is perhaps the earliest result in game theory, due to Zermelo in 1913 (see the historical notes at the end of the chapter). The intuition here should be clear; since players take turns, and everyone gets to see everything that happened thus far before making a move, it is never necessary to introduce randomness into action selection in order to find an equilibrium. We will see this plainly when we discuss *backward induction* below. Both this intuition and the theorem will cease to hold when we discuss more general classes of games such as imperfect-information games in extensive form. First, however, we discuss an important refinement of the concept of Nash equilibrium.

backward
induction

5.1.3 *Subgame-perfect equilibrium*

As we have discussed, the notion of Nash equilibrium is as well defined in perfect-information games in extensive form as it is in the normal form. However, as the following example shows, the Nash equilibrium can be too weak a notion for the extensive form. Consider again the perfect-information extensive-form game shown in Figure 5.2. There are three pure-strategy Nash equilibria in this game: $\{(A, G), (C, F)\}, \{(A, H), (C, F)\}$, and $\{(B, H), (C, E)\}$. This can be determined by examining the normal form image of the game, as indicated in Figure 5.4.

However, examining the normal form image of an extensive-form game obscures the game's temporal nature. To illustrate a problem that can arise in certain equilibria of extensive-form games, in Figure 5.5 we contrast the equilibria $\{(A, G), (C, F)\}$ and $\{(B, H), (C, E)\}$ by drawing them on the extensive-form game tree.

First consider the equilibrium $\{(A, G), (C, F)\}$. If player 1 chooses A then player 2 receives a higher payoff by choosing C than by choosing D. If player 2

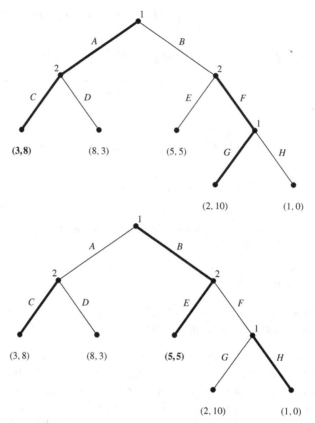

Figure 5.5 Two out of the three equilibria of the game from Figure 5.2: $\{(A, G), (C, F)\}$ and $\{(B, H), (C, E)\}$. Bold edges indicate players' choices at each node.

played the strategy (C, E) rather than (C, F) then player 1 would prefer to play B at the first node in the tree; as it is, player 1 gets a payoff of 3 by playing A rather than a payoff of 2 by playing B. Hence we have an equilibrium.

The second equilibrium $\{(B, H), (C, E)\}$ is less intuitive. First, note that $\{(B, G), (C, E)\}$ is *not* an equilibrium: player 2's best response to (B, G) is (C, F). Thus, the only reason that player 2 chooses to play the action E is that he knows that player 1 would play H at his second decision node. This behavior by player 1 is called a *threat*: by committing to choose an action that is harmful to player 2 in his second decision node, player 1 can cause player 2 to avoid that part of the tree. (Note that player 1 benefits from making this threat: he gets a payoff of 5 instead of 2 by playing (B, H) instead of (B, G).) So far so good. The problem, however, is that player 2 may not consider player 1's threat to be credible: if player 1 did reach his final decision node, actually choosing H over G would also reduce player 1's own utility. If player 2 played F, would player 1 really follow through on his threat and play H, or would he relent and pick G instead?

To formally capture the reason why the $\{(B, H), (C, E)\}$ equilibrium is un-satisfying, and to define an equilibrium refinement concept that does not suffer from this problem, we first define the notion of a subgame.

Definition 5.1.4 (Subgame) *Given a perfect-information extensive-form game G, the* subgame *of G rooted at node h is the restriction of G to the descendants of h. The set of subgames of G consists of all of subgames of G rooted at some node in G.*

Now we can define the notion of a *subgame-perfect equilibrium*, a refinement of the Nash equilibrium in perfect-information games in extensive form, which eliminates those unwanted Nash equilibria.[1]

subgame-perfect equilibrium (SPE)
Definition 5.1.5 (Subgame-perfect equilibrium) *The* subgame-perfect equilibria *(SPE) of a game G are all strategy profiles s such that for any subgame G′ of G, the restriction of s to G′ is a Nash equilibrium of G′.*

Since *G* is its own subgame, every SPE is also a Nash equilibrium. Furthermore, although SPE is a stronger concept than Nash equilibrium (i.e., every SPE is a NE, but not every NE is a SPE) it is still the case that every perfect-information extensive-form game has at least one subgame-perfect equilibrium.

This definition rules out "noncredible threats" of the sort illustrated in the above example. In particular, note that the extensive-form game in Figure 5.2 has only one subgame-perfect equilibrium, $\{(A, G), (C, F)\}$. Neither of the other Nash equilibria is subgame perfect. Consider the subgame rooted at player 1's second choice node. The unique Nash equilibrium of this (trivial) game is for player 1 to play *G*. Thus the action *H*, the restriction of the strategies (A, H) and (B, H) to this subgame, is not optimal in this subgame, and cannot be part of a subgame-perfect equilibrium of the larger game.

5.1.4 *Computing equilibria: backward induction*

n-player, general-sum games: the backward induction algorithm

backward induction
Inherent in the concept of subgame-perfect equilibrium is the principle of *backward induction*. One identifies the equilibria in the "bottom-most" subgame trees, and assumes that those equilibria will be played as one backs up and considers increasingly larger trees. We can use this procedure to compute a sample Nash equilibrium. This is good news: not only are we guaranteed to find a subgame-perfect equilibrium (rather than possibly finding a Nash equilibrium that involves noncredible threats), but also this procedure is computationally simple. In particular, it can be implemented as a single depth-first traversal of the game tree and thus requires time linear in the size of the game representation. Recall in contrast that the best known methods for finding Nash equilibria of general games require time exponential in the size of the normal form; remember as well that the induced normal form of an extensive-form game is exponentially larger than the original representation.

The algorithm BACKWARDINDUCTION is described in Figure 5.6. The variable *util_at_child* is a vector denoting the utility for each player at the child node; *util_at_child*$_{\rho(h)}$ denotes the element of this vector corresponding to the utility for

---

1. Note that the word "perfect" is used in two different senses here.

function BACKWARDINDUCTION (node h) **returns** $u(h)$
if $h \in Z$ **then**
   | **return** $u(h)$ // h is a terminal node
$best\_util \leftarrow -\infty$
forall $a \in \chi(h)$ **do**
   | $util\_at\_child \leftarrow$ BACKWARDINDUCTION$(\sigma(h, a))$
   | **if** $util\_at\_child_{\rho(h)} > best\_util_{\rho(h)}$ **then**
   | | $best\_util \leftarrow util\_at\_child$
return $best\_util$

Figure 5.6 Procedure for finding the value of a sample (subgame-perfect) Nash equilibrium of a perfect-information extensive-form game.

player $\rho(h)$ (the player who gets to move at node h). Similarly, *best_util* is a vector giving utilities for each player.

Observe that this procedure does not return an equilibrium strategy for each of the n players, but rather describes how to label each node with a vector of n real numbers. This labeling can be seen as an extension of the game's utility function to the nonterminal nodes H. The players' equilibrium strategies follow straightforwardly from this extended utility function: every time a given player i has the opportunity to act at a given node $h \in H$ (i.e., $\rho(h) = i$), that player will choose an action $a_i \in \chi(h)$ that solves $\arg\max_{a_i \in \chi(h)} u_i(\sigma(a_i, h))$. These strategies can also be returned by BACKWARDINDUCTION given some extra bookkeeping.

While the procedure demonstrates that in principle a sample SPE is effectively computable, in practice many game trees are not enumerated in advance and are hence unavailable for backward induction. For example, the extensive-form representation of chess has around 10^{150} nodes, which is vastly too large to represent explicitly. For such games it is more common to discuss the size of the game tree in terms of the average branching factor b (the average number of actions which are possible at each node) and a maximum depth m (the maximum number of sequential actions). A procedure which requires time linear in the size of the representation thus expands $O(b^m)$ nodes. Unfortunately, we can do no better than this on arbitrary perfect-information games.

Two-player, zero-sum games: minimax and alpha-beta pruning

We *can* make some computational headway in the widely applicable case of two-player, zero-sum games. We first note that BACKWARDINDUCTION has another name in the two-player, zero-sum context: the *minimax algorithm*. Recall that in such games, only a single payoff number is required to characterize any outcome. Player 1 wants to maximize this number, while player 2 wants to minimize it. In this context BACKWARDINDUCTION can be understood as propagating these single payoff numbers from the leaves of the tree up to the root. Each decision node for player 1 is labeled with the maximum of the labels of its child nodes (representing the fact that player 1 would choose the corresponding action), and each decision node for player 2 is labeled with the minimum of that node's

minimax algorithm

function ALPHABETAPRUNING (node h, real α, real β) **returns** $u_1(h)$
if $h \in Z$ **then**
  $\lfloor$ **return** $u_1(h)$ // h is a terminal node
$best\_util \leftarrow (2\rho(h) - 3) \times \infty$ // $-\infty$ for player 1; ∞ for player 2
forall $a \in \chi(h)$ **do**
  $\mid$ **if** $\rho(h) = 1$ **then**
  $\mid$ $\mid$ $best\_util \leftarrow \max(best\_util, \text{ALPHABETAPRUNING}(\sigma(h, a), \alpha, \beta))$
  $\mid$ $\mid$ **if** $best\_util \geq \beta$ **then**
  $\mid$ $\mid$ $\lfloor$ **return** $best\_util$
  $\mid$ $\mid$ $\alpha \leftarrow \max(\alpha, best\_util)$
  $\mid$ **else**
  $\mid$ $\mid$ $best\_util \leftarrow \min(best\_util, \text{ALPHABETAPRUNING}(\sigma(h, a), \alpha, \beta))$
  $\mid$ $\mid$ **if** $best\_util \leq \alpha$ **then**
  $\mid$ $\mid$ $\lfloor$ **return** $best\_util$
  $\mid$ $\lfloor$ $\beta \leftarrow \min(\beta, best\_util)$
return $best\_util$

Figure 5.7 The alpha-beta pruning algorithm. It is invoked at the root node h as ALPHABETAPRUN-
ING($h, -\infty, \infty$).

children's labels. The label on the root node is the value of the game: player 1's
payoff in equilibrium.

pruning   How can we improve on the minimax algorithm? The fact that player 1 and
player 2 always have strictly opposing interests means that we can *prune* away
some parts of the game tree: we can recognize that certain subtrees will never
be reached in equilibrium, even without examining the nodes in these subtrees.
This leads us to a new algorithm called ALPHABETAPRUNING, which is given in
Figure 5.7.

There are several ways in which ALPHABETAPRUNING differs from BACK-
WARDINDUCTION. Some concern the fact that we have now restricted ourselves
to a setting where there are only two players, and one player's utility is the negative of the other's. We thus deal only with the utility for player 1. This is why
we treat the two players separately, maximizing for player 1 and minimizing for
player 2.

At each node h either α or β is updated. These variables take the value of the
previously encountered node that their corresponding player (player 1 for α and
player 2 for β) would most prefer to choose *instead* of h. For example, consider
the variable β at some node h. Now consider all the different choices that player
2 could make at ancestors of h that would prevent h from ever being reached,
and that would ultimately lead to previously encountered terminal nodes. β is
the best value that player 2 could obtain at any of these terminal nodes. Because
the players do not have any alternative to starting at the root of the tree, at the
beginning of the search $\alpha = -\infty$ and $\beta = \infty$.

We can now concentrate on the important difference between BACKWARDIN-
DUCTION and ALPHABETAPRUNING: in the latter procedure, the search can backtrack at a node that is not terminal. Let us think about things from the point
of view of player 1, who is considering what action to play at node h. (As we

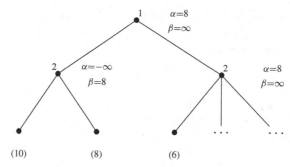

Figure 5.8 An example of alpha-beta pruning. We can backtrack after expanding the first child of the right choice node for player 2.

encourage you to check for yourself, a similar argument holds when it is player 2's turn to move at node h.) For player 1, this backtracking occurs on the line that reads "if *best_util* $\geq \beta$ then return *best_util*." What is going on here? We have just explored some, but not all, of the children of player 1's decision node h; the highest value among these explored nodes is *best_util*. The value of node h is therefore lower bounded by *best_util* (it is *best_util* if h has no children with larger values, and is some larger amount otherwise). Either way, if *best_util* $\geq \beta$ then player 1 knows that player 2 prefers choosing his best alternative (at some ancestor node of h) rather than allowing player 1 to act at node h. Thus node h cannot be on the equilibrium path[2] and so there is no need to continue exploring the game tree below h.

A simple example of ALPHABETAPRUNING in action is given in Figure 5.8. The search begins by heading down the left branch and visiting both terminal nodes, and eventually setting $\beta = 8$. (Do you see why?) It then returns the value 8 as the value of this subgame, which causes α to be set to 8 at the root node. In the right subgame the search visits the first terminal node and so sets *best_util* = 6 at the shaded node, which we will call h. Now at h we have *best_util* $\leq \alpha$, which means that we can backtrack. This is safe to do because we have just shown that player 1 would never choose this subgame: he can guarantee himself a payoff of 8 by choosing the left subgame, whereas his utility in the right subgame would be no more than 6.

The effectiveness of the alpha-beta pruning algorithm depends on the order in which nodes are considered. For example, if player 1 considers nodes in increasing order of their value, and player 2 considers nodes in decreasing order of value, then no nodes will ever be pruned. In the best case (where nodes are ordered in decreasing value for player 1 and in increasing order for player 2), alpha-beta pruning has complexity of $O(b^{\frac{m}{2}})$. We can rewrite this expression as $O(\sqrt{b}^m)$, making more explicit the fact that the game's branching factor would effectively be cut to the square root of its original value. If nodes are examined in random order then the analysis becomes somewhat more complicated;

---

2. In fact, in the case *best_util* $= \beta$, it *is* possible that h could be reached on an equilibrium path; however, in this case there is still always an equilibrium in which player 2 plays his best alternative and h is not reached.

when b is fairly small, the complexity of alpha-beta pruning is $O(b^{\frac{3m}{4}})$, which is still an exponential improvement. In practice, it is usually possible to achieve performance somewhere between the best case and the random case. This technique thus offers substantial practical benefit over straightforward backward induction in two-player, zero-sum games for which the game tree is represented implicitly.

Techniques like alpha-beta pruning are commonly used to build strong computer players for two-player board games such as chess. (However, they perform poorly on games with extremely large branching factors, such as go.) Of course, building a good computer player involves a great deal of engineering, and requires considerable attention to game-specific heuristics such as those used to order actions. One general technique is required by many such systems, however, and so is worth discussing here. The game tree in practical games can be so large that it is infeasible to search all the way down to leaf nodes. Instead, the search proceeds to some shallower depth (which is chosen either statically or dynamically). Where do we get the node values to propagate up using backward induction? The trick is to use an *evaluation function* to estimate the value of the deepest node reached (taking into account game-relevant features such as board position, number of pieces for each player, who gets to move next, etc., and either built by hand or learned). When the search has reached an appropriate depth, the node is treated as terminal with a call to the evaluation function replacing the evaluation of the utility function at that node. This requires a small change to the beginning of ALPHABETAPRUNING; otherwise, the algorithm works unchanged.

evaluation
function

Two-player, general-sum games: computing all subgame-perfect equilibria

While the BACKWARDINDUCTION procedure identifies one subgame-perfect equilibrium in linear time, it does not provide an efficient way of finding all of them. One might wonder how there could even *be* more than one SPE in a perfect-information game. Multiple subgame-perfect equilibria can exist when there exist one or more decision nodes at which a player chooses between subgames in which he receives the same utility. In such cases BACKWARDINDUCTION simply chooses the first subgame it encountered. It could be useful to find the set of all subgame-perfect equilibria if we wanted to find a specific SPE (as we did with Nash equilibria of normal-form games in Section 4.2.4) such as the one that maximizes social welfare.

Here let us restrict ourselves to two-player perfect-information extensive-form games, but lift our previous restriction that the game be zero-sum. A somewhat more complicated algorithm can find the set of *all* subgame-perfect equilibrium values in worst-case cubic time.

Theorem 5.1.6 *Given a two-player perfect-information extensive-form game with ℓ leaves, the set of subgame-perfect equilibrium payoffs can be represented as the union of $O(\ell^2)$ axis-aligned rectangles and can be computed in time $O(\ell^3)$.*

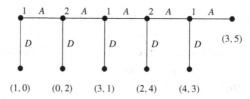

Figure 5.9 The Centipede game.

Intuitively, the algorithm works much like BACKWARDINDUCTION, but the variable *util_at_child* holds a representation of all equilibrium values instead of just one. The "max" operation we had previously implemented through *best_util* is replaced by a subroutine that returns a representation of all the values that can be obtained in subgame-perfect equilibria of the node's children. This can include mixed strategies if multiple children are simultaneously best responses. More information about this algorithm can be found in the reference cited in the chapter notes.

An example and criticisms of backward induction

Despite the fact that strong arguments can be made in its favor, the concept of backward induction is not without controversy. To see why this is, consider the well-known *Centipede game*, depicted in Figure 5.9. (The game starts at the node at the upper left.) In this game two players alternate in making decisions, at each turn choosing between going "down" and ending the game or going "across" and continuing it (except at the last node where going "across" also ends the game). The payoffs are constructed in such a way that the only SPE is for each player to always choose to go down. To see why, consider the last choice. Clearly at that point the best choice for the player is to go down. Since this is the case, going down is also the best choice for the other player in the previous choice point. By induction the same argument holds for all choice points.

This would seem to be the end of this story, except for two pesky factors. The first problem is that the SPE prediction in this case flies in the face of intuition. Indeed, in laboratory experiments subjects in fact continue to play "across" until close to the end of the game. The second problem is theoretical. Imagine that you are the second player in the game, and in the first step of the game the first player actually goes across. What should you do? The SPE suggests you should go down, but the same analysis suggests that you would not have gotten to this choice point in the first place. In other words, you have reached a state to which your analysis has given a probability of zero. How should you amend your beliefs and course of action based on this measure-zero event? It turns out this seemingly small inconvenience actually raises a fundamental problem in game theory. We will not develop the subject further here, but let us only mention that there exist different accounts of this situation, and they depend on the probabilistic assumptions made, on what is common knowledge (in particular, whether there is common knowledge of rationality), and on exactly how one revises one's beliefs in the face of measure-zero events. The last question is intimately related to the subject of belief revision discussed in Chapter 14.

Centipede game

5.2 Imperfect-information extensive-form games

Up to this point, in our discussion of extensive-form games we have allowed players to specify the action that they would take at every choice node of the game. This implies that players know the node they are in, and—recalling that in such games we equate nodes with the histories that led to them—all the prior choices, including those of other agents. For this reason we have called these *perfect-information games.*

We might not always want to make such a strong assumption about our players and our environment. In many situations we may want to model agents needing to act with partial or no knowledge of the actions taken by others, or even agents with limited memory of their own past actions. The sequencing of choices allows us to represent such ignorance to a limited degree; an "earlier" choice might be interpreted as a choice made without knowing the "later" choices. However, so far we could not represent two choices made in the same play of the game in mutual ignorance of each other.

5.2.1 *Definition*

Imperfect-information games in extensive form address this limitation. An imperfect-information game is an extensive-form game in which each player's choice nodes are partitioned into information sets; intuitively, if two choice nodes are in the same information set then the agent cannot distinguish between them.[3]

Definition 5.2.1 (Imperfect-information game) *An imperfect-information game (in extensive form) is a tuple* $(N, A, H, Z, \chi, \rho, \sigma, u, I)$, *where:*

- $(N, A, H, Z, \chi, \rho, \sigma, u)$ *is a perfect-information extensive-form game; and*
- $I = (I_1, \ldots, I_n)$, *where* $I_i = (I_{i,1}, \ldots, I_{i,k_i})$ *is a set of equivalence classes on (i.e., a partition of)* $\{h \in H : \rho(h) = i\}$ *with the property that* $\chi(h) = \chi(h')$ *and* $\rho(h) = \rho(h')$ *whenever there exists a* j *for which* $h \in I_{i,j}$ *and* $h' \in I_{i,j}$.

Note that in order for the choice nodes to be truly indistinguishable, we require that the set of actions at each choice node in an information set be the same (otherwise, the player would be able to distinguish the nodes). Thus, if $I_{i,j} \in I_i$ is an equivalence class, we can unambiguously use the notation $\chi(I_{i,j})$ to denote the set of actions available to player i at any node in information set $I_{i,j}$.

Consider the imperfect-information extensive-form game shown in Figure 5.10. In this game, player 1 has two information sets: the set including the top choice node, and the set including the bottom choice nodes. Note that the two bottom choice nodes in the second information set have the same set of possible actions. We can regard player 1 as not knowing whether player 2 chose A or B when he makes his choice between ℓ and r.

---

3. From the technical point of view, imperfect-information games are obtained by overlaying a partition structure, as defined in Chapter 13 in connection with models of knowledge, over a perfect-information game.

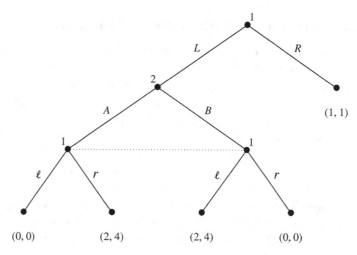

Figure 5.10 An imperfect-information game.

5.2.2 *Strategies and equilibria*

A pure strategy for an agent in an imperfect-information game selects one of the available actions in each information set of that agent.

Definition 5.2.2 (Pure strategies) *Let* $G = (N, A, H, Z, \chi, \rho, \sigma, u, I)$ *be an imperfect-information extensive-form game. Then the pure strategies of player i consist of the Cartesian product* $\prod_{I_{i,j} \in I_i} \chi(I_{i,j})$.

Thus perfect-information games can be thought of as a special case of imperfect-information games, in which every equivalence class of each partition is a singleton.

Consider again the Prisoner's Dilemma game, shown as a normal-form game in Figure 3.3. An equivalent imperfect-information game in extensive form is given in Figure 5.11.

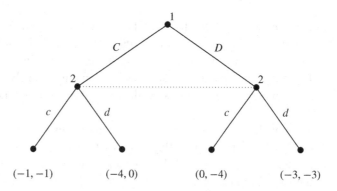

Figure 5.11 The Prisoner's Dilemma game in extensive form.

Note that we could have chosen to make player 2 choose first and player 1 choose second.

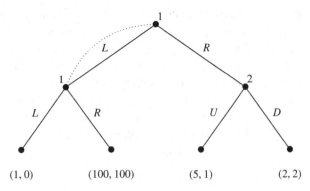

Figure 5.12 A game with imperfect recall.

Recall that perfect-information games were not expressive enough to capture the prisoner's dilemma game and many other ones. In contrast, as is obvious from this example, any normal-form game can be trivially transformed into an equivalent imperfect-information game. However, this example is also special in that the Prisoner's Dilemma is a game with a dominant strategy solution, and thus in particular a pure-strategy Nash equilibrium. This is not true in general for imperfect-information games. To be precise about the equivalence between a normal-form game and its extensive-form image we must consider mixed strategies, and this is where we encounter a new subtlety.

As we did for perfect-information games, we can define the normal-form game corresponding to any given imperfect-information game; this normal game is again defined by enumerating the pure strategies of each agent. Now, we define the set of mixed strategies of an imperfect-information game as simply the set of mixed strategies in its image normal-form game; in the same way, we can also define the set of Nash equilibria.[4] However, we can also define the set of *behavioral*
behavioral *strategies* in the extensive-form game. These are the strategies in which, rather than randomizing over complete pure strategies, the agent randomizes independently at each information set. And so, whereas a mixed strategy is a distribution over vectors (each vector describing a pure strategy), a behavioral strategy is a vector of distributions.

In general, the expressive power of behavioral strategies and the expressive power of mixed strategies are noncomparable; in some games there are outcomes that are achieved via mixed strategies but not any behavioral strategies, and in some games it is the other way around.

Consider for example the game in Figure 5.12. In this game, when considering mixed strategies (but not behavioral strategies), R is a strictly dominant strategy for agent 1, D is agent 2's strict best response, and thus (R, D) is the unique Nash equilibrium. Note in particular that in a mixed strategy, agent 1 decides

---

4. Note that we have defined two transformations—one from any normal-form game to an imperfect-information game, and one in the other direction. However the first transformation is not one to one, and so if we transform a normal-form game to an extensive-form one and then back to normal form, we will not in general get back the same game we started out with. However, we will get a game with identical strategy spaces and equilibria.

probabilistically whether to play L or R in his information set, but once he decides he plays that pure strategy consistently. Thus the payoff of 100 is irrelevant in the context of mixed strategies. On the other hand, with behavioral strategies agent 1 gets to randomize afresh each time he finds himself in the information set. Noting that the pure strategy D is weakly dominant for agent 2 (and in fact is the unique best response to all strategies of agent 1 other than the pure strategy L), agent 1 computes the best response to D as follows. If he uses the behavioral strategy $(p, 1 - p)$ (i.e., choosing L with probability p each time he finds himself in the information set), his expected payoff is

$$1 * p^2 + 100 * p(1 - p) + 2 * (1 - p).$$

The expression simplifies to $-99p^2 + 98p + 2$, whose maximum is obtained at $p = 98/198$. Thus $(R, D) = ((0, 1), (0, 1))$ is no longer an equilibrium in behavioral strategies, and instead we get the equilibrium $((98/198, 100/198), (0, 1))$.

There is, however, a broad class of imperfect-information games in which the expressive power of mixed and behavioral strategies coincides. This is the class of games of *perfect recall*. Intuitively speaking, in these games no player forgets any information he knew about moves made so far; in particular, he remembers precisely all his own moves. A formal definition follows.

perfect recall

Definition 5.2.3 (Perfect recall) *Player i has* perfect recall *in an imperfect-information game G if for any two nodes h, h' that are in the same information set for player i, for any path $h_0, a_0, h_1, a_1, h_2, \ldots, h_m, a_m, h$ from the root of the game to h (where the h_j are decision nodes and the a_j are actions) and for any path $h_0, a'_0, h'_1, a'_1, h'_2, \ldots, h'_{m'}, a'_{m'}, h'$ from the root to h' it must be the case that:*

1. *$m = m'$;*
2. *for all $0 \leq j \leq m$, h_j and h'_j are in the same equivalence class for player i; and*
3. *for all $0 \leq j \leq m$, if $\rho(h_j) = i$ (i.e., h_j is a decision node of player i), then $a_j = a'_j$.*

G is a game of perfect recall if every player has perfect recall in it.

Clearly, every perfect-information game is a game of perfect recall.

Theorem 5.2.4 (Kuhn, 1953) *In a game of perfect recall, any mixed strategy of a given agent can be replaced by an equivalent behavioral strategy, and any behavioral strategy can be replaced by an equivalent mixed strategy. Here two strategies are equivalent in the sense that they induce the same probabilities on outcomes, for any fixed strategy profile (mixed or behavioral) of the remaining agents.*

As a corollary we can conclude that the set of Nash equilibria does not change if we restrict ourselves to behavioral strategies. This is true only in games of perfect recall, and thus, for example, in perfect-information games. We stress again, however, that in general imperfect-information games, mixed and behavioral strategies yield noncomparable sets of equilibria.

5.2.3 *Computing equilibria: the sequence form*

Because any extensive-form game can be converted into an equivalent normal-form game, an obvious way to find an equilibrium of an extensive-form game is to first convert it into a normal-form game and then find the equilibria using, for example the Lemke–Howson algorithm. This method is inefficient, however, because the number of actions in the normal-form game is *exponential* in the size of the extensive-form game. The normal-form game is created by considering all combinations of information set actions for each player, and the payoffs that result when these strategies are employed.

One way to avoid this problem is to operate directly on the extensive-form representation. This can be done by employing behavioral strategies to express a game using a description called the sequence form.

Defining the sequence form

The sequence form is (primarily) useful for representing imperfect-information extensive-form games of perfect recall. Definition 5.2.5 describes the elements of the sequence-form representation of such games; we then go on to explain what each of these elements means.

Definition 5.2.5 (Sequence-form representation) *Let G be an imperfect-information game of perfect recall. The sequence-form representation of G is a tuple (N, Σ, g, C), where:*

- *N is a set of agents;*
- *$\Sigma = (\Sigma_1, \ldots, \Sigma_n)$, where Σ_i is the set of sequences available to agent i;*
- *$g = (g_1, \ldots, g_n)$, where $g_i : \Sigma \mapsto \mathbb{R}$ is the payoff function for agent i; and*
- *$C = (C_1, \ldots, C_n)$, where C_i is a set of linear constraints on the realization probabilities of agent i.*

Now let us define all these terms. To begin with, what is a sequence? The key insight of the sequence form is that, while there are exponentially many pure strategies in an extensive-form game, there are only a small number of nodes in the game tree. Rather than building a player's strategy around the idea of pure strategies, the sequence form builds it around paths in the tree from the root to each node.

Definition 5.2.6 (Sequence) *A sequence of actions of player $i \in N$, defined by a node $h \in H \cup Z$ of the game tree, is the ordered set of player i's actions that lie on the path from the root to h. Let $\emptyset$ denote the sequence corresponding to the root node. The set of sequences of player i is denoted Σ_i, and $\Sigma = \Sigma_1 \times \cdots \times \Sigma_n$ is the set of all sequences.*

A sequence can thus be thought of as a string listing the action choices that player i would have to take in order to get from the root to a given node h. Observe that h may or may not be a leaf node; observe also that the other players' actions that form part of this path are not part of the sequence.

	Ø	A	B
Ø	0, 0	0, 0	0, 0
L	0, 0	0, 0	0, 0
R	1, 1	0, 0	0, 0
Lℓ	0, 0	0, 0	2, 4
Lr	0, 0	2, 4	0, 0

Figure 5.13 The sequence form of the game from Figure 5.10.

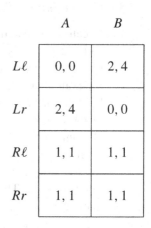

	A	B
Lℓ	0, 0	2, 4
Lr	2, 4	0, 0
Rℓ	1, 1	1, 1
Rr	1, 1	1, 1

Figure 5.14 The induced normal form of the game from Figure 5.10.

sequence-form payoff function

Definition 5.2.7 (Payoff function) *The* payoff function $g_i : \Sigma \mapsto \mathbb{R}$ *for agent i is given by* $g(\sigma) = u(z)$ *if a leaf node* $z \in Z$ *would be reached when each player played his sequence* $\sigma_i \in \sigma$, *and by* $g(\sigma) = 0$ *otherwise.*

Given the set of sequences Σ and the payoff function g, we can think of the sequence form as defining a tabular representation of an imperfect-information extensive-form game, much as the induced normal form does. Consider the game given in Figure 5.10 (see p. 126). The sets of sequences for the two players are $\Sigma_1 = \{\emptyset, L, R, L\ell, Lr\}$ and $\Sigma_2 = \{\emptyset, A, B\}$. The payoff function is given in Figure 5.13. For comparison, the induced normal form of the same game is given in Figure 5.14. Written this way, the sequence form is larger than the induced normal form. However, many of the entries in the game matrix in Figure 5.13 correspond to cases where the payoff function is defined to be zero because the given pair of sequences does not correspond to a leaf node in the game tree. These entries are shaded in gray to indicate that they could not arise in play. Each payoff that *is* defined at a leaf in the game tree occurs exactly once in the sequence-form table. Thus, if g was represented using a sparse encoding, only five values would have to be stored. Compare this to the induced normal form, where all of the eight entries correspond to leaf nodes from the game tree.

We now have a set of players, a set of sequences, and a mapping from sequences to payoffs. At first glance this may look like everything we need to describe our game. However, sequences do not quite take the place of actions. In particular, a player cannot simply select a single sequence in the way that he would select a pure strategy—the other player(s) might not play in a way that would allow him to follow it to its end. Put another way, players still need to define what they would do in every information set that could be reached in the game tree.

What we want is for agents to select behavioral strategies. (Since we have assumed that our game G has perfect recall, Theorem 5.2.4 tells us that any equilibrium will be expressible using behavioral strategies.) However, it turns out that it is not a good idea to work with behavioral strategies directly—if we did so, the optimization problems we develop later would be computationally harder to solve. Instead, we will develop the alternate concept of a *realization plan*, which corresponds to the probability that a given sequence would arise under a given behavioral strategy.

Consider an agent i following a behavioral strategy that assigned probability $\beta_i(h, a_i)$ to taking action a_i at a given decision node h. Then we can construct a *realization plan* that assigns probabilities to sequences in a way that recovers i's behavioral strategy β.

realization plan
of β_i

realization
probability

Definition 5.2.8 (Realization plan of β_i) *The* realization plan of β_i *for player* $i \in N$ *is a mapping* $r_i : \Sigma_i \mapsto [0, 1]$ *defined as* $r_i(\sigma_i) = \prod_{c \in \sigma_i} \beta_i(c)$. *Each value* $r_i(\sigma_i)$ *is called a* realization probability.

Definition 5.2.8 is not the most useful way of defining realization probabilities. There is a second, equivalent definition with the advantage that it involves a set of linear equations, although it is a bit more complicated. This definition relies on two functions that we will make extensive use of in this section.

To define the first function, we make use of our assumption that G is a game of perfect recall. This entails that, given an information set $I \in I_i$, there must be one single sequence that player i can play to reach all of his nonterminal choice nodes $h \in I$. We denote this mapping as $seq_i : I_i \mapsto \Sigma_i$, and call $seq_i(I)$ the sequence

$seq_i(I)$: the
sequence
leading to I

leading to information set I. Note that while there is only one sequence that leads to a given information set, a given sequence can lead to multiple different information sets. For example, if player 1 moves first and player 2 observes his move, then the sequence $\emptyset$ will lead to multiple information sets for player 2.

The second function considers ways that sequences can be built from other sequences. By $\sigma_i a_i$ denote a sequence that consists of the sequence σ_i followed by the single action a_i. As long as the new sequence still belongs to Σ_i, we say that the sequence $\sigma_i a_i$ *extends* the sequence σ_i. A sequence can often be extended in multiple ways—for example, perhaps agent i could have chosen an action a_i'

$Ext_i(\sigma_i)$:
sequences
extending σ_i

instead of a_i after playing sequence σ_i. We denote by $Ext_i : \Sigma_i \mapsto 2^{\Sigma_i}$ a function mapping from sequences to sets of sequences, where $Ext_i(\sigma_i)$ denotes the set of sequences that extend the sequence σ_i. We define $Ext_i(\emptyset)$ to be the set of all single-action sequences. Note that extension always refers to playing a *single* action beyond a given sequence; thus, $\sigma_i a_i a_i'$ does not belong to $Ext_i(\sigma_i)$, even if it is a valid sequence. (It *does* belong to $Ext_i(\sigma_i a_i)$.) Also note that not all sequences have extensions; one example is sequences leading to leaf nodes. In such cases $Ext_i(\sigma)$ returns the empty set. Finally, to reduce notation we introduce

$Ext_i(I) =$
$Ext_i(seq_i(I))$

the shorthand $Ext_i(I) = Ext_i(seq_i(I))$: the sequences extending an information set are the sequences extending the (unique) sequence leading to that information set.

realization plan   **Definition 5.2.9 (Realization plan)** *A realization plan for player* $i \in N$ *is a function* $r_i : \Sigma_i \mapsto [0, 1]$ *satisfying the following constraints.*

$$r_i(\emptyset) = 1 \qquad\qquad\qquad (5.1)$$

$$\sum_{\sigma_i' \in \text{Ext}_i(I)} r_i(\sigma_i') = r_i(\text{seq}_i(I)) \qquad\qquad \forall I \in I_i \qquad (5.2)$$

$$r_i(\sigma_i) \geq 0 \qquad\qquad\qquad \forall \sigma_i \in \Sigma_i \qquad (5.3)$$

If a player i follows a realization plan r_i, we must be able to recover a behavioral strategy β_i from it. For a decision node h for player i that is in information set $I \in I_i$, and for any sequence $(\text{seq}_i(I)a_i) \in \text{Ext}_i(I)$, $\beta_i(h, a_i)$ is defined as $\frac{r_i(\text{seq}_i(I)a_i)}{r_i(\text{seq}_i(I))}$, as long as $r_i(\text{seq}_i(I)) > 0$. If $r_i(\text{seq}_i(I)) = 0$ then we can assign $\beta_i(h, a_i)$ an arbitrary value from $[0, 1]$: here β_i describes the player's behavioral strategy at a node that could never be reached in play because of the player's own previous decisions, and so the value we assign to β_i is irrelevant.

Let C_i be the set of constraints (5.2) on realization plans of player i. Let $C = (C_1, \ldots, C_n)$. We have now defined all the elements[5] of a sequence-form representation $G = (N, \Sigma, g, C)$, as laid out in Definition 5.2.5.

What is the space complexity of the sequence-form representation? Unlike the normal form, the size of this representation is linear in the size of the extensive-form game. There is one sequence for each node in the game tree, plus the $\emptyset$ sequence for each player. As argued previously, the payoff function g can be represented sparsely, so that each payoff corresponding to a leaf node is stored only once, and no other payoffs are stored at all. There is one version of constraint (5.2) for each edge in the game tree. Each such constraint for player i references only $|\text{Ext}_i(I)| + 1$ variables, again allowing sparse encoding.

Computing best responses in two-player games

The sequence-form representation can be leveraged to allow the computation of equilibria far more efficiently than can be done using the induced normal form. Here we will consider the case of two-player games, as it is these games for which the strongest results hold. First we consider the problem of determining player 1's best response to a fixed behavioral strategy of player 2 (represented as a realization plan). This problem can be written as the following linear program.

$$\text{maximize} \quad \sum_{\sigma_1 \in \Sigma_1} \left(\sum_{\sigma_2 \in \Sigma_2} g_1(\sigma_1, \sigma_2) r_2(\sigma_2) \right) r_1(\sigma_1) \qquad (5.4)$$

$$\text{subject to} \quad r_1(\emptyset) = 1 \qquad\qquad\qquad (5.5)$$

$$\sum_{\sigma_1' \in \text{Ext}_1(I)} r_1(\sigma_1') = r_1(\text{seq}_1(I)) \qquad\qquad \forall I \in I_1 \qquad (5.6)$$

$$r_1(\sigma_1) \geq 0 \qquad\qquad\qquad \forall \sigma_1 \in \Sigma_1 \qquad (5.7)$$

---

5. We do not need to explicitly store constraints (5.1) and (5.3), because they are always the same for every sequence-form representation.

This linear program is straightforward. First, observe that $g_1(\cdot)$ and $r_2(\cdot)$ are constants, while $r_1(\cdot)$ are variables. The LP states that player 1 should choose r_1 to maximize his expected utility (given in the objective function (5.4)) subject to constraints (5.5)–(5.7) which require that r_1 corresponds to a valid realization plan.

In an equilibrium, player 1 and player 2 best respond simultaneously. However, if we treated both r_1 and r_2 as variables in Equations (5.4)–(5.7) then the objective function (5.4) would no longer be linear. Happily, this problem does not arise in the dual of this linear program.[6] Denote the variables of our dual LP as v; there will be one v_I for every information set $I \in I_1$ (corresponding to constraint (5.6) from the primal) and one additional variable v_0 (corresponding to constraint (5.5)). For notational convenience, we define a "dummy" information set 0 for player 1; thus, we can consider every dual variable to correspond to an information set.

$\mathcal{I}_i(\sigma_i)$: the last information set encountered in σ_i

We now define one more function. Let $\mathcal{I}_i : \Sigma_i \mapsto I_i \cup \{0\}$ be a mapping from player i's sequences to information sets. We define $\mathcal{I}_i(\sigma_i)$ to be 0 iff $\sigma_i = \emptyset$, and to be the information set $I \in I_i$ in which the final action in σ_i was taken otherwise. Note that the information set in which each action in a sequence was taken is unambiguous because of our assumption that the game has perfect recall. Finally, we again overload notation to simplify the expressions that follow. Given a set of sequences Σ', let $\mathcal{I}_i(\Sigma')$ denote $\{\mathcal{I}_i(\sigma')|\sigma'_i \in \Sigma'_i\}$. Thus, for example, $\mathcal{I}_i(\mathrm{Ext}_i(\sigma_1))$ is the (possibly empty) set of final information sets encountered in the (possibly empty) set of extensions of σ_i.

$\mathcal{I}_i(\mathrm{Ext}_i(\sigma_1)) = \{\mathcal{I}_i(\sigma')|\sigma' \in \mathrm{Ext}_i(\sigma_1)\}$

The dual LP follows.

$$\text{minimize} \quad v_0 \tag{5.8}$$

$$\text{subject to} \quad v_{\mathcal{I}_1(\sigma_1)} - \sum_{I' \in \mathcal{I}_1(\mathrm{Ext}_1(\sigma_1))} v_{I'} \geq \sum_{\sigma_2 \in \Sigma_2} g_1(\sigma_1, \sigma_2) r_2(\sigma_2) \quad \forall \sigma_1 \in \Sigma_1 \tag{5.9}$$

The variable v_0 represents player 1's expected utility under the realization plan he chooses to play, given player 2's realization plan. In the optimal solution v_0 will correspond to player 1's expected utility when he plays his best response. (This follows from LP duality—primal and dual linear programs always have the same optimal solutions.) Each other variable v_I can be understood as the portion of this expected utility that player 1 will achieve under his best-response realization plan in the subgame starting from information set I, again given player 2's realization plan r_2.

There is one version of constraint (5.9) for every sequence σ_1 of player 1. Observe that there is always exactly one positive variable on the left-hand side of the inequality, corresponding to the information set of the last action in the sequence. There can also be zero or more negative variables, each of which corresponds to a different information set in which player 1 can end up after playing the given sequence. To understand this constraint, we will consider three different cases.

First, there are zero of these negative variables when the sequence cannot be extended—that is, when player 1 never gets to move again after $\mathcal{I}_1\sigma_1$, no

---

6. The dual of a linear program is defined in Appendix B.

matter what player 2 does. In this case, the right-hand side of the constraint will evaluate to player 1's expected payoff from the subgame beyond σ_1, given player 2's realization probabilities r_2. (This subgame is either a terminal node or one or more decision nodes for player 2 leading ultimately to terminal nodes.) Thus, here the constraint states that the expected utility from a decision at information set $\mathcal{I}_1(\sigma_1)$ must be at least as large as the expected utility from making the decision according to σ_1. In the optimal solution this constraint will be realized as equality if σ_1 is played with positive probability; contrapositively, if the inequality is strict, σ_1 will never be played.

The second case is when the structure of the game is such that player 1 will face another decision node no matter how he plays at information set $\mathcal{I}_1(\sigma_1)$. For example, this occurs if $\sigma_1 = \emptyset$ and player 1 moves at the root node: then $\mathcal{I}_1(\text{Ext}_1(\sigma_1)) = \{1\}$ (the first information set). As another example, if player 2 takes one of two moves at the root node and player 1 observes this move before choosing his own move, then for $\sigma_1 = \emptyset$ we will have $\mathcal{I}_1(\text{Ext}_1(\sigma_1)) = \{1, 2\}$. Whenever player 1 is guaranteed to face another decision node, the right-hand side of constraint (5.9) will evaluate to zero because $g_1(\sigma_1, \sigma_2)$ will equal 0 for all σ_2. Thus the constraint can be interpreted as stating that player 1's expected utility at information set $\mathcal{I}_1(\sigma_1)$ must be equal to the sum of the expected utilities at the information sets $\mathcal{I}_1(\text{Ext}_1(\sigma_1))$. In the optimal solution, where v_0 is minimized, these constraints are always be realized as equality.

Finally, there is the case where there exist extensions of sequence σ_1, but where it is also possible that player 2 will play in a way that will deny player 1 another move. For example, consider the game in Figure 5.2 from earlier in the chapter. If player 1 adopts the sequence B at his first information set, then he will reach his second information set if player 2 plays F, and will reach a leaf node otherwise. In this case there will be both negative terms on the left-hand side of constraint (5.9) (one for every information set that player 1 could reach beyond sequence σ_1) *and* positive terms on the right-hand side (expressing the expected utility player 1 achieves for reaching a leaf node). Here the constraint can be interpreted as asserting that i's expected utility at $\mathcal{I}_1(\sigma_1)$ can only exceed the sum of the expected utilities of i's successor information sets by the amount of the expected payoff due to reaching leaf nodes from player 2's move(s).

Computing equilibria of two-player zero-sum games

For two-player zero-sum games the sequence form allows us to write a linear program for computing a Nash equilibrium that can be solved in time polynomial in the size of the extensive form. Note that in contrast, the methods described in Section 4.1 would require time exponential in the size of the extensive form, because they require construction of an LP with a constraint for each pure strategy of each player and a variable for each pure strategy of one of the players.

This new linear program for games in sequence form can be constructed quite directly from the dual LP given in Equations (5.8)–(5.9). Intuitively, we simply treat the terms $r_2(\cdot)$ as variables rather than constants, and add in the constraints from Definition 5.2.9 to ensure that r_2 is a valid realization plan. The program follows.

$$\text{minimize} \quad v_0 \tag{5.10}$$

$$\text{subject to} \quad v_{\mathcal{I}_1(\sigma_1)} - \sum_{I' \in \mathcal{I}_1(\text{Ext}_1(\sigma_1))} v_{I'} \geq \sum_{\sigma_2 \in \Sigma_2} g_1(\sigma_1, \sigma_2) r_2(\sigma_2) \qquad \forall \sigma_1 \in \Sigma_1 \tag{5.11}$$

$$r_2(\emptyset) = 1 \tag{5.12}$$

$$\sum_{\sigma_2' \in \text{Ext}_2(I)} r_2(\sigma_2') = r_2(\text{seq}_2(I)) \qquad \forall I \in I_2 \tag{5.13}$$

$$r_2(\sigma_2) \geq 0 \qquad \forall \sigma_2 \in \Sigma_2 \tag{5.14}$$

The fact that r_2 is now a variable means that player 2's realization plan will now be selected to minimize player 1's expected utility when player 1 best responds to it. In other words, we find a minmax strategy for player 2 against player 1, and since we have a two-player zero-sum game it is also a Nash equilibrium by Theorem 3.4.4. Observe that if we had tried this same trick with the primal LP given in Equations (5.4)–(5.7) we would have ended up with a quadratic objective function, and hence not a linear program.

Computing equilibria of two-player general-sum games

For two-player general-sum games, the problem of finding a Nash equilibrium can be formulated as a linear complementarity problem as follows.

$$r_1(\emptyset) = 1 \tag{5.15}$$

$$r_2(\emptyset) = 1 \tag{5.16}$$

$$\sum_{\sigma_1' \in \text{Ext}_1(I)} r_1(\sigma_1') = r_1(\text{seq}_1(I)) \qquad \forall I \in I_1 \tag{5.17}$$

$$\sum_{\sigma_2' \in \text{Ext}_2(I)} r_2(\sigma_2') = r_2(\text{seq}_2(I)) \qquad \forall I \in I_2 \tag{5.18}$$

$$r_1(\sigma_1) \geq 0 \qquad \forall \sigma_1 \in \Sigma_1 \tag{5.19}$$

$$r_2(\sigma_2) \geq 0 \qquad \forall \sigma_2 \in \Sigma_2 \tag{5.20}$$

$$\left(v_{\mathcal{I}_1(\sigma_1)}^1 - \sum_{I' \in \mathcal{I}_1(\text{Ext}_1(\sigma_1))} v_{I'}^1 \right) - \left(\sum_{\sigma_2 \in \Sigma_2} g_1(\sigma_1, \sigma_2) r_2(\sigma_2) \right) \geq 0 \qquad \forall \sigma_1 \in \Sigma_1 \tag{5.21}$$

$$\left(v_{\mathcal{I}_2(\sigma_2)}^2 - \sum_{I' \in \mathcal{I}_2(\text{Ext}_2(\sigma_2))} v_{I'}^2 \right) - \left(\sum_{\sigma_1 \in \Sigma_1} g_2(\sigma_1, \sigma_2) r_1(\sigma_1) \right) \geq 0 \qquad \forall \sigma_2 \in \Sigma_2 \tag{5.22}$$

$$r_1(\sigma_1) \left[\left(v_{\mathcal{I}_1(\sigma_1)}^1 - \sum_{I' \in \mathcal{I}_1(\text{Ext}_1(\sigma_1))} v_{I'}^1 \right) - \left(\sum_{\sigma_2 \in \Sigma_2} g_1(\sigma_1, \sigma_2) r_2(\sigma_2) \right) \right] = 0 \quad \forall \sigma_1 \in \Sigma_1 \tag{5.23}$$

$$r_2(\sigma_2) \left[\left(v_{\mathcal{I}_2(\sigma_2)}^2 - \sum_{I' \in \mathcal{I}_2(\text{Ext}_2(\sigma_2))} v_{I'}^2 \right) - \left(\sum_{\sigma_1 \in \Sigma_1} g_2(\sigma_1, \sigma_2) r_1(\sigma_1) \right) \right] = 0 \quad \forall \sigma_2 \in \Sigma_2 \tag{5.24}$$

Like the linear complementarity problem for two-player games in normal form given in Equations (4.14)–(4.19) on Page 91, this is a feasibility problem consisting of linear constraints and complementary slackness conditions. The linear constraints are those from the primal LP for player 1 (constraints (5.15), (5.17), and (5.19)), from the dual LP for player 1 (constraint (5.21)), and from the corresponding versions of these primal and dual programs for player 2 (constraints (5.16), (5.18), (5.20), and (5.22)). Note that we have rearranged some of these constraints by moving all terms to the left side, and have superscripted the v's with the appropriate player number.

If we stopped at constraint (5.22) we would have a linear program, but the variables v would be allowed to take arbitrarily large values. The complementary slackness conditions (constraints (5.23) and (5.24)) fix this problem at the expense of shifting us from a linear program to a linear complementarity problem. Let us examine constraint (5.23). It states that either sequence σ_1 is never played (i.e., $r_1(\sigma_1) = 0$) or that

$$v^1_{\mathcal{I}_1(\sigma_1)} - \sum_{I' \in \mathcal{I}_1(\mathrm{Ext}_1(\sigma_1))} v^1_{I'} = \sum_{\sigma_2 \in \Sigma_2} g_1(\sigma_1, \sigma_2) r_2(\sigma_2). \qquad (5.25)$$

What does it mean for Equation (5.25) to hold? The short answer is that this equation requires a property that we previously observed of the optimal solution to the dual LP given in Equations (5.8)–(5.9): that the weak inequality in constraint (5.9) will be realized as strict equality whenever the corresponding sequence is played with positive probability. We were able to achieve this property in the dual LP by minimizing v_0; however, this does not work in the two-player general-sum case where we have both v^1_0 and v^2_0. Instead, we use the complementary slackness idea that we previously applied in the LCP for normal-form games (constraint (4.19)).

This linear complementarity program cannot be solved using the Lemke–Howson algorithm, as we were able to do with our LCP for normal-form games. However, it *can* be solved using the Lemke algorithm, a more general version of Lemke–Howson. Neither algorithm is polynomial time in the worst case. However, it is exponentially faster to run the Lemke algorithm on a game in sequence form than it is to run the Lemke–Howson algorithm on the game's induced normal form. We omit the details of how to apply the Lemke algorithm to sequence-form games, but refer the interested reader to the reference given at the end of the chapter.

5.2.4 *Sequential equilibrium*

We have already seen that the Nash equilibrium concept is too weak for perfect-information games, and how the more selective notion of subgame-perfect equilibrium can be more instructive. The question is whether this essential idea can be applied to the broader class of imperfect-information games; it turns out that it can, although the details are considerably more involved.

Recall that in a subgame-perfect equilibrium we require that the strategy of each agent be a best response in every subgame, not only (rather than L). It is immediately apparent that the definition does not apply in imperfect-information games, if for no other reason than we no longer have a well-defined notion of a subgame. What we have instead at each information set is a "subforest" or a

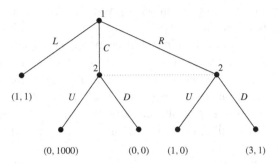

Figure 5.15 Player 2 knows where in the information set he is.

collection of subgames. We could require that each player's strategy be a best response in each subgame in each forest, but that would be both too strong a requirement and too weak. To see why it is too strong, consider the game in Figure 5.15.

The pure strategies of player 1 are $\{L, C, R\}$ and of player 2 $\{U, D\}$. Note also that the two pure Nash equilibria are (L, U) and (R, D). But should either of these be considered "subgame perfect?" On the face of it the answer is ambiguous, since in one subtree U (dramatically) dominates D and in the other D dominates U. However, consider the following argument. R dominates C for player 1, and player 2 knows this. So although player 2 does not have explicit information about which of the two nodes he is in within his information set, he can deduce that he is in the rightmost one based on player 1's incentives, and hence will go D. Furthermore player 1 knows that player 2 can deduce this, and therefore player 1 should go R. Thus, (R, D) is the only subgame-perfect equilibrium.

This example shows how a requirement that a sub strategy be a best response in all subgames is too simplistic. However, in general it is not the case that subtrees of an information set can be pruned as in the previous example so that all remaining ones agree on the best strategy for the player. In this case the naive application of the SPE intuition would rule out all strategies.

There have been several related proposals that apply the intuition underlying subgame-perfection in more sophisticated ways. One of the more influential notions has been that of *sequential equilibrium* (SE). It shares some features with the notion of trembling-hand perfection, discussed in Section 3.4.6. Note that indeed trembling-hand perfection, which was defined for normal-form games, applies here just as well; just think of the normal form induced by the extensive-form game. However, this notion makes no reference to the tree structure of the game. SE does, but at the expense of additional complexity.

sequential
equilibrium

Sequential equilibrium is defined for games of perfect recall. As we have seen, in such games we can restrict our attention to behavioral strategies. Consider for the moment a fully mixed-strategy profile.[7] Such a strategy profile induces a positive probability on every node in the game tree. This means in particular

_____

7. Again, recall that a strategy is fully mixed if, at every information set, each action is given some positive probability.

that every information set is given a positive probability. Therefore, for a given fully mixed-strategy profile, one can meaningfully speak of i's expected utility, given that he finds himself in any particular information set. (The expected utility of starting at any node is well defined, and since each node is given positive probability, one can apply Bayes' rule to aggregate the expected utilities of the different nodes in the information set.) If the fully mixed-strategy profile constitutes an equilibrium, it must be that each agent's strategy maximizes his expected utility in each of his information sets, holding the strategies of the other agents fixed.

All of the preceding discussion is for a fully mixed-strategy profile. The problem is that equilibria are rarely fully mixed, and strategy profiles that are not fully mixed do *not* induce a positive probability on every information set. The expected utility of starting in information sets whose probability is zero under the given strategy profile is simply not well defined. This is where the ingenious device of SE comes in. Given any strategy profile s (not necessarily fully mixed), imagine a probability distribution $\mu(h)$ over each information set. μ has to be *consistent* with s, in the sense that for sets whose probability is nonzero under their parents' conditional distribution s, this distribution is precisely the one defined by Bayes' rule. However, for other information sets, it can be any distribution. Intuitively, one can think of these distributions as the new beliefs of the agents, if they are surprised and find themselves in a situation they thought would not occur.[8] This means that each agent's expected utility is now well defined in any information set, including those having measure zero. For information set h belonging to agent i, with the associated probability distribution $\mu(h)$, the expected utility under strategy profile s is denoted by $u_i(s \mid h, \mu(h))$.

With this, the precise definition of SE is as follows.

Definition 5.2.10 (Sequential equilibrium) *A strategy profile s is a sequential equilibrium of an extensive-form game G if there exist probability distributions $\mu(h)$ for each information set h in G, such that the following two conditions hold:*

1. *$(s, \mu) = \lim_{n \to \infty}(s^m, \mu^m)$ for some sequence $(s^1, \mu^1), (s^2, \mu^2), \ldots,$ where s^m is fully mixed, and μ^m is consistent with s^m (in fact, since s^m is fully mixed, μ^m is uniquely determined by s^m); and*

2. *For any information set h belonging to agent i, and any alternative strategy s_i' of i, we have that*

$$u_i(s \mid h, \mu(h)) \geq u_i((s', s_{-i}) \mid h, \mu(h)).$$

Analogous to subgame perfect equilibria in games of perfect information, sequential equilibria are guaranteed to always exist.

Theorem 5.2.11 *Every finite game of perfect recall has a sequential equilibrium.*

Finally, while sequential equilibria are defined for games of imperfect information, they are obviously also well defined for the special case of games of

---

8. This construction is essentially that of an LPS, discussed in Chapter 13.

perfect information. This raises the question of what relationship holds between the two solution concepts in games of perfect information.

Theorem 5.2.12 *In extensive-form games of perfect information, the sets of subgame-perfect equilibria and sequential equilibria are always equivalent.*

5.3 History and references

As in Chapter 3, much of the material in this chapter is covered in modern game theory textbooks. Some of the historical references are as follows. The earliest game-theoretic publication is arguably that of Zermelo, who in 1913 introduced the notions of a game tree and backward induction and argued that in principle chess admits a trivial solution [Zermelo, 1913]. It was already mentioned in Chapter 3 that extensive-form games were discussed explicitly in von Neumann and Morgenstern [1944], as was backward induction. Subgame perfection was introduced by Selten [1965], who received a Nobel Prize in 1994. The material on computing all subgame-perfect equilibria is based on Littman et al. [2006]. The Centipede game was introduced by Rosenthal [1981]; many other papers discuss the rationality of backward induction in such games [Aumann, 1995, 1996; Binmore, 1996].

In 1953 Kuhn introduced extensive-form games of imperfect information, including the distinction and connection between mixed and behavioral strategies [Kuhn, 1953]. The sequence form, and its application to computing the equilibria of zero-sum games of imperfect information with perfect recall, is due to von Stengel [1996]. Many of the same ideas were developed earlier by Koller and Megiddo [1992]; see [von Stengel, 1996, pp. 242–243] for the distinctions. The use of the sequence form for computing the equilibria of general-sum two-player games of imperfect information is explained by Koller et al. [1996]. Sequential equilibria were introduced by Kreps and Wilson [1982]. Here, as in normal-form games, the full list of alternative solution concepts and connection among them is long, and the interested reader is referred to Hillas and Kohlberg [2002] and Govindan and Wilson [2005b] for a more extensive survey than is possible here.

6

Richer Representations: Beyond the Normal and Extensive Forms

In this chapter we will go beyond the normal and extensive forms by considering a variety of richer game representations. These further representations are important because the normal and extensive forms are not always suitable for modeling large or realistic game-theoretic settings.

First, we may be interested in games that are not finite and that therefore cannot be represented in normal or extensive form. For example, we may want to consider what happens when a simple normal-form game such as the Prisoner's Dilemma is repeated infinitely. We might want to consider a game played by an uncountably infinite set of agents. Or we may want to use an interval of the real numbers as each player's action space.[1]

Second, both of the representations we have studied so far presume that agents have perfect knowledge of everyone's payoffs. This seems like a poor model of many realistic situations, where, for example, agents might have private information that affects their own payoffs and other agents might have only probabilistic information about each others' private information. An elaboration like this can have a big impact, because one agent's actions can depend on what he knows about another agent's payoffs.

Finally, as the numbers of players and actions in a game grow—even if they remain finite—games can quickly become far too large to reason about or even to write down using the representations we have studied so far. Luckily, we are not usually interested in studying arbitrary strategic situations. The sorts of noncooperative settings that are most interesting in practice tend to involve highly structured payoffs. This can occur because of constraints imposed by the fact that the play of a game actually unfolds over time (e.g., because a large game actually corresponds to finitely repeated play of a small game). It can also occur because of the nature of the problem domain (e.g., while the world may involve many agents, the number of agents who are able to directly affect any given agent's payoff is small). If we understand the way in which agents' payoffs are structured, we can represent them much more compactly than we would be able to do using

---

1. We will explore the first example in detail in this chapter. A thorough treatment of infinite sets of players or action spaces is beyond the scope of this book; nevertheless, we will consider certain games with infinite sets of players in Section 6.4.4 and with infinite action spaces in Chapters 10 and 11.

the normal or extensive forms. Often, these compact representations also allow us to reason more efficiently about the games they describe (e.g., the computation of Nash equilibria can be provably faster, or pure-strategy Nash equilibria can be proved to always exist).

In this chapter we will present various different representations that address these limitations of the normal and extensive forms. In Section 6.1 we will begin by considering the special case of extensive-form games that are constructed by repeatedly playing a normal-form game and then we will extend our consideration to the case where the normal form is repeated infinitely. This will lead us to stochastic games in Section 6.2, which are like repeated games but do not require that the same normal-form game is played in each time step. In Section 6.3 we will consider structure of a different kind: instead of considering time, we will consider games involving uncertainty. Specifically, in Bayesian games agents face uncertainty—and hold private information—about the game's payoffs. Section 6.4 describes congestion games, which model situations in which agents contend for scarce resources. Finally, in Section 6.5 we will consider representations that are motivated primarily by compactness and by their usefulness for permitting efficient computation (e.g., of Nash equilibria). Such compact representations can extend any other existing representation, such as normal-form games, extensive-form games, or Bayesian games.

6.1 Repeated games

In repeated games, a given game (often thought of in normal form) is played multiple times by the same set of players. The game being repeated is called the *stage game*. For example, Figure 6.1 depicts two players playing the Prisoner's Dilemma exactly twice in a row.

stage game

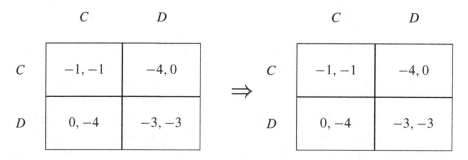

Figure 6.1 Twice-played Prisoner's Dilemma.

This representation of the repeated game, while intuitive, obscures some key factors. Do agents see what the other agents played earlier? Do they remember what they knew? And, while the utility of each stage game is specified, what is the utility of the entire repeated game?

We answer these questions in two steps. We first consider the case in which the game is repeated a finite and commonly-known number of times. Then we

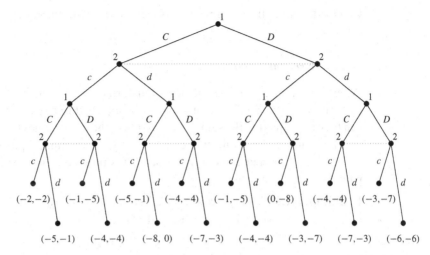

Figure 6.2 Twice-played Prisoner's Dilemma in extensive form.

consider the case in which the game is repeated infinitely often, or a finite but unknown number of times.

6.1.1 *Finitely repeated games*

One way to completely disambiguate the semantics of a finitely repeated game is to specify it as an imperfect-information game in extensive form. Figure 6.2 describes the twice-played Prisoner's Dilemma game in extensive form. Note that it captures the assumption that at each iteration the players do not know what the other player is playing, but afterward they do. Also note that the payoff function of each agent is additive; that is, it is the sum of payoffs in the two-stage games.

The extensive form also makes it clear that the strategy space of the repeated game is much richer than the strategy space in the stage game. Certainly one strategy in the repeated game is to adopt the same strategy in each stage game; clearly, this memory less strategy, called a *stationary strategy*, is a behavioral strategy in the extensive-form representation of the game. But in general, the action (or mixture of actions) played at a stage game can depend on the history of play thus far. Since this fact plays a particularly important role in infinitely repeated games, we postpone further discussion of it to the next section. Indeed, in the finite, known repetition case, we encounter again the phenomenon of backward induction, which we first encountered when we introduced subgame-perfect equilibria. Recall that in the Centipede game, discussed in Section 5.1.3, the unique SPE was to go down and terminate the game at every node. Now consider a finitely repeated Prisoner's Dilemma game. Again, it can be argued, in the last round it is a dominant strategy to defect, no matter what happened so far. This is common knowledge, and no choice of action in the preceding rounds will impact the play in the last round. Thus in the second-to-last round too it is a dominant strategy to defect. Similarly, by induction, it can be argued that the only equilibrium in this case is to always defect. However, as in the case of the

stationary
strategy

Centipede game, this argument is vulnerable to both empirical and theoretical criticisms.

6.1.2 *Infinitely repeated games*

When the infinitely repeated game is transformed into extensive form, the result is an infinite tree. So the payoffs cannot be attached to any terminal nodes, nor can they be defined as the sum of the payoffs in the stage games (which in general will be infinite). There are two common ways of defining a player's payoff in an infinitely repeated game to get around this problem. The first is the average payoff of the stage game in the limit.[2]

Definition 6.1.1 (Average reward) *Given an infinite sequence of payoffs $r_i^{(1)}$, $r_i^{(2)}, \ldots$ for player i, the* average reward *of i is*

$$\lim_{k \to \infty} \frac{\sum_{j=1}^{k} r_i^{(j)}}{k}.$$

The *future discounted reward* to a player at a certain point of the game is the sum of his payoff in the immediate stage game, plus the sum of future rewards discounted by a constant factor. This is a recursive definition, since the future rewards again give a higher weight to early payoffs than to later ones.

Definition 6.1.2 (Discounted reward) *Given an infinite sequence of payoffs $r_i^{(1)}, r_i^{(2)}, \ldots$ for player i, and a discount factor β with $0 \le \beta \le 1$, the* future discounted reward *of i is $\sum_{j=1}^{\infty} \beta^j r_i^{(j)}$.*

The discount factor can be interpreted in two ways. First, it can be taken to represent the fact that the agent cares more about his well-being in the near term than in the long term. Alternatively, it can be assumed that the agent cares about the future just as much as he cares about the present, but with some probability the game will be stopped any given round; $1 - \beta$ represents that probability. The analysis of the game is not affected by which perspective is adopted.

Now let us consider strategy spaces in an infinitely repeated game. In particular, consider the infinitely repeated Prisoner's Dilemma game. As we discussed, there are many strategies other than stationary ones. One of the most famous is *Tit-for-Tat*. TfT is the strategy in which the player starts by cooperating and thereafter chooses in round $j + 1$ the action chosen by the other player in round j. Beside being both simple and easy to compute, this strategy is notoriously hard to beat; it was the winner in several repeated Prisoner's Dilemma competitions for computer programs.

Since the space of strategies is so large, a natural question is whether we can characterize all the Nash equilibria of the repeated game. For example, if the discount factor is large enough, both players playing TfT is a Nash equilibrium. But there is an infinite number of others. For example, consider the *trigger strategy*. This is a draconian version of TfT; in the trigger strategy, a player starts

average reward

future discounted reward

Tit-for-Tat (TfT)

trigger strategy

---

2. The observant reader will notice a potential difficulty in this definition, since the limit may not exist. One can extend the definition to cover these cases by using the lim sup operator in Definition 6.1.1 rather than lim.

by cooperating, but if ever the other player defects then the first defects forever. Again, for sufficiently large discount factor, the trigger strategy forms a Nash equilibrium not only with itself but also with TfT.

The folk theorem—so-called because it was part of the common lore before it was formally written down—helps us understand the space of all Nash equilibria of an infinitely repeated game, by answering a related question. It does not characterize the equilibrium strategy profiles, but rather the payoffs obtained in them. Roughly speaking, it states that in an infinitely repeated game the set of average rewards attainable in equilibrium are precisely those pairs attainable under mixed strategies in a single-stage game, with the constraint on the mixed strategies that each player's payoff is at least the amount he would receive if the other players adopted minmax strategies against him.

More formally, consider any n-player game $G = (N, A, u)$ and any payoff profile $r = (r_1, r_2, \ldots, r_n)$. Let

$$v_i = \min_{s_{-i} \in S_{-i}} \max_{s_i \in S_i} u_i(s_{-i}, s_i).$$

In words, v_i is player i's minmax value: his utility when the other players play minmax strategies against him, and he plays his best response.

Before giving the theorem, we provide some more definitions.

Definition 6.1.3 (Enforceable) *A payoff profile* $r = (r_1, r_2, \ldots, r_n)$ *is enforceable if* $\forall i \in N, r_i \geq v_i$.

Definition 6.1.4 (Feasible) *A payoff profile* $r = (r_1, r_2, \ldots, r_n)$ *is feasible if there exist rational, nonnegative values* α_a *such that for all* i*, we can express* r_i *as* $\sum_{a \in A} \alpha_a u_i(a)$*, with* $\sum_{a \in A} \alpha_a = 1$.

In other words, a payoff profile is feasible if it is a convex, rational combination of the outcomes in G.

folk theorem   **Theorem 6.1.5 (Folk Theorem)** *Consider any* n*-player normal-form game* G *and any payoff profile* $r = (r_1, r_2, \ldots, r_n)$.

1. *If* r *is the payoff profile for any Nash equilibrium* s *of the infinitely repeated* G *with average rewards, then for each player* i*,* r_i *is enforceable.*
2. *If* r *is both feasible and enforceable, then* r *is the payoff profile for some Nash equilibrium of the infinitely repeated* G *with average rewards.*

This proof is both instructive and intuitive. The first part uses the definition of minmax and best response to show that an agent can never receive less than his minmax value in any equilibrium. The second part shows how to construct an equilibrium that yields each agent the average payoffs given in any feasible and enforceable payoff profile r. This equilibrium has the agents cycle in perfect lock-step through a sequence of game outcomes that achieve the desired average payoffs. If any agent deviates, the others punish him forever by playing their minmax strategies against him.

Proof. Part 1: Suppose r is not enforceable, that is, $r_i < v_i$ for some i. Then consider an alternative strategy for i: playing $BR(s_{-i}(h))$, where $s_{-i}(h)$ is the equilibrium strategy of other players given the current history h and $BR(s_{-i}(h))$ is a function that returns a best response for i to a given strategy profile s_{-i} in the (unrepeated) stage game G. By definition of a minmax strategy, player i receives a payoff of at least v_i in every stage game if he plays $BR(s_{-i}(h))$, and so i's average reward is also at least v_i. Thus, if $r_i < v_i$ then s cannot be a Nash equilibrium.

Part 2: Since r is a feasible enforceable payoff profile, we can write it as $r_i = \sum_{a \in A} (\frac{\beta_a}{\gamma}) u_i(a)$, where β_a and γ are nonnegative integers. (Recall that α_a were required to be rational. So we can take γ to be their common denominator.) Since the combination was convex, we have $\gamma = \sum_{a \in A} \beta_a$.

We are going to construct a strategy profile that will cycle through all outcomes $a \in A$ of G with cycles of length γ, each cycle repeating action a exactly β_a times. Let (a^t) be such a sequence of outcomes. Let us define a strategy s_i of player i to be a trigger version of playing (a^t): if nobody deviates, then s_i plays a_i^t in period t. However, if there was a period t' in which some player $j \neq i$ deviated, then s_i will play $(p_{-j})_i$, where (p_{-j}) is a solution to the minimization problem in the definition of v_j.

First observe that if everybody plays according to s_i, then, by construction, player i receives average payoff of r_i (look at averages over periods of length γ). Second, this strategy profile is a Nash equilibrium. Suppose everybody plays according to s_i, and player j deviates at some point. Then, forever after, player j will receive his min max payoff $v_j \leq r_j$, rendering the deviation unprofitable. ∎

The reader might wonder why this proof appeals to i's minmax value rather than his maxmin value. First, notice that the trigger strategies in Part 2 of the proof use minmax strategies to punish agent i. This makes sense because even in cases where i's minmax value is strictly greater than his maxmin value,[3] i's minmax value is the smallest amount that the other agents can guarantee that i will receive. When i best responds to a minmax strategy played against him by $-i$, he receives exactly his minmax value; this is the deviation considered in Part 1.

Theorem 6.1.5 is actually an instance of a large family of folk theorems. As stated, Theorem 6.1.5 is restricted to infinitely repeated games, to average reward, to the Nash equilibrium, and to games of complete information. However, there are folk theorems that hold for other versions of each of these conditions, as well as other conditions not mentioned here. In particular, there are folk theorems for infinitely repeated games with discounted reward (for a large enough discount factor), for finitely repeated games, for subgame-perfect equilibria (i.e., where agents only administer finite punishments to deviators), and for games of incomplete information. We do not review them here, but the message of each of them

---

3. This can happen in games with more than two players, as discussed in Section 3.4.1.

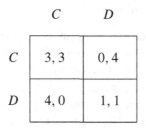

	C	D
C	3, 3	0, 4
D	4, 0	1, 1

Figure 6.3 Prisoner's Dilemma game.

is fundamentally the same: the payoffs in the equilibria of a repeated game are essentially constrained only by enforceability and feasibility.

6.1.3 *"Bounded rationality": repeated games played by automata*

Until now we have assumed that players can engage in arbitrarily deep reasoning and mutual modeling, regardless of their complexity. In particular, consider the fact that we have tended to rely on equilibrium concepts as predictions of—or prescriptions for—behavior. Even in the relatively uncontroversial case of two-player zero-sum games, this is a questionable stance in practice; otherwise, for example, there would be no point in chess competitions. While we will continue to make this questionable assumption in much of the remainder of the book, we pause here to revisit it. We ask what happens when agents are not perfectly rational expected-utility maximizers. In particular, we ask what happens when we impose specific computational limitations on them.

Consider (yet again) an instance of the Prisoner's Dilemma, which is reproduced in Figure 6.3. In the finitely repeated version of this game, we know that each player's dominant strategy (and thus the only Nash equilibrium) is to choose the strategy D in each iteration of the game. In reality, when people actually play the game, we typically observe a significant amount of cooperation, especially in the earlier iterations of the game. While much of game theory is open to the criticism that it does not match well with human behavior, this is a particularly stark example of this divergence. What models might explain this fact?

One early proposal in the literature is based on the notion of an ϵ-equilibrium, defined in Section 3.4.7. Recall that this is a strategy profile in which no agent can gain more than ϵ by changing his strategy; a Nash equilibrium is thus the special case of a 0-equilibrium. This equilibrium concept is motivated by the idea that agents' rationality may be bounded in the sense that they are willing to settle for payoffs that are slightly below their best response payoffs. In the finitely repeated Prisoner's Dilemma game, as the number of repetitions increases, the corresponding sets of ϵ-equilibria include outcomes with longer and longer sequences of the "cooperate" strategy.

Various other models of bounded rationality exist, but we will focus on what has proved to be the richest source of results so far, namely, restricting agents' strategies to those implemented by automata of the sort investigated in computer science.

Finite-state automata

The motivation for using automata becomes apparent when we consider the representation of a strategy in a repeated game. Recall that a finitely repeated game is an imperfect-information extensive-form game, and that a strategy for player i in such a game is a specification of an action for every information set belonging to that player. A strategy for k repetitions of an m-action game is thus a specification of $\frac{m^k-1}{m-1}$ different actions. However, a naive encoding of a strategy as a table mapping each possible history to an action can be extremely inefficient. For example, the strategy of choosing D in every round can be represented using just the single-stage strategy D, and the *Tit-for-Tat* strategy can be represented simply by specifying that the player mimic what his opponent did in the previous round. One representation that exploits this structure is the *finite-state automaton*, or *Moore machine*. The formal definition of a finite-state automaton in the context of a repeated game is as follows.

finite-state
automaton

Moore machine

Definition 6.1.6 (Automaton) *Given a game* $G = (N, A, u)$ *that will be played repeatedly, an automaton* M_i *for player* i *is a four-tuple* $(Q_i, q_i^0, \delta_i, f_i)$, *where:*

- Q_i *is a set of states;*
- q_i^0 *is the start state;*
- $\delta_i : Q_i \times A \mapsto Q_i$ *is a transition function mapping the current state and an action profile to a new state; and*
- $f_i : Q_i \mapsto A_i$ *is a strategy function associating with every state an action for player* i.

An automaton is used to represent each player's repeated game strategy as follows. The machine begins in the start state q_i^0, and in the first round plays the action given by $f_i(q_i^0)$. Using the transition function and the actions played by the other players in the first round, it then transitions automatically to the new state $\delta_i(q_i^0, a_1, \ldots, a_n)$ before the beginning of round 2. It then plays the action $f_i(\delta_i(q_i^0, a_1, \ldots, a_n))$ in round two, and so on. More generally, we can specify the current strategy and state at round t using the following recursive definitions.

$$a_i^t = f_i(q_i^t)$$

$$q_i^{t+1} = \delta_i(q_i^t, a_1^t, \ldots, a_n^t)$$

Automaton representations of strategies are very intuitive when viewed graphically. The following figures show compact automaton representations of some common strategies for the repeated Prisoner's Dilemma game. Each circle is a state in the automaton and its label is the action to play at that state. The transitions are represented as labeled arrows. From the current state, we transition along the arrow labeled with the move the opponent played in the current game. The unlabeled arrow enters the initial state.

Figure 6.4 An automaton representing the repeated *Defect* action.

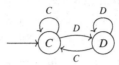

Figure 6.5 An automaton representing the *Tit-for-Tat* strategy.

The automaton represented by Figure 6.4 plays the constant D strategy, while Figure 6.5 encodes the more interesting *Tit-for-Tat* strategy. It starts in the C state, and the transitions are constructed so that the automaton always mimics the opponent's last action.

machine game

We can now define a new class of games, called *machine games*, in which each player selects an automaton representing a repeated game strategy.

Definition 6.1.7 (Machine game) *A two-player machine game $G^M = (\{1, 2\}, \mathcal{M}, G)$ of the k-period repeated game G is defined by:*

- *a pair of players $\{1, 2\}$;*
- *$\mathcal{M} = \mathcal{M}_1, \mathcal{M}_2$, where $\mathcal{M}_1$ is a set of available automata for player i; and*
- *a normal-form game $G = (\{1, 2\}, A, u)$.*

A pair $M_1 \in \mathcal{M}_1$ and $M_2 \in \mathcal{M}_2$ deterministically yield an outcome $o^t(M_1, M_2)$ at each iteration t of the repeated game. Thus, G^M induces a normal-form game $(\{1, 2\}, \mathcal{M}, U)$, in which each player i chooses an automaton $M_i \in \mathcal{M}_i$, and obtains utility $U_i(M_1, M_2) = \sum_{t=1}^{k} u_i(o^t(M_1, M_2))$.

Note that we can easily replace the k-period repeated game with a discounted (or limit of means) infinitely repeated game, with a corresponding change to $U_i(M_1, M_2)$ in the induced normal-form game.

In what follows, the function $s : M \mapsto \mathbb{Z}$ represents the number of states of an automaton M, and the function $S(\mathcal{M}_i) = \max_{M \in \mathcal{M}_i} s(M)$ represents the size of the largest automaton among a set of automata $\mathcal{M}_i$.

Automata of bounded size

Intuitively, automata with fewer states represent *simpler* strategies. Thus, one way to bound the rationality of the player is by limiting the number of states in the automaton.

Placing severe restrictions on the number of states not only induces an equilibrium in which cooperation always occurs, but also causes the always-defect equilibrium to disappear. This equilibrium in a finitely repeated Prisoner's Dilemma game depends on the assumption that each player can use *backward induction* (see Section 5.1.4) to find his dominant strategy. In order to perform backward induction in a k-period repeated game, each player needs to keep track of at least k distinct states: one state to represent the choice of strategy in each repetition of the game. In the Prisoner's Dilemma, it turns out that if $2 < \max(S(\mathcal{M}_1), S(\mathcal{M}_2)) < k$, then the constant-defect strategy does not yield a symmetric equilibrium, while the Tit-for-Tat automaton does.

When the size of the automaton is not restricted to be less than k, the constant-defect equilibrium does exist. However, there is still a large class of machine games in which other equilibria exist in which some amount of cooperation occurs, as shown in the following result.

Theorem 6.1.8 *For any integer x, there exists an integer k_0 such that for all $k > k_0$, any machine game $G^M = (\{1, 2\}, \mathcal{M}, G)$ of the k-period repeated Prisoner's Dilemma game G, in which $k^{1/x} \leq \min\{S(\mathcal{M}_1), S(\mathcal{M}_2)\} \leq \max\{S(\mathcal{M}_1), S(\mathcal{M}_2)\} \leq k^x$ holds has a Nash equilibrium in which the average payoffs to each player are at least $3 - \frac{1}{x}$.*

Thus the average payoffs to each player can be much higher than $(1, 1)$; in fact they can be arbitrarily close to $(3, 3)$, depending on the choice of x. While this result uses pure strategies for both players, a stronger result can be proved through the use of a mixed-strategy equilibrium.

Theorem 6.1.9 *For every $\epsilon > 0$, there exists an integer k_0 such that for all $k > k_0$, any machine game $G^M = (\{1, 2\}, \mathcal{M}, G)$ of the k-period repeated Prisoner's Dilemma game G, in which $\min\{S(\mathcal{M}_1), S(\mathcal{M}_2)\} < 2^{\frac{\epsilon k}{12(1+\epsilon)}}$ has a Nash equilibrium in which the average payoffs to each player are at least $3 - \epsilon$.*

Thus, if even one of the players' automata has a size that is less than exponential in the length of the game, an equilibrium with some degree of cooperation exists.

Automata with a cost of complexity

Now, instead of imposing constraints on the complexity of the automata, we will incorporate this complexity as a cost into the agent's utility function. This could reflect, for example, the implementation cost of a strategy or the cost to learn it. While we cannot show theorems similar to those in the preceding section, it turns out that we can get mileage out of this idea even when we incorporate it in a minimal way. Specifically, an agent's disutility for complexity will only play a tie-breaking role.

lexicographic
disutility for
complexity

Definition 6.1.10 (Lexicographic disutility for complexity) *Agents have lexicographic disutility for complexity in a machine game if their utility functions $U_i(\cdot)$ in the induced normal-form game are replaced by preference orderings $\succeq_i$ such that $(M_1, M_2) \succ_i (M_1', M_2')$ whenever either $U_i(M_1, M_2) > U_i(M_1', M_2')$ or $U_i(M_1, M_2) = U_i(M_1', M_2')$ and $s(M_i) < s(M_i')$.*

Consider a machine game G^M of the discounted infinitely repeated Prisoner's Dilemma in which both players have a lexicographic disutility for complexity. The trigger strategy is an equilibrium strategy in the infinitely repeated Prisoner's Dilemma game with discounting. When the discount factor β is large enough, if player 2 is using the trigger strategy, then player 1 cannot achieve a higher payoff

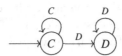

Figure 6.6 An automaton representing the trigger strategy.

by using any strategy other than the trigger strategy himself. We can represent the trigger strategy using the machine M shown in Figure 6.6. However, while no other machine can give player 1 a higher payoff, there does exist another machine that achieves the *same* payoff and is less complex. Player 1's machine M never enters the state D during play; it is designed only as a threat to the other player. Thus the machine which contains only the state C will achieve the same payoff as the machine M, but with less complexity. As a result, the outcome (M, M) is not a Nash equilibrium of the machine game G^M when agents have a lexicographic disutility for complexity.

We can also show several interesting properties of the equilibria of machine games in which agents have a lexicographic disutility for complexity. First, because machines in equilibrium must minimize complexity, they have no unused states. Thus we know that in an infinite game, every state must be visited in some period. Second, the strategies represented by the machines in a Nash equilibrium of the machine game also form a Nash equilibrium of the infinitely repeated game.

Computing best-response automata

In the previous sections we limited the rationality of agents in repeated games by bounding the number of states that they can use to represent their strategies. However, it could be the case that the number of states used by the equilibrium strategies is small, but the time required to compute them is prohibitively large. Furthermore, one can argue (by introspection, for example) that bounding the computation of an agent is a more appropriate means of capturing bounded rationality than bounding the number of states.

It seems reasonable that an equilibrium must be at least verifiable by agents. But this does not appear to be the case for finite automata. (The results that follow are for the limit-average case, but can be adapted to the discounted case as well.)

Theorem 6.1.11 *Given a two-player machine game $G^M = (N, \mathcal{M}, G)$ of a limit average infinitely repeated two-player game $G = (N, A, u)$ with unknown N, and a choice of automata $M_1, \ldots, M_n$ for all players, there does not exist a polynomial time algorithm for verifying whether M_i is a best-response automaton for player i.*

The news is not all bad; if we hold N fixed, then the problem does belong to P. We can explain this informally by noting that player i does not have to scan all of his possible strategies in order to decide whether automaton M_i is the best response; since he knows the strategies of the other players, he merely needs to scan the actual path taken on the game tree, which is bounded by the length of the game tree.

Notice that the previous result held even when the other players were assumed to play pure strategies. The following result shows that the verification problem is hard even in the two-player case when the players can randomize over machines.

Theorem 6.1.12 *Given a two-player machine game $G^M = (\{1, 2\}, \mathcal{M}, G)$ of a limit-average infinitely repeated game $G = (\{1, 2\}, A, u)$, and a mixed strategy for player 2 in which the set of automata that are played with positive probability is finite, the problem of verifying that an automaton M_1 is a best-response automaton for player 1 is NP-complete.*

So far we have abandoned the bounds on the number of states in the automata, and one might wonder whether such bounds could improve the worst-case complexity. However, for the repeated Prisoner's Dilemma game, it has the opposite effect: limiting the size of the automata under consideration increases the complexity of computing a best response. By Theorem 6.1.11 we know that when the size of the automata under consideration are unbounded and the number of agents is two, the problem of computing the best response is in the class P. The following result shows that when the automata under consideration are instead bounded, the problem becomes NP-complete.

Theorem 6.1.13 *Given a machine game $G^M = (\{1, 2\}, \mathcal{M}, G)$ of the limit average infinitely repeated Prisoner's Dilemma game G, an automaton M_2, and an integer k, the problem of computing a best-response automaton M_1 for player 1, such that $s(M_1) \leq k$, is NP-complete.*

From finite automata to Turing machines

Turing machines are more powerful than finite-state automata due to their infinite memories. One might expect that in this richer model, unlike with finite automata, game-theoretic results will be preserved. But they are not. For example, there is strong evidence (if not yet proof) that a Prisoner's Dilemma game of two Turing machines can have equilibria that are arbitrarily close to the repeated C payoff. Thus cooperative play can be approximated in equilibrium even if the machines memorize the entire history of the game and are capable of counting the number of repetitions.

The problem of computing a best response yields another unintuitive result. Even if we restrict the opponent to strategies for which the best-response Turing machine is computable, the general problem of finding the best response for any such input is not Turing computable when the discount factor is sufficiently close to one.

Theorem 6.1.14 *For the discounted, infinitely-repeated Prisoner's Dilemma game G, there exists a discount factor $\underline{\beta} > 0$ such that for any rational discount factor $\beta \in (\underline{\beta}, 1)$ there is no Turing-computable procedure for computing a best response to a strategy drawn from the set of all computable strategies that admit a computable best response.*

Finally, even before worrying about computing a best response, there is a more basic challenge: the best response to a Turing machine may not be a Turing machine!

Theorem 6.1.15 *For the discounted, infinitely-repeated Prisoner's Dilemma game G, there exists a discount factor $\beta > 0$ such that for any rational discount factor $\beta \in (\underline{\beta}, 1)$ there exists an equilibrium profile (s_1, s_2) such that s_2 can be implemented by a Turing machine, but no best response to s_2 can be implemented by a Turing machine.*

6.2 Stochastic games

Intuitively speaking, a stochastic game is a collection of normal-form games; the agents repeatedly play games from this collection, and the particular game played at any given iteration depends probabilistically on the previous game played and on the actions taken by all agents in that game.

6.2.1 *Definition*

Stochastic games are very broad framework, generalizing both Markov decision processes (MDPs; see Appendix C) and repeated games. An MDP is simply a stochastic game with only one player, while a repeated game is a stochastic game in which there is only one stage game.

stochastic game   **Definition 6.2.1 (Stochastic game)** *A* stochastic game *(also known as a* Markov game*) is a tuple (Q, N, A, P, r), where:*

Markov game

- *Q is a finite set of games;*
- *N is a finite set of n players;*
- *$A = A_1 \times \cdots \times A_n$, where A_i is a finite set of actions available to player i;*
- *$P : Q \times A \times Q \mapsto [0, 1]$ is the transition probability function; $P(q, a, \hat{q})$ is the probability of transitioning from state q to state $\hat{q}$ after action profile a; and*
- *$R = r_1, \ldots, r_n$, where $r_i : Q \times A \mapsto \mathbb{R}$ is a real-valued payoff function for player i.*

In this definition we have assumed that the strategy space of the agents is the same in all games, and thus that the difference between the games is only in the payoff function. Removing this assumption adds notation, but otherwise presents no major difficulty or insights. Restricting Q and each A_i to be finite is a substantive restriction, but we do so for a reason; the infinite case raises a number of complications that we wish to avoid.

We have specified the payoff of a player at each stage game (or in each state), but not how these payoffs are aggregated into an overall payoff. To solve this problem, we can use solutions already discussed earlier in connection with infinitely repeated games (Section 6.1.2). Specifically, the two most commonly used aggregation methods are *average reward* and *future discounted reward*.

6.2.2 *Strategies and equilibria*

We now define the strategy space of an agent. Let $h_t = (q^0, a^0, q^1, a^1, \ldots, a^{t-1}, q^t)$ denote a history of t stages of a stochastic game, and let H_t be the set of all possible histories of this length. The set of deterministic strategies is the Cartesian product $\prod_{t, H_t} A_i$, which requires a choice for each possible history at each point in time. As in the previous game forms, an agent's strategy can consist of any mixture over deterministic strategies. However, there are several restricted classes of strategies that are of interest, and they form the following hierarchy. The first restriction is the requirement that the mixing take place at each history independently; this is the restriction to behavioral strategies seen in connection with extensive-form games.

Definition 6.2.2 (Behavioral strategy) *A behavioral strategy $s_i(h_t, a_{i_j})$ returns the probability of playing action a_{i_j} for history h_t.*

A Markov strategy further restricts a behavioral strategy so that, for a given time t, the distribution over actions depends only on the current state.

Markov strategy **Definition 6.2.3 (Markov strategy)** *A Markov strategy s_i is a behavioral strategy in which $s_i(h_t, a_{i_j}) = s_i(h'_t, a_{i_j})$ if $q_t = q'_t$, where q_t and q'_t are the final states of h_t and h'_t, respectively.*

The final restriction is to remove the possible dependence on the time t.

stationary **Definition 6.2.4 (Stationary strategy)** *A stationary strategy s_i is a Markov strat-*
strategy *egy in which $s_i(h_{t_1}, a_{i_j}) = s_i(h'_{t_2}, a_{i_j})$ if $q_{t_1} = q'_{t_2}$, where q_{t_1} and q'_{t_2} are the final states of h_{t_1} and h'_{t_2}, respectively.*

Now we can consider the equilibria of stochastic games, a topic that turns out to be fraught with subtleties. The discounted-reward case is the less problematic one. In this case it can be shown that a Nash equilibrium exists in every stochastic game. In fact, we can state a stronger property. A strategy profile is called a
Markov perfect *Markov perfect equilibrium* if it consists of only Markov strategies, and is a
equilibrium Nash equilibrium regardless of the starting state. In a sense, MPE plays a role
(MPE) analogous to the subgame-perfect equilibrium in perfect-information games.

Theorem 6.2.5 *Every n-player, general-sum, discounted-reward stochastic game has a Markov perfect equilibrium.*

The case of average rewards presents greater challenges. For one thing, the limit average may not exist (i.e., although the stage-game payoffs are bounded, their average may cycle and not converge). However, there is a class of stochastic
irreducible games that is well behaved in this regard. This is the class of *irreducible* stochastic
stochastic game games. A stochastic game is irreducible if every strategy profile gives rise to an irreducible Markov chain over the set of games, meaning that every game can be reached with positive probability regardless of the strategy adopted. In such games the limit averages are well defined, and we have the following theorem.

Theorem 6.2.6 *Every two-player, general-sum, average reward, irreducible stochastic game has a Nash equilibrium.*

Indeed, under the same condition we can state a folk theorem similar to that presented for repeated games in Section 6.1.2. That is, as long as we give each player an expected payoff that is at least as large as his minmax value, any feasible payoff pair can be achieved in equilibrium through the use of threats.

Theorem 6.2.7 *For every two-player, general-sum, irreducible stochastic game, and every feasible outcome with a payoff vector r that provides to each player at least his minmax value, there exists a Nash equilibrium with a payoff vector r. This is true for games with average rewards, as well as games with large enough discount factors (or, with players that are sufficiently patient).*

6.2.3   *Computing equilibria*

The algorithms and results for stochastic games depend greatly on whether we use discounted reward or average reward for the agent utility function. We will discuss both separately, starting with the discounted reward case. The first question to ask about the problem of finding a Nash equilibrium is whether a polynomial procedure is available. The fact that there exists an linear programming formulation for solving MDPs (for both the discounted reward and average reward cases) gives us a reason for optimism, since stochastic games are a generalization of MDPs. While such a formulation does not exist for the full class of stochastic games, it does for several nontrivial subclasses.

One such subclass is the set of two-player, general-sum, discounted-reward stochastic games in which the transitions are determined by a single player. The *single-controller* condition is formally defined as follows.

single-controller
stochastic game

Definition 6.2.8 (Single-controller stochastic game) *A stochastic game is single-controller if there exists a player i such that $\forall q, q' \in Q, \forall a \in A$, $P(q, a, q') = P(q, a', q')$ if $a_i = a_i'$.*

The same results hold when we replace the single-controller restriction with the following pair of restrictions: that the state and action profile have independent effects on the reward achieved by each agent, and that the transition function only depends on the action profile. Formally, this pair is called the *separable reward state independent transition* condition.

Definition 6.2.9 (SR-SIT stochastic game) *A stochastic game is separable reward state independent transition (SR-SIT) if the following two conditions hold:*

- *there exist functions α, γ such that $\forall i, q \in Q, \forall \in A$ it is the case that $r_i(q, a) = \alpha(q) + \gamma(a)$; and*
- *$\forall q, q', q'' \in Q, \forall \in A$ it is the case that $P(q, a, q'') = P(q', a, q'')$.*

Even when the problem does not fall into one of these subclasses, practical solutions still exist for the discounted case. One such solution is to apply a modified version of Newton's method to a nonlinear program formulation of the problem. An advantage of this method is that no local minima exist. For zero-sum games, an alternative is to use an algorithm developed by Shapley that is related to value iteration, a commonly-used method for solving MDPs (see Appendix C).

Moving on to the average reward case, we have to impose more restrictions in order to use a linear program than we did for the discounted reward case. Specifically, for the class of two-player, general-sum, average-reward stochastic games, the single-controller assumption no longer suffices—we also need the game to be zero sum.

Even when we cannot use a linear program, irreducibility allows us to use an algorithm that is guaranteed to converge. This algorithm is a combination of policy iteration (another method used for solving MDPs) and successive approximation.

6.3   Bayesian games

Bayesian game

All of the game forms discussed so far assumed that all players know what game is being played. Specifically, the number of players, the actions available to each player, and the payoff associated with each action vector have all been assumed to be common knowledge among the players. Note that this is true even of imperfect-information games; the actual moves of agents are not common knowledge, but the game itself is. In contrast, *Bayesian games*, or games of incomplete information, allow us to represent players' uncertainties about the very game being played.[4] This uncertainty is represented as a probability distribution over a set of possible games. We make two assumptions.

1. All possible games have the same number of agents and the same strategy space for each agent; they differ only in their payoffs.
2. The beliefs of the different agents are posteriors, obtained by conditioning a common prior on individual private signals.

The second assumption is substantive, and we return to it shortly. The first is not particularly restrictive, although at first it might seem to be. One can imagine many other potential types of uncertainty that players might have about the game—how many players are involved, what actions are available to each player, and perhaps other aspects of the situation. It might seem that we have severely limited the discussion by ruling these out. However, it turns out that these other types of uncertainty can be reduced to uncertainty only about payoffs via problem reformulation.

For example, imagine that we want to model a situation in which one player is uncertain about the number of actions available to the other players. We can reduce this uncertainty to uncertainty about payoffs by padding the game with irrelevant actions. For example, consider the following two-player game, in which the row player does not know whether his opponent has only the two strategies L and R or also the third one C:

---

4. It is easy to confuse the term "incomplete information" with "imperfect information"; don't...

	L	R
U	1, 1	1, 3
D	0, 5	1, 13

	L	C	R
U	1, 1	0, 2	1, 3
D	0, 5	2, 8	1, 13

Now consider replacing the leftmost, smaller game by a padded version, in which we add a new C column.

	L	C	R
U	1, 1	0, -100	1, 3
D	0, 5	2, -100	1, 13

Clearly the newly added column is dominated by the others and will not participate in any Nash equilibrium (or any other reasonable solution concept). Indeed, there is an isomorphism between Nash equilibria of the original game and the padded one. Thus the uncertainty about the strategy space can be reduced to uncertainty about payoffs.

Using similar tactics, it can be shown that it is also possible to reduce uncertainty about other aspects of the game to uncertainty about payoffs only. This is not a mathematical claim, since we have given no mathematical characterization of all the possible forms of uncertainty, but it is the case that such reductions have been shown for all the common forms of uncertainty.

common-prior assumption

The second assumption about Bayesian games is the *common-prior assumption*, addressed in more detail in our discussion of multiagent probabilities and KP-structures in Chapter 13. As discussed there, a Bayesian game thus defines not only the uncertainties of agents about the game being played, but also their beliefs about the beliefs of other agents about the game being played, and indeed an entire infinite hierarchy of nested beliefs (the so-called epistemic type space). As also discussed in Chapter 13, the common-prior assumption is a substantive assumption that limits the scope of applicability. We nonetheless make this assumption since it allows us to formulate the main ideas in Bayesian games, and without the assumption the subject matter becomes much more involved than is appropriate for this text. Indeed, most (but not all) work in game theory makes this assumption.

6.3.1 Definition

There are several ways of presenting Bayesian games; we will offer three different definitions. All three are equivalent, modulo some subtleties that lie outside the

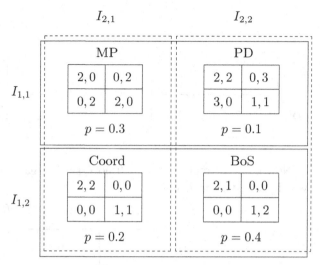

Figure 6.7 A Bayesian game.

scope of this book. We include all three since each formulation is useful in different settings and offers different intuition about the underlying structure of this family of games.

Information sets

First, we present a definition that is based on information sets. Under this definition, a Bayesian game consists of a set of games that differ only in their payoffs, a common prior defined over them, and a partition structure over the games for each agent.[5]

Bayesian game **Definition 6.3.1 (Bayesian game: information sets)** *A* Bayesian game *is a tuple* (N, G, P, I) *where:*

- *N is a set of agents;*
- *G is a set of games with N agents each such that if $g, g' \in G$ then for each agent $i \in N$ the strategy space in g is identical to the strategy space in g';*
- *$P \in \Pi(G)$ is a common prior over games, where $\Pi(G)$ is the set of all probability distributions over G; and*
- *$I = (I_1, ..., I_N)$ is a tuple of partitions of G, one for each agent.*

Figure 6.7 gives an example of a Bayesian game. It consists of four 2×2 games (Matching Pennies, Prisoner's Dilemma, Coordination and Battle of the Sexes), and each agent's partition consists of two equivalence classes.

Extensive form with chance moves

A second way of capturing the common prior is to hypothesize a special agent called Nature who makes probabilistic choices. While we could have Nature's

---

5. This combination of a common prior and a set of partitions over states of the world turns out to correspond to a KP-structure, defined in Chapter 13.

choice be interspersed arbitrarily with the agents' moves, without loss of generality we assume that Nature makes all its choices at the outset. Nature does not have a utility function (or, alternatively, can be viewed as having a constant one), and has the unique strategy of randomizing in a commonly known way. The agents receive individual signals about Nature's choice, and these are captured by their information sets in a standard way. The agents have no additional information; in particular, the information sets capture the fact that agents make their choices without knowing the choices of others. Thus, we have reduced games of incomplete information to games of imperfect information, albeit ones with chance moves. These chance moves of Nature require minor adjustments of existing definitions, replacing payoffs by their expectations given Nature's moves.[6]

For example, the Bayesian game of Figure 6.7 can be represented in extensive form as depicted in Figure 6.8.

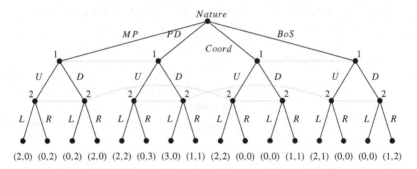

Figure 6.8 The Bayesian game from Figure 6.7 in extensive form.

Although this second definition of Bayesian games can be initially more intuitive than our first definition, it can also be more cumbersome to work with. This is because we use an extensive-form representation in a setting where players are unable to observe each others' moves. (Indeed, for the same reason we do not routinely use extensive-form games of imperfect information to model simultaneous interactions such as the Prisoner's Dilemma, though we could do so if we wished.) For this reason, we will not make further use of this definition. We close by noting one advantage that it does have, however: it extends very naturally to Bayesian games in which players move sequentially and do (at least sometimes) learn about previous players' moves.

Epistemic types

Recall that a game may be defined by a set of players, actions, and utility functions. In our first definition agents are uncertain about which game they are playing; however, each possible game has the same sets of actions and players, and so agents are really only uncertain about the game's utility function.

---

6. Note that the special structure of this extensive-form game means that we do not have to agonize over the refinements of Nash equilibrium; since agents have no information about prior choices made other than by Nature, all Nash equilibria are also sequential equilibria.

a_1	a_2	θ_1	θ_2	u_1	u_2		a_1	a_2	θ_1	θ_2	u_1	u_2
U	L	$\theta_{1,1}$	$\theta_{2,1}$	2	0		D	L	$\theta_{1,1}$	$\theta_{2,1}$	0	2
U	L	$\theta_{1,1}$	$\theta_{2,2}$	2	2		D	L	$\theta_{1,1}$	$\theta_{2,2}$	3	0
U	L	$\theta_{1,2}$	$\theta_{2,1}$	2	2		D	L	$\theta_{1,2}$	$\theta_{2,1}$	0	0
U	L	$\theta_{1,2}$	$\theta_{2,2}$	2	1		D	L	$\theta_{1,2}$	$\theta_{2,2}$	0	0
U	R	$\theta_{1,1}$	$\theta_{2,1}$	0	2		D	R	$\theta_{1,1}$	$\theta_{2,1}$	2	0
U	R	$\theta_{1,1}$	$\theta_{2,2}$	0	3		D	R	$\theta_{1,1}$	$\theta_{2,2}$	1	1
U	R	$\theta_{1,2}$	$\theta_{2,1}$	0	0		D	R	$\theta_{1,2}$	$\theta_{2,1}$	1	1
U	R	$\theta_{1,2}$	$\theta_{2,2}$	0	0		D	R	$\theta_{1,2}$	$\theta_{2,2}$	1	2

Figure 6.9 Utility functions u_1 and u_2 for the Bayesian game from Figure 6.7.

epistemic type

Our third definition uses the notion of an *epistemic type*, or simply a *type* as a way of defining uncertainty directly over a game's utility function.

Bayesian game

Definition 6.3.2 (Bayesian game: types) *A* Bayesian game *is a tuple* (N, A, Θ, p, u) *where:*

- *N is a set of agents;*
- *$A = A_1 \times \cdots \times A_n$, where A_i is the set of actions available to player i;*
- *$\Theta = \Theta_1 \times \ldots \times \Theta_n$, where Θ_i is the type space of player i;*
- *$p : \Theta \mapsto [0, 1]$ is a common prior over types; and*
- *$u = (u_1, \ldots, u_n)$, where $u_i : A \times \Theta \mapsto \mathbb{R}$ is the utility function for player i.*

The assumption is that all of the above is common knowledge among the players, and that each agent knows his own type. This definition can seem mysterious, because the notion of type can be rather opaque. In general, the type of agent encapsulates all the information possessed by the agent that is not common knowledge. This is often quite simple (e.g., the agent's knowledge of his private payoff function), but can also include his beliefs about other agents' payoffs, about their beliefs about his own payoff, and any other higher-order beliefs.

We can get further insight into the notion of a type by relating it to the formulation at the beginning of this section. Consider again the Bayesian game in Figure 6.7. For each of the agents we have two types, corresponding to his two information sets. Denote player 1's actions as U and D, player 2's actions as L and R. Call the types of the first agent $\theta_{1,1}$ and $\theta_{1,2}$, and those of the second agent $\theta_{2,1}$ and $\theta_{2,2}$. The joint distribution on these types is as follows: $p(\theta_{1,1}, \theta_{2,1}) = .3$, $p(\theta_{1,1}, \theta_{2,2}) = .1$, $p(\theta_{1,2}, \theta_{2,1}) = .2$, $p(\theta_{1,2}, \theta_{2,2}) = .4$. The conditional probabilities for the first player are $p(\theta_{2,1} \mid \theta_{1,1}) = 3/4$, $p(\theta_{2,2} \mid \theta_{1,1}) = 1/4$, $p(\theta_{2,1} \mid \theta_{1,2}) = 1/3$, and $p(\theta_{2,2} \mid \theta_{1,2}) = 2/3$. Both players' utility functions are given in Figure 6.9.

6.3.2 *Strategies and equilibria*

Now that we have defined Bayesian games, we must explain how to reason about them. We will do this using the epistemic type definition, because that is the definition most commonly used in mechanism design (discussed in Chapter 10), one of the main applications of Bayesian games. All of the concepts defined below can also be expressed in terms of the first two Bayesian game definitions as well.

The first task is to define an agent's strategy space in a Bayesian game. Recall that in an imperfect-information extensive-form game a pure strategy is a mapping from information sets to actions. The definition is similar in Bayesian games: a pure strategy $\alpha_i : \Theta_i \mapsto A_i$ is a mapping from every type agent i could have to the action he would play if he had that type. We can then define mixed strategies in the natural way as probability distributions over pure strategies. As before, we denote a mixed strategy for i as $s_i \in S_i$, where S_i is the set of all i's mixed strategies. Furthermore, we use the notation $s_j(a_j|\theta_j)$ to denote the probability under mixed strategy s_j that agent j plays action a_j, given that j's type is θ_j.

Next, since we have defined an environment with multiple sources of uncertainty, we will pause to reconsider the definition of an agent's expected utility. In a Bayesian game setting, there are three meaningful notions of expected utility: *ex post*, *ex interim* and *ex ante*. The first is computed based on all agents' actual types, the second considers the setting in which an agent knows his own type but not the types of the other agents, and in the third case the agent does not know anybody's type.

ex post expected utility **Definition 6.3.3 (*Ex post* expected utility)** *Agent i's* ex post *expected utility in a Bayesian game (N, A, Θ, p, u), where the agents' strategies are given by s and the agent' types are given by θ, is defined as*

$$EU_i(s, \theta) = \sum_{a \in A} \left(\prod_{j \in N} s_j(a_j|\theta_j) \right) u_i(a, \theta). \tag{6.1}$$

In this case, the only uncertainty concerns the other agents' mixed strategies, since agent i's *ex post* expected utility is computed based on the other agents' actual types. Of course, in a Bayesian game no agent *will* know the others' types; while that does not prevent us from offering the definition given, it might make the reader question its usefulness. We will see that this notion of expected utility is useful both for defining the other two and also for defining a specialized equilibrium concept.

Definition 6.3.4 (*Ex interim* expected utility) *Agent i's* ex interim *expected utility in a Bayesian game (N, A, Θ, p, u), where i's type is θ_i and where the agents' strategies are given by the mixed-strategy profile s, is defined as*

$$EU_i(s, \theta_i) = \sum_{\theta_{-i} \in \Theta_{-i}} p(\theta_{-i}|\theta_i) \sum_{a \in A} \left(\prod_{j \in N} s_j(a_j|\theta_j) \right) u_i(a, \theta_{-i}, \theta_i), \tag{6.2}$$

or equivalently as

$$EU_i(s, \theta_i) = \sum_{\theta_{-i} \in \Theta_{-i}} p(\theta_{-i}|\theta_i) EU_i(s, (\theta_i, \theta_{-i})). \tag{6.3}$$

Thus, i must consider every assignment of types to the other agents θ_{-i} and every pure action profile a in order to evaluate his utility function $u_i(a, \theta_i, \theta_{-i})$. He must weight this utility value by two amounts: the probability that the other players' types would be θ_{-i} given that his own type is θ_i, and the probability that the pure action profile a would be realized given all players' mixed strategies and types. (Observe that agents' types may be correlated.) Because uncertainty over mixed strategies was already handled in the *ex post* case, we can also write *ex interim* expected utility as a weighted sum of $EU_i(s, \theta)$ terms.

Finally, there is the *ex ante* case, where we compute i's expected utility under the joint mixed strategy s without observing any agents' types.

ex ante expected utility   **Definition 6.3.5 (*Ex ante* expected utility)** *Agent i's* ex ante *expected utility in a Bayesian game* (N, A, Θ, p, u), *where the agents' strategies are given by the mixed-strategy profile s, is defined as*

$$EU_i(s) = \sum_{\theta \in \Theta} p(\theta) \sum_{a \in A} \left(\prod_{j \in N} s_j(a_j | \theta_j) \right) u_i(a, \theta), \tag{6.4}$$

or equivalently as

$$EU_i(s) = \sum_{\theta \in \Theta} p(\theta) EU_i(s, \theta), \tag{6.5}$$

or again equivalently as

$$EU_i(s) = \sum_{\theta_i \in \Theta_i} p(\theta_i) EU_i(s, \theta_i). \tag{6.6}$$

Next, we define best response.

best response in a Bayesian game   **Definition 6.3.6 (Best response in a Bayesian game)** *The set of agent i's best responses to mixed-strategy profile s_{-i} are given by*

$$BR_i(s_{-i}) = \arg\max_{s_i' \in S_i} EU_i(s_i', s_{-i}). \tag{6.7}$$

Note that BR_i is a set because there may be many strategies for i that yield the same expected utility. It may seem odd that BR is calculated based on i's *ex ante* expected utility. However, write $EU_i(s)$ as $\sum_{\theta_i \in \Theta_i} p(\theta_i) EU_i(s, \theta_i)$ and observe that $EU_i(s_i', s_{-i}, \theta_i)$ does not depend on strategies that i would play if his type were not θ_i. Thus, we are in fact performing independent maximization of i's *ex interim* expected utilities conditioned on each type that he could have. Intuitively speaking, if a certain action is best after the signal is received, it is also the best conditional plan devised ahead of time for what to do should that signal be received.

We are now able to define the Bayes–Nash equilibrium.

Bayes–Nash equilibrium   **Definition 6.3.7 (Bayes–Nash equilibrium)** *A* Bayes–Nash equilibrium *is a mixed-strategy profile s that satisfies $\forall i \; s_i \in BR_i(s_{-i})$.*

This is exactly the definition we gave for the Nash equilibrium in Definition 3.3.4: each agent plays a best response to the strategies of the other players. The difference from Nash equilibrium, of course, is that the definition of Bayes–Nash equilibrium is built on top of the Bayesian game definitions of best response and expected utility. Observe that we would not be able to define equilibrium in this way if an agent's strategies were not defined for every possible type. In order for a given agent i to play a best response to the other agents $-i$, i must know what strategy each agent would play for each of his possible types. Without this information, it would be impossible to evaluate the term $EU_i(s_i', s_{-i})$ in Equation (6.7).

6.3.3 *Computing equilibria*

Despite its similarity to the Nash equilibrium, the Bayes–Nash equilibrium may seem conceptually more complicated. However, as we did with extensive-form games, we can construct a normal-form representation that corresponds to a given Bayesian game.

As with games in extensive form, the induced normal form for Bayesian games has an action for every pure strategy. That is, the actions for an agent i are the distinct mappings from Θ_i to A_i. Each agent i's payoff given a pure-strategy profile s is his *ex ante* expected utility under s. Then, as it turns out, the Bayes–Nash equilibria of a Bayesian game are precisely the Nash equilibria of its induced normal form. This fact allows us to note that Nash's theorem applies directly to Bayesian games, and hence that Bayes–Nash equilibria always exist.

An example will help. Consider the Bayesian game from Figure 6.9. Note that in this game each agent has four possible pure strategies (two types and two actions). Then player 1's four strategies in the Bayesian game can be labeled UU, UD, DU, and DD: UU means that 1 chooses U regardless of his type, UD that he chooses U when he has type $\theta_{1,1}$ and D when he has type $\theta_{1,2}$, and so forth. Similarly, we can denote the strategies of player 2 in the Bayesian game by RR, RL, LR, and LL.

We now define a 4×4 normal-form game in which these are the four strategies of the two agents, and the payoffs are the expected payoffs in the individual games, given the agents' common prior beliefs. For example, player 2's *ex ante* expected utility under the strategy profile (UU, LL) is calculated as follows:

$$u_2(UU, LL)$$
$$= \sum_{\theta \in \Theta} p(\theta) u_2(U, L, \theta)$$
$$= p(\theta_{1,1}, \theta_{2,1}) u_2(U, L, \theta_{1,1}, \theta_{2,1}) + p(\theta_{1,1}, \theta_{2,2}) u_2(U, L, \theta_{1,1}, \theta_{2,2})$$
$$\quad + p(\theta_{1,2}, \theta_{2,1}) u_2(U, L, \theta_{1,2}, \theta_{2,1}) + p(\theta_{1,2}, \theta_{2,2}) u_2(U, L, \theta_{1,2}, \theta_{2,2})$$
$$= 0.3(0) + 0.1(2) + 0.2(2) + 0.4(1) = 1.$$

Continuing in this manner, the complete payoff matrix can be constructed as shown in Figure 6.10.

	LL	LR	RL	RR
UU	2, 1	1, 0.7	1, 1.2	0, 0.9
UD	0.8, 0.2	1, 1.1	0.4, 1	0.6, 1.9
DU	1.5, 1.4	0.5, 1.1	1.7, 0.4	0.7, 0.1
DD	0.3, 0.6	0.5, 1.5	1.1, 0.2	1.3, 1.1

Figure 6.10 Induced normal form of the game from Figure 6.9.

Now the game may be analyzed straightforwardly. For example, we can determine that player 1's best response to RL is DU.

Given a particular signal, the agent can compute the posterior probabilities and recompute the expected utility of any given strategy vector. Thus in the previous example once the row agent gets the signal $\theta_{1,1}$ he can update the expected payoffs and compute the new game shown in Figure 6.11.

	LL	LR	RL	RR
UU	2, 0.5	1.5, 0.75	0.5, 2	0, 2.25
UD	2, 0.5	1.5, 0.75	0.5, 2	0, 2.25
DU	0.75, 1.5	0.25, 1.75	2.25, 0	1.75, 0.25
DD	0.75, 1.5	0.25, 1.75	2.25, 0	1.75, 0.25

Figure 6.11 *Ex interim* induced normal-form game, where player 1 observes type $\theta_{1,1}$.

Note that for the row player, DU is still a best response to RL; what has changed is how much better it is compared to the other three strategies. In particular, the row player's payoffs are now independent of his choice of which action to take upon observing type $\theta_{1,2}$; in effect, conditional on observing type $\theta_{1,1}$ the player needs only to select a single action U or D. (Thus, we could have written the *ex interim* induced normal form in Figure 6.11 as a table with four columns but only two rows.)

Although we can use this matrix to find best responses for player 1, it turns out to be meaningless to analyze the Nash equilibria in this payoff matrix. This is because these expected payoffs are not common knowledge; if the column player were to condition on his signal, he would arrive at a different set of numbers (though, again, for him best responses would be preserved). Ironically, it is only in the induced normal form, in which the payoffs do not correspond to any *ex interim* assessment of any agent, that the Nash equilibria are meaningful.

expectimax
algorithm

Other computational techniques exist for Bayesian games that also have temporal structure—that is, for Bayesian games written using the "extensive form with chance moves" formulation, for which the game tree is smaller than its induced normal form. First, there is an algorithm for Bayesian games of perfect information that generalizes backward induction (defined in Section 5.1.4), is called *expectimax*. Intuitively, this algorithm is very much like the standard backward induction algorithm given in Figure 5.6. Like that algorithm, expectimax recursively explores a game tree, labeling each non-leaf node h with a payoff vector by examining the labels of each of h's child nodes—the actual payoffs when these child nodes are leaf nodes—and keeping the payoff vector in which the agent who moves at h achieves maximal utility. The new wrinkle is that chance nodes must also receive labels. Expectimax labels a chance node h with a weighted sum of the labels of its child nodes, where the weights are the probabilities that each child node will be selected. The same idea of labeling chance nodes with the expected value of the next node's label can also be applied to extend the minimax algorithm (from which expectimax gets its name) and alpha-beta pruning (see Figure 5.7) in order to solve zero-sum games. This is a popular algorithmic framework for building computer players for perfect-information games of chance such as Backgammon.

There are also efficient computational techniques for computing sample equilibria of imperfect-information extensive-form games with chance nodes. In particular, all the computational results for computing with the sequence form that we discussed in Section 5.2.3 still hold when chance nodes are added. Intuitively, the only change we need to make is to replace our definition of the payoff function (Definition 5.2.7) with an expected payoff that supplies the expected value, ranging over Nature's possible actions, of the payoff the agent would achieve by following a given sequence. This means that we can sometimes achieve a substantial computational savings by working with the extensive-form representation of a Bayesian game, rather than considering the game's induced normal form.

6.3.4 Ex post *equilibrium*

Finally, working with *ex post* utilities allows us to define an equilibrium concept that is stronger than the Bayes–Nash equilibrium.

ex post
equilibrium

Definition 6.3.8 (*Ex post* equilibrium) *An* ex post *equilibrium is a mixed-strategy profile s that satisfies $\forall \theta$, $\forall i$, $s_i \in \arg\max_{s_i' \in S_i} EU_i(s_i', s_{-i}, \theta)$.*

Observe that this definition does not presume that each agent actually *does* know the others' types; instead, it says that no agent would ever want to deviate from his mixed strategy *even if* he knew the complete type vector θ. This form of equilibrium is appealing because it is unaffected by perturbations in the type distribution $p(\theta)$. Said another way, an *ex post* equilibrium does not ever require any agent to believe that the others have accurate beliefs about his own type distribution. (Note that a standard Bayes–Nash equilibrium *can* imply this requirement.) The *ex post* equilibrium is thus similar in flavor to equilibria in dominant strategies, which do not require agents to believe that other agents act rationally.

Indeed, many dominant strategy equilibria are also *ex post* equilibria, making it easy to believe that this relationship always holds. In fact, it does not, as the following example shows. Consider a two-player Bayesian game where each agent has two actions and two corresponding types ($\forall_{i \in N}$, $A_i = \Theta_i = \{H, L\}$) distributed uniformly ($\forall_{i \in N}$, $P(\theta_i = H) = 0.5$), and with the same utility function for each agent i:

$$u_i(a, \theta) = \begin{cases} 10 & a_i = \theta_{-i} = \theta_i; \\ 2 & a_i = \theta_{-i} \neq \theta_i; \\ 0 & \text{otherwise.} \end{cases}$$

In this game, each agent has a dominant strategy of choosing the action that corresponds to his type, $a_i = \theta_i$. An equilibrium in these dominant strategies is not *ex post* because if either agent knew the other's type, he would prefer to deviate to playing the strategy that corresponds to the other agent's type, $a_i = \theta_{-i}$.

Unfortunately, another sense in which *ex post* equilibria are in fact similar to equilibria in dominant strategies is that neither kind of equilibrium is guaranteed to exist.

Finally, we note that the term "*ex post* equilibrium" has been used in several different ways in the literature. One alternate usage requires that each agent's strategy constitute a best response not only to every possible *type* of the others, but also to every *pure strategy profile* that can be realized given the others' mixed strategies. (Indeed, this solution concept has also been applied in settings where there is no uncertainty about agents' types.) A third usage even more stringently requires that no agent ever play a mixed strategy. Both of these definitions can be useful, e.g., in the context of mechanism design (see Chapter 10). However, the advantage of Definition 6.3.8 is that of the three, it describes the most general prior-free equilibrium concept for Bayesian games.

6.4 Congestion games

Congestion games are a restricted class of games that are useful for modeling some important real-world settings and that also have attractive theoretical properties. Intuitively, they simplify the representation of a game by imposing constraints on the effects that a single agent's action can have on other agents' utilities.

6.4.1 *Definition*

Intuitively, in a congestion game each player chooses some subset from a set of resources, and the cost of each resource depends on the number of other agents who select it. Formally, a congestion game is single-shot n-player game, defined as follows.

congestion game **Definition 6.4.1 (Congestion game)** *A congestion game is a tuple (N, R, A, c), where*

- *N is a set of n agents;*
- *R is a set of r resources;*

- $A = A_1 \times \cdots \times A_n$, where $A_i \subseteq 2^R \setminus \{\emptyset\}$ *is the set of* actions *for agent i;* *and*
- $c = (c_1, \ldots, c_r)$, *where* $c_k : \mathbb{N} \mapsto \mathbb{R}$ *is a* cost function *for resource* $k \in R$.

The players' utility functions are defined in terms of the cost functions c_k. Define $\# : R \times A \mapsto \mathbb{N}$ as a function that counts the number of players who took any action that involves resource r under action profile a. For each resource k, define a cost function $c_k : \mathbb{N} \mapsto \mathbb{R}$. Now we are ready to state the utility function,[7] which is the same for all players. Given a pure-strategy profile $a = (a_i, a_{-i})$,

$$u_i(a) = - \sum_{r \in R | r \in a_i} c_r(\#(r, a)).$$

Observe that while the agents can have different actions available to them, they all have the same utility function. Furthermore, observe that congestion games have an *anonymity* property: players care about *how may* others use a given resource, but they do not care about *which* others do so.

anonymity

One motivating example for this formulation is a computer network in which several users want to send a message from one node to another at approximately the same time. Each link connecting two nodes is a resource, and an action for a user is to select a path of links connecting their source and target node. The cost function for each resource expresses the latency on each link as a function of its congestion.

As the name suggests, a congestion game typically features functions $c_k(\cdot)$ that are increasing in the number of people who choose that resource, as would be the case in the network example. However, congestion games can just as easily handle positive externalities (or even cost functions that oscillate). A popular formulation that captures both types of externalities is the *Santa Fe (or, El Farol) Bar problem*, in which each of a set of people independently selects whether or not to go to the bar. The utility of attending increases with the number of other people who select the same night, up to the capacity of the bar. Beyond this point, utility decreases because the bar gets too crowded. Deciding not to attend yields a baseline utility that does not depend on the actions of the participants.[8]

Santa Fe Bar problem

6.4.2 *Computing equilibria*

Congestion games are interesting for reasons beyond the fact that they can compactly represent realistic n-player games like the examples given earlier. One particular example is the following result.

Theorem 6.4.2 *Every congestion game has a pure-strategy Nash equilibrium.*

We defer the proof for the moment, though we note that the property is important because mixed-strategy equilibria are open to criticisms that they are

---

7. This utility function is negated because the cost functions are historically understood as penalties that the agents want to minimize. We note that the c_r functions are also permitted to be negative.
8. Incidentally, this problem is typically studied in a repeated game context, in which (possibly boundedly rational) agents must learn to play an equilibrium. It is famous partly for not having a symmetric pure-strategy equilibrium, and has been generalized with the concept of *minority games*, in which agents get the highest payoff for choosing a minority action.

less likely than pure-strategy equilibria to arise in practice. Furthermore, this theorem tells us that if we want to compute a sample Nash equilibrium of a congestion game, we can look for a pure-strategy equilibrium. Consider the *myopic best-response* process, described in Figure 6.12.

myopic best-response

function MYOPICBESTRESPONSE (game G, action profile a) **returns** a
while *there exists an agent i for whom a_i is not a best response to a_{-i}* **do**
$\quad\quad a_i' \leftarrow$ some best response by i to a_{-i}
$\quad\quad a \leftarrow (a_i', a_{-i})$
return a

Figure 6.12 Myopic best response algorithm. It is invoked starting with an arbitrary (e.g., random) action profile a.

By the definition of equilibrium, MYOPICBESTRESPONSE returns a pure-strategy Nash equilibrium if it terminates. Because this procedure is so simple, it is an appealing way to search for an equilibrium. However, in general games MYOPICBESTRESPONSE can get caught in a cycle, even when a pure-strategy Nash equilibrium exists. For example, consider the game in Figure 6.13.

	L	C	R
U	$-1, 1$	$1, -1$	$-2, -2$
M	$1, -1$	$-1, 1$	$-2, -2$
D	$-2, -2$	$-2, -2$	$2, 2$

Figure 6.13 A game on which MYOPICBESTRESPONSE can fail to terminate.

This game has one pure-strategy Nash equilibrium, (D, R). However, if we run MYOPICBESTRESPONSE with $a = (L, U)$ the procedure will cycle forever. (Do you see why?) This suggests that MYOPICBESTRESPONSE may be too simplistic to be useful in practice. Interestingly, it *is* useful for congestion games.

Theorem 6.4.3 *The* MYOPICBESTRESPONSE *procedure is guaranteed to find a pure-strategy Nash equilibrium of a congestion game.*

6.4.3 Potential games

To prove the two theorems from the previous section, it is useful to introduce the concept of potential games.[9]

---

9. The potential games we discuss here are more formally known as *exact potential games*, though it is correct to shorten their name to the term *potential games*. There are other variants with somewhat different

potential game

Definition 6.4.4 (Potential game) *A game* $G = (N, A, u)$ *is a* potential game *if there exists a function* $P : A \mapsto \mathbb{R}$ *such that, for all* $i \in N$, *all* $a_{-i} \in A_{-i}$ *and* $a_i, a_i' \in A_i$, $u_i(a_i, a_{-i}) - u_i(a_i', a_{-i}) = P(a_i, a_{-i}) - P(a_i', a_{-i})$.

It is easy to prove the following property.

Theorem 6.4.5 *Every (finite) potential game has a pure-strategy Nash equilibrium.*

Proof. Let $a^* = \arg\max_{a \in A} P(a)$. Clearly for any other action profile a', $P(a^*) \geq P(a')$. Thus by the definition of a potential function, for any agent i who can change the action profile from a^* to a' by changing his own action, $u_i(a^*) \geq u_i(a')$. $\blacksquare$

Let $\mathbb{I}_{r \in a_i}$ be an indicator function that returns 1 if $r \in a_i$ for a given action a_i, and 0 otherwise. We also overload the notation # to give the expression $\#(r, a_{-i})$ its obvious meaning. Now we can show the following result.

Theorem 6.4.6 *Every congestion game is a potential game.*

Proof. We demonstrate that every congestion game has the potential function $P(a) = \sum_{r \in R} \sum_{j=1}^{\#(r,a)} c_r(j)$. To accomplish this, we must show that for any agent i and any action profiles (a_i, a_{-i}) and (a_i', a_{-i}), the difference between the potential function evaluated at these action profiles is the same as i's difference in utility.

$$P(a_i, a_{-i}) - P(a_i', a_{-i})$$

$$= \left[\sum_{r \in R} \sum_{j=1}^{\#(r,(a_i, a_{-i}))} c_r(j) \right] - \left[\sum_{r \in R} \sum_{j=1}^{\#(r,(a_i', a_{-i}))} c_r(j) \right]$$

$$= \left[\sum_{r \in R} \left(\left(\sum_{j=1}^{\#(r,(a_{-i}))} c_r(j) \right) + \mathbb{I}_{r \in a_i} c_r(j+1) \right) \right]$$

$$\quad - \left[\sum_{r \in R} \left(\left(\sum_{j=1}^{\#(r,(a_{-i}))} c_r(j) \right) + \mathbb{I}_{r \in a_i'} c_r(j+1) \right) \right]$$

$$= \left[\sum_{r \in R} \mathbb{I}_{r \in a_i} c_r(\#(r, a_{-i}) + 1) \right] - \left[\sum_{r \in R} \mathbb{I}_{r \in a_i'} c_r(\#(r, a_{-i}) + 1) \right]$$

$$= \left[\sum_{r \in R | r \in a_i} c_r(\#(r, (a_i, a_{-i}))) \right] - \left[\sum_{r \in R | r \in a_i'} c_r(\#(r, (a_i', a_{-i}))) \right]$$

$$= u_i(a_i, a_{-i}) - u_i(a_i', a_{-i}) \qquad \blacksquare$$

---

properties, such as *weighted potential games* and *ordinal potential games*. These variants differ in the expression that appears in Definition 6.4.4; for example, ordinal potential games generalize potential games with the condition $u_i(a_i, a_{-i}) - u_i(a_i', a_{-i}) > 0$ iff $P(a_i, a_{-i}) - P(a_i', a_{-i}) > 0$. More can be learned about these distinctions by consulting the reference given in the chapter notes; most importantly, potential games of all these variants are still guaranteed to have pure-strategy Nash equilibria.

Now that we have this result, the proof to Theorem 6.4.2 (stating that every congestion game has a pure-strategy Nash equilibrium) follows directly from Theorems 6.4.5 and 6.4.6. Furthermore, though we do not state this result formally, it turns out that the mapping given in Theorem 6.4.6 also holds in the other direction: every potential game can be represented as a congestion game.

Potential games (along with their equivalence to congestion games) also make it easy to prove Theorem 6.4.3 (stating that MYOPICBESTRESPONSE will always find a pure-strategy Nash equilibrium), which we had previously deferred.

> **Proof of Theorem 6.4.3.** By Theorem 6.4.6 it is sufficient to show that MYOPICBESTRESPONSE finds a pure-strategy Nash equilibrium of any potential game. With every step of the while loop, $P(a)$ strictly increases, because by construction $u_i(a_i', a_{-i}) > u_i(a_i, a_{-i})$, and thus by the definition of a potential function $P(a_i', a_{-i}) > P(a_i, a_{-i})$. Since there are only a finite number of action profiles, the algorithm must terminate. ∎

Thus, when given a congestion game MYOPICBESTRESPONSE will converge regardless of the cost functions (e.g., they do not need to be monotonic), the action profile with which the algorithm is initialized, and which agent we choose as agent i in the while loop (when there is more than one agent who is not playing a best response). Furthermore, we can see from the proof that it is not even necessary that agents *best respond* at every step. The algorithm will still converge to a pure-strategy Nash equilibrium by the same argument as long as agents deviate to a *better* response. On the other hand, it has recently been shown that the problem of finding a pure Nash equilibrium in a congestion game is PLS-complete: as hard to find as any other object whose existence is guaranteed by a potential function argument. Intuitively, this means that our problem is as hard as finding a local minimum in a traveling salesman problem using local search. This cautions us to expect that MYOPICBESTRESPONSE will be inefficient in the worst case.

6.4.4 *Nonatomic congestion games*

A nonatomic congestion game is a congestion game that is played by an uncountably infinite number of players. These games are used to model congestion scenarios in which the number of agents is very large, and each agent's effect on the level of congestion is very small. For example, consider modeling traffic congestion in a freeway system.

nonatomic
congestion
games

Definition 6.4.7 (Nonatomic congestion game) *A nonatomic congestion game is a tuple (N, μ, R, A, ρ, c), where:*

- $N = \{1, \ldots, n\}$ *is a set of* types *of players;*
- $\mu = (\mu_1, \ldots, \mu_n)$; *for each $i \in N$ there is a continuum of players represented by the interval $[0, \mu_i]$;*
- *R is a set of k resources;*
- *$A = A_1 \times \cdots \times A_n$, where $A_i \subseteq 2^R \setminus \{\emptyset\}$ is the set of actions for agents of type i;*

- $\rho = (\rho_1, \ldots, \rho_n)$, where for each $i \in N$, $\rho_i : A_i \times R \mapsto \mathbb{R}_+$ denotes the amount of congestion contributed to a given resource $r \in R$ by players of type i selecting a given action $a_i \in A_i$; and
- $c = (c_1, \ldots, c_k)$, where $c_r : \mathbb{R}_+ \mapsto \mathbb{R}$ is a cost function for resource $r \in R$, and c_r is nonnegative, continuous and nondecreasing.

To simplify notation, assume that $A_1, \ldots, A_n$ are disjoint; denote their union as $\mathcal{A}$. Let $\mathcal{S} = \mathbb{R}_+^{|\mathcal{A}|}$. An *action distribution* $s \in \mathcal{S}$ indicates how many players choose each action; by $s(a_i)$, denote the element of s that corresponds to the measure of the set of players of type i who select action $a_i \in A_i$. An action distribution s must have the properties that all entries are nonnegative real numbers and that $\sum_{a_i \in A_i} s(a_i) = \mu_i$. Note that $\rho_i(a_i, r) = 0$ when $r \notin a_i$. Overloading notation, we write as s_r the amount of congestion induced on resource $r \in R$ by action distribution s:

$$s_r = \sum_{i \in N} \sum_{a_i \in A_i} \rho_i(a_i, r) s(a_i).$$

We can now express the utility function. As in (atomic) congestion games, all agents have the same utility function, and the function depends only on how many agents choose each action rather than on these agents' identities. By $c_{a_i, s}$ we denote the cost, under an action distribution s, to agents of type i who choose action a_i. Then

$$c_{a_i}(s) = \sum_{r \in a_i} \rho(a_i, r) c_r(s_r),$$

and so we have $u_i(a_i, s) = -c_{a_i}(s)$. Finally, we can define the *social cost* of an action profile as the total cost born by all the agents,

$$C(s) = \sum_{i \in N} \sum_{a_i \in A_i} s(a_i) c_{a_i}(s).$$

Despite the fact that we have an uncountably infinite number of agents, we can still define a Nash equilibrium in the usual way.

Definition 6.4.8 (Pure-strategy Nash equilibrium of a nonatomic congestion game) *An action distribution s arises in a pure-strategy equilibrium of a nonatomic congestion game if for each player type $i \in N$ and each pair of actions $a_1, a_2 \in A_i$ with $s(a_1) > 0$, $u_i(a_1, s) \geq u_i(a_2, s)$ (and hence $c_{a_1}(s) \leq c_{a_2}(s)$).*

A couple of warnings are in order. First, the attentive reader will have noticed that we have glossed over the difference between actions and strategies. This is to simplify notation, and because we will only be concerned with pure-strategy equilibria. We do note that results exist concerning mixed-strategy equilibria of nonatomic congestion games; see the references cited at the end of the chapter. Second, we say only that an action distribution *arises in* an equilibrium because an action distribution does not identify the action taken by every individual agent, and hence cannot *constitute* an equilibrium. Nevertheless, from this point on we will ignore these issues.

We can now state some properties of nonatomic congestion games.

Theorem 6.4.9 *Every nonatomic congestion game has a pure-strategy Nash equilibrium.*

Furthermore, limiting ourselves by considering only pure-strategy equilibria is in some sense not restrictive.

Theorem 6.4.10 *All equilibria of a nonatomic congestion game have equal social cost.*

Intuitively, because the players are nonatomic, any mixed-strategy equilibrium corresponds to an "equivalent" pure-strategy equilibrium where the number of agents playing a given action is the expected number under the original equilibrium.

6.4.5 *Selfish routing and the price of anarchy*

selfish routing

Selfish routing is a model of how self-interested agents would route traffic through a congested network. This model was studied as early as 1920—long before game theory developed as a field. Today, we can understand these problems as nonatomic congestion games.

Defining selfish routing

First, let us formally define the problem. Let $G = (V, E)$ be a directed graph having n source–sink pairs $(s_1, t_1), \ldots, (s_n, t_n)$. Some volume of traffic must be routed from each source to each sink. For a given source–sink pair (s_i, t_i) let $\mathcal{P}_i$ denote the set of simple paths from s_i to t_i. We assume that $\mathcal{P} \neq \emptyset$ for all i; it is permitted for there to be multiple "parallel" edges between the same pair of nodes in V, and for paths from $\mathcal{P}_i$ and $\mathcal{P}_j$ $(j \neq i)$ to share edges. Let $\mu \in \mathbb{R}^n_+$ denote a vector of *traffic rates*; μ_i denotes the amount of traffic that must be routed from s_i to t_i. Finally, every edge $e \in E$ is associated with a cost function $c_e : \mathbb{R}_+ \mapsto \mathbb{R}$ (think of it an amount of delay) that can depend on the amount of traffic carried by the edge. The problem in selfish routing is to determine how the given traffic rates will lead traffic to flow along each edge, assuming that agents are selfish and will direct their traffic to minimize the sum of their own costs.

Selfish routing problems can be encoded as nonatomic congestion games as follows:

- N is the set of source–sink pairs;
- μ is the set of traffic rates;
- R is the set of edges E;
- A_i is the set of paths $\mathcal{P}_i$ from s_i to t_i;
- ρ_i is always 1; and
- c_r is the edge cost function c_e.

The price of anarchy

From the above reduction to nonatomic congestion games and from Theorems 6.4.9 and 6.4.10 we can conclude that every selfish routing problem has at least one pure-strategy Nash equilibrium,[10] and that all of a selfish routing problem's equilibria have equal social cost. These properties allow us to ask an interesting question: how similar is the optimal social cost to the social cost under an equilibrium action distribution?

price of anarchy

Definition 6.4.11 (Price of anarchy) *The* price of anarchy *of a nonatomic congestion game* (N, μ, R, A, ρ, c) *having equilibrium s and social cost minimizing action distribution* s^* *is defined as* $\frac{C(s)}{C(s^*)}$ *unless* $C(s^*) = 0$, *in which case the price of anarchy is defined to be* 1.

Intuitively, the price of anarchy is the proportion of additional social cost that is incurred because of agents' self-interested behavior. When this ratio is close to 1 for a selfish routing problem, one can conclude that the agents are routing traffic about as well as possible, given the traffic rates and network structure. When this ratio is large, however, the agents' selfish behavior is causing significantly suboptimal network performance. In this latter case one might want to seek ways of changing either the network or the agents' behavior in order to reduce the social cost.

To gain a better understanding of the price of anarchy, and to lay the groundwork for some theoretical results, consider the examples in Figure 6.14.

Figure 6.14 Pigou's example: a selfish routing problem with an interesting price of anarchy. Left: linear version; right: nonlinear version.

In this example there is only one type of agent ($n = 1$) and the rate of traffic is 1 ($\mu_1 = 1$). There are two paths from s to t, one of which is relatively slow but immune to congestion, and the other of which has congestion-dependent cost.

Consider first the linear version of the problem given in Figure 6.14 (left). It is not hard to see that the Nash equilibrium is for all agents to choose the lower edge—indeed, this is a Nash equilibrium in dominant strategies. The social cost of this Nash equilibrium is 1. Consider what would happen if we required half of the agents to choose the upper edge, and the other half of the agents to choose the lower edge. In this case the social cost would be 3/4, because half the agents would continue to pay a cost of 1, while half the agents would now pay a cost of only 1/2. It is easy to show that this is the smallest social cost that can be achieved in this example, meaning that the price of anarchy here is 4/3.

Now consider the nonlinear problem given in Figure 6.14 (right), where p is some large value. Again in the Nash equilibrium all agents will choose the lower edge, and again the social cost of this equilibrium is 1. Social cost is minimized when the marginal costs of the two edges are equalized; this occurs when a $(p + 1)^{-1/p}$ fraction of the agents choose the lower edge. In this case the social cost is $1 - p \cdot (p + 1)^{-(p+1)/p}$, which approaches 0 as $p \to \infty$. Thus we can

---

10. In the selfish routing literature these equilibria are known as Wardrop equilibria, after the author who first proposed their use. For consistency we avoid that term here.

see that the price of anarchy tends to infinity in the nonlinear version of Pigou's example as p grows.

Bounding the price of anarchy

These examples illustrate that the price of anarchy is unbounded for unrestricted cost functions. On the other hand, it turns out to be possible to offer bounds in the case where cost functions are restricted to a particular set $\mathcal{C}$. First, we must define the so-called Pigou bound:

$$\alpha(\mathcal{C}) = \sup_{c \in \mathcal{C}} \sup_{x, \mu \geq 0} \frac{r \cdot c(r)}{x \cdot c(x) + (r - x)c(r)}.$$

When $\alpha(\mathcal{C})$ evaluates to $\frac{0}{0}$, we define it to be 1. We can now state a surprisingly strong result.

Theorem 6.4.12 *The price of anarchy of a selfish routing problem whose cost functions are taken from the set $\mathcal{C}$ is never more than $\alpha(\mathcal{C})$.*

Observe that Theorem 6.4.12 makes a very broad statement, bounding a selfish routing problem's price of anarchy regardless of network structure and for any given family of cost functions. Because α appears difficult to evaluate, one might find it hard to get excited about this result. However, α can be evaluated for a variety of interesting sets of cost functions. For example, when $\mathcal{C}$ is the set of linear functions $ax + b$ with $a, b \geq 0$, $\alpha(\mathcal{C}) = 4/3$. Indeed, $\alpha(\mathcal{C})$ takes the same value when $\mathcal{C}$ is the set of all convex functions. This means that the bound from Theorem 6.4.12 is tight for this set of functions: Pigou's linear example from Figure 6.14 (left) uses only convex cost functions and we have already shown that this problem has a price of anarchy of precisely 4/3. The linear version of Pigou's example thus serves as a worst case for the price of anarchy among all selfish routing problems with convex cost functions. Because the price of anarchy is relatively close to 1 for networks with convex edge costs, this result indicates that centralized control of traffic offers limited benefit in this case.

What about other families of cost functions, such as polynomials with non-negative coefficients and bounded degree? It turns out that the Pigou bound is also tight for this family and that the nonlinear variant of Pigou's example offers the worst-possible price of anarchy in this case (where p is the bound on the polynomials' degree). For this family $\alpha(\mathcal{C}) = [1 - p \cdot (p + 1)^{-(p+1)/p}]^{-1}$. To give some examples, this means that the price of anarchy is about 1.6 for $p = 2$, about 2.2 for $p = 4$, about 18 for $p = 100$ and—as it was earlier—unbounded as $p \to \infty$.

Results also exist bounding the price of anarchy for general nonatomic congestion games. It is beyond the scope of this section to state these results formally, but we note that they are qualitatively similar to the results given above. More information can be found in the references cited in the chapter notes.

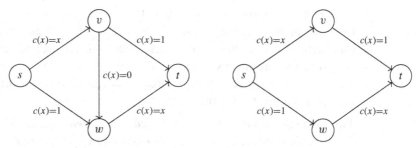

Figure 6.15 Braess' paradox: removing an edge that has zero cost can improve social welfare. Left: original network; right: after edge removal.

Reducing the social cost of selfish routing

Braess' paradox

When the equilibrium social cost is undesirably high, a network operator might want to intervene in some way in order to reduce it. First, we give an example to show that such interventions are possible, known as *Braess' paradox*.

Consider first the example in Figure 6.15 (Left). This selfish routing problem is essentially a more complicated version of the linear version of Pigou's example from Figure 6.14 (left). Again $n = 1$ and $\mu_1 = 1$. Agents have a weakly dominant strategy of choosing the path s-v-w-t, and so in equilibrium all traffic will flow along this path. The social cost in equilibrium is therefore 1. Minimal social cost is achieved by having half of the agents choose the path s-v-t and having the other half of the agents choose the path s-w-t; the social cost in this case is 3/4. Like the linear version of Pigou's example, therefore, the price of anarchy is 4/3.

The interesting thing about this new example is the role played by the edge v-w. One might intuitively believe that zero-cost edges can only help in routing problems, because they provide agents with a costless way of routing traffic from one node to another. At worst, one might reason, such edges would be ignored. However, this intuition is wrong. Consider the network in Figure 6.15 (right). This network was constructed from the network in Figure 6.15 (left) by removing the zero-cost edge v-w. In this modified problem agents no longer have a dominant strategy; the equilibrium is for half of them to choose each path. This is also the optimal action distribution, and hence the price of anarchy in this case is 1. We can now see the (apparent) paradox: removing even a zero-cost edge can transform a selfish routing problem from the very worst (a network having the highest price of anarchy possible given its family of cost functions) to the very best (a network in which selfish agents will choose to route themselves optimally).

A network operator facing a high price of anarchy might therefore want to remove one or more edges in order to improve the network's social cost in equilibrium. Unfortunately, however, the problem of determining which edges to remove is computationally hard.

Theorem 6.4.13 *It is NP-complete to determine whether there exists any set of edges whose removal from a selfish routing problem would reduce the social cost in equilibrium.*

In particular, this result implies that identifying the optimal set of edges to remove from a selfish routing problem in order to minimize the social cost in equilibrium is also NP-complete.

Of course, it is always possible to reduce a network's social cost in equilibrium by reducing all of the edge costs. (This could be done in an electronic network, for example, by installing faster routers.) Interestingly, even in the case where the edge functions are unconstrained and the price of anarchy is therefore unbounded, a relatively modest reduction in edge costs can outperform the imposition of centralized control in the original network.

Theorem 6.4.14 *Let Γ be a selfish routing problem, and let Γ' be identical to Γ except that each edge cost $c_e(x)$ is replaced by $c'_e(x) = c_e(x/2)/2$. The social cost in equilibrium of Γ' is less than or equal to the* optimal *social cost in Γ.*

This result suggests that when it is relatively inexpensive to speed up a network, doing so can have more significant benefits than getting agents to change their behavior.

Finally, we will briefly mention two other methods of reducing social cost in
Stackelberg routing
equilibrium. First, in so-called *Stackelberg routing* a small fraction of agents are routed centrally, and the remaining population of agents is free to choose their own actions. It should already be apparent from the example in Figure 6.14 (right) that such an approach can be very effective in certain networks. Second, taxes can be imposed on certain edges in the graph in order to encourage agents to adopt more socially beneficial behavior. The dominant idea here is to charge agents according to "marginal cost pricing"—each agent pays the amount his presence cost other agents who are using the same edge.[11] Under certain assumptions taxes can be set up in a way that induces optimal action distributions; however, the taxes themselves can be very large. Various papers in the literature elaborate on and refine both of these ideas.

6.5   Computationally motivated compact representations

So far we have examined game representations that are motivated primarily by the goals of capturing relevant details of real-world domains and of showing that all games expressible in the representation share useful theoretical properties. Many of these representations—especially the normal and extensive forms— suffer from the problem that their encodings of interesting games are so large as to be impractical. For example, when you describe to someone the rules of poker, you do not give them a normal or extensive-form description; such a description would fill volumes and be almost unintelligible. Instead, you describe the rules of the game in a very compact form, which is possible because of the inherent structure of the game. In this section we explore some computationally motivated alternative representations that allow certain large games to be compactly described and also make it possible to efficiently find an equilibrium. The first two representations, graphical games and action-graph games, apply to normal-form games, while the following two, multiagent influence diagrams and the GALA language, apply to extensive-form games.

---

11. Here we anticipate the idea of *mechanism design*, introduced in Chapter 10, and especially the VCG mechanism from Section 10.4.

6.5.1 *The expected utility problem*

We begin by defining a problem that is fundamental to the discussion of computationally motivated compact representations.

Definition 6.5.1 (EXPECTEDUTILITY) *Given a game (possibly represented in a compact form), a mixed-strategy profile s, and i ∈ N, the* EXPECTEDUTILITY *problem is to compute $EU_i(s)$, the expected utility of player i under mixed-strategy profile s.*

EXPECTEDUTILITY problem

Our chief interest in this section will be in the computational complexity of the EXPECTEDUTILITY problem for different game representations. When we considered normal-form games, we showed (in Definition 3.2.7) that EXPECTEDUTILITY can be computed as

$$EU_i(s) = \sum_{a \in A} u_i(a) \prod_{j=1}^{n} s_j(a_j). \tag{6.8}$$

If we interpret Equation (6.8) as a simple algorithm, we have a way of solving EXPECTEDUTILITY in time exponential in the number of agents. This algorithm is exponential because, assuming for simplicity that all agents have the same number of actions, the size of A is $|A_i|^n$. However, since the representation size of a normal-form game is itself exponential in the number of agents (it is $O(|A_i|^n)$), the problem can in fact be solved in time linear in the size of the representation. Thus EXPECTEDUTILITY does not appear to be very computationally difficult.

Interestingly though, as game representations become exponentially more compact than the normal form, it grows more challenging to solve the EXPECTEDUTILITY problem efficiently. This is because our simple algorithm given by Equation (6.8) requires time exponential in the size of such more compact representations. The trick with compact representations, therefore, will not be simply finding some way of representing payoffs compactly—indeed, there are any number of schemes from the compression literature that could achieve this goal. Rather, we will want the additional property that the compactness of the representation can be leveraged by an efficient algorithm for computing EXPECTEDUTILITY.

The first challenge is to ensure that the inputs to EXPECTEDUTILITY can be specified compactly.

polynomial type

Definition 6.5.2 (Polynomial type) *A game representation has* polynomial type *if the number of agents n and the sizes of the action sets $|A_i|$ are polynomially bounded in the size of the representation.*

Representations always have polynomial type when their action sets are specified explicitly. However, some representations—such as the extensive form—implicitly specify action spaces that are exponential in the size of the representation and so do not have polynomial type.

When we combine the polynomial type requirement with a further requirement about EXPECTEDUTILITY being efficiently computable, we obtain the following theorem.

Theorem 6.5.3 *If a game representation satisfies the following properties:*

1. *the representation has polynomial type; and*
2. EXPECTEDUTILITY *can be computed using an arithmetic binary circuit consisting of a polynomial number of nodes, where each node evaluates to a constant value or performs addition, subtraction or multiplication on its inputs;*

then the problem of finding a Nash equilibrium in this representation can be reduced to the problem of finding a Nash equilibrium in a two-player normal-form game that is only polynomially larger.

We know from Theorem 4.2.1 in Section 4.2 that the problem of finding a Nash equilibrium in a two-player normal-form game is PPAD-complete. Therefore this theorem implies that if the above condition holds, the problem of finding a Nash equilibrium for a compact game representation is in PPAD. This should be understood as a positive result: if a game in its compact representation is exponentially smaller than its induced normal form, and if computing an equilibrium for this representation belongs to the same complexity class as computing an equilibrium of a normal-form game, then equilibria can be computed exponentially more quickly using the compact representation.

Observe that the second condition in Theorem 6.5.3 implies that the EXPECTEDUTILITY algorithm takes polynomial time; however, not every polynomial-time algorithm will satisfy this condition. Congestion games are an example of games that do meet the conditions of Theorem 6.5.3. We will see two more such representations in the next sections.

What about extensive-form games, which do not have polynomial type—might it be harder to compute their Nash equilibria? Luckily we can use behavioral strategies, which can be represented linearly in the size of the game tree. Then we obtain the following result.

Theorem 6.5.4 *The problem of computing a Nash equilibrium in behavioral strategies in an extensive-form game can be polynomially reduced to finding a Nash equilibrium in a two-player normal-form game.*

This shows that the speedups we achieved by using the sequence form in Section 5.2.3 were not achieved simply because of inefficiency in our algorithms for normal-form games. Instead, there is a fundamental computational benefit to working with extensive-form games, at least when we restrict ourselves to behavioral strategies.

Fast algorithms for solving EXPECTEDUTILITY are useful for more than just demonstrating the worst-case complexity of finding a Nash equilibrium for a game representation. EXPECTEDUTILITY is also a bottleneck step in several practical algorithms for computing Nash equilibria, such as the Govindan–Wilson algorithm or simplicial subdivision methods (see Section 4.3). Plugging a fast method for solving EXPECTEDUTILITY into one of these algorithms offers a simple way of more quickly computing a Nash equilibrium of a compactly represented game.

The complexity of the EXPECTEDUTILITY problem is also relevant to the computation of solution concepts other than the Nash equilibrium.

Theorem 6.5.5 *If a game representation has polynomial type and has a polynomial algorithm for computing* EXPECTEDUTILITY, *then a correlated equilibrium can be computed in polynomial time.*

The attentive reader may recall that we have already showed (in Section 4.6) that correlated equilibria can be identified in polynomial time by solving a linear program (Equations (4.52)–(4.54)). Thus, Theorem 6.5.5 may not seem very interesting. The catch, as with expected utility, is that while this LP has size polynomial in size of the normal form, its size would be exponential in the size of many compact representations. Specifically, there is one variable in the linear program for each action profile, and so overall the linear program has size exponential in any representation for which the simple EXPECTEDUTILITY algorithm discussed earlier is inadequate. Indeed, in these cases even *representing* a correlated equilibrium using these probabilities of action profiles would be exponential. Theorem 6.5.5 is thus a much deeper result than it may first seem. Its proof begins by showing that there exists a correlated equilibrium of every compactly represented game that can be written as the mixture of a polynomial number of *product distributions*, where a product distribution is a joint probability distribution over action profiles arising from each player independently randomizing over his actions (i.e., adopting a mixed-strategy profile). Since the theorem requires that the game representation has polynomial type, each of these product distributions can be compactly represented. Thus a polynomial mixture of product distributions can also be represented polynomially. The rest of the proof appeals to linear programming duality and to properties of the ellipsoid algorithm.

6.5.2 *Graphical games*

Graphical games are a compact representation of normal-form games that use graphical models to capture the *payoff independence* structure of the game. Intuitively, a player's payoff matrix can be written compactly if his payoff is affected only by a subset of the other players.

Let us begin with an example, which we call the Road game. Consider n agents, each of whom has purchased a piece of land alongside a road. Each agent has to decide what to build on his land. His payoff depends on what he builds himself, what is built on the land to either side of his own, and what is built across the road. Intuitively, the payoff relationships in this situation can be understood using the graph shown in Figure 6.16, where each node represents an agent.

Now let us define the representation formally. First, we define a neighborhood relation on a graph: the set of nodes connected to a given node, plus the node itself.

Definition 6.5.6 (Neighborhood relation) *For a graph defined on a set of nodes* neighborhood relation *N and edges E, for every $i \in N$ define the* neighborhood relation *$\nu : N \mapsto 2^N$ as $\nu(i) = \{i\} \cup \{j | (j, i) \in E\}$.*

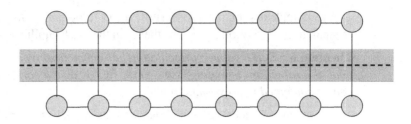

Figure 6.16 Graphical game representation of the Road game.

Now we can define the graphical game representation.

graphical game   **Definition 6.5.7 (Graphical game)** *A graphical game is a tuple (N, E, A, u), where:*

- *N is a set of n vertices, representing agents;*
- *E is a set of undirected edges connecting the nodes N;*
- *$A = A_1 \times \cdots \times A_n$, where A_i is the set of actions available to agent i; and*
- *$u = (u_1, \ldots, u_n)$, $u_i : A^{(i)} \mapsto \mathbb{R}$, where $A^{(i)} = \prod_{j \in v(i)} A_j$.*

An edge between two vertices in the graph can be interpreted as meaning that the two agents are able to affect each other's payoffs. In other words, whenever two nodes i and j are *not* connected in the graph, agent i must always receive the same payoff under any action profiles (a_j, a_{-j}) and (a'_j, a_{-j}), $a_j, a'_j \in A_j$. Graphical games can represent any game, but of course they are not always compact. The space complexity of the representation is exponential in the size of the largest $v(i)$. In the example above the size of the largest $v(i)$ is 4, and this is independent of the total number of agents. As a result, the graphical game representation of the example requires space polynomial in n, while a normal-form representation would require space exponential in n.

The following is sufficient to show that the properties we discussed above in Section 6.5.1 hold for graphical games.

Lemma 6.5.8 *The* EXPECTEDUTILITY *problem can be computed in polynomial time for graphical games, and such an algorithm can be translated to an arithmetic circuit as required by Theorem 6.5.3.*

The way that graphical games capture payoff independence in games is similar to the way that Bayesian networks and Markov random fields capture conditional independence in multivariate probability distributions. It should therefore be unsurprising that many computations on graphical games can be performed efficiently using algorithms similar to those proposed in the graphical models literature. For example, when the graph (N, E) defines a tree, a message-passing algorithm called NASHPROP can compute an ϵ-Nash equilibrium in time polynomial in $1/\epsilon$ and the size of the representation. NASHPROP consists of two phases: a "downstream" pass in which messages are passed from the leaves to the root and then an "upstream" pass in which messages are passed from the root to the leaves. When the graph is a path, a similar algorithm can find an exact equilibrium in polynomial time.

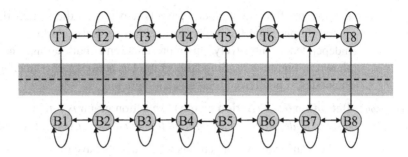

Figure 6.17 Modified Road game.

We may also be interested in finding pure-strategy Nash equilibria. Determining whether a pure-strategy equilibrium exists in a graphical game is NP-complete. However, the problem can be formulated as a constraint satisfaction problem (or alternatively as a Markov random field) and solved using standard algorithms. In particular, when the graph has constant *treewidth*,[12] the problem can be solved in polynomial time.

Graphical games have also been useful as a theoretical tool. For example, they are instrumental in the proof of Theorem 4.2.1, which showed that finding a sample Nash equilibrium of a normal-form game is PPAD-complete. Intuitively, graphical games are important to this proof because such games can be constructed to simulate arithmetic circuits in their equilibria.

6.5.3 *Action-graph games*

Consider a scenario similar to the Road game given in Section 6.5.2, the same or with one major difference: instead of deciding what to build, here agents need to decide *where* to build. Suppose each of the n agents is interested in opening a business (say a coffee shop), and can choose to locate in any block along either side of a road. Multiple agents can choose the same block. Agent i's payoff depends on the number of agents who chose the same block as he did, as well as the numbers of agents who chose each of the adjacent blocks of land. This game has an obvious graphical structure, which is illustrated in Figure 6.17. Here nodes correspond to actions, and each edge indicates that an agent who takes one action affects the payoffs of other agents who take some second action.

Notice that any pair of agents can potentially affect each other's payoffs by choosing the same or adjacent locations. This means that the graphical game representation of this game is a clique, and the space complexity of this representation is the same as that of the normal form (exponential in n). The problem is that graphical games are only compact for games with *strict payoff independencies*: that is, where there exist pairs of players who can never (directly) affect each other. This game exhibits *context-specific independence* instead: whether two agents

---

12. A graph's *treewidth* is a measure of how similar the graph is to a tree. It is defined using the *tree decomposition* of the graph. Many NP-complete problems on graphs can be solved efficiently when a graph has small treewidth.

are able to affect each other's payoffs depends on the actions they choose. The action-graph game (AGG) representation exploits this kind of context-specific independence. Intuitively, this representation is built around the graph structure shown in Figure 6.17. Since this graph has actions rather than agents serving as the nodes, it is referred to as an action graph.

action graph **Definition 6.5.9 (Action graph)** *An* action graph *is a tuple* $(\mathcal{A}, E)$*, where* $\mathcal{A}$ *is a set of nodes corresponding to actions and E is a set of directed edges.*

We want to allow for settings where agents have different actions available to them, and hence where an agent's action set is not identical to $\mathcal{A}$. (For example, no two agents could be able to take the "same" action, or every agent could have the same action set as in Figure 6.17.) We thus define as usual a set of action profiles $A = A_1 \times \cdots \times A_n$, and then let $\mathcal{A} = \bigcup_{i \in N} A_i$. If two actions by different agents have the same name, they will collapse to the same element of $\mathcal{A}$; otherwise they will correspond to two different elements of $\mathcal{A}$.

Given an action graph and a set of agents, we can further define a *configuration*, which is a possible arrangement of agents over nodes in an action graph.

Definition 6.5.10 (Configuration) *Given an action graph $(\mathcal{A}, E)$ and a set of*
configuration (of *action profiles A, a* configuration *c is a tuple of $|\mathcal{A}|$ nonnegative integers, where*
an action-graph *the kth element c_k is interpreted as the number of agents who chose the kth action*
game) *$\alpha_k \in \mathcal{A}$, and where there exists some $a \in A$ that would give rise to c. Denote the set of all configurations as C.*

Observe that multiple action profiles might give rise to the same configuration, because configurations simply count the number of agents who took each action without worrying about which agent took which action. For example, in the Road game all action profiles in which exactly half of the agents take action $T1$ and exactly half the agents take action $B8$ give rise to the same configuration.
anonymity Intuitively, configurations will allow AGGs to compactly represent *anonymity* structure: cases where an agent's payoffs depend on the aggregate behavior of other agents, but not on which particular agents take which actions. Recall that we saw such structure in congestion games (Section 6.4).

Intuitively, we will use the edges of the action graph to denote context-specific independence relations in the game. Just as we did with graphical games, we will define a utility function that depends on the actions taken in some local neighborhood. As it was for graphical games, the neighborhood ν will be defined by the edges E; indeed, we will use exactly the same definition (Definition 6.5.6). In action graph games the idea will be that the payoff of a player playing an action $\alpha \in \mathcal{A}$ only depends on the configuration over the neighbors of α.[13] We must therefore define notation for such a configuration over a neighborhood. Let $C^{(\alpha)}$ denote the set of all restrictions of configurations to the elements corresponding to the neighborhood of $\alpha \in \mathcal{A}$. (That is, each $c \in C^{(\alpha)}$ is a tuple of length $|\nu(\alpha)|$.) Then u_α, the utility for *any* agent who takes action $\alpha \in \mathcal{A}$, is a mapping from $C^{(\alpha)}$ to the real numbers.

---

13. We use the notation α rather than a to denote an element of $\mathcal{A}$ in order to emphasize that we speak about a single action rather than an action profile.

Summing up, we can state the formal definition of action-graph games as follows.

Definition 6.5.11 *An* action-graph game (AGG) *is a tuple* $(N, A, (\mathcal{A}, E), u)$, *where:*

- *N is the set of agents;*
- *$A = A_1 \times \cdots \times A_n$, where A_i is the set of actions available to agent i;*
- *$(\mathcal{A}, E)$ is an action graph, where $\mathcal{A} = \bigcup_{i \in N} A_i$ is the set of distinct actions; and*
- *$u = \{u_\alpha | \alpha \in \mathcal{A}\}$, $u_\alpha : C^{(\alpha)} \mapsto \mathbb{R}$.*

Since each utility function is a mapping only from the possible configurations over the neighborhood of a given action, the utility function can be represented concisely. In the Road game, since each node has at most four incoming edges, we only need to store $O(n^4)$ numbers for each node, and $O(|\mathcal{A}|n^4)$ numbers for the entire game. In general, when the in-degree of the action graph is bounded by a constant, the space complexity of the AGG representation is polynomial in n.

Like graphical games, AGGs are fully expressive. Arbitrary normal-form games can be represented as AGGs with nonoverlapping action sets. Graphical games can be encoded in the same way, but with a sparser edge structure. Indeed, the AGG encoding of a graphical game is just as compact as the original graphical game.

Although it is somewhat involved to show why this is true, AGGs have the theoretical properties we have come to expect from a compact representation.

Theorem 6.5.12 *Given an AGG,* EXPECTEDUTILITY *can be computed in time polynomial in the size of the AGG representation by an algorithm represented as an arithmetic circuit as required by Theorem 6.5.3. In particular, if the in-degree of the action graph is bounded by a constant, the time complexity is polynomial in n.*

The AGG representation can be extended to include *function nodes*, which are special nodes in the action graph that do not correspond to actions. For each function node p, c_p is defined as a deterministic function of the configuration of its neighbors $\nu(p)$. Function nodes can be used to represent a utility function's intermediate parameters, allowing the compact representation of games with additional forms of independence structure. Computationally, when a game with function nodes has the property that each player affects the configuration c independently, EXPECTEDUTILITY can still be computed in polynomial time. AGGs can also be extended to exploit additivity in players' utility functions. Given both of these extensions, AGGs are able to compactly represent a broad array of realistic games, including congestion games.

6.5.4 *Multiagent influence diagrams*

Multiagent influence diagrams (MAIDs) are a generalization of *influence diagrams* (IDs), a compact representation for decision-theoretic reasoning in the single-agent case. Intuitively, MAIDs can be seen as a combination of graphical

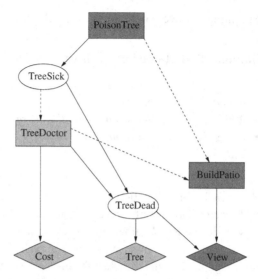

Figure 6.18 A multiagent influence diagram. Nodes for Alice are in dark gray, while Bob's are in light gray.

games and extensive-form games with chance moves (see Section 6.3). Not all variables (moves by nature) and action nodes depend on all other variables and action nodes, and only the dependencies need to be represented and reasoned about.

We will give a brief overview of MAIDs using the following example. Alice is considering building a patio behind her house, and the patio would be more valuable to her if she could get a clear view of the ocean. Unfortunately, there is a tree in her neighbor Bob's yard that blocks her view. Being somewhat unscrupulous, Alice considers poisoning Bob's tree, which might cause it to become sick. Bob cannot tell whether Alice has poisoned his tree, but he can tell if the tree is getting sick, and he has the option of calling in a tree doctor (at some cost). The attention of a tree doctor reduces the chance that the tree will die during the coming winter. Meanwhile, Alice must make a decision about building her patio before the weather gets too cold. When she makes this decision, she knows whether a tree doctor has come, but she cannot observe the health of the tree directly. A MAID for this scenario is shown in Figure 6.18.

Chance variables are represented as ovals, decision variables as rectangles, and utility variables as diamonds. Each variable has a set of parents, which may be chance variables or decision variables. Each chance node is characterized by a conditional probability distribution, which defines a distribution over the variable's domain for each possible instantiation of its parents. Similarly, each utility node records the conditional value for the corresponding agent. If multiple utility nodes exist for the same agent, as they do for in this example for Bob, the total utility is simply the sum of the values from each node. Decision variables differ in that their parents (connected by dotted arrows) are the variables that an agent observes when making his decision. This allows us to represent the information sets in a compact way.

For each decision node, the corresponding agent constructs a decision rule, which is a distribution over the domain of the decision variable for each possible instantiation of this node's parents. A strategy for an agent consists of a decision rule for each of his decision nodes. Since a decision node acts as a chance node once its decision rule is set, we can calculate the expected utility of an agent given a strategy profile. As you would expect, a strategy profile is a Nash equilibrium in a MAID if no agent can improve its expected utility by switching to a different set of decision rules.

This example shows several of the advantages of the MAID representation over the equivalent extensive-form game representation. Since there are a total of five chance and decision nodes and all variables are binary, the game tree would have 32 leaves, each with a value for both agents. In the MAID, however, we only need four values for each agent to fill tables for the utility nodes. Similarly, redundant chance nodes of the game tree are replaced by small conditional probability tables. In general, the space savings of MAIDs can be exponential (although it is possible that this relationship is reversed if the game tree is sufficiently asymmetric).

The most important advantage of MAIDs is that they allow more efficient algorithms for computing equilibria, as we will informally show for the example. strategic The efficiency of the algorithm comes from exploiting the property of *strategic* relevance *relevance* in a way that is related to backward induction in perfect-information games. A decision node D_2 is strategically relevant to another decision node D_1 if, to optimize the rule at D_1, the agent needs to consider the rule at D_2. We omit a formal definition of strategic relevance, but point out that it can be computed in polynomial time.

No decision nodes are strategically relevant to *BuildPatio* for Alice, because she observes both of the decision nodes (*PoisonTree* and *TreeDoctor*) that could affect her utility before she has to make this decision. Thus, when finding an equilibrium, we can optimize this decision rule independently of the others and effectively convert it into a chance node. Next, we observe that *PoisonTree* is not strategically relevant to *TreeDoctor*, because any influence that *PoisonTree* has on a utility node for Bob must go through *TreeSick*, which is a parent of *TreeDoctor*. After optimizing this decision node, we can obviously optimize *PoisonTree* by itself, yielding an equilibrium strategy profile.

Obviously not all games allow such a convenient decomposition. However, as long as there exists some subset of the decision nodes such that no node outside of this subset is relevant to any node in the subset, then we can achieve some computational savings by jointly optimizing the decision rules for this subset before tackling the rest of the problem. Using this general idea, an equilibrium can often be found exponentially more quickly than in standard extensive-form games.

An efficient algorithm also exists for computing EXPECTEDUTILITY for MAIDs.

Theorem 6.5.13 *The* EXPECTEDUTILITY *problem for MAIDs can be computed in time polynomial in the size of the MAID representation.*

Unfortunately the only known algorithm for efficiently solving EXPEC-TEDUTILITY in MAIDS uses division and so cannot be directly translated to an arithmetic circuit as required in Theorem 6.5.3, which does not allow division

```
game(blind_tic_tac_toe,                                             (1)
  [ players : [a,b],                                                (2)
    objects : [grid_board : array('$size', '$size')],              (3)
    params : [size],                                                (4)
    flow : (take_turns(mark,unless(full),until(win))),             (5)
    mark : (choose('$player', (X, Y, Mark),                        (6)
                (empty(X,Y), member(Mark, [x,o]))),                (7)
              reveal('$opponent',(X,Y)),                            (8)
              place((X,Y),Mark)),                                   (9)
    full : (\+(empty(_,_)) → outcome(draw)),                       (10)
    win : (straight_line(_,_,length = 3,                           (11)
              contains(Mark)) → outcome(wins('$player')))]).       (12)
```

Figure 6.19 A GALA description of Blind Tic-Tac-Toe.

operations. It is unknown whether the problem of finding a Nash equilibrium in a MAID can be reduced to finding a Nash equilibrium in a two-player game. Nevertheless many other applications for computing EXPECTEDUTILITY we discussed in Section 6.5.1 apply to MAIDs. For example, the EXPECTEDUTILITY algorithm can be used as a subroutine to Govindan and Wilson's algorithm for computing Nash equilibria in extensive-form games (see Section 4.3).

6.5.5 *GALA*

While MAIDs allow us to capture exactly the relevant information needed to make a decision at each point in the game, we still need to explicitly record each choice point of the game. When, instead of modeling real-world setting, we are modeling a board or card game, this task would be rather cumbersome, if not impossible. The key property of these games that is not being exploited is their repetitive nature—the game alternates between the opponents whose possible moves are independent of the depth of the game tree, and can instead be defined in terms of the current state of the game and an unchanging set of rules. The

GALA Prolog-based language *GALA* exploits this fact to allow concise specifications of large, complex games.

We present the main ideas of the language using the code in Figure 6.19 for an imperfect-information variant of Tic-Tac-Toe. Each player can mark a square with either an "x" or an "o," but the opponent sees only the position of the mark, not its type. A player wins if his move creates a line of the same type of mark.

Lines 3 and 5 define the central components of the representation—the object `grid_board` that records all marks, and the flow of the game, which is defined as two players alternating moves until either the board is full or one of the them wins the game. Lines 6–12 then provide the definitions of the terms used in line 5. Three of the functions found in these lines are particularly important because of their relation to the corresponding extensive-form game: `choose` (line 8) defines the available actions at each node, `reveal` (line 6) determines the information sets of the players, and `outcome` (lines 10 and 12) defines the payoffs at the leaves.

Reading through the code in Figure 6.19, one finds not only primitives like `array`, but also several high-level modules, like `straight_line`, that are not defined. The GALA language contains many such predicates, built up from primitives, that were added to handle conditions common to games people play. For example, the high-level predicate `straight_line` is defined using the intermediate-level predicate `chain`, which in turn is defined to take a predicate and a set as input and return true if the predicate holds for the entire set. The idea behind intermediate-level predicates is that they make it easier to define the high-level predicates specific to a game. For example, `chain` can be used in poker to define a flush.

On top of the language, the GALA system was implemented to take a description of a game in the GALA language, generate the corresponding game tree, and then solve the game using the sequence form of the (defined in Section 5.2.3).

Since we lose the space savings of the GALA language when we actually solve the game, the main advantage of the language is the ease with which it allows a human to describe a game to the program that will solve it.

6.6 History and references

Some of the earliest and most influential work on repeated games is Luce and Raiffa [1957a] and Aumann [1959]. Of particular note is that the former provided the main ideas behind the folk theorem and that the latter explored the theoretical differences between finitely and infinitely repeated games. Aumann's work on repeated games led to a Nobel Prize in 2005. Our proof of the folk theorem is based on Osborne and Rubinstein [1994]. For an extensive discussion of the Tit-for-Tat strategy in repeated Prisoner's Dilemma, and in particular this strategy's strong performance in a tournament of computer programs, see Axelrod [1984].

While most game theory textbooks have material on so-called bounded rationality, the most comprehensive repository of results in the area was assembled by Rubinstein [1998]. Some of the specific references are as follows. Theorem 6.1.8 is due to Neyman [1985], while Theorem 6.1.9 is due to Papadimitriou and Yannakakis [1994]. Theorem 6.1.11 is due to Gilboa [1988], and Theorem 6.1.12 is due to Ben-Porath [1990]. Theorem 6.1.13 is due to Papadimitriou [1992]. Finally, Theorems 6.1.14 and 6.1.15 are due to Nachbar and Zame [1996].

Stochastic games were introduced in Shapley [1953]. The state of the art regarding them circa 2003 appears in the edited collection Neyman and Sorin [2003]. Filar and Vrieze [1997] provide a rigorous introduction to the topic, integrating MDPs (or single-agent stochastic games) and two-person stochastic games.

Bayesian games were introduced by Harsanyi [1967–1968]; in 1994 he received a Nobel Prize, largely because of this work.

Congestion games were first defined by Rosenthal [1973]; later potential games were introduced by Monderer and Shapley [1996a] and were shown to be equivalent to congestion games (up to isomorphism). The PLS-completeness result is due to Fabrikant et al. [2004]. Nonatomic congestion games are due to Schmeidler [1973]. Selfish routing was first studied as early as 1920 [Pigou, 1920; Beckmann et al., 1956]. Pigou's example comes from the former reference; Braess' paradox was introduced in Braess [1968]. The Wardrop equilibrium is due to Wardrop [1952]. The concept of the price of anarchy is due to Koutsoupias and Papadimitriou [1999]. Most of the results in Section 6.4.5 are due to Roughgarden and his coauthors; see his recent book Roughgarden [2005]. Similar results have also been shown for broader classes of nonatomic congestion games; see Roughgarden and Tardos [2004] and Correa et al. [2005].

Theorems 6.5.3 and 6.5.4 are due to Daskalakis et al. [2006a]. Theorem 6.5.5 is due to Papadimitriou [2005]. Graphical games were introduced in Kearns et al. [2001]. The problem of finding pure Nash equilibria in graphical games was analyzed in Gottlob et al. [2003] and Daskalakis and Papadimitriou [2006]. Action graph games were defined in Bhat and Leyton-Brown [2004] and extended in Jiang and Leyton-Brown [2006]. Multiagent influence diagrams were introduced in Koller and Milch [2003], which also contains the running example we used

game network   for that section. A related notion of *game networks* was concurrently developed by La Mura [2000]. Theorem 6.5.13 is due to Blum et al. [2006]. GALA is described in Koller and Pfeffer [1995], which also contained the sample code for the Tic-Tac-Toe example.

7

Learning and Teaching

The capacity to learn is a key facet of intelligent behavior, and it is no surprise that much attention has been devoted to the subject in the various disciplines that study intelligence and rationality. We will concentrate on techniques drawn primarily from two such disciplines—artificial intelligence and game theory—although those in turn borrow from a variety of disciplines, including control theory, statistics, psychology and biology, to name a few. We start with an informal discussion of the various subtle aspects of learning in multiagent systems and then discuss representative theories in this area.

7.1 Why the subject of "learning" is complex

The subject matter of this chapter is fraught with subtleties, and so we begin with an informal discussion of the area. We address three issues—the interaction between learning and teaching, the settings in which learning takes place and what constitutes learning in those settings, and the yardsticks by which to measure this or that theory of learning in multiagent systems.

7.1.1 *The interaction between learning and teaching*

Most work in artificial intelligence concerns the learning performed by an individual agent. In that setting the goal is to design an agent that learns to function successfully in an environment that is unknown and potentially also changes as the agent is learning. A broad range of techniques have been developed, and learning rules have become quite sophisticated.

In a multiagent setting, however, an additional complication arises, since the environment contains (or perhaps consists entirely of) other agents. The problem is not only that the other agents' learning will change the environment for our protagonist agent—dynamic environments feature already in the single-agent case—but that these changes will depend in part on the actions of the protagonist agent. That is, the learning of the other agents will be impacted by the learning performed by our protagonist.

The simultaneous learning of the agents means that every learning rule leads to a dynamical system, and sometimes even very simple learning rules can lead to complex global behaviors of the system. Beyond this mathematical fact, however, lies a conceptual one. In the context of multiagent systems one cannot separate

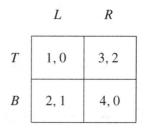

Figure 7.1 Stackelberg game: player 1 must teach player 2.

learning and
teaching

the phenomenon of *learning* from that of *teaching*; when choosing a course of action, an agent must take into account not only what he has learned from other agents' past behavior, but also how he wishes to influence their future behavior.

The following example illustrates this point. Consider the infinitely repeated game with average reward (i.e., where the payoff to a given agent is the limit average of his payoffs in the individual stage games, as in Definition 6.1.1), in which the stage game is the normal-form game shown in Figure 7.1.

Stackelberg
game

First note that player 1 (the row player) has a dominant strategy, namely *B*. Also note that (B, L) is the unique Nash equilibrium of the game. Indeed, if player 1 were to play *B* repeatedly, it is reasonable to expect that player 2 would always respond with *L*. Of course, if player 1 were to choose *T* instead, then player 2's best response would be *R*, yielding player 1 a payoff of 3 which is greater than player 1's Nash equilibrium payoff. In a single-stage game it would be hard for player 1 to convince player 2 that he (player 1) will play *T*, since it is a strictly dominated strategy.[1] However, in a repeated-game setting agent 1 has an opportunity to put his payoff where his mouth is, and adopt the role of a teacher. That is, player 1 could repeatedly play *T*; presumably, after a while player 2, if he has any sense at all, would get the message and start responding with *R*.

In the preceding example it is pretty clear who the natural candidate for adopting the teacher role is. But consider now the repetition of the Coordination game, reproduced in Figure 7.2. In this case, either player could play the teacher with equal success. However, if both decide to play teacher and happen to select uncoordinated actions (Left, Right) or (Right, Left) then the players will receive a payoff of zero forever.[2] Is there a learning rule that will enable them to coordinate without an external designation of a teacher?

7.1.2   *What constitutes learning?*

In the preceding examples the setting was a repeated game. We consider this a "learning" setting because of the temporal nature of the domain, and the regularity across time (at each time the same players are involved, and they play the same game as before). This allows us to consider strategies in which future action is

---

1. See related discussion on signaling and cheap talk in Chapter 8.
2. This is reminiscent of the "sidewalk shuffle," that awkward process of trying to get by the person walking toward you while he is doing the same thing, the result being that you keep blocking each other.

	Left	Right
Left	1, 1	0, 0
Right	0, 0	1, 1

Figure 7.2 Who's the teacher here?

selected based on the experience gained so far. When discussing repeated games in Chapter 6 we mentioned a few simple strategies. For example, in the context of repeated Prisoner's Dilemma, we mentioned the Tit-for-Tat (TfT) and trigger strategies. These, in particular TfT, can be viewed as very rudimentary forms of learning strategies. But one can imagine much more complex strategies, in which an agent's next choice depends on the history of play in more sophisticated ways. For example, the agent could guess that the frequency of actions played by his opponent in the past might be his current mixed strategy, and play a best response to that mixed strategy. As we shall see in Section 7.2, this basic learning rule is called *fictitious play*.

Repeated games are not the only context in which learning takes place. Certainly the more general category of stochastic games (also discussed in Chapter 6) is also one in which regularity across time allows meaningful discussion of learning. Indeed, most of the techniques discussed in the context of repeated games are applicable more generally to stochastic games, though specific results obtained for repeated games do not always generalize.

In both cases—repeated and stochastic games—there are additional aspects of the settings worth discussing. These have to do with whether the (e.g., repeated) game is commonly known by the players. If it is, any "learning" that takes place is only about the strategies employed by the other. If the game is not known, the agent can in addition learn about the structure of the game itself. For example, in a stochastic game setting, the agent may start out not knowing the payoff functions at a given stage game or the transition probabilities, but learn those over time in the course of playing the game. It is most interesting to consider the case in which the game being played is unknown; in this case there is a genuine process of discovery going on. (Such a setting could be modeled as a Bayesian game, as described in Section 6.3, though the formal modeling details are not necessary for the discussion in this chapter.) Some of the remarkable results are that, with certain learning strategies, agents can sometimes converge to an equilibrium of the game even without knowing the game being played. Additionally, there is observability   the question of whether the game is *observable*; do the players see each others' actions, and/or each others' payoffs? (Of course, in the case of a known game, the actions also reveal the payoffs.)

While repeated and stochastic games constitute the main setting in which we will investigate learning, there are other settings as well. Chief among them are models of large populations. These models, which were largely inspired by

	Yield	Dare
Yield	2, 2	1, 3
Dare	3, 1	0, 0

Figure 7.3 The game of Chicken.

evolutionary models in biology, are superficially quite different from the setting of repeated or stochastic games. Unlike the latter, which involve a small number of players, the evolutionary models consist of a large number of players, who repeatedly play a given game among themselves (e.g., pairwise in the case of two-player games). A closer look, however, shows that these models are in fact closely related to the models of repeated games. We discuss this further in the last section of this chapter.

7.1.3 *If learning is the answer, what is the question?*

It is very important to be clear on why we study learning in multiagent systems, and how we judge whether a given learning theory is successful or not. These might seem like trivial questions, but in fact the answers are not obvious and not unique.

First, note that in the following, when we speak about learning strategies, these should be understood as complete strategies, which involve learning in the sense of choosing action as well as updating beliefs. One consequence is that learning in the sense of "accumulated knowledge" is not always beneficial. In the abstract, accumulating knowledge never hurts, since one can always ignore what has been learned. But when one precommits to a particular strategy for acting on accumulated knowledge, sometimes less is more.

This point is related to the inseparability of learning from teaching, discussed earlier. For example, consider a protagonist agent planning to play an infinitely
Chicken game repeated game of *Chicken*, depicted in Figure 7.3. In the presence of any opponent who attempts to learn the protagonist agent's strategy and play a best response, an optimal strategy is to play the stationary policy of always daring; this is the "watch out: I'm crazy" policy. The opponent will learn to always yield, a worse outcome for him than learning anything.[3]

descriptive   Broadly speaking, we can divide theories of learning in multiagent systems
theory   into two categories—*descriptive theories* and *prescriptive theories*.

prescriptive
theory

_____

3. The literary-minded reader may be reminded of the quote from Oscar Wilde's *A Woman of No Importance*: "[...] the worst tyranny the world has ever known; the tyranny of the weak over the strong. It is only tyranny that ever lasts." Except here it is the tyranny of the simpleton over the sophisticated.

Descriptive theories

Descriptive theories attempt to study the way learning takes place in real life—usually by people, but sometimes by other entities such as organizations or animal species. The goal here is to show experimentally that a certain model of learning agrees with behavior (typically, in laboratory experiments) and then to identify interesting properties of the formal model.

The ideal descriptive theory would have two properties.

realism **Property 7.1.1 (Realism)** *There should be a good match between the formal theory and the natural phenomenon being studied.*

convergence **Property 7.1.2 (Convergence)** *The formal theory should exhibit interesting behavioral properties, in particular convergence of the strategy profile being played to some solution concept (e.g., equilibrium) of the game being played.*

One approach to demonstrating realism is to apply the experimental methodology of the social sciences. While we will not focus on this approach, there are several good examples of it in economics and game theory. But there can be other reasons for studying a given learning process. For example, to the extent that one accepts the Bayesian model as at least an idealized model of human decision making, this model provides support for the idea of *rational learning*, which we discuss later.

Convergence properties come in various flavors. Here we survey four of them.

First of all, the holy grail has been showing convergence to stationary strategies which form a Nash equilibrium of the stage game. In fact often this is the hidden motive of the research. It has been noted that game theory is somewhat unusual in having the notion of an equilibrium without associated dynamics that give rise to the equilibrium. Showing that the equilibrium arises naturally would correct this anomaly.[4]

A second approach recognizes that actual convergence to Nash equilibria is a rare occurrence under many learning processes. It pursues an alternative: not requiring that the agents converge to a strategy profile that is a Nash equilibrium, but rather requiring that the empirical frequency of play converge to such an equilibrium. For example, consider a repeated game of Matching Pennies. If both agents repeatedly played (H,H) and (T,T), the frequency of both their plays would converge to (.5, .5), the strategy in the unique Nash equilibrium, even though the payoffs obtained would be very different from the equilibrium payoffs.

Third and yet more radically, we can give up entirely on Nash equilibrium as the relevant solution concept. One alternative is to seek convergence to a *correlated equilibrium* of the stage game. This is interesting in a number of ways. No-regret learning, which we discuss later, can be shown to converge to correlated equilibria in certain cases. Indeed, convergence to a correlated equilibrium provides a

---

4. However, recent theoretical progress on the complexity of computing a Nash equilibrium (see Section 4.2.1) raises doubts about whether any such procedure could be guaranteed to converge to an equilibrium, at least within polynomial time.

justification for the no-regret learning concept; the "correlating device" in this case is not an abstract notion, but the prior history of play.

Finally, we can give up on convergence to stationary policies, but require that the non-stationary policies converge to an interesting state. In particular, learning strategies that include building an explicit model of the opponents' strategies (as we shall see, these are called *model-based* learning rules) can be required to converge to correct models of the opponents' strategies.

Prescriptive theories

In contrast with descriptive theories, prescriptive theories ask how agents—people, programs, or otherwise—*should* learn. A such they are not required to show a match with real-world phenomena. By the same token, their main focus is not on behavioral properties, though they may investigate convergence issues as well. For the most part, we will concentrate on *strategic* normative theories, in which individual agents are self-motivated.

In zero-sum games, and even in repeated or stochastic zero sum games, it is meaningful to ask whether an agent is learning in an optimal fashion. But in general this question is not meaningful, since the answer depends not only on the learning being done but also on the behavior of other agents in the system. When all agents adopt the same strategy (e.g., they all adopt TfT, or all adopt self-play   reinforcement learning, to be discussed shortly), this is called *self-play*. One way to judge learning procedures is based on their performance in self-play. However, learning agents can be judged also by how they do in the context of other types of agents; a TfT agent may perform well against another TfT agent, but less well against an agent using reinforcement learning.

No learning procedure is optimal against all possible opponent behaviors. This observation is simply an instance of the general move in game theory away from the notion of "optimal strategy" and toward "best response" and equilibrium. Indeed, in the broad sense in which we use the term, a "learning strategy" is simply a strategy in a game that has a particular structure (namely, the structure of a repeated or stochastic game) that happens to have a component that is naturally viewed as adaptive.

So how do we evaluate a prescriptive learning strategy? There are several answers. The first is to adopt the standard game-theoretic stance: give up on judging a strategy in isolation, and instead ask which learning rules are in equilibrium with each other. Note that requiring that repeated-game learning strategies be in equilibrium with each other is very different from the convergence requirements discussed above; those speak about equilibrium in the stage game, not in the repeated game. For example, TfT is in equilibrium with itself in an infinitely repeated Prisoner's Dilemma game, but does not lead to the repeated Defect play, the only Nash equilibrium of the stage game. This "equilibrium of learning strategies" approach is not common, but we shall see one example of it later on.

A more modest, but by far more common and perhaps more practical approach is to ask whether a learning strategy achieves payoffs that are "high enough." This approach is both stronger and weaker than the requirement of "best response."

Best response requires that the strategy yield the highest possible payoff against a particular strategy of the opponent(s). A focus on "high enough" payoffs can consider a broader class of opponents, but makes weaker requirements regarding the payoffs, which are allowed to fall short of best response.

There are several different versions of such high-payoff requirements, each adopting and/or combining different basic properties.

safety of a learning rule

Property 7.1.3 (Safety) *A learning rule is safe if it guarantees the agent at least its maxmin payoff, or "security value." (Recall that this is the payoff the agent can guarantee to himself regardless of the strategies adopted by the opponents; see Definition 3.4.1.)*

rationality of a learning rule

Property 7.1.4 (Rationality) *A learning rule is rational if whenever the opponent settles on a stationary strategy of the stage game (i.e., the opponent adopts the same mixed strategy each time, regardless of the past), the agent settles on a best response to that strategy.*

universal consistency

Hannan consistency

no-regret

Property 7.1.5 (No-regret, informal) *A learning rule is* universally consistent, *or* Hannan consistent, *or exhibits* no regret *(these are all synonymous terms), if, loosely speaking, against any set of opponents it yields a payoff that is no less than the payoff the agent could have obtained by playing any one of his pure strategies throughout. We give a more formal definition of this condition later in the chapter.*

Some of these basic requirements are quite strong, and can be weakened in a variety of ways. One way is to allow slight deviations, either in terms of the magnitude of the payoff obtained, or the probability of obtaining it, or both. For example, rather than require optimality, one can require ϵ, δ-optimality, meaning that with probability of at least $1 - \delta$ the agent's payoff comes within ϵ of the payoff obtained by the best response. Another way of weakening the requirements is to limit the class of opponents against which the requirement holds. For example, attention can be restricted to the case of self play, in which the agent plays a copy of itself. (Note that while the learning strategies are identical, the game being played may not be symmetric.) For example, one might require that the learning rule guarantee convergence in self play. More broadly, as in the case of *targeted optimality*, which we discuss later, one might require a best response only against a particular class of opponents.

In the next sections, as we discuss several learning rules, we will encounter various versions of these requirements and their combinations. For the most part we will concentrate on repeated, two-player games, though in some cases we will broaden the discussion and discuss stochastic games and games with more than two players.

7.2 Fictitious play

fictitious play

Fictitious play is one of the earliest learning rules. It was actually not proposed initially as a learning model at all, but rather as an iterative method for computing

Nash equilibria in zero-sum games. It happens to not be a particularly effective way of performing this computation, but since it employs an intuitive update rule, it is usually viewed as a model of learning, albeit a simplistic one, and subjected to convergence analyses of the sort discussed above.

Fictitious play is an instance of model-based learning, in which the learner explicitly maintains beliefs about the opponent's strategy. The structure of such techniques is straightforward.

> Initialize beliefs about the opponent's strategy
> **repeat**
> > Play a best response to the assessed strategy of the opponent
> > Observe the opponent's actual play and update beliefs accordingly

Note that in this scheme the agent is oblivious to the payoffs obtained or obtainable by other agents. We do however assume that the agent knows his own payoff matrix in the stage game (i.e., the payoff he would get in each action profile, whether or not encountered in the past).

In fictitious play, an agent believes that his opponent is playing the mixed strategy given by the empirical distribution of the opponent's previous actions. That is, if A is the set of the opponent's actions, and for every $a \in A$ we let $w(a)$ be the number of times that the opponent has played action a, then the agent assesses the probability of a in the opponent's mixed strategy as

$$P(a) = \frac{w(a)}{\sum_{a' \in A} w(a')}.$$

For example, in a repeated Prisoner's Dilemma game, if the opponent has played C, C, D, C, D in the first five games, before the sixth game he is assumed to be playing the mixed strategy $(0.6, 0.4)$. Note that we can represent a player's beliefs with either a probability measure or with the set of counts $(w(a_1), \ldots, w(a_k))$.

We have not fully specified fictitious play. There exist different versions of fictitious play which differ on the tie-breaking method used to select an action when there is more than one best response to the particular mixed strategy induced by an agent's beliefs. In general the tie-breaking rule chosen has little effect on the results of fictitious play.

On the other hand, fictitious play is very sensitive to the players' initial beliefs. This choice, which can be interpreted as action counts that were observed before the start of the game, can have a radical impact on the learning process. Note that one must pick some nonempty prior belief for each agent; the prior beliefs cannot be $(0, \ldots, 0)$ since this does not define a meaningful mixed strategy.

Fictitious play is somewhat paradoxical in that each agent assumes a stationary policy of the opponent, yet no agent plays a stationary policy except when the process happens to converge to one. The following example illustrates the operation of fictitious play. Recall the Matching Pennies game from Chapter 3, reproduced here as Figure 7.4. Two players are playing a repeated game of Matching Pennies. Each player is using the fictitious play learning rule to update

	Heads	Tails
Heads	1, −1	−1, 1
Tails	−1, 1	1, −1

Figure 7.4 Matching Pennies game.

Round	1's action	2's action	1's beliefs	2's beliefs
0			(1.5,2)	(2,1.5)
1	T	T	(1.5,3)	(2,2.5)
2	T	H	(2.5,3)	(2,3.5)
3	T	H	(3.5,3)	(2,4.5)
4	H	H	(4.5,3)	(3,4.5)
5	H	H	(5.5,3)	(4,4.5)
6	H	H	(6.5,3)	(5,4.5)
7	H	T	(6.5,4)	(6,4.5)
⋮	⋮	⋮	⋮	⋮

Table 7.1 Fictitious play of a repeated game of Matching Pennies.

his beliefs and select actions. Player 1 begins the game with the prior belief that player 2 has played heads 1.5 times and tails 2 times. Player 2 begins with the prior belief that player 1 has played heads 2 times and tails 1.5 times. How will the players play?

The first seven rounds of play of the game is shown in Table 7.1.

As you can see, each player ends up alternating back and forth between playing heads and tails. In fact, as the number of rounds tends to infinity, the empirical distribution of the play of each player will converge to $(0.5, 0.5)$. If we take this distribution to be the mixed strategy of each player, the play converges to the unique Nash equilibrium of the normal form stage game, that in which each player plays the mixed strategy $(0.5, 0.5)$.

Fictitious play has several nice properties. First, connections can be shown to pure-strategy Nash equilibria, when they exist.

steady state

absorbing state

Definition 7.2.1 (Steady state) *An action profile a is a* steady state *(or* absorbing state*) of fictitious play if it is the case that whenever a is played at round t it is also played at round t + 1 (and hence in all future rounds as well).*

The following two theorems establish a tight connection between steady states and pure-strategy Nash equilibria.

Theorem 7.2.2 *If a pure-strategy profile is a strict Nash equilibrium of a stage game, then it is a steady state of fictitious play in the repeated game.*

Note that the pure-strategy profile must be a *strict* Nash equilibrium, which means that no agent can deviate to another action without strictly decreasing its payoff. We also have a converse result.

Theorem 7.2.3 *If a pure-strategy profile is a steady state of fictitious play in the repeated game, then it is a (possibly weak) Nash equilibrium in the stage game.*

Of course, one cannot guarantee that fictitious play always converges to a Nash equilibrium, if only because agents can only play pure strategies and a pure-strategy Nash equilibrium may not exist in a given game. However, while the stage game strategies may not converge, the empirical distribution of the stage game strategies over multiple iterations may. And indeed this was the case in the Matching Pennies example given earlier, where the empirical distribution of the each player's strategy converged to their mixed strategy in the (unique) Nash equilibrium of the game. The following theorem shows that this was no accident.

Theorem 7.2.4 *If the empirical distribution of each player's strategies converges in fictitious play, then it converges to a Nash equilibrium.*

This seems like a powerful result. However, notice that although the theorem gives sufficient conditions for the empirical distribution of the players' actions to converge to a mixed-strategy equilibrium, we have not made any claims about the distribution of the particular outcomes played.

To better understand this point, consider the following example. Consider the *Anti-Coordination game* shown in Figure 7.5.

Anti-Coordination game

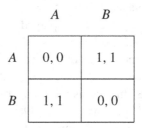

Figure 7.5 The Anti-Coordination game.

Clearly there are two pure Nash equilibria of this game, (A, B) and (B, A), and one mixed Nash equilibrium, in which each agent mixes A and B with probability 0.5. Either of the two pure-strategy equilibria earns each player a payoff of 1, and the mixed-strategy equilibrium earns each player a payoff of 0.5.

Now let us see what happens when we have agents play the repeated Anti-Coordination game using fictitious play. Let us assume that the weight function for each player is initialized to $(1, 0.5)$. The play of the first few rounds is shown in Table 7.2.

As you can see, the play of each player converges to the mixed strategy $(0.5, 0.5)$, which is the mixed strategy Nash equilibrium. However, the payoff received by each player is 0, since the players never hit the outcomes with positive payoff. Thus, although the empirical distribution of the strategies converges to

Round	1's action	2's action	1's beliefs	2's beliefs
0			(1,0.5)	(1,0.5)
1	B	B	(1,1.5)	(1,1.5)
2	A	A	(2,1.5)	(2,1.5)
3	B	B	(2,2.5)	(2,2.5)
4	A	A	(3,2.5)	(3,2.5)
⋮	⋮	⋮	⋮	⋮

Table 7.2 Fictitious play of a repeated Anti-Coordination game.

	Rock	Paper	Scissors
Rock	0, 0	0, 1	1, 0
Paper	1, 0	0, 0	0, 1
Scissors	0, 1	1, 0	0, 0

Figure 7.6 Shapley's Almost-Rock-Paper-Scissors game.

the mixed strategy Nash equilibrium, the players may not receive the expected payoff of the Nash equilibrium, because their actions are miscorrelated.

Finally, the empirical distributions of players' actions need not converge at all. Consider the game in Figure 7.6. Note that this example, due to Shapley, is a modification of the rock-paper-scissors game; this game is not constant sum.

The unique Nash equilibrium of this game is for each player to play the mixed strategy $(1/3, 1/3, 1/3)$. However, consider the fictitious play of the game when player 1's weight function has been initialized to $(0, 0, 0.5)$ and player 2's weight function has been initialized to $(0, 0.5, 0)$. The play of this game is shown in Table 7.3. Although it is not obvious from these first few rounds, it can be shown that the empirical play of this game never converges to any fixed distribution.

For certain restricted classes of games we *are* guaranteed to reach convergence.

Theorem 7.2.5 *Each of the following is a sufficient condition for the empirical frequencies of play to converge in fictitious play:*

- *The game is zero sum;*
- *The game is solvable by iterated elimination of strictly dominated strategies;*
- *The game is a potential game;*[5]
- *The game is $2 \times n$ and has generic payoffs.*[6]

---

5. Actually an even more more general condition applies here, that the players have "identical interests," but we will not discuss this further here.

6. Full discussion of genericity in games lies outside the scope of this book, but here is the essential idea, at least for games in normal form. Roughly speaking, a game in normal form is generic if it does

Round	1's action	2's action	1's beliefs	2's beliefs
0			(0,0,0.5)	(0,0.5,0)
1	Rock	Scissors	(0,0,1.5)	(1,0.5,0)
2	Rock	Paper	(0,1,1.5)	(2,0.5,0)
3	Rock	Paper	(0,2,1.5)	(3,0.5,0)
4	Scissors	Paper	(0,3,1.5)	(3,0.5,1)
5	Scissors	Paper	(0,1.5,0)	(1,0,0.5)
⋮	⋮	⋮	⋮	⋮

Table 7.3 Fictitious play of a repeated game of the Almost-Rock-Paper-Scissors game.

Overall, fictitious play is an interesting model of learning in multiagent systems not because it is realistic or because it provides strong guarantees, but because it is very simple to state and gives rise to nontrivial properties. But it is very limited; its model of beliefs and belief update is mathematically constraining, and is clearly implausible as a model of human learning. There exist various variants of fictitious play that score somewhat better on both fronts. We will mention one of them—called *smooth fictitious play*—when we discuss no-regret learning methods.

7.3 Rational learning

rational learning

Bayesian learning

Rational learning (also sometimes called *Bayesian learning*) adopts the same general model-based scheme as fictitious play. Unlike fictitious play, however, it allows players to have a much richer set of beliefs about opponents' strategies. First, the set of strategies of the opponent can include repeated-game strategies such as TfT in the Prisoner's Dilemma game, not only repeated stage-game strategies. Second, the beliefs of each player about his opponent's strategies may be expressed by any probability distribution over the set of all possible strategies.

Bayesian updating

As in fictitious play, each player begins the game with some prior beliefs. After each round, the player uses *Bayesian updating* to update these beliefs. Let S be the set of the opponent's strategies considered possible by player i, and H be the set of possible histories of the game. Then we can use Bayes' rule to express the probability assigned by player i to the event in which the opponent is playing a particular strategy $s_{-i} \in S^i_{-i}$ given the observation of history $h \in H$, as

$$P_i(s_{-i}|h) = \frac{P_i(h|s_{-i})P_i(s_{-i})}{\sum_{s'_{-i} \in S^i_{-i}} P_i(h|s'_{-i})P_i(s'_{-i})}.$$

not have any interesting property that does not also hold with probability 1 when the payoffs are selected independently from a sufficiently rich distribution (e.g., the uniform distribution over a fixed interval). Of course, to make this precise we would need to define "interesting" and "sufficiently." Intuitively, though, this means that the payoffs do not have accidental properties. A game whose payoffs are all distinct is necessarily generic.

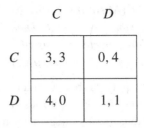

Figure 7.7 Prisoner's Dilemma game.

For example, consider two players playing the infinitely repeated Prisoner's Dilemma game, reproduced in Figure 7.7.

Suppose that the support of the prior belief of each player (i.e., the strategies of the opponent to which the player ascribes nonzero probability; see Definition 3.2.6) consists of the strategies $g_1, g_2, \ldots g_\infty$, defined as follows. g_∞ is the *trigger strategy* that was presented in Section 6.1.2. A player using the trigger strategy begins the repeated game by cooperating, and if his opponent defects in any round, he defects in every subsequent round. For $T < \infty$, g_T coincides with g_∞ at all histories shorter than T but prescribes unprovoked defection starting from time T on. Following this convention, strategy g_0 is the strategy of constant defection.

Suppose furthermore that each player happens indeed to select a best response from among $g_0, g_1, \ldots, g_\infty$. (There are of course infinitely many additional best responses outside this set.) Thus each round of the game will be played according to some strategy profile (g_{T_1}, g_{T_2}).

After playing each round of the repeated game, each player performs Bayesian updating. For example, if player i has observed that player j has always cooperated, the Bayesian updating after history $h_t \in H$ of length t reduces to

$$P_i(g_T | h_t) = \begin{cases} 0 & \text{if } T \le t; \\ \frac{P_i(g_T)}{\sum_{k=t+1}^{\infty} P_i(g_k)} & \text{if } T > t. \end{cases}$$

Rational learning is a very intuitive model of learning, but its analysis is quite involved. The formal analysis focuses on self-play, that is, on properties of the repeated game in which all agents employ rational learning (though they may start with different priors). Broadly, the highlights of this model are as follows:

- Under some conditions, in self-play rational learning results in agents having close to correct beliefs about the observable portion of their opponent's strategy.
- Under some conditions, in self-play rational learning causes the agents to converge toward a Nash equilibrium with high probability.
- Chief among these "conditions" *absolute continuity*, a strong assumption.

In the remainder of this section we discuss these points in more detail, starting with the notion of absolute continuity.

(margin note: trigger strategy)

Definition 7.3.1 (Absolute continuity) *Let X be a set and let μ, $\mu' \in \Pi(X)$ be probability distributions over X. Then the distribution μ is said to be* absolutely continuous *with respect to the distribution μ' iff for $x \subset X$ that is measurable[7] it is the case that if $\mu(x) > 0$ then $\mu'(x) > 0$.*

absolute continuity

Note that the players' beliefs and the actual strategies each induce probability distributions over the set of histories H. Let $s = (s_1, \ldots, s_n)$ be a strategy profile. If we assume that these strategies are used by the players, we can calculate the probability of each history of the game occurring, thus inducing a distribution over H. We can also induce such a distribution with a player's beliefs about players' strategies. Let S_j^i be a set of strategies that i believes possible for j, and $P_j^i \in \Pi(S_j^i)$ be the distribution over S_j^i believed by player i. Let $P_i = (P_1^i, \ldots, P_n^i)$ be the tuple of beliefs about the possible strategies of every player. Now, if player i assumes that all players (including himself) will play according to his beliefs, he can also calculate the probability of each history of the game occurring, thus inducing a distribution over H. The results that follow all require that the distribution over histories induced by the actual strategies is absolutely continuous with respect to the distribution induced by a player's beliefs; in other words, if there is a positive probability of some history given the actual strategies, then the player's beliefs should also assign the history positive probability. (Colloquially, it is sometimes said that the beliefs of the players must contain a *grain of truth*.) Although the results that follow are very elegant, it must be said that the absolute continuity assumption is a significant limitation of the theoretical results associated with rational learning.

grain of truth

In the Prisoner's Dilemma example discussed earlier, it is easy to see that the distribution of histories induced by the actual strategies is absolutely continuous with respect to the distribution predicted by the prior beliefs of the players. All positive probability histories in the game are assigned positive probability by the original beliefs of both players: if the true strategies are g_{T_1}, g_{T_2}, players assign positive probability to the history with cooperation up to time $t < \min(T_1, T_2)$ and defection in all times exceeding the $\min(T_1, T_2)$.

The rational learning model is interesting because it has some very desirable properties. Roughly speaking, players satisfying the assumptions of the rational learning model will have beliefs about the play of the other players that converge to the truth, and furthermore, players will in finite time converge to play that is arbitrarily close to the Nash equilibrium. Before we can state these results we need to define a measure of the similarity of two probability measures.

Definition 7.3.2 (ϵ-closeness) *Given an $\epsilon > 0$ and two probability measures μ and μ' on the same space, we say that μ is ϵ-close to μ' if there is a measurable set Q satisfying:*

- *$\mu(Q)$ and $\mu'(Q)$ are each greater than $1 - \epsilon$; and*
- *for every measurable set $A \subseteq Q$, we have that*

$$(1 + \epsilon)\mu'(A) \geq \mu(A) \geq (1 - \epsilon)\mu'(A).$$

---

7. Recall that a probability distribution over a domain X does not necessarily give a value for all subsets of X, but only over some σ-algebra of X, the collection of measurable sets.

Now we can state a result about the accuracy of the beliefs of a player using rational learning.

Theorem 7.3.3 (Rational learning and belief accuracy) *Let s be a repeated-game strategy profile for a given n-player game[8], and let $P = P_1, \ldots, P_n$ be a tuple of probability distributions over such strategy profiles (P_i is interpreted as player i's beliefs). Let μ_s and μ_P be the distributions over infinite game histories induced by the strategy profile s and the belief tuple P, respectively. If we have that:*

- *at each round, each player i plays a best response strategy given his beliefs P_i;*
- *after each round each player i updates P_i using Bayesian updating; and*
- *μ_s is absolutely continuous with respect to μ_{P_i},*

then for every $\epsilon > 0$ and for almost every history in the support of μ_s (i.e., every possible history given the actual strategy profile s), there is a time T such that for all $t \geq T$, the play μ_{P_i} predicted by the player i's beliefs is ϵ-close to the distribution of play μ_s predicted by the actual strategies.

Thus a player's beliefs will eventually converge to the truth if he is using Bayesian updating, is playing a best response strategy, and the play predicted by the other players' real strategies is absolutely continuous with respect to that predicted by his beliefs. In other words, he will correctly predict the on-path portions of the other players' strategies.

Note that this result does *not* state that players will learn the true strategy being played by their opponents. As stated earlier, there are an infinite number of possible strategies that their opponent could be playing, and each player begins with a prior distribution that assigns positive probability to only some subset of the possible strategies. Instead, players' beliefs will accurately predict the play of the game, and no claim is made about their accuracy in predicting the off-path portions of the opponents' strategies.

Consider again the two players playing the infinitely repeated Prisoner's Dilemma game, as described in the previous example. Let us verify that, as Theorem 7.3.3 dictates, the future play of this game will be correctly predicted by the players. If $T_1 < T_2$ then from time $T_1 + 1$ on, player 2's posterior beliefs will assign probability 1 to player 1's strategy, g_{T_1}. On the other hand, player 1 will never fully know player 2's strategy, but will know that $T_2 > T_1$. However, this is sufficient information to predict that player 2 will always choose to defect in the future.

A player's beliefs must converge to the truth even when his strategy space is incorrect (does not include the opponent's actual strategy), as long as they satisfy the absolute continuity assumption. Suppose, for instance, that player 1 is playing the trigger strategy g_∞, and player 2 is playing tit-for-tat, but that player 1 believes that player 2 is also playing the trigger strategy. Thus player 1's beliefs about player 2's strategy are incorrect. Nevertheless, his beliefs will correctly predict the future play of the game.

---

8. That is, a tuple of repeated-game strategies, one for each player.

We have so far spoken about the accuracy of beliefs in rational learning. The following theorem addresses convergence to equilibrium. Note that the conditions of this theorem are identical to those of Theorem 7.3.3, and that the definition refers to the concept of an ϵ-Nash equilibrium from Section 3.4.7, as well as to ϵ-closeness as defined earlier.

Theorem 7.3.4 (Rational Learning and Nash) *Let s be a repeated-game strategy profile for a given n-player game, and let $P = P_1, \ldots, P_n$ be a a tuple of probability distributions over such strategy profiles. Let μ_s and μ_P be the distributions over infinite game histories induced by the strategy profile s and the belief tuple P, respectively. If we have that:*

- *at each round, each player i plays a best response strategy given his beliefs P_i;*
- *after each round each player i updates P_i using Bayesian updating; and*
- *μ_s is absolutely continuous with respect to μ_{P_i},*

then for every $\epsilon > 0$ and for almost every history in the support of μ_s there is a time T such that for every $t \geq T$ there exists an ϵ-equilibrium s^* of the repeated game in which the play μ_{P_i} predicted by player i's beliefs is ϵ-close to the play μ_{s^*} of the equilibrium.

In other words, if utility-maximizing players start with individual subjective beliefs with respect to which the true strategies are absolutely continuous, then in the long run, their behavior must be essentially the same as a behavior described by an ϵ-Nash equilibrium.

Of course, the space of repeated-game equilibria is huge, which leaves open the question of which equilibrium will be reached. Here notice a certain self-fulfilling property: players' optimism can lead to high rewards, and likewise pessimism can lead to low rewards. For example, in a repeated Prisoner's Dilemma game, if both players begin believing that their opponent will likely play the TfT strategy, they each will tend to cooperate, leading to mutual cooperation. If, on the other hand, they each assign high prior probability to constant defection, or to the grim-trigger strategy, they will each tend to defect.

7.4 Reinforcement learning

reinforcement learning

In this section we look at multiagent extensions of learning in MDPs, that is, in single-agent stochastic games (see Appendix C for a review of MDP essentials). Unlike the first two learning techniques discussed, and with one exception discussed in section 7.4.4, *reinforcement learning* does not explicitly model the opponent's strategy. The specific family of techniques we look at are derived from the Q-learning algorithm for learning in unknown (single-agent) MDPs. Q-learning is described in the next section, after which we present its extension to zero-sum stochastic games. We then briefly discuss the difficulty in extending the methods to general-sum stochastic games.

7.4.1 *Learning in unknown MDPs*

First, consider (single-agent) MDPs. Value iteration, as described in Appendix C, assumes that the MDP is known. What if we do not know the rewards or transition probabilities of the MDP? It turns out that, if we always know what state[9] we are in and the reward received in each iteration, we can still converge to the correct Q-values.

Q-learning **Definition 7.4.1 (Q-learning)** *Q-learning is the following procedure:*

Initialize the Q-function and V values (arbitrarily, for example)
repeat *until convergence*
> Observe the current state s_t.
> Select action a_t and and take it.
> Observe the reward $r(s_t, a_t)$
> Perform the following updates (and do not update any other Q-values):
> $Q_{t+1}(s_t, a_t) \leftarrow (1 - \alpha)Q_t(s_t, a_t) + \alpha_t(r(s_t, a_t) + \beta V_t(s_{t+1}))$
> $V_{t+1}(s) \leftarrow \max_a Q_t(s, a)$

Theorem 7.4.2 *Q-learning guarantees that the Q and V values converge to those of the optimal policy, provided that each state-action pair is sampled an infinite number of times, and that the time-dependent learning rate α_t obeys $0 \le \alpha_t < 1$, $\sum_0^\infty \alpha_t = \infty$ and $\sum_0^\infty \alpha_t^2 < \infty$.*

The intuition behind this approach is that we approximate the unknown transition probability by using the actual distribution of states reached in the game itself. Notice that this still leaves us a lot of room in designing the order in which the algorithm selects actions.

Note that this theorem says nothing about the rate of convergence. Furthermore, it gives no assurance regarding the accumulation of optimal future discounted rewards by the agent; it could well be, depending on the discount factor, that by the time the agent converges to the optimal policy it has paid too high a cost, which cannot be recouped by exploiting the policy going forward. This is not a concern if the learning takes place during training sessions, and only when learning has converged sufficiently is the agent unleashed on the world (e.g., think of a fighter pilot being trained on a simulator before going into combat). But in general Q-learning should be thought of as guaranteeing good learning, but neither quick learning nor high future discounted rewards.

7.4.2 *Reinforcement learning in zero-sum stochastic games*

In order to adapt the method presented from the setting of MDPs to stochastic games, we must make a few modifications. The simplest possible modification is to have each agent ignore the existence of the other agent (recall that zero-sum games involve only two agents). We then define $Q_i^\pi : S \times A_i \mapsto \mathbb{R}$ to be the

---

9. For consistency with the literature on reinforcement learning, in this section we use the notation s and S for a state and set of states respectively, rather than for a strategy profile and a set of strategy profiles as elsewhere in the book.

value for player i if the two players follow strategy profile π after starting in state s and player i chooses the action a. We can now apply the Q-learning algorithm. As mentioned earlier in the chapter, the multiagent setting forces us to forego our search for an "optimal" policy, and instead to focus on one that performs well against its opponent. For example, we might require that it satisfy Hannan consistency (Property 7.1.5). Indeed, the Q-learning procedure can be shown to be Hannan-consistent for an agent in a stochastic game against opponents playing stationary policies. However, against opponents using more complex strategies, such as Q-learning itself, we do not obtain such a guarantee.

The above approach, assuming away the opponent, seems unmotivated. Instead, if the agent is aware of what actions its opponent selected at each point in its history, we can use a modified Q-function, $Q_i^\pi : S \times A \mapsto \mathbb{R}$, defined over states and action profiles, where $A = A_1 \times A_2$. The formula to update Q is simple to modify and would be the following for a two-player game.

$$Q_{i,t+1}(s_t, a_t, o_t) = (1 - \alpha_t)Q_{i,t}(s_t, a_t, o_t) + \alpha_t(r_i(s_t, a_t, o_t) + \beta V_t(s_{t+1}))$$

Now that the actions range over both our agent's actions and that of its competitor, how can we calculate the value of a state? Recall that for (two-player) zero-sum games, the policy profile where each agent plays its maxmin strategy forms a Nash equilibrium. The payoff to the first agent (and thus the negative of the payoff to the second agent) is called the *value* of the game, and it forms the basis for our revised value function for Q-learning,

value of a zero-sum game

$$V_t(s) = \max_{\Pi_i} \min_o Q_{i,t}(s, \Pi_i(s), o).$$

minimax-Q Like the basic Q-learning algorithm, the above *minimax-Q* learning algorithm is guaranteed to converge in the limit of infinite samples of each state and action profile pair. While this will guarantee the agent a payoff at least equal to that of its maxmin strategy, it no longer satisfies Hannan consistency. If the opponent is playing a suboptimal strategy, minimax-Q will be unable to exploit it in most games.

The minimax-Q algorithm is described in Figure 7.8. Note that this algorithm specifies not only how to update the Q and V values, but also how to update the strategy Π. There are still some free parameters, such as how to update the learning parameter, α. One way of doing so is to simply use a decay rate, so that α is set to $\alpha * decay$ after each Q-value update, for some value of $decay < 1$. Another possibility from the Q-learning literature is to keep separate α's for each state and action profile pair. In this case, a common method is to use $\alpha = 1/k$, where k equals the number of times that particular Q-value has been updated including the current one. So, when first encountering a reward for a state s where an action profile a was played, the Q-value is set entirely to the observed reward plus the discounted value of the successor state ($\alpha = 1$). On the next time that state–action profile pair is encountered, it will be set to be half of the old Q-value plus half of the new reward and discounted successor state value.

We now look at an example demonstrating the operation of minimax-Q learning in a simple repeated game: repeated Matching Pennies (see Figure 7.4) against an unknown opponent. Note that the convergence results for Q-learning impose

```
// Initialize:
forall s ∈ S, a ∈ A, and o ∈ O do
    L  Q(s, a, o) ← 1
forall s in S do
    L  V(s) ← 1
forall s ∈ S and a ∈ A do
    L  Π(s, a) ← 1/|A|
α ← 1.0
// Take an action:
```

when in state s, with probability *explor* choose an action uniformly at random, and with probability $(1 - explor)$ choose action a with probability $\Pi(s, a)$

// Learn:

after receiving reward rew for moving from state s to s' via action a and opponent's action o

$$Q(s, a, o) \leftarrow (1 - \alpha) * Q(s, a, o) + \alpha * (rew + \gamma * V(s'))$$
$$\Pi(s, \cdot) \leftarrow \arg\max_{\Pi'(s, \cdot)} (\min_{o'} \sum_{a'} (\Pi(s, a') * Q(s, a', o')))$$

// The above can be done, for example, by linear programming

$$V(s) \leftarrow \min_{o'} (\sum_{a'} (\Pi(s, a') * Q(s, a', o')))$$

Update α

Figure 7.8 The minimax-Q algorithm.

only weak constraints on how to select actions and visit states. In this example, we follow the given algorithm and assume that the agent chooses an action randomly some fraction of the time (denoted *explor*), and plays according to his current best strategy otherwise. For updating the learning rate, we have chosen the second method discussed earlier, with $\alpha = 1/k$, where k is the number of times the state and action profile pair has been observed. Assume that the Q-values are initialized to 1 and that the discount factor of the game is 0.9.

Table 7.4 shows the values of player 1's Q-function in the first few iterations of this game as well as his best strategy at each step. We see that the value of the game, 0, is being approached, albeit slowly. This is not an accident.

Theorem 7.4.3 *Under the same conditions that assure convergence of Q-learning to the optimal policy in MDPs, in zero-sum games Minimax-Q converges to the value of the game in self play.*

Here again, no guarantee is made about the rate of convergence or about the accumulation of optimal rewards. We can achieve more rapid convergence if we are willing to sacrifice the guarantee of finding a perfectly optimal maxmin strategy. In particular, we can consider the framework of *probably approximately correct (PAC) learning*. In this setting, choose some $\epsilon > 0$ and $1 > \delta > 0$, and seek an algorithm that can guarantee—regardless of the opponent—a payoff of at least that of the maxmin strategy minus ϵ, with probability $(1 - \delta)$. If we are willing to settle for this weaker guarantee, we gain the property that it will always hold after a polynomially-bounded number of time steps.

probably approximately correct (PAC) learning

t	Actions	Reward$_1$	$Q_t(H,H)$	$Q_t(H,T)$	$Q_t(T,H)$	$Q_t(T,T)$	$V(s)$	$\pi_1(H)$
0			1	1	1	1	1	0.5
1	(H*,H)	1	1.9	1	1	1	1	0.5
2	(T,H)	-1	1.9	1	-0.1	1	1	0.55
3	(T,T)	1	1.9	1	-0.1	1.9	1.279	0.690
4	(H*,T)	-1	1.9	0.151	-0.1	1.9	0.967	0.534
5	(T,H)	-1	1.9	0.151	-0.115	1.9	0.964	0.535
6	(T,T)	1	1.9	0.151	-0.115	1.884	0.960	0.533
7	(T,H)	-1	1.9	0.151	-0.122	1.884	0.958	0.534
8	(H,T)	-1	1.9	0.007	-0.122	1.884	0.918	0.514
⋮	⋮	⋮	⋮	⋮	⋮	⋮	⋮	⋮
100	(H,H)	1	1.716	-0.269	-0.277	1.730	0.725	0.503
⋮	⋮	⋮	⋮	⋮	⋮	⋮	⋮	⋮
1000	(T,T)	1	1.564	-0.426	-0.415	1.564	0.574	0.500
⋮	⋮	⋮	⋮	⋮	⋮	⋮	⋮	⋮

Table 7.4 Minimax-Q learning in a repeated Matching Pennies game.

One example of such an algorithm is the model-based learning algorithm *R-max*. It first initializes its estimate of the value of each state to be the highest reward that can be returned in the game (hence the name). This philosophy has been referred to as *optimism in the face of uncertainty* and helps guarantee that the agent will explore its environment to the best of its ability. The agent then uses these optimistic values to calculate a maxmin strategy for the game. Unlike normal Q-learning, the algorithm does not update its values for any state and action profile pair until it has visited them "enough" times to have a good estimate of the reward and transition probabilities. Using a theoretical method called *Chernoff bounds*, it is possible to polynomially bound the number of samples necessary to guarantee that the accuracy of the average over the samples deviates from the true average by at most ϵ with probability $(1 - \delta)$ for any selected value of ϵ and δ. The polynomial is in Σ, k, T, $1/\epsilon$, and $1/\delta$, where Σ is the number of states (or games) in the stochastic game, k is the number of actions available to each agent in a game (without loss of generally we can assume that this is the same for all agents and all games), and T is the ϵ-return *mixing time* of the optimal policy, that is, the smallest length of time after which the optimal policy is guaranteed to yield an expected payoff at most ϵ away from optimal. The notes at the end of the chapter point to further reading on R-max, and a predecessor algorithm called *E3* (pronounced "E cubed").

R-max algorithm

Chernoff bounds

mixing time

E3 algorithm

7.4.3 *Beyond zero-sum stochastic games*

So far we have shown results for the class of zero-sum stochastic games. Although the algorithms discussed, in particular minimax-Q, are still well defined in the general-sum case, the guarantee of achieving the maxmin strategy payoff is less compelling. Another subclass of stochastic games that has been addressed is that of common-payoff (pure coordination) games, in which all agents receive the same reward for an outcome. This class has the advantage of reducing the problem to identifying an optimal action profile and coordinating with the other agents to play it. In many ways this problem can really be seen as a single-agent problem of distributed control. This is a relatively well-understood problem, and various algorithms exist for it, depending on precisely how the problem is defined.

Expanding reinforcement learning algorithms to the general-sum case is quite problematic, on the other hand. There have been attempts to generalize Q-learning to general-sum games, but they have not yet been truly successful. As was discussed at the beginning of this chapter, the question of what it means to learn in general-sum games is subtle. One yardstick we have discussed is convergence to Nash equilibrium of the stage game during self play. No generalization of Q-learning has been put forward that has this property.

7.4.4 Belief-based reinforcement learning

There is also a version of reinforcement learning that includes explicit modeling of the other agent(s), given by the following equations.

$$Q_{t+1}(s_t, a_t) \leftarrow (1 - \alpha)Q_t(s_t, a_t) + \alpha_t(r(s_t, a_t) + \beta V_t(s_{t+1}))$$

$$V_t(s) \leftarrow \max_{a_i} \sum_{a_{-i} \subset A_{-i}} Q_t(s, (a_i, a_{-i})) Pr_i(a_{-i})$$

In this version, the agent updates the value of the game using the probability he assigns to the opponent(s) playing each action profile. Of course, the belief function must be updated after each play. How it is updated depends on what the function is. Indeed, belief-based reinforcement learning is not a single procedure but a family, each member characterized by how beliefs are formed and updated. For example, in one version the beliefs are of the kind considered in fictitious play, and in another they are Bayesian in the style of rational learning. There are some experimental results that show convergence to equilibrium in self-play for some versions of belief-based reinforcement learning and some classes of games, but no theoretical results.

7.5 No-regret learning and universal consistency

As discussed above, learning rule is universally consistent or (equivalently) exhibits no regret if, loosely speaking, against any set of opponents it yields a payoff that is no less than the payoff the agent could have obtained by playing any one of his pure strategies throughout.

More precisely, let α^t be the average per-period reward the agent received up until time t, and let $\alpha^t(s_i)$ be the average per-period reward the agent *would have* received up until time t had he played pure strategy s instead, assuming all other agents continue to play as they did.

regret **Definition 7.5.1 (Regret)** *The* regret *an agent experiences at time t for not having played s is* $R^t(s) = \alpha^t - \alpha^t(s)$.

Observe that this is conceptually the same as the definition of regret we offered in Section 3.4 (Definition 3.4.5).

A learning rule is said to exhibit *no regret*[10] if it guarantees that with high probability the agent will experience no positive regret.

no-regret   **Definition 7.5.2 (No-regret learning rule)** *A learning rule exhibits* no regret *if for any pure strategy of the agent s it holds that* $Pr([\liminf R^t(s)] \le 0) = 1$.

The quantification is over all of the agent's pure strategies of the stage game, but note that it would make no difference if instead one quantified over all mixed strategies of the stage game. (Do you see why?) Note also that this guarantee is only in expectation, since the agent's strategy will in general be mixed, and thus the payoff obtained at any given time—u_i^t—is uncertain.

It is important to realize that this "in hindsight" requirement ignores the possibility that the opponents' play might change as a result of the agent's own play. This is true for stationary opponents, and might be a reasonable approximation in the context of a large number of opponents (such as in a public securities market), but less in the context of a small number of agents, of the sort game theory tends to focus on. For example, in the finitely-repeated Prisoner's Dilemma game, the only strategy exhibiting no regret is to always defect. This precludes strategies that capitalize on cooperative behavior by the opponent, such as Tit-for-Tat. In this connection see our earlier discussion of the inseparability of learning and teaching.

regret matching   Over the years, a variety of no-regret learning techniques have been developed. Here are two, *regret matching* and *smooth fictitious play*.

smooth fictitious
play

- *Regret matching*: At each time step each action is chosen with probability proportional to its regret. That is,

$$\sigma_i^{t+1}(s) = \frac{R^t(s)}{\sum_{s' \in S_i} R^t(s')},$$

where $\sigma_i^{t+1}(s)$ is the probability that agent i plays pure strategy s at time $t + 1$.

- *Smooth fictitious play*: Instead of playing the best response to the empirical frequency of the opponent's play, as fictitious play prescribes, one introduces a perturbation that gradually diminishes over time. That is, rather than adopt at time $t + 1$ a pure strategy s_i that maximizes $u_i(s_i, P^t)$ where P^t is the empirical distribution of opponent's play until time t, agent i adopts a mixed strategy σ_i that maximizes $u_i(s_i, P^t) + \lambda v_i(\sigma_i)$. Here λ is any constant, and v_i is a smooth, concave function with boundaries at the unit simplex. For example, v_i can be the entropy function, $v_i(\sigma_i) = -\sum_{S_i} \sigma_i(s_i) \log \sigma_i(s_i)$.

Regret matching can be shown to exhibit no regret, and smooth fictitious play approaches no regret as λ tends to zero. The proofs are based on Blackwell's Approachability Theorem; the notes at the end of the chapter provide pointers for further reading on it, as well as on other no-regret techniques.

---

10. There are actually several versions of regret. The one described here is called *external regret* in computer science, and *unconditional regret* in game theory.

7.6   Targeted learning

No-regret learning was one approach to ensuring good rewards, but as we discussed this sense of "good" has some drawbacks. Here we discuss an alternative sense of "good," which retains the requirement of best response, but limits it to a particular class of opponents. The intuition guiding this approach is that in any strategic setting, in particular a multiagent learning setting, one has *some* sense of the agents in the environment. A chess player has studied previous plays of his opponent, a skipper in a sailing competition knows a lot about his competitors, and so on. And so it makes sense to try to optimize against this set of opponents, rather than against completely unknown opponents.

targeted learning   Technically speaking, the model of *targeted learning* takes as a parameter a class—the "target class"—of likely opponents and is required to perform particularly well against these likely opponents. At the same time one wants to ensure at least the maxmin payoff against opponents outside the target class. Finally, an additional desirable property is for the algorithm to perform well in self-play; the algorithm should be designed to "cooperate" with itself.

For games with only two agents, these intuitions can be stated formally as follows.

targeted   **Property 7.6.1 (Targeted optimality)** *Against any opponent in the target class,*
optimality   *the expected payoff is the best-response payoff.*[11]

safety   **Property 7.6.2 (Safety)** *Against any opponent, the expected payoff is at least the individual security (or maxmin) value for the game.*

Property 7.6.3 (Autocompatibility) *Self-play—in which both agents adopt the*
autocompatibility   *learning procedure in question—is strictly Pareto efficient.*[12]

We introduce one additional twist. Since we are interested in quick learning, not only learning in the limit, we need to allow some departure from the ideal. And so we amend the requirements as follows.

Definition 7.6.4 (Efficient targeted learning) *A learning rule exhibits* efficient
efficient targeted   *targeted learning if for every $\epsilon > 0$ and $1 > \delta > 0$, there exists an M polynomial*
learning   *in $1/\epsilon$ and $1/\delta$ such that after M time steps, with probability greater than $1 - \delta$, all three payoff requirements listed previously are achieved within ϵ.*

Note the difference from no-regret learning. For example, consider learning in a repeated Prisoner's Dilemma game. Suppose that the target class consists of all opponents whose strategies rely on the past iteration; note this includes the Tit-for-Tat strategy. In this case successful targeted learning will result in constant cooperation, while no-regret learning prescribes constant defection.

---

11. Note: the expectation is over the mixed-strategy profiles, but not over opponents; this requirement is for any fixed opponent.

12. Recall that strict Pareto efficiency means that one agent's expected payoff cannot increase without the other's decreasing; see Definition 3.3.2. Also note that we do not restrict the discussion to symmetric games, and so self play does not in general mean identical play by the agents, nor identical payoffs. We abbreviate "strictly Pareto efficient" as "Pareto efficient."

How hard is it to achieve efficient targeted learning? The answer depends of course on the target class. Provably correct (with respect to this criterion) learning procedures exist for the class of stationary opponents, and the class of opponents whose memory is limited to a finite window into the past. The basic approach is to construct a number of building blocks and then specialize and combine them differently depending on the precise setting. The details of the algorithms can get involved, especially in the interesting case of nonstationary opponents, but the essential flow is as follows.

1. Start by assuming that the opponent is in the target set and learn a best response to the particular agent under this assumption. If the payoffs you obtain stray too much from your expectation, move on.
2. Signal to the opponent to find out whether he is employing the same learning strategy. If he is, coordinate to a Pareto-efficient outcome. If your payoffs stray too far off, move on.
3. Play your security-level strategy.

Note that so far we have restricted the discussion to two-player games. Can we generalize the criteria—and the algorithms—to games with more players? The answer is yes, but various new subtleties creep in. For example, in the two-agent case we needed to worry about three cases, corresponding to whether the opponent is in the target set, is a self-play agent, or is neither. We must now consider three sets of agents—self play agents (i.e., agents using the algorithm in question), agents in the target set, and unconstrained agents, and ask how agents in the first set can jointly achieve a Pareto-efficient outcome against the second set and yet protect themselves from exploitation by agents in the third set. This raises questions about possible coordination among the agents:

- Can self-play agents coordinate other than implicitly through their actions?
- Can opponents—whether in the target set or outside—coordinate other than through the actions?

The section at the end of the chapter points to further reading on this topic.

7.7 Evolutionary learning and other large-population models

In this section we shift our focus from models of the learning of individual agents to models of the learning of populations of agents (although, as we shall see, we will not abandon the single-agent perspective altogether). When we speak about learning in a population of agents, we mean the change in the constitution and behavior of that population over time. These models were originally developed by population biologists to model the process of biological evolution, and later adopted and adapted by other fields.

In the first subsection we present the model of the *replicator dynamic*, a simple model inspired by evolutionary biology. In the second subsection we present the concept of *evolutionarily stable strategies*, a stability concept that is related to the replicator dynamic. We conclude with a somewhat different model of *agent-based simulation* and the concept of *emergent conventions*.

7.7.1 *The replicator dynamic*

replicator
dynamic

The *replicator dynamic* models a population undergoing frequent interactions. We will concentrate on the symmetric, two-player case, in which the agents repeatedly play a two-player symmetric normal-form stage game[13] against each other.

Definition 7.7.1 (Symmetric 2×2 **game)** *Let a two-player two-action normal-*

symmetric game   *form game be called a* symmetric game *if it has the following form:*

	A	B
A	x, x	u, v
B	v, u	y, y

Intuitively, this requirement says that the agents do not have distinct roles in the game, and the payoff for agents does not depend on their identities. We have already seen several instances of such games, including the Prisoner's Dilemma.[14]

The replicator dynamic describes a population of agents playing such a game in an ongoing fashion. At each point in time, each agent only plays a pure strategy. Informally speaking, the model then pairs all agents and has them play each other, each obtaining some payoff. This payoff is called the agent's *fitness*. At this point

fitness   the biological inspiration kicks in—each agent now "reproduces" in a manner proportional to this fitness, and the process repeats. The question is whether the process converges to a fixed proportion of the various pure strategies within the population, and if so to which fixed proportions.

The verbal description above is only meant to be suggestive. The actual mathematical model is a little different. First, we never explicitly model the play of the game between particular sets of players; we only model the proportions of the populations associated with a given strategy. Second, the model is not one of discrete repetitions of play, but rather one of continuous evolution. Third, beyond the fitness-based reproduction, there is also a random element that impacts the proportions in the population. (Again, because of the biological inspiration, this

mutation   random element is called *mutation*.)

The formal model is as follows. Given a normal-form game $G = (\{1, 2\}, A, u)$, let $\varphi_t(a)$ denote the number of players playing action a at time t. Also, let

$$\theta_t(a) = \frac{\varphi_t(a)}{\sum_{a' \in A} \varphi_t(a')}$$

---

13. There exist much more general notions of symmetric normal-form games with multiple actions and players, but the following is sufficient for our purposes.
14. This restriction to symmetric games is very convenient, simplifying both the substance and notation of what follows. However, there exist more complicated evolutionary models, including ones allowing both different strategy spaces for different agents and nonsymmetric payoffs. At the end of the chapter we point the reader to further reading on these models.

be the proportion of players playing action a at time t. We denote with φ_t the vector of measures of players playing each action, and with θ_t the vector of population shares for each action.

The expected payoff to any individual player for playing action a at time t is

$$u_t(a) = \sum_{a'} \theta_t(a')u(a, a').$$

The change in the number of agents playing action a at time t is defined to be proportional to his fitness, that is, his average payoff at the current time,

$$\dot{\varphi}_t(a) = \varphi_t(a)u_t(a).$$

The absolute numbers of agents of each type are not important; only the relative ratios are. Defining the average expected payoff of the whole population as

$$u_t^* = \sum_a \theta_t(a)u_t(a),$$

we have that the change in the fraction of agents playing action a at time t is

$$\dot{\theta}_t(a) = \frac{\left[\dot{\varphi}_t(a)\sum_{a'\in A}\varphi_t(a')\right] - \left[\varphi_t(a)\sum_{a'\in A}\dot{\varphi}_t(a')\right]}{\left[\sum_{a'\in A}\varphi_t(a')\right]^2} = \theta_t(a)[u_t(a) - u_t^*].$$

The system we have defined has a very intuitive quality. If an action does better than the population average then the proportion of the population playing this action increases, and vice versa. Note that even an action that is not a best response to the current population state can grow as a proportion of the population when its expected payoff is better than the population average.

How should we interpret this evolutionary model? A straightforward interpretation is that it describes agents repeatedly interacting and replicating within a large population. However, we can also interpret the fraction of agents playing a certain strategy as the mixed strategy of a single agent, and the process as that of two identical agents repeatedly updating their identical mixed strategies based on their previous interaction. Seen in this light, except for its continuous-time nature, the evolutionary model is not as different from the repeated-game model as it seems at first glance.

We would like to examine the equilibrium points in this system. Before we do, we need a definition of stability.

steady state   **Definition 7.7.2 (Steady state)** *A steady state of a population using the replicator dynamic is a population state θ such that for all $a \in A$, $\dot{\theta}(a) = 0$.*

In other words, a steady state is a state in which the population shares of each action are constant. This stability concept has a major flaw. Any state in which all players play the same action is a steady state. The population shares of the actions will remain constant because the replicator dynamic does not allow the "entry" of strategies that are not already being played. To disallow these states, we will often require that our steady states are *stable*.

stable steady state

Definition 7.7.3 (Stable steady state) *A steady state θ of a replicator dynamic is stable if there exists an $\epsilon > 0$ such that for every ϵ-neighborhood U of θ there exists another neighborhood U' of θ such that if $\theta_0 \in U'$ then $\theta_t \in U$ for all $t > 0$.*

That is, if the system starts close enough to the steady state, it remains nearby.

Finally, we might like to define an equilibrium state which, if perturbed, will eventually return back to the state. We call this *asymptotic stability*.

asymptotically stable state

Definition 7.7.4 (Asymptotically stable state) *A steady state θ of a replicator dynamic is* asymptotically stable *if it is stable, and in addition there exists an $\epsilon > 0$ such that for every ϵ-neighborhood U of θ it is the case that if $\theta_0 \in U$ then $\lim_{t \to \infty} \theta_t = \theta$.*

The following example illustrates some of these concepts. Consider a homogeneous population playing the Anti-Coordination game, repeated in Figure 7.9.

$$
\begin{array}{c c c}
 & A & B \\
A & \boxed{\;0,0\;} & \boxed{\;1,1\;} \\
B & \boxed{\;1,1\;} & \boxed{\;0,0\;}
\end{array}
$$

Figure 7.9 The Anti-Coordination game.

The game has two pure-strategy Nash equilibria, (A, B) and (B, A), and one mixed-strategy equilibrium in which both players select actions from the distribution $(0.5, 0.5)$. Because of the symmetric nature of the setting, there is no way for the replicator dynamic to converge to the pure-strategy equilibria. However, note that the state corresponding to the mixed-strategy equilibrium is a steady state, because when half of the players are playing A and half are playing B, both strategies have equal expected payoff (0.5) and the population shares of each are constant. Moreover, notice that this state is also asymptotically stable. The replicator dynamic, when started in any other state of the population (where the share of players playing A is more or less than 0.5) will converge back to the state $(0.5, 0.5)$. More formally we can express this as

$$
\begin{aligned}
\dot{\theta}(A) &= \theta(A)(1 - \theta(A) - 2\theta(A)(1 - \theta(A))) \\
&= \theta(A)(1 - 3\theta(A) + 2\theta(A)^2).
\end{aligned}
$$

This expression is positive for $\theta(A) < 0.5$, exactly 0 at 0.5, and negative for $\theta(A) > 0.5$, implying that the state $(0.5, 0.5)$ is asymptotically stable.

This example suggests that there may be a special relationship between Nash equilibria and states in the replicator dynamic. Indeed, this is the case, as the following results indicate.

Theorem 7.7.5 *Given a normal-form game $G = (\{1, 2\}, A = \{a_1, \ldots, a_k\}, u)$, if the strategy profile (s, s) is a (symmetric) mixed strategy Nash equilibrium of G then the population share vector $\theta = (s(a_1), \ldots, s(a_k))$ is a steady state of the replicator dynamic of G.*

In other words, every symmetric Nash equilibrium is a steady state. The reason for this is quite simple. In a state corresponding to a mixed Nash equilibrium, all strategies being played have the same average payoff, so the population shares remain constant.

As mentioned above, however, it is not the case that every steady state of the replicator dynamic is a Nash equilibrium. In particular, states in which not all actions are played may be steady states because the replicator dynamic cannot introduce new actions, even when the corresponding mixed-strategy profile is not a Nash equilibrium. On the other hand, the relationship between Nash equilibria and *stable* steady states is much tighter.

Theorem 7.7.6 *Given a normal-form game* $G = (\{1, 2\}, A\{a_1, \ldots, a_k\}, u)$ *and a mixed strategy* s, *if the population share vector* $\theta = (s(a_1), \ldots, s(a_k))$ *is a stable steady state of the replicator dynamic of* G, *then the strategy profile* (s, s) *is a mixed strategy Nash equilibrium of* G.

In other words, every stable steady state is a Nash equilibrium. It is easier to understand the contrapositive of this statement. If a mixed-strategy profile is not a Nash equilibrium, then some action must have a higher payoff than some of the actions in its support. Then in the replicator dynamic the share of the population using this better action will increase, once it exists. Then it is not possible that the population state corresponding to this mixed-strategy profile is a stable steady state.

Finally, we show that asymptotic stability corresponds to a notion that is stronger than Nash equilibrium. Recall the definition of trembling-hand perfection (Definition 3.4.14), reproduced here for convenience.

Definition 7.7.7 (Trembling-hand perfect equilibrium) *A mixed strategy profile* s *is a* (trembling-hand) *perfect equilibrium of a normal-form game* G *if there exists a sequence* $s^0, s^1, \ldots$ *of fully mixed-strategy profiles such that* $\lim_{n \to \infty} s^n = s$, *and such that for each* s^k *in the sequence and each player* i, *the strategy* s_i *is a best response to the strategies* s^k_{-i}.

Furthermore, we say informally that an equilibrium strategy profile is *isolated* if there does not exist another equilibrium strategy profile in the neighborhood (i.e., reachable via small perturbations of the strategies) of the original profile. Then we can relate trembling-hand perfection to the replicator dynamic as follows.

Theorem 7.7.8 *Given a normal-form game* $G = (\{1, 2\}, A, u)$ *and a mixed strategy* s, *if the population share vector* $\theta = (s(a_1), \ldots, s(a_k))$ *is an asymptotically stable steady state of the replicator dynamic of* G, *then the strategy profile* (s, s) *is a Nash equilibrium of* G *that is trembling-hand perfect and isolated.*

7.7.2   *Evolutionarily stable strategies*

evolutionarily
stable strategy
(ESS)

An *evolutionarily stable strategy (ESS)* is a stability concept that was inspired by the replicator dynamic. However, unlike the steady states discussed earlier, it does not require the replicator dynamic, or any dynamic process, explicitly;

	H	D
H	$-2, -2$	$6, 0$
D	$0, 6$	$3, 3$

Figure 7.10 Hawk–Dove game.

rather it is a static solution concept. Thus in principle it is not inherently linked to learning.

Roughly speaking, an evolutionarily stable strategy is a mixed strategy that is "resistant to invasion" by new strategies. Suppose that a population of players is playing a particular mixed strategy in the replicator dynamic. Then suppose that a small population of "invaders" playing a different strategy is added to the population. The original strategy is considered to be an ESS if it gets a higher payoff against the resulting mixture of the new and old strategies than the invaders do, thereby "chasing out" the invaders.

More formally, we have the following.

Definition 7.7.9 (Evolutionarily stable strategy (ESS)) *Given a symmetric two-player normal-form game $G = (\{1, 2\}, A, u)$ and a mixed strategy s, we say that s is an evolutionarily stable strategy if and only if for some $\epsilon > 0$ and for all other strategies s' it is the case that*

$$u(s, (1 - \epsilon)s + \epsilon s') > u(s', (1 - \epsilon)s + \epsilon s').$$

We can use properties of expectation to state this condition equivalently as

$$(1 - \epsilon)u(s, s) + \epsilon u(s, s') > (1 - \epsilon)u(s', s) + \epsilon u(s', s').$$

Note that, since this only needs to hold for small ϵ, this is equivalent to requiring that either $u(s, s) > u(s', s)$ holds, or else both $u(s, s) = u(s', s)$ and $u(s, s') > u(s', s')$ hold. Note that this is a strict definition. We can also state a weaker definition of ESS.

Definition 7.7.10 (Weak ESS) *s is a* weak evolutionarily stable strategy *if and only if for some $\epsilon > 0$ and for all s' it is the case that either $u(s, s) > u(s', s)$ holds, or else both $u(s, s) = u(s', s)$ and $u(s, s') \geq u(s', s')$ hold.*

weak evolutionarily stable strategy

This weaker definition includes strategies in which the invader does just as well against the original population as it does against itself. In these cases the population using the invading strategy will not grow, but it will also not shrink.

We illustrate the concept of ESS with the instance of the *Hawk–Dove* game shown in Figure 7.10. The story behind this game might be as follows. Two animals are fighting over a prize such as a piece of food. Each animal can choose between two behaviors: an aggressive hawkish behavior H, or an accommodating dovish behavior D. The prize is worth 6 to each of them. Fighting costs each player 5. When a hawk meets a dove he gets the prize without a fight, and hence the payoffs are 6 and 0, respectively. When two doves meet they split the prize without a fight, hence a payoff of 3 to each one. When two hawks meet a fight breaks out, costing each player 5 (or, equivalently, yielding −5). In addition, each player has a 50% chance of ending up with the prize, adding an expected benefit of 3, for an overall payoff of −2.

It is not hard to verify that the game has a unique symmetric Nash equilibrium (s, s), where $s = (\frac{3}{5}, \frac{2}{5})$, and that s is also the unique ESS of the game. To confirm that s is an ESS, we need that for all $s' \neq s$, $u(s, s) = u(s', s)$ and $u(s, s') > u(s', s')$. The equality condition is true of any mixed strategy equilibrium with full support, so follows directly. To demonstrate that the inequality holds, it is sufficient to find the s'—or equivalently, the probability of playing H—that minimizes $f(s') = u(s, s') - u(s', s')$. Expanding $f(s')$ we see that it is a quadratic equation with the (unique) maximum $s' = s$, proving our result.

This connection between an ESS and a Nash equilibrium is not accidental. The following two theorems capture this connection.

Theorem 7.7.11 *Given a symmetric two-player normal-form game $G = (\{1, 2\}, A, u)$ and a mixed strategy s, if s is an evolutionarily stable strategy then (s, s) is a Nash equilibrium of G.*

This is easy to show. Note that by definition an ESS S must satisfy

$$u(s, s) \geq u(s', s).$$

In other words, it is a best response to itself and thus must be a Nash equilibrium. However, not every Nash equilibrium is an ESS; this property is guaranteed only for strict equilibria.

Theorem 7.7.12 *Given a symmetric two-player normal-form game $G = (\{1, 2\}, A, u)$ and a mixed strategy s, if (s, s) is a strict (symmetric) Nash equilibrium of G, then s is an evolutionarily stable strategy.*

This is also easy to show. Note that for any strict Nash equilibrium s it must be the case that

$$u(s, s) > u(s', s).$$

But this satisfies the first criterion of an ESS.

The ESS also is related to the idea of stability in the replicator dynamic.

Theorem 7.7.13 *Given a symmetric two-player normal-form game $G = (\{1, 2\}, A, u)$ and a mixed strategy s, if s is an evolutionarily stable strategy then it is an asymptotically stable steady state of the replicator dynamic of G.*

Intuitively, if a state is an ESS then we know that it will be resistant to invasions by other strategies. Thus, when this strategy is represented by a population in the replicator dynamic, it will be resistant to small perturbations. What is interesting, however, is that the converse is *not* true. The reason for this is that in the replicator dynamic, only pure strategies can be inherited. Thus some states that are asymptotically stable would actually not be resistant to invasion by a mixed strategy, and thus not an ESS.

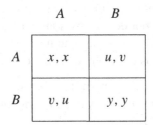

	A	B
A	x, x	u, v
B	v, u	y, y

Figure 7.11 A game for agent-based simulation models.

7.7.3 *Agent-based simulation and emergent conventions*

It was mentioned in Section 7.7.1 that, while motivated by a notion of dynamic process within a population, in fact the replicator dynamic only models the gross statistics of the process, not its details. There are other large-population models that provide a more fine-grained model of the process, with many parameters that can impact the dynamics. We call such models, which explicitly model the individual agents, *agent-based simulation* models.

agent-based simulation In this we look at one such model, geared toward the investigation of how conventions emerge in a society. In Section 2.4 we saw how in any realistic multi-agent system it is crucial that the agents agree on certain *social laws*, in order to decrease conflicts among them and promote cooperative behavior. Without such laws even the simplest goals might become unattainable by any of the agents, or at least not efficiently attainable (just imagine driving in the absence of traffic rules). A social law restricts the options available to each agent. A special case of social laws are *social conventions*, which limit the agents to exactly one option from the many available ones (e.g., always driving on the right side of the road). A good social law or convention strike as balance between on the one hand allowing agents sufficient freedom to achieve their goals, and on the other hand restricting them so that they do not interfere too much with one another.

social law

social convention

In Section 2.4 we asked how social laws and conventions can be designed by a social designer, but here we ask how such conventions can emerge organically. Roughly speaking, the process we aim to study is one in which individual agents occasionally interact with one another, and as a result gain some new information. Based on his personal accumulated information, each agent updates his behavior over time. This process is reminiscent of the replicator dynamic, but there are crucial differences. We start in the same way, and restrict the discussion to symmetric, two-player-two-choices games. Here too one can look at much more general settings, but we will restrict ourselves to the game schema in Figure 7.11.

However, unlike the replicator dynamic, here we assume a discrete process, and furthermore assume that at each stage exactly one pair of agents—selected at random from the population—play. This contrasts sharply with the replicator dynamic, which can be interpreted as implicitly assuming that almost all pairs of agents play before updating their choices of action. In this discrete model each agent is tracked individually, and indeed different agents end up possessing very different information.

Most importantly, in contrast with the replicator dynamic, the evolution of the system is not defined by some global statistics of the system. Instead, each agent decides how to play the next game based on his individual accumulated experience thus far. There are two constraints we impose on such rules.

anonymous learning rule

Property 7.7.14 (Anonymity) *The selection function cannot be based on the identities of agents or the names of actions.*

local learning rule

Property 7.7.15 (Locality) *The selection function is purely a function of the agent's personal history; in particular, it is not a function of global system properties.*

The requirement of anonymity deserves some discussion. We are interested in how social conventions emerge when we cannot anticipate in advance the games that will be played. For example, if we know that the coordination problem will be that of deciding whether to drive on the left of the road or on the right, we can very well use the names "left" and "right" in the action-selection rule; in particular, we can admit the trivial update rule that has all agents drive on the right immediately. Instead, the type of coordination problem we are concerned with is better typified by the following example. Consider a collection of manufacturing robots that have been operating at a plant for five years, at which time a new collection of parts arrive that must be assembled. The assembly requires using one of two available attachment widgets, which were introduced three years ago (and hence were unknown to the designer of the robots five years ago). Either of the widgets will do, but if two robots use different ones then they incur the high cost of conversion when it is time for them to mate their respective parts. Our goal is that the robots learn to use the same kind of widget. The point to emphasize about this example is that five years ago the designer could have stated rules of the general form "if in the future you have several choices, each of which has been tried this many times and has yielded this much payoff, then next time make the following choice"; the designer could not, however, have referred to the specific choices of widget, since those were only invented two years later.

The prohibition on using agent identities in the rules (e.g., "if you see Robot 17 use a widget of a certain type then do the same, but if you see Robot 5 do it then never mind") is similarly motivated. In a dynamic society agents appear and disappear, denying the designer the ability to anticipate membership in advance. One can sometimes refer to the *roles* of agents (such as Head Robot), and have them treated in a special manner, but we will not discuss this interesting aspect here.

Finally, the notion of "personal history" can be further honed. We will assume that the agent has access to the action he has taken and the reward he received at each instance. One could assume further that the agent observes the choices of others in the games in which he participated, and perhaps also their payoffs. But we will look specifically at an action-selection rule that does not make this

highest cumulative reward (HCR)

assumption. This rule, called the *highest cumulative reward (HCR)* rule, is the following learning procedure:

1. Initialize the cumulative reward for each action (e.g., to zero).
2. Pick an initial action.

3. Play according to the current action and update its cumulative reward.
4. Switch to a new action iff the total payoff obtained from that action in the latest *m* iterations is greater than the payoff obtained from the currently chosen action in the same time period.
5. Go to step 3.

The parameter *m* in the procedure denotes a finite bound, but the bound may vary. HCR is a simple and natural procedure, but it admits many variants. One can consider rules that use a weighted accumulation of feedback rather than simple accumulation, or ones that normalize the reward somehow rather than looking at absolute numbers. However even this basic rule gives rise to interesting properties. In particular, under certain conditions it guarantees convergence to a "good" convention.

Theorem 7.7.16 *Let g be a symmetric game as defined earlier, with $x > 0$ or $y > 0$ or $x = y > 0$, and either $u < 0$ or $v < 0$ or $x < 0$ or $y < 0$. Then if all agents employ the HCR rule, it is the case that for every $\epsilon > 0$ there exists an integer δ such that after δ iterations of the process the probability that a social convention is reached is greater than $1 - \epsilon$. Once a convention is reached, it is never left. Furthermore, this convention guarantees to the agent a payoff which is no less than the maxmin value of g.*

There are many more questions to ask about the evolution of conventions: How quickly does a convention evolve? How does this time depend on the various parameters, for example *m*, the history remembered? How does it depend on the initial choices of action? How does the particular convention reached— since there are many—depend on these variables? The discussion below points the reader to further reading on this topic.

7.8   History and references

There are quite a few broad introductions to, and textbooks on, single-agent learning. In contrast, there are few general introductions to the area of *multiagent* learning. Fudenberg and Levine [1998] provide a comprehensive survey of the area from a game-theoretic perspective, as does Young [2004]. A special issue of the *Journal of Artificial Intelligence* [Vohra and Wellman, 2007] looked at the foundations of the area. Parts of this chapter are based on Shoham et al. [2007] from that special issue. Some of the specific references are as follows.

Fictitious play was introduced by Brown [1951] and Robinson [1951]. The convergence results for fictitious play in Theorem 7.2.5 are taken respectively from Robinson [1951], Nachbar [1990], Monderer and Shapley [1996b] and Berger [2005]. The *non*-convergence example appeared in Shapley [1964].

Rational learning was introduced and analyzed by Kalai and Lehrer [1993]. A rich literature followed, but this remains the seminal paper on the topic.

Single-agent reinforcement learning is surveyed in Kaelbling et al. [1996]. Some key publications in the literature include Bellman [1957] on value iteration in known MDPs, and Watkins [1989] and Watkins and Dayan [1992] on Q-learning in unknown MDPs. The literature on multiagent reinforcement learning begins with Littman [1994]. Some other milestones in this line of research are as follows. Littman and Szepesvari [1996] completed the story regarding zero-sum games, Claus and Boutilier [1998] defined belief-based reinforcement learning and showed experimental results in the case of pure coordination (or team) games, and Hu and Wellman [1998], Bowling and Veloso [2001], and Littman [2001] attempted to generalize the approach to general-sum games. The R-max algorithm was introduced by Brafman and Tennenholtz [2002], and its predecessor, the E3 algorithm, by Kearns and Singh [1998].

The notion of no-regret learning can be traced to Blackwell's approachability theorem [Blackwell, 1956] and Hannan's notion of Universal Consistency [Hannan, 1957]. A good review of the history of this line of thought is provided in Foster and Vohra [1999]. The regret-matching algorithm and the analysis of its convergence to correlated equilibria appears in Hart and Mas-Colell [2000]. Modifications of fictitious play that exhibit no regret are discussed in Fudenberg and Levine [1995] and Fudenberg and Levine [1999].

Targeted learning was introduced in Powers and Shoham [2005b], and further refined and extended in Powers and Shoham [2005a] and Vu et al. [2006]. (However, the term *targeted learning* was invented later to apply to this approach to learning.)

The replicator dynamic is borrowed from biology. While the concept can be traced back at least to Darwin, work that had the most influence on game theory is perhaps Taylor and Jonker [1978]. The specific model of replicator dynamics discussed here appears in Schuster and Sigmund [1982]. The concept of evolutionarily stable strategies (ESSs) again has a long history, but was most explicitly put forward in Maynard Smith and Price [1973]—which also introduced the Hawk–Dove game—and figured prominently a decade later in the seminal Maynard Smith [1982]. Experimental work on learning and the evolution of cooperation appears in Axelrod [1984]. It includes discussion of a celebrated tournament among computer programs that played a finitely repeated Prisoner's Dilemma game and in which the simple Tit-for-Tat strategy emerged victorious. Emergent conventions and the HCR rule were introduced in Shoham and Tennenholtz [1997].

8

Communication

Agents communicate; this is one of the defining characteristics of a multiagent system. In traditional linguistic analysis, the communication is taken to have a certain form (syntax), to carry a certain meaning (semantics), and to be influenced by various circumstances of the communication (pragmatics). As we shall see, a closer look at communication adds to the complexity of the story. We can distinguish between purely *informational* theories of communication and *motivational* ones. In informational communication, agents simply inform each other of different facts. The theories of belief change, introduced in Chapter 14, look at ways in which beliefs change in the face of new information—depending on whether the beliefs are logical or probabilistic, consistent with prior beliefs or not. In this chapter we broaden the discussion and consider motivational theories of communication, involving agents with individual motivations and possible courses of actions.

We divide the discussion into three parts. The first concerns *cheap talk* and describes a situation in which self-motivated agents can engage in costless communication before taking action. As we see, in some situations this talk influences future behavior, and in some it does not. Cheap talk can be viewed as "doing by talking"; in contrast, *signaling games* can be viewed as "talking by doing." In signaling games an agent can take actions that, by virtue of the underlying incentives, communicate to the other agent something new. Since these theories draw on game theory, cheap talk and signaling both apply in cooperative as well as in competitive situations. In contrast, *speech-act theory*, which draws on philosophy and linguistics, applies in purely cooperative situations. It describes pragmatic ways in which language is used not only to convey information but to effect change; as such, it too has the flavor of "doing by talking."

8.1 "Doing by talking" I: cheap talk

Consider the Prisoner's Dilemma game, reproduced here in Figure 8.1. Recall that the game has a unique equilibrium in dominant strategies, the strategy profile (D, D), which is ironically also the only outcome that is not Pareto optimal; both players would do better if they both choose C instead. Suppose now that the prisoners are allowed to communicate before they play; will this change the outcome of the game? Intuitively, the answer is no. Regardless of the other agent's action, the given agent's best action is still D; the other agent's talk is indeed

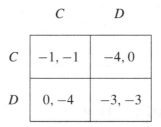

Figure 8.1 The Prisoner's Dilemma game.

cheap. Furthermore, regardless of his true intention, it is the interest of a given agent to get the other agent to play C; his talk is not only cheap, but also not credible (or, as the saying goes, the talk is free—and worth every penny).

Contrast this with cheap talk prior to the Coordination game given in Figure 8.2.

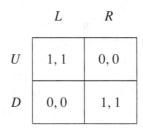

Figure 8.2 Coordination game.

self-committing
utterance

self-revealing
utterance

Here, if the row player declares "I will play U" prior to playing the game, the column player should take this seriously. Indeed, this utterance by the row player is both *self-committing* and *self-revealing*. These two notions are related but subtly different. A declaration of intent is self-committing if, once uttered, and assuming it is believed, the optimal course of action for the player is indeed to act as declared. In this example, if the column player believes the utterance "I will play U," then his best response is to play L. But then the row player's best response is indeed to play U. In contrast, an utterance is self-revealing if, assuming that it is uttered with the expectation that it will be believed, it is uttered only when indeed the intention was to act that way. In our case, a row player intending to play D will never announce the intention to play U, and so the utterance is self-revealing.

It must be mentioned that the precise analysis of this example, as well as the later examples, is subtle in a number of ways. In particular, the equilibrium analysis reveals other, less desirable equilibria than the ones in which a meaningful message is transmitted and received. For example, this example has another, less obvious equilibrium. The column player could ignore anything the row player says, allowing its beliefs to be unaffected by signals. In this case, the row player has no incentive to say anything in particular, and he might as well "babble," that is, send signals that are uncorrelated with his type. For this

babbling
equilibrium

reason, we call this a *babbling equilibrium*. In theory, every cheap talk game has a babbling equilibrium; there is always an equilibrium in which one party sends a meaningless signal and the other party ignores it. An equilibrium that

revealing
equilibrium

focal point

Stag Hunt game

is not a babbling equilibrium is called a *revealing equilibrium*. In a similar fashion one can have odd equilibria in which messages are not ignored but are used in a nonstandard way. For example, the row player might send the signal U when she means D and vice versa, so long as the column player adopts the same convention. However, going forward we will ignore these complications, and assume a meaningful and straightforward communication among the parties.

It might seem that self-commitment and self-revelation are inseparable, but this is an artifact of the pure coordination nature of the game. In such games the utterance creates a so-called *focal point*, a signal on which the agents can coordinate their actions. But now consider the well-known *Stag Hunt game*, whose payoff matrix is shown in Figure 8.3.

	Stag	Hare
Stag	9, 9	0, 8
Hare	8, 0	7, 7

Figure 8.3 Payoff matrix for the Stag Hunt game.

In the story behind this game, Artemis and Calliope are about to go hunting, and are trying to decide whether they want to hunt stag or hare. If both hunt stag, they do very well; if one tries to hunt stag alone, she fails completely. On the other hand, if one hunts rabbits alone, she will do well, for there is no competition; if both hunt rabbits together, they only do OK, for they each have competition.

In each cell of the matrix, Artemis' payoff is listed first and Calliope's payoff is listed second. This game has a symmetric mixed-strategy equilibrium, in which each player hunts stag with probability $\frac{7}{8}$, yielding an expected utility of $7\frac{7}{8}$. But now suppose Artemis can speak to Calliope before the game; can he do any better? The answer is arguably yes. Consider the message "I plan to hunt stag." It is not self-revealing; Artemis would like Calliope to believe this, even if she does not actually plan to hunt stag. However, it *is* self-committing; if Artemis were to think that Calliope believes her, then Artemis would actually prefer to hunt stag. There is however the question of whether Calliope would believe the utterance, knowing that it is not self-revealing on the part of Artemis.

For this reason, some view self-commitment without self-revelation as a notion lacking force. To gain further insight into this issue, let us define the Stag Hunt game more generally. Consider the game in Figure 8.4. Here, if x is less than 7, then the message "I plan to hunt stag" is possibly credible. However, if x is greater than 7, then that message is not at all credible, because it is in Artemis' best interest to get Calliope to hunt stag, no matter what Artemis actually intends to play.

We have so far spoken about communication in the context of games of perfect information. In such games all that can possibly be revealed in the intention to

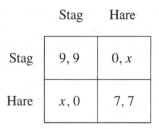

Figure 8.4 More general payoff matrix for the Stag Hunt game.

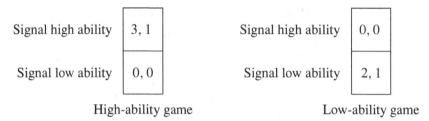

Figure 8.5 Payoff matrix for the Job Hunt game.

act a certain way. In games of incomplete information, however, there is an opportunity to reveal one's own private information prior to acting.

Consider the following example. The Acme Corporation wants to hire Sally into one of two positions: a demanding and an undemanding position. Sally may have high or low ability. Sally prefers the demanding position if she has high ability (because of salary and intellectual challenge) and she prefers the undemanding positions if she instead has low ability (because it will be more manageable). Acme too prefers that Sally be in the demanding position if she has high ability, and that she be in the undemanding position if she is of low ability. The actual game being played is determined by Nature; for concreteness, let us assume that selection is done with uniform probability. Importantly, however, only Sally knows what her true ability level is. However, before they play the game, Sally can send Acme a signal about her ability level. Suppose for the sake of simplicity that Sally can only choose from two signals: "My ability is low," and "My ability is high." Note that Sally may choose to be either sincere or insincere. The situation is modeled by the two games in Figure 8.5; in each cell of the matrix, Sally's payoff is listed first, and Acme's payoff is listed second.

What signal should Sally send? It seems obvious that she should tell the truth. She has no incentive to lie about her ability. If she were to lie, and Acme were to believe her, then she would receive a lower payoff than if she had told the truth. Acme knows that she has no reason to lie and so will believe her. Thus there in an equilibrium in which when Sally has low ability she says so, and Acme gives her an undemanding job, and when Sally has high ability she also says so, and Acme gives her a demanding job. The message is therefore self-signaling; assuming she will be believed, Sally will send the message only if it is true.

8.2 "Talking by doing": signaling games

We have so far discussed the situation in which talk preceded action. But sometimes actions speak louder than words. In this section we consider a class of imperfect-information games called *signaling games*.

signaling game   **Definition 8.2.1 (Signaling game)** *A signaling game is a two-player game in which Nature selects a game to be played according to a commonly known distribution, player 1 is informed of that choice and chooses an action, and player 2 then chooses an action without knowing Nature's choice, but knowing player 1's choice.*

In other words, a signaling game is an extensive-form game in which player 2 has incomplete information.

It is tempting to model player 2's decision problem as follows. Since each of the possible games has a different set of payoffs, player 2 must first calculate the posterior probability distribution over possible games, given the message that she received from player 1. She can calculate this using Bayes rule with the prior distribution over games and the conditional probabilities of player 1's message given the game. More precisely, the expected payoff for each action is as follows.

$$
\begin{aligned}
u_2(a, m) &= \mathbb{E}(u_2(g, m, a)|m, a) \\
&= \sum_{g \in G} u_2(g, m, a) P(g|m, a) \\
&= \sum_{g \in G} u_2(g, m, a) P(g|m) \\
&= \sum_{g \in G} u_2(g, m, a) \frac{P(m|g)P(g)}{P(m)} \\
&= \sum_{g \in G} u_2(g, m, a) \frac{P(m|g)P(g)}{\sum_{g \in G} P(m|g)P(g)}
\end{aligned}
$$

One problem with this formulation is that the use of Bayes' rule requires that the probabilities involved be nonzero. But more acutely, how does player 2 calculate the probability of player 1's message given a certain game? This is not at all obvious in light of the fact that player 2 knows that player 1 knows that player 2 will go through such reasoning, et cetera. Indeed, even if player 1 has a dominant strategy in the game being played the situation is not straightforward. Consider the following signaling game. Nature chooses with equal probability one of the two zero-sum normal-form games given in Figure 8.6.

Recall that player 1 knows which game is being played, and will choose his message first (U or D), and then player 2, who does not know which game is being played, will choose his action (L or R). What should player 1 do?

Note that in the leftmost game (U, R) is an equilibrium in dominant strategies, and in rightmost game (D, L) is an equilibrium in dominant strategies. Since player 2's preferred action depends entirely on the game being played, and he is confident that player 1 will play his dominant strategy, his best response is R if

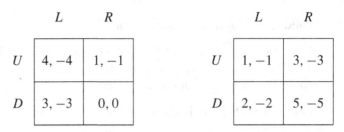

Figure 8.6 A signaling setting: Nature chooses randomly between the two games.

player 1 chooses U, and L if player 1 chooses D. If player 2 plays in this fashion, we can calculate the expected payoff to player 1 as

$$\mathbb{E}(u_1) = (0.5)1 + (0.5)2 = 1.5.$$

This seems like an optimal strategy. However, consider a different strategy for player 1. If player 1 always chooses D, regardless of what game he is playing, then his payoff is independent of player 2's action. We calculate the expected payoff to player 1 as follows, assuming that player 2 plays L with probability p and R with probability $(1 - p)$:

$$\mathbb{E}(u_1) = (0.5)(3p + 0(1 - p)) + (0.5)(2p + 5(1 - p)) = 2.5.$$

Thus, player 1 has a higher expected payoff if he always chooses the message D.

The example highlights an interesting property of signaling games. Although player 1 has privileged information, it may not always be to his advantage to exploit it. This is because by exploiting the advantage, he is effectively telling player 2 what game is being played and thereby losing his advantage. Thus, in some cases player 1 can receive a higher payoff by ignoring his information.

Signaling games fall under the umbrella term of *games of asymmetric information*. One of the best-known examples is the so-called *Spence signaling game*, which offers a rationale for enrolling in a difficult academic program. Consider the situation in which an employer is trying to decide how much to pay a new worker. The worker may or may not be talented, and can signal this to the employer by choosing to get either a high or low level of education. Specifically, we can model the setting as a Bayesian game between an employer and a worker in which *Nature* first chooses the level of the worker's talent, θ, to be either θ_L or θ_H, such that $\theta_L < \theta_H$. This value of θ defines two different possible games. In each possible game, the worker's strategy space is the level of education e to get for both possible types, or level of talent. We use e_L to refer to the level of education chosen by the worker if his talent level is θ_L and e_H for the education chosen if his talent level is θ_H. We assume that the worker knows his talent.

Finally, the employer's strategy specifies two wages, w_H and w_L, to offer a worker based on whether his signal is e_H or e_L. We assume that the employer does not know the level of talent of the worker, but does get to observe his level of education. The employer is assumed to have two choices. One is to ignore the signal and set $w_H = w_L = p_H\theta_H + p_L\theta_L$, where $p_L + p_H = 1$ are the probabilities with which *Nature* chooses a high and low talent for the worker.

games of asymmetric information

Spence signaling game

The other is to pay a worker with a high education w_H and a worker with a low education w_L.

The payoff to the employer is $\theta - w$, the difference between the talent of the worker and the payment to him. The payoff to the worker is $w - e/\theta$, reflecting the assumption that education is easier when talent is higher.

pooling
equilibrium

This game has two equilibria. The first is a *pooling equilibrium*, in which the worker will choose the same level of education regardless of his type ($e_L = e_H = e^*$), and the employer pays all workers the same amount. The other is a *separating equilibrium*, in which the worker will choose a different level of education depending on his type. In this case a low-talent worker will choose to get no education, $e_L = 0$, because the wage paid to this worker is w_L, independent of e_L. The education chosen by a high-talent worker is set in such a way as to make it unprofitable for either type of worker to mimic the other. This is the case only if the following two inequalities are satisfied.

separating
equilibrium

$$\theta_L \geq \theta_H - e_H/\theta_L$$
$$\theta_L \leq \theta_H - e_H/\theta_H$$

These inequalities can be rewritten in terms of e_H as

$$\theta_L(\theta_H - \theta_L) \leq e_H \leq \theta_H(\theta_H - \theta_L).$$

Note that since $\theta_H > \theta_L$, a separating equilibrium always exists.

8.3 "Doing by talking" II: speech-act theory

Human communication is as rich and imprecise as natural language, tone, affect, and body language permit, and human motivations are similarly complex. It is not surprising that philosophers and linguists have attempted to model such communication. As mentioned at the very start of the chapter, human communication is analyzed on many different levels of abstraction, among them the *syntactic*, *semantic*, and *pragmatic* levels. The discussion of speech acts lies squarely within the pragmatic level, although it should be noted that there are legitimate arguments against a crisp separation among these layers.

8.3.1 *Speech acts*

The traditional view of communication is that it is the sharing of information. Speech-act theory, due to the philosopher J. L. Austin, embodies the insight that some communications can instead be viewed as actions, intended to achieve some goal.

Speech-act theory distinguishes between three different kinds of speech acts, or, if you wish, three levels at which an utterance can be analyzed. The *locutionary act* is merely the emission of a signal carrying a certain meaning. When I say "there's a car coming your way," the locution refers to the content transmitted. Locutions establish a proposition, which may be true or false. However, the

locutionary act

illocutionary act   utterance can also be viewed as an *illocutionary act*, which in this case is a
warning. In general, an illocution is the invocation of a conventional force on the
receiver through the utterances. Other illocutions can be making a request, telling
a joke, or, indeed, simply informing.

perlocutionary   Finally, if the illocution captures the intention of the speaker, the *perlocution-*
act   *ary act* is bringing about an effect on the hearer as a result of an utterance. Although the illocutionary and perlocutionary acts may seem similar, it is important
to distinguish between an illocutionary act and its perlocutionary consequences.
Illocutionary acts do something *in* saying something, while perlocutionary acts do
something *by* saying something. Perlocutionary acts include scaring, convincing,
and saddening. In our car example, the perlocution would be an understanding
by the hearer of the imminent danger causing him to jump from in front of the
car.

performative   Illocutions thus may or may not be successful. *Performatives* constitute a type
of act that is inherently successful. Merely saying something achieves the desired
effect. For example, the utterance "please get off my foot" (or, somewhat more
stiffly, "I hereby request you to get off my foot") is a performative. The speaker
asserts that the utterance is a request, and is thereby successful in communicating
the request to the listener, because the listener assumes that the speaker is an
expert on his own mental state. Some utterances are performatives only under
some circumstances. For example, the statement "I hereby pronounce you man
and wife" is a performative only if the speaker is empowered to conduct marriage
ceremonies in that time and place, if the rest of the ceremony follows protocol, if
the bride and groom are eligible for marriage, and so on.[1]

8.3.2   *Rules of conversation*

Building on the notion of speech acts as a foundation, another important contribu-
rules of   tion to language pragmatics takes the form of *rules of conversation*, as developed
conversation   by P. Grice, another philosopher. The simple observation is that humans seem
to undertake the act of conversation cooperatively. Humans generally seek to
understand and be understood when engaging in conversation, even when other
motivations may be at odds. It is in both parties' best interest to communicate
cooperative   clearly and efficiently. This is called the *cooperative principle*.
principle   It is also the case that humans generally follow some basic rules when con-
versing, which presumably help them to achieve the larger shared goal of the
Cooperative Principle. These rules have come to be known as the *Gricean max-*
Gricean maxims   *ims*. The four Gricean maxims are *quantity*, *quality*, *relation*, and *manner*. We
discuss each one in turn.

The rule of *quantity* states that humans tend to provide listeners with exactly
the amount of information required in the current conversation, even when they
have access to more information. As an example, imagine that a waitress asks
you, "how do you like your coffee?" You would probably answer, "Cream, no
sugar, please," or something similar. You would probably not answer, "I like

---

1. It is however interesting to contemplate a world in which any such utterance results in a marriage.

arabica beans, grown in the mountains of Guatemala. I prefer the medium roast from Peet's Coffee. I like to buy whole beans, which I keep in the freezer, and grind them just before brewing. I like the coffee strong, and served with a dash of cream." The latter response clearly provides the waitress with much more information than she needs. You also probably would not respond, "no sugar," because this does not give the waitress enough information to do her job.

The rule of *quality* states that humans usually only say things that they actually believe. More specifically, humans do not say things they know to be false, and do not say things for which they lack adequate evidence. For example, if someone asks you about the weather outside, you respond that it is raining only if in fact you believe that it is raining, and if you have evidence to support that belief.

The rule of *relation* states that humans tend to say things that are relevant to the current conversation. If a stranger approaches you on the street to ask for directions to the nearest gas station, they would be quite surprised if you began to tell them a story about your grandmother's cooking.

Finally, the rule of *manner* states that humans generally say things in a manner that is brief and clear. When you are asked at the airport whether anyone unknown to you has asked you to carry something in your luggage, the appropriate answer is either "yes" or "no," not "many people assume that they know their family members, but what does that really mean?" In general, humans tend to avoid obscurity, ambiguity, prolixity, and disorganization.

These maxims help explain a surprising phenomenon about human speech, namely that we often succeed in communicating much more meaning than is contained directly in the words they say. This phenomenon is called *implicature*. For example, suppose that *A* and *B* are talking about a mutual friend, *C*, who is now working in a bank. *A* asks *B* how *C* is getting on in his job, and *B* replies, "Oh quite well, I think; he likes his colleagues, and he has not been to prison yet." Clearly, by stating the simple fact that *C* hasn't been to prison yet, which is a truism for most people, *B* is implying, suggesting, or meaning something else. He may mean that *C* is the kind of person who is likely to yield to temptation or that *C*'s colleagues are really very treacherous people, for example. In this case the implicature may be clear from the context of their conversation, or *A* may have to ask *B* what he means.

Grice distinguished between *conventional* and *nonconventional* implicature. The former refers to the case in which the conventional meaning of the words used determines what is implicated. In the latter, the implication does not follow directly from the conventional meaning of the words, but instead follows from context, or from the structure of the conversation, as is the case in *conversational implicatures*.

In conversational implicatures, the implied meaning relies on the fact that the hearer assumes that the speaker is following the Gricean maxims. Let us begin with an example. *A* is standing by an immobilized car, and is approached by *B*. *A* says, "I am out of gas." *B* says, "There is a garage around the corner." Although *B* does not explicitly say it, she implicates effectively that she thinks that the garage is open and sells gasoline. This follows immediately from the assumption that *B* is following the Gricean maxims of relation and quality. If she were not following

(margin notes:) implicature

conversational implicature

the maxim of relation, her utterance about the garage could be a *non sequitur*; if she were not following the maxim of quality, she could be lying. In order for a conversational implicature to occur, (1) the hearer must assume that the speaker is following the maxims, (2) this assumption is necessary for the hearer to get the implied meaning, and (3) it is common knowledge that the hearer can work out the implication.

Grice offers three types of conversational implicature. In the first, no maxim is violated, as in the aforementioned example. In the second, a maxim is violated, but the hearer assumes that the violation is because of a clash with another maxim. For example, if *A* asks, "Where does *C* live?" and *B* responds, "Somewhere in the South of France," *A* can presume that *B* does not know more and thus violates the maxim of quantity in order to obey the maxim of quality. Finally, in the third type of conversational implicature, a maxim is flouted, and the hearer assumes that there must be another reason for it. For example, when a recommendation letter says very little about the candidate in question, the maxim of quantity is flouted, and the reader can safely assume that there is very little positive to say.

We give some examples of commonly-occurring conversational implicatures. Humans often use an *if* statement to implicate an *if and only if* statement. Suppose *A* says to *B*, "If you teach me speech act theory I'll kiss you." In this case, if *A* did not mean *if and only if*, then *A* might kiss *B* whether or not *B* teaches *A* speech act theory. Then *A* would have been violating the maxim of quantity, telling the *B* something that did not contain any useful information.

In another common case, people often make a direct statement as a way to implicate that they believe the statement. When *A* says to *B*, "Austin was right," *B* is meant to implicate, "*A* believes Austin was right." Otherwise, *A* would have been violating the maxim of quality.

Finally, humans use a presupposition to implicate that the presupposition is true. When *A* says to *B*, "Grice's maxims are incomplete," *A* intends *B* to assume that Grice has axioms. Otherwise, *A* would have been violating the maxim of quality.

indirect speech   Note that conversational implicatures enable *indirect* speech acts. Consider
act   the classic Eddie Murphy skit in which his mother says to him, "It's cold in here, Eddie." Although her utterance is on the surface merely an informational locution, it is in fact implicating a request for Eddie to do something to warm up the room.

8.3.3   *A game-theoretic view of speech acts*

The discussion of speech acts so far has clearly been relatively discursive and informal as compared to the discussion in the other sections, and indeed to most of the book. This reflects the nature of the work in the field. There are advantages to the relative laxness; it enables a very broad and multifaceted theory. Indeed, quite a number of researchers and practitioners in several disciplines have drawn inspiration from speech act theory. But it also comes at a price, as the theory can be pushed only so far before the missing details halt progress.

One could look in a number of directions for such formal foundations. Since the definition of speech acts appeals to the mental state of the speaker and hearer, one could plausibly try to apply the formal theories of mental state discussed later in the book, and in particular theories of attitudes such as belief, desire and intention. Section 14.4 outlines one such theory, but also makes it clear that so-called BDI theories are not yet fully developed. Here we will explore a different direction. Our starting point is the fact that there are at least two agents involved in communication, the speaker and the hearer. So why not model this as a game between them, in the sense of game theory, and analyze that game?

disambiguation

Although this direction too is not yet well developed, we shall see that some insights can be gleaned from the game-theoretic perspective. We illustrate this via the phenomenon of *disambiguation* in language. One of the factors that render natural language understanding so hard is that speech is rife with ambiguities at all levels, from the phonemic through the lexical to the sentence and whole text level. We will analyze the following sentence-level ambiguity:

Every ten minutes a person gets mugged in New York City.

The intended interpretation is of course that every ten minutes some different person gets mugged. The unintended, but still permissible, alternative interpretation that the same person gets mugged over and over again. (Indeed, if one adds the sentence "I feel very bad for him," the implausible interpretation becomes the only permissible one.) How do the hearer and speaker implicitly understand which interpretation is intended?

One way is to set this up as a common-payoff game of incomplete information between the speaker and hearer (indeed, as we shall see, in this example we end up with a signaling game as defined in Section 8.2, albeit a purely cooperative one). The game proceeds as follows:

1. There exist two situations:

   s: Muggings of different people take place at ten-minute intervals in NYC.

   t: The same person is repeatedly mugged every ten minutes in NYC.

2. Nature selects between s and t according to a distribution known commonly to A and B.

3. Nature's choice is revealed to A but not to B.

4. A decides between uttering one of three possible sentences:

   p: "Every ten minutes a person gets mugged in New York City."

   q: "Every ten minutes some person or another gets mugged in New York City."

   r: "There is a person who gets mugged every ten minutes in New York City."

5. B hears A, and must decide whether s or t obtain.

This is a simplified view of the world (more on this shortly), but let us simplify it even further. Let us assume that A cannot utter r when t obtains, and cannot utter q when s obtains (i.e., he can be ambiguous, but not deceptive). Let us furthermore assume that when B hears either r or q he has no interpretation decision, and knows exactly which situation obtains (s or t, respectively).

In order to analyze the game, we must supply some numbers. Let us assume that the probability of s is much higher that that of t. Say, $P(s) = .99$ and $P(t) = .01$. Finally, we need to decide on the payoffs. We assume that this is a game of pure coordination, that is a common-payoff game. A and B jointly have the goal that B correctly have the right interpretation. In addition, though, both A and B have a preference for simple sentences, since long sentences place a cognitive burden on them and waste time. And so the payoffs are as follows: If the sentence used is p and a correct interpretation is reached, the payoff is 10. If either q or r are uttered (after which by assumption a correct interpretation is reached), the payoff is 7; and if an incorrect interpretation is reached the payoff is -10.

The resulting game is depicted in Figure 8.7.

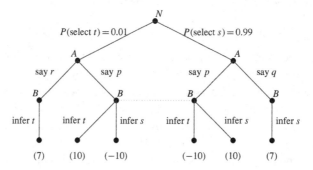

Figure 8.7 Communication as a signaling game.

What are the equilibria of this game? Here are two.

1. A's strategy: say q in s and r in t. B's strategy: When hearing p, select between the s and t interpretations with equal probability.
2. A's strategy: say p in s and r in t. B's strategy: When hearing p, select the s interpretation.

First, you should persuade yourself that these are in fact Nash equilibria (and even subgame-perfect ones, or, to be more precise since this is a game of imperfect information, sequential equilibria). But then we might ask, is there a reason to prefer one over the other? Well, one way to select is based on the expected payoff to the players. After all, this is a cooperative game, and it makes sense to expect the players to coordinate on the equilibrium with the highest payoff. Indeed, this would be one way to implement Grice's cooperative principle. Note that in the first equilibrium the (common) payoff is 7, while in the second equilibrium the expected payoff is $0.99 \cdot 10 + 0.01 \cdot 7 = 9.97$. And so it would seem that we have a winner on our hands, and a particularly pleasing one since this use of language accords well with real-life usage. Intuitively, to economize we use shorthand for commonly-occurring situations. This allows the hearer to make some default assumptions, but use more verbose language in the relatively rare situations in which those defaults are misleading.

This example can be extended in various ways. *A* can be given the freedom to say other sentences, and *B* can be given greater freedom to interpret them. Not only could *A* say *q* in *s*, but *A* could even say "I like cucumbers" in *s*. This is no less useful a sentence than *p*, so long as *B* conditions its interpretation correctly on it. The problem is of course that we end up with infinitely many good equilibria, and payoff maximization cannot distinguish between them. And focal point so language can be seen to have evolved so as to provide *focal points* among these equilibria; the "straightforward interpretation" of the sentence is a device to coordinate on one of the optimal equilibria.

Although we are still far from being able to account for the entire pragmatics of language in this fashion, one can apply similar analysis to more complex linguistic phenomena, and it remains an interesting area of investigation.

8.3.4 *Applications*

The framework of speech-act theory has been put to practical use in a number of computer science and artificial intelligence applications. We give a brief description of some of these applications below.

Intelligent dialog systems

dialog system One obvious application of speech act theory is a *dialog system*, which communicates with human users through a natural language dialog interface. In order to communicate efficiently and naturally with the user, dialog systems must obey the principles of human conversation, including those from Austin and Grice presented in this chapter.

TRAINS/TRIPS is a well-known dialog system, and is to assist the user in accomplishing tasks in a transportation domain. The system has access to information about the state of the transportation network, and the user makes decisions about what actions to take. The system maintains an ongoing conversation with the user about possible actions and the state of the network.

The TRAINS/TRIPS dialog system both uses and extends the principles of speech act theory. It incorporates a *Speech Act Interpreter*, which hypothesizes what speech acts the user is making, and a *Dialog Manager*, which uses knowledge of those acts to maintain the dialog. It extends speech act theory by creating a hierarchy of *conversation acts*, as shown in Table 8.1. As you can see, speech acts appear in this framework as the conversation acts that occur at the discourse level.

Workflow systems

Another useful application of speech act theory is in workflow software, software used to track and manage complex interactions within and between human organizations. These interactions range from simple business transactions to long-term collaborative projects, and each requires the involvement of many different human participants. To track and manage the interactions effectively, workflow software provides a medium for structured communications between all of the participants.

Discourse level	Act type	Sample acts
Multidiscourse	Argumentation acts	elaborate, summarize, clarify, convince
Discourse	Speech acts	inform, accept, request, suggest, offer, promise
Utterance	Grounding acts	initiate, continue, acknowledge, repair
Subutterance	Turn-taking acts	take-turn, keep-turn, release-turn, assign-turn

Table 8.1 Conversation acts used by the TRAINS/TRIPS system.

Many workflow applications are designed around an information processing framework, in which, for example, interactions may be modeled as assertions and queries to a database. This perspective is useful, but lacks an explicit understanding and representation of the pragmatic structure of human communications. An alternative is to view each communication as an illocutionary speech act, which states an intention on the part of the sender and places constraints on the possible responses of the recipient. Instead of generic messages, as in the case of email communications, users must choose from a set of communication types when composing messages to other participants. Within this framework, they can write freely. For example, when responding to a request, users might be given the following options.

- Acknowledge
- Promise
- Free form
- Counter offer
- Commit-to-commit
- Decline
- Interim report
- Report completion

The speech act framework confers a number of advantages to developers and users of workflow software. Because the basic unit of communication is a conversation, rather than a message, the organization of communications is straightforward, and retrieval simple. Furthermore, the status and urgency of messages is clear. Users can ask "In which conversations is someone waiting for me to do something?" or "In which conversations have I promised to do things?". Finally, access to messages can be organized and controlled easily, depending on project involvement and authorization levels. The downside is that it involves additional overhead in the communication, which may not be justified by the benefits, especially if the conversational structures implemented in the system do not capture well the rich set of communications that takes place in the workplace.

Agent communication languages

Perhaps the most widespread use of speech act theory within the field of computer science is for communication between software applications. Increasingly, computer systems are structured in such a way that individual applications can act as agents (e.g., with the popularization of the Internet and electronic commerce), each with its own goals and planning mechanisms. In such a system, software applications must communicate with each other and with their human users to enlist the support of other agents to achieve goals, to commit to helping another agent, to report their own status, to request a status report from another, and so on.

KQML

Not surprisingly, several proposals have been made for artificial languages to serve as the medium for this interapplication communication. A relatively simple example is presented by *Knowledge Query and Manipulation Language (KQML)*, which was developed in the early 1990s. KQML incorporates some ideas from speech-act theory, especially the idea of performatives. It has a built-in set of performatives, such as ACHIEVE, ADVERTISE, BROKER, REGISTER, and TELL.

The following is an example of a KQML message, taken from a communication between two applications operating in the blocks world domain.

```
(tell
        :sender      Agent1
        :receiver    Agent2
        :language    KIF
        :ontology    Blocks-World
        :content     (AND (Block A) (Block B) (On A B)))
```

Note that the message is a performative. The content of the message uses blocks world semantics, which are completely independent of the semantics of the performative itself.

XML

Semantic Web

KQML is no longer an influential standard, but the ideas of structured interactions among software agents that are based in part on speech acts live on in more modern protocols defined on top of abstract markup languages such as *XML* and the so-called *Semantic Web*.

Rational programming

We have described how speech act theory can be used in communication between software applications. Some authors have also proposed to use it directly in the development of software applications, that is, as part of a programming language itself. This proposal is part of a more general effort to introduce elements of rationality into programming languages. This new programming paradigm has

rational
programming

been termed *rational programming*. Just as object-oriented programming shifted the paradigm from writing procedures to creating objects, rational programming shifts the paradigm from creating informational objects to creating motivational agents.

So where does communication come in? The motivational agents created by rational programming must act in the world, and because the agents are not likely to have a physical embodiment, their actions consist of sending and receiving signals; in other words, their actions will be speech acts. Of course, as shown in the previous section, it is possible to construct communicating agents within existing programming paradigms. However, by incorporating speech acts as primitives, rational programming constructs make such programs more powerful, easier to create, and more readable.

We give a few examples for clarity. *Elephant2000* is a programming language described by McCarthy which explicitly incorporates speech acts. Thus, for example, an Elephant2000 program can make a promise to another and cannot renege on a promise. The following is an example statement from Elephant2000, taken from a hypothetical travel agency program:

```
if ¬ full (flight)
    then accept.request(
        make (commitment (admit (psgr, flight))))
```

The intuitive reading of this statement is "if a passenger has requested to reserve a seat on a given flight, and that flight is not full, then make the reservation."

Agent-Oriented Programming (AOP)

Agent-Oriented Programming (AOP) is a separate proposal that is similar in several respects. It too embraces speech acts as the form and meaning of the communication among agents. The most significant difference from Elephant2000 is that AOP also embraces the notion of mental state, consisting of beliefs and commitments. Thus the result of an inform speech act is a new belief. AOP is not actually a single language, but a general design that allows multiple languages; one particular simple language, *Agent0* was defined and implemented. The following is an example statement in Agent0, taken from a print server application.

```
IF
    MSG COND: (?msgId ?someone REQUEST
                    ((FREE-PRINTER 5min) ?time))
    MENTAL COND:
                    ((NOT (CMT ?other (PRINT ?doc (?time+10min))))
                     (B (FRIENDLY ?someone)))
    THEN COMMIT
        ((?someone (FREE-PRINTER 5min) ?time)
         (myself (INFORM ?someone (ACCEPT ?msgId)) now))
```

The approximate reading of this statement is "if you get a request to free the printer for five minutes at a future time, if you are not committed to finishing a print job within ten minutes of that time, and if you believe the requester to be friendly, then accept the request and tell them that you did."

8.4 History and references

The literature on language and natural language understanding is of course vast and we cannot do it justice here. We will focus on the part of the literature that bears most directly on the material presented in the chapter.

Two early seminal discussions on cheap talk are due to Crawford and Sobel [1982] and Farrell [1987]. Later references include Rabin [1990] and Farrell [1993]. Good overviews are given by Farrell [1995] and Farrell and Rabin [1996].

The literature on signaling games dates considerably farther back. The Stackelberg leadership model, later couched in game-theoretic terminology (as we do in the book), was introduced by Heinrich von Stackelberg, a German economist, as a model of duopoly in economics [von Stackelberg, 1934]. The literature on information economics, and in particular on asymmetric information, continued to flourish, culminating in the 2001 Nobel Prize awarded to three pioneers in the area (Akerlof, Spence, and Stiglitz). The Spence signaling game, which we cover in the chapter, appeared in Spence [1973].

Austin's seminal book, *How to Do Things with Words*, was published in 1962 [Austin, 1962], but is also available in a more recent second edition [Austin, 2006]. Grice's ideas were developed in several publications starting in the late 1960s, for example, Grice [1969]. This and many other of his relevant publications were collected in Grice [1989]. Another important reference is Searl [1979]. The game-theoretic perspective on speech acts is more recent. The discussion here for the most part follows Parikh [2001]. Another recent reference covering a number of issues at the interface of language and economics is Rubinstein [2000].

The TRAINS dialog system is described by Allen et al. [1995], and the TRIPS system is described by Ferguson and Allen [1998]. The speech-act-based approach to workflow systems follows the ideas of Winograd and Flores [1986] and Flores et al. [1988]. The KQML language is described by Finin et al. [1997]. The term *rational programming* was coined by Shoham [1997]. Elements of the Elephant2000 programming language are described by McCarthy [1994]. The AOP framework was described by Shoham [1993].

9

Aggregating Preferences: Social Choice

In the preceding chapters we adopted what might be called the "agent perspective": we asked what an agent believes or wants, and how an agent should or would act in a given situation. We now adopt a complementary, "designer perspective": we ask what rules should be put in place by the authority (the "designer") orchestrating a set of agents. In this chapter this will take us away from game theory, but before too long (in the next two chapters) it will bring us right back to it.

9.1 Introduction

A simple example of the designer perspective is voting. How should a central authority pool the preferences of different agents so as to best reflect the wishes of the population as a whole? It turns out that voting, the kind familiar from our political and other institutions, is only a special case of the general class of *social choice problems*. Social choice is a motivational but nonstrategic theory—agents have preferences, but do not try to camouflage them in order to manipulate the outcome (of voting, for example) to their personal advantage.[1] This problem is thus analogous to the problem of belief fusion that we present in Section 14.2.1, which is also nonstrategic; here, however, we examine the problem of aggregating preferences rather than beliefs.

social choice problem

We start with a brief and informal discussion of the most familiar voting scheme, plurality. We then give the formal model of social choice theory, consider other voting schemes, and present two seminal results about the sorts of preference aggregation rules that it is possible to construct. Finally, we consider the problem of building ranking systems, where agents rate each other.

9.1.1 *Example: plurality voting*

To get a feel for social choice theory, consider an example in which you are babysitting three children—Will, Liam, Vic—and need to decide on an activity for them. You can choose among going to the video arcade (a), playing basketball (b), and driving around in a car (c). Each kid has a different preference over these

1. Some sources use the term "social choice" to refer to both strategic and nonstrategic theories; we do not follow that usage here.

activities, which is represented as a strict total ordering over the activities and which he reveals to you truthfully. By $a \succ b$ denote the proposition that outcome a is preferred to outcome b.

$$
\begin{aligned}
\text{Will:} \quad & a \succ b \succ c \\
\text{Liam:} \quad & b \succ c \succ a \\
\text{Vic:} \quad & c \succ b \succ a
\end{aligned}
$$

What should you do? One straightforward approach would be to ask each kid to vote for his favorite activity and then to pick the activity that received the largest number of votes. This amounts to what is called the *plurality* method. While quite standard, this method is not without problems. For one thing, we need to select a tie-breaking rule (e.g., we could select the candidate ranked first alphabetically). A more disciplined way is to hold a runoff election among the candidates tied at the top.

plurality voting

Even absent a tie, however, the method is vulnerable to the criticism that it does not meet the *Condorcet condition*. This condition states that if there exists a candidate x such that if for all other candidates y at least half the voters prefer x to y, then x must be chosen. If each child votes for his top choice, the plurality method would declare a tie between all three candidates and, in our example, would choose a. However, the Condorcet condition would choose b, since two of the three children prefer b to a, and likewise prefer b to c.

Condorcet condition

Based on this example the Condorcet rule might seem unproblematic (and actually useful since it breaks the tie without resorting to an arbitrary choice such as alphabetical ordering), but now consider a similar example in which the preferences are as follows.

$$
\begin{aligned}
\text{Will:} \quad & a \succ b \succ c \\
\text{Liam:} \quad & b \succ c \succ a \\
\text{Vic:} \quad & c \succ a \succ b
\end{aligned}
$$

In this case the Condorcet condition does not tell us what to do, illustrating the fact that it does not tell us how to aggregate arbitrary sets of preferences. We will return to the question of what properties can be guaranteed in social choice settings; for the moment, we aim simply to illustrate that social choice is not a straightforward matter. In order to study it precisely, we must establish a formal model. Our definition will cover voting, but will also handle more general situations in which agents' preferences must be aggregated.

9.2 A formal model

Let $N = \{1, 2, \ldots, n\}$ denote a set of agents, and let O denote a finite set of outcomes (or alternatives, or candidates). Making a multiagent extension to the preference notation introduced in Section 3.1.2, denote the proposition that agent i weakly prefers outcome o_1 to outcome o_2 by $o_1 \succeq_i o_2$. We use the notation $o_1 \succ_i o_2$ to capture strict preference (shorthand for $o_1 \succeq_i o_2$ and not $o_2 \succeq_i o_1$) and $o_1 \sim_i o_2$ to capture indifference (shorthand for $o_1 \succeq_i o_2$ and $o_2 \succeq_i o_1$).

preference
ordering

Because preferences are transitive, an agent's preference relation induces a *preference ordering*, a (nonstrict) total ordering on O. Let $L.$ be the set of nonstrict total orders; we will understand each agent's preference ordering as an element of $L.$. Overloading notation, we denote an element of $L.$ using the same symbol we used for the relational operator: $\succeq_i \in L.$. Likewise, we define

preference
profile

a *preference profile* $[\succeq] \in L.^n$ as a tuple giving a preference ordering for each agent.

Note that the arguments in Section 3.1.2 show that preference orderings and utility functions are tightly related. We can define an ordering $\succeq_i \in L.$ in terms of a given utility function $u_i : O \mapsto \mathbb{R}$ for an agent i by requiring that o_1 is weakly preferred to o_2 if and only if $u_i(o_1) \geq u_i(o_2)$.

In what follows, we define two kinds of social functions. In both cases, the input is a preference profile. Both classes of functions aggregate these preferences, but in a different way.

Social choice functions simply select one of the alternatives (or, in a more general version, some subset).

social choice
function

Definition 9.2.1 (Social choice function) *A* social choice function *(over N and O) is a function* $C : L.^n \mapsto O$.

social choice
correspondence

A *social choice correspondence* differs from a social choice function only in that it can return a set of candidates, instead of just a single one.

social choice
correspondence

Definition 9.2.2 (Social choice correspondence) *A* social choice correspondence *(over N and O) is a function* $C : L.^n \mapsto 2^O$.

In our babysitting example there were three agents (Will, Liam, and Vic) and three possible outcomes (a, b, c). The social choice correspondence defined by plurality voting of course picks the subset of candidates with the most votes; in this example either the subset must be the singleton consisting of one of the candidates or else it must include all candidates. Plurality is turned into a social choice function by any deterministic tie-breaking rule (e.g., alphabetical).[2]

Let $\#(o_i \succ o_j)$ denote the number of agents who prefer outcome o_i to outcome o_j under preference profile $[\succeq] \in L.^n$. We can now give a formal statement of the Condorcet condition.

Condorcet
winner

Definition 9.2.3 (Condorcet winner) *An outcome* $o \in O$ *is a* Condorcet winner *if* $\forall o' \in O$, $\#(o \succ o') \geq \#(o' \succ o)$.

A social choice function satisfies the *Condorcet condition* if it always picks a Condorcet winner when one exists. We saw earlier that for some sets of preferences there does *not* exist a Condorcet winner. (Indeed, under reasonable conditions the probability that there will exist a Condorcet winner approaches zero as the number of candidates approaches infinity.) Thus, the Condorcet condition does not always tell us anything about which outcome to choose.

---

2. One can also define probabilistic versions of social choice functions; however, we will focus on the deterministic variety.

An alternative is to find a rule that identifies a *set* of outcomes among which we can choose. Extending on the idea of the Condorcet condition, a variety of other conditions have been proposed that are guaranteed to identify a nonempty set of outcomes. We will not describe such rules in detail; however, we give one prominent example here.

Smith set **Definition 9.2.4 (Smith set)** *The* Smith set *is the smallest set $S \subseteq O$ having the property that $\forall o \in S, \forall o' \notin S, \#(o \succ o') \geq \#(o' \succ o)$.*

That is, every outcome *in* the Smith set is preferred by at least half of the agents to every outcome *outside* the set. This set always exists. When there is a Condorcet winner then that candidate is also the only member of the Smith set; otherwise, the Smith set is the set of candidates who participate in a "stalemate" (or "top cycle").

The other important flavor of social function is the *social welfare function*. These are similar to social choice functions, but produce richer objects, total orderings on the set of alternatives.

social welfare **Definition 9.2.5 (Social welfare function)** *A* social welfare function *(over N function and O) is a function $W : L_{\cdot}^{n} \mapsto L_{\cdot\cdot}$*

Although the usefulness of these functions is somewhat less intuitive, they are very important to social choice theory. We will discuss them further in Section 9.4.1, in which we present Arrow's famous impossibility theorem.

9.3 Voting

We now survey some important voting methods and discuss their properties. Then we demonstrate that the problem of voting is not as easy as it might appear, showing some counterintuitive ways in which these methods can behave.

9.3.1 *Voting methods*

nonranking The most standard class of voting methods is called *nonranking voting*, in which voting each agent votes for one of the candidates. We have already discussed plurality voting.

Definition 9.3.1 (Plurality voting) *Each voter casts a single vote. The candidate with the most votes is selected.*

As discussed earlier, ties must be broken according to a tie-breaking rule (e.g., based on a lexicographic ordering of the candidates; through a runoff election between the first-place candidates, etc.). Since the issue arises in the same way for all the voting methods we discuss, we will not belabor it in what follows.

Plurality voting gives each voter a very limited way of expressing his preferences. Various other rules are more generous in this regard. Consider *cumulative* cumulative *voting*. voting

Definition 9.3.2 (Cumulative voting) *Each voter is given k votes, which can be cast arbitrarily (e.g., several votes could be cast for one candidate, with the remainder of the votes being distributed across other candidates). The candidate with the most votes is selected.*

approval voting *Approval voting* is similar.

Definition 9.3.3 (Approval voting) *Each voter can cast a single vote for as many of the candidates as he wishes; the candidate with the most votes is selected.*

We have presented cumulative voting and approval voting to give a sense of the range of voting methods. We will defer discussion of such rules to Section 9.5, however, since in the (nonstrategic) voting setting as we have defined it so far, it is not clear how agents should choose when to vote for more than one candidate. Furthermore, although it is more expressive than plurality, approval voting still fails to allow voters to express their full preference orderings. Voting methods ranking voting that do so are called *ranking voting* methods. Among them, one of the best known is *plurality with elimination*; for example, this method is used for some plurality voting with elimination political elections. When preference orderings are elicited from agents before any elimination has occurred, the method is also known as *instant runoff*.

Definition 9.3.4 (Plurality with elimination) *Each voter casts a single vote for their most-preferred candidate. The candidate with the fewest votes is eliminated. Each voter who cast a vote for the eliminated candidate casts a new vote for the candidate he most prefers among the candidates that have not been eliminated. This process is repeated until only one candidate remains.*

Borda voting Another method which has been widely studied is *Borda voting*.

Definition 9.3.5 (Borda voting) *Each voter submits a full ordering on the candidates. This ordering contributes points to each candidate; if there are n candidates, it contributes $n - 1$ points to the highest ranked candidate, $n - 2$ points to the second highest, and so on; it contributes no points to the lowest ranked candidate. The winners are those whose total sum of points from all the voters is maximal.*

Nanson's method *Nanson's method* is a variant of Borda that eliminates the candidate with the lowest Borda score, recomputes the remaining candidates' scores, and repeats. This method has the property that it always chooses a member of the Condorcet set if it is nonempty, and otherwise chooses a member of the Smith set.

pairwise elimination Finally, there is *pairwise elimination*.

Definition 9.3.6 (Pairwise elimination) *In advance, voters are given a schedule for the order in which pairs of candidates will be compared. Given two candidates (and based on each voter's preference ordering) determine the candidate that each voter prefers. The candidate who is preferred by a minority of voters is eliminated, and the next pair of noneliminated candidates in the schedule is considered. Continue until only one candidate remains.*

9.3.2 *Voting paradoxes*

At this point it is reasonable to wonder why so many voting schemes have been invented. What are their strengths and weaknesses? For that matter, is there one voting method that is appropriate for all circumstances? We will give a more formal (and more general) answer to the latter question in Section 9.4. First, however, we will consider the first question by considering some sets of preferences for which our voting methods exhibit undesirable behavior. Our aim is not to point out every problem that exists with every voting method defined above; rather, it is to illustrate the fact that voting schemes that seem reasonable can often fail in surprising ways.

Condorcet condition

Let us start by revisiting the Condorcet condition. Earlier, we saw two examples: one in which plurality voting chose the Condorcet winner, and another in which a Condorcet winner did not exist. Now consider a situation in which there are 1,000 agents with three different sorts of preferences.

$$
\begin{array}{ll}
499 \text{ agents:} & a \succ b \succ c \\
3 \text{ agents:} & b \succ c \succ a \\
498 \text{ agents:} & c \succ b \succ a
\end{array}
$$

Observe that 501 people out of 1,000 prefer b to a, and 502 prefer b to c; this makes b the Condorcet winner. However, many of our voting methods would fail to select b as the winner. Plurality would pick a, as a has the largest number of first-place votes. Plurality with elimination would first eliminate b and would subsequently pick c as the winner. In this example Borda does select b, but there are other cases where it fails to select the Condorcet winner—can you construct one?

Sensitivity to a losing candidate

Consider the following preferences by 100 agents.

$$
\begin{array}{ll}
35 \text{ agents:} & a \succ c \succ b \\
33 \text{ agents:} & b \succ a \succ c \\
32 \text{ agents:} & c \succ b \succ a
\end{array}
$$

Plurality would pick candidate a as the winner, as would Borda. (To confirm the latter claim, observe that Borda assigns $a, b,$ and c the scores 103, 98, and 99 respectively.) However, if the candidate c did not exist, then plurality would pick b, as would Borda. (With only two candidates, Borda is equivalent to plurality.) A third candidate who stands no chance of being selected can thus act as a "spoiler," changing the selected outcome.

Another example demonstrates that the inclusion of a least-preferred candidate can even cause the Borda method to *reverse* its ordering on the other candidates.

$$
\begin{array}{ll}
3 \text{ agents:} & a \succ b \succ c \succ d \\
2 \text{ agents:} & b \succ c \succ d \succ a \\
2 \text{ agents:} & c \succ d \succ a \succ b
\end{array}
$$

Given these preferences, the Borda method ranks the candidates $c \succ b \succ a \succ d$, with scores of 13, 12, 11, and 6 respectively. If the lowest-ranked candidate d is dropped, however, the Borda ranking is $a \succ b \succ c$ with scores of 8, 7, and 6.

Sensitivity to the agenda setter

Finally, we examine the pairwise elimination method, and consider the influence that the agenda setter can have on the selected outcome. Consider the following preferences, which we discussed previously.

$$
\begin{array}{ll}
35 \text{ agents:} & a \succ c \succ b \\
33 \text{ agents:} & b \succ a \succ c \\
32 \text{ agents:} & c \succ b \succ a
\end{array}
$$

First, consider the order a, b, c. a is eliminated in the pairing between a and b; then c is chosen in the pairing between b and c. Second, consider the order a, c, b. a is chosen in the pairing between a and c; then b is chosen in the pairing between a and b. Finally, under the order b, c, a, we first eliminate b and ultimately choose a. Thus, given these preferences, the agenda setter can select whichever outcome he wants by selecting the appropriate elimination order!

Next, consider the following preferences.

$$
\begin{array}{ll}
1 \text{ agent:} & b \succ d \succ c \succ a \\
1 \text{ agent:} & a \succ b \succ d \succ c \\
1 \text{ agent:} & c \succ a \succ b \succ d
\end{array}
$$

Consider the elimination ordering a, b, c, d. In the pairing between a and b, a is preferred; c is preferred to a and then d is preferred to c, leaving d as the winner. However, *all* of the agents prefer b to d—the selected candidate is Pareto dominated!

Last, we give an example showing that Borda is fundamentally different from pairwise elimination, *regardless* of the elimination ordering. Consider the following preferences.

$$
\begin{array}{ll}
3 \text{ agents:} & a \succ b \succ c \\
2 \text{ agents:} & b \succ c \succ a \\
1 \text{ agent:} & b \succ a \succ c \\
1 \text{ agent:} & c \succ a \succ b
\end{array}
$$

Regardless of the elimination ordering, pairwise elimination will select the candidate a. The Borda method, on the other hand, selects candidate b.

9.4   Existence of social functions

The previous section has illustrated several senses in which some popular voting methods exhibit undesirable or unfair behavior. In this section, we consider this state of affairs from a more formal perspective, examining both social welfare functions and social choice functions.

In this section only, we introduce an additional assumption to simplify the exposition. Specifically, we will assume that all agents' preferences are *strict* total orderings on the outcomes, rather than nonstrict total orders; denote the set of such orders as L, and denote an agent i's preference ordering as $\succ_i \in L$. Denote a preference profile (a tuple giving a preference ordering for each agent) as $[\succ'] \in L^n$, and denote agent i's preferences from preference profile $[\succ']$ as $\succ_i'$. We also redefine social welfare functions to return a strict total ordering over the outcomes, $W : L^n \mapsto L$. In other words, we assume that no agent is ever indifferent between outcomes and that the social welfare function is similarly decisive. We stress that this assumption is *not required* for the results that follow; analysis of the general case can be found in the works cited at the end of the chapter.[3]

Finally, let us introduce some new notation. Social welfare functions take preference profiles as input; denote the preference ordering selected by the social welfare function W, given preference profile $[\succ'] \in L^n$, as $\succ_{W([\succ'])}$. When the input ordering $[\succ']$ is understood from context, we abbreviate our notation for the social ordering as $\succ_W$.

9.4.1 *Social welfare functions*

Arrow's impossibility theorem

In this section we examine *Arrow's impossibility theorem*, without a doubt the most influential result in social choice theory. Its surprising conclusion is that fairness is multifaceted and that it is *impossible* to achieve all of these kinds of fairness simultaneously.

Now, let us review these multifaceted notions of fairness.

Pareto efficiency (PE)

Definition 9.4.1 (Pareto efficiency (PE)) W *is* Pareto efficient *if for any* $o_1, o_2 \in O$, $\forall i\ o_1 \succ_i o_2$ *implies that* $o_1 \succ_W o_2$.

In words, PE means that when all agents agree on the ordering of two outcomes, the social welfare function must select that ordering. Observe that this definition is effectively the same as *strict Pareto efficiency* as defined in Definition 3.3.2.[4]

independence of irrelevant alternatives (IIA)

Definition 9.4.2 (Independence of irrelevant alternatives (IIA)) W *is* independent of irrelevant alternatives *if, for any* $o_1, o_2 \in O$ *and any two preference profiles* $[\succ'], [\succ''] \in L^n$, $\forall i\ (o_1 \succ_i' o_2$ *if and only if* $o_1 \succ_i'' o_2)$ *implies that* $(o_1 \succ_{W([\succ'])} o_2$ *if and only if* $o_1 \succ_{W([\succ''])} o_2)$.

That is, the selected ordering between two outcomes should depend only on the relative orderings they are given by the agents.

---

3. Intuitively, because we will be looking for social functions that work given *any* preferences the agents might have, when we show that desirable social welfare and social choice functions cannot exist even when agents are assumed to have strict preferences, we will also have shown that the claim holds when we relax this restriction.

4. One subtle difference does arise from our assumption in this section that all preferences are strict.

nondictatorship **Definition 9.4.3 (Nondictatorship)** W *does not have a* dictator *if* $\neg \exists i \, \forall o_1,$
$o_2(o_1 \succ_i o_2 \Rightarrow o_1 \succ_W o_2)$.

Nondictatorship means that there does not exist a single agent whose preferences always determine the social ordering. We say that W is *dictatorial* if it fails to satisfy this property.

Surprisingly, it turns out that there exists no social welfare function W that satisfies these three properties for all of its possible inputs. This result relies on our previous assumption that N is finite.

Theorem 9.4.4 (Arrow, 1951) *If $|O| \geq 3$, any social welfare function W that is Pareto efficient and independent of irrelevant alternatives is dictatorial.*

> **Proof.** We will assume that W is both PE and IIA and show that W must be dictatorial. The argument proceeds in four steps.
>
> **Step 1:** *If every voter puts an outcome b at either the very top or the very bottom of his preference list, b must be at either the very top or very bottom of $\succ_W$ as well.*
>
> Consider an arbitrary preference profile $[\succ]$ in which every voter ranks some $b \in O$ at either the very bottom or very top, and assume for contradiction that the preceding claim is not true. Then, there must exist some pair of distinct outcomes $a, c \in O$ for which $a \succ_W b$ and $b \succ_W c$.
>
> Now let us modify $[\succ]$ so that every voter moves c just above a in his preference ranking, and otherwise leaves the ranking unchanged; let us call this new preference profile $[\succ']$. We know from IIA that for $a \succ_W b$ or $b \succ_W c$ to change, the pairwise relationship between a and b and/or the pairwise relationship between b and c would have to change. However, since b occupies an extremal position for all voters, c can be moved above a without changing either of these pairwise relationships. Thus in profile $[\succ']$ it is also the case that $a \succ_W b$ and $b \succ_W c$. From this fact and from transitivity, we have that $a \succ_W c$. However, in $[\succ']$, every voter ranks c above a and so PE requires that $c \succ_W a$. We have a contradiction.
>
> **Step 2:** *There is some voter n^* who is* extremely pivotal *in the sense that by changing his vote at some profile, he can move a given outcome b from the bottom of the social ranking to the top.*
>
> Consider a preference profile $[\succ]$ in which every voter ranks b last, and in which preferences are otherwise arbitrary. By PE, W must also rank b last. Now let voters from 1 to n successively modify $[\succ]$ by moving b from the bottom of their rankings to the top, preserving all other relative rankings. Denote as n^* the first voter whose change causes the social ranking of b to change. There must clearly be some such voter: when the voter n moves b to the top of his ranking, PE will require that b be ranked at the top of the social ranking.
>
> Denote by $[\succ^1]$ the preference profile just before n^* moves b, and denote by $[\succ^2]$ the preference profile just after n^* has moved b to the top of his ranking. (These preference profiles are illustrated in Figures 9.1a and 9.1b, with the indicated positions of outcomes a and c in each agent's ranking

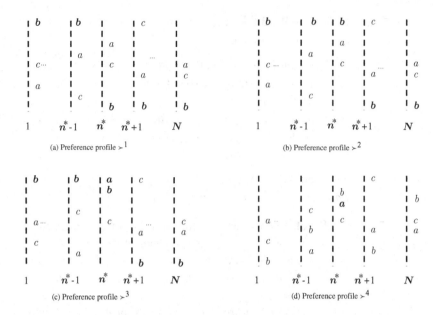

Figure 9.1 The four preference profiles used in the proof of Arrow's theorem. A higher position along the dotted line indicates a higher position in an agent's preference ordering. The outcomes indicated in bold (i.e., b in profiles $[\succ^1]$, $[\succ^2]$, and $[\succ^3]$ and a for voter n^* in profiles $[\succ^3]$ and $[\succ^4]$) must be in the exact positions shown. (In profile $[\succ^4]$, a must simply be ranked above c.) The outcomes not indicated in bold are simply examples and can occur in any relative ordering that is consistent with the placement of the bold outcomes.

serving only as examples.) In $[\succ^1]$, b is at the bottom in $\succ_w$. In $[\succ^2]$, b has changed its position in $\succ_w$, and every voter ranks b at either the top or the bottom. By the argument from Step 1, in $[\succ^2]$ b must be ranked at the top of $\succ_w$.

Step 3: n^* *(the agent who is extremely pivotal on outcome b) is a dictator over any pair ac not involving b.*

We begin by choosing one element from the pair ac; without loss of generality, let us choose a. We will construct a new preference profile $[\succ^3]$ from $[\succ^2]$ by making two changes. (Profile $[\succ^3]$ is illustrated in Figure 9.1c.) First, we move a to the top of n^*'s preference ordering, leaving it otherwise unchanged; thus $a \succ_{n^*} b \succ_{n^*} c$. Second, we arbitrarily rearrange the relative rankings of a and c for all voters other than n^*, while leaving b in its extremal position.

In $[\succ^1]$ we had $a \succ_w b$, as b was at the very bottom of $\succ_w$. When we compare $[\succ^1]$ to $[\succ^3]$, relative rankings between a and b are the same for all voters. Thus, by IIA, we must have $a \succ_w b$ in $[\succ^3]$ as well. In $[\succ^2]$ we had $b \succ_w c$, as b was at the very top of $\succ_w$. Relative rankings between b and c are the same in $[\succ^2]$ and $[\succ^3]$. Thus in $[\succ^3]$, $b \succ_w c$. Using the two aforementioned facts about $[\succ^3]$ and transitivity, we can conclude that $a \succ_w c$ in $[\succ^3]$.

Now construct one more preference profile, $[\succ^4]$, by changing $[\succ^3]$ in two ways. First, arbitrarily change the position of b in each voter's ordering while keeping all other relative preferences the same. Second, move a to an arbitrary position in n^*'s preference ordering, with the constraint that a remains ranked higher than c. (Profile $[\succ^4]$ is illustrated in Figure 9.1d.) Observe that all voters other than n^* have entirely arbitrary preferences in $[\succ^4]$, while n^*'s preferences are arbitrary except that $a \succ_{n^*} c$. In $[\succ^3]$ and $[\succ^4]$, all agents have the same relative preferences between a and c; thus, since $a \succ_W c$ in $[\succ^3]$ and by IIA, $a \succ_W c$ in $[\succ^4]$. Thus we have determined the social preference between a and c without assuming anything except that $a \succ_{n^*} c$.

Step 4: n^* *is a dictator over all pairs ab.*

Consider some third outcome c. By the argument in Step 2, there is a voter n^{**} who is extremely pivotal for c. By the argument in Step 3, n^{**} is a dictator over any pair $\alpha\beta$ not involving c. Of course, ab is such a pair $\alpha\beta$. We have already observed that n^* is able to affect W's ab ranking—for example, when n^* was able to change $a \succ_W b$ in profile $[\succ^1]$ into $b \succ_W a$ in profile $[\succ^2]$. Hence, n^{**} and n^* must be the same agent.

We have now shown that n^* is a dictator over all pairs of outcomes. ∎

9.4.2 *Social choice functions*

Arrow's theorem tells us that we cannot hope to find a voting scheme that satisfies all of the notions of fairness that we find desirable. However, maybe the problem is that Arrow's theorem considers too hard a problem—the identification of a social ordering over *all* outcomes. We now consider the setting of social choice functions, which are required only to identify a single top-ranked outcome. First, we define concepts analogous to Pareto efficiency, independence of irrelevant alternatives and nondictatorship for social choice functions.

weak Pareto efficiency
Definition 9.4.5 (Weak Pareto efficiency) *A social choice function C is weakly Pareto efficient if, for any preference profile $[\succ] \in L^n$, if there exist a pair of outcomes o_1 and o_2 such that $\forall i \in N$, $o_1 \succ_i o_2$, then $C([\succ]) \neq o_2$.*

This definition prohibits the social choice function from selecting any outcome that is dominated by another alternative for all agents. (That is, if all agents prefer o_1 to o_2, the social choice rule does not have to choose o_1, but it cannot choose o_2.) The definition implies that the social choice rule must respect agents' unanimous choices: if outcome o is the top choice according to each $\succ_i$, then we must have $C([\succ]) = o$. Thus, the definition is less demanding than *strict Pareto efficiency* as defined in Definition 3.3.2—a strictly Pareto efficient choice rule would also always satisfy weak Pareto efficiency, but the reverse is not true.

monotonicity
Definition 9.4.6 (Monotonicity) *C is monotonic if, for any $o \in O$ and any preference profile $[\succ] \in L^n$ with $C([\succ]) = o$, then for any other preference profile*

$[\succ']$ *with the property that* $\forall i \in N, \forall o' \in O, o \succ'_i o'$ *if* $o \succ_i o'$, *it must be that* $C([\succ']) = o$.

Monotonicity says that when a social choice rule C selects the outcome o for a preference profile $[\succ]$, then for any second preference profile $[\succ']$ in which, for every agent i, the set of outcomes to which o is preferred under $\succ'_i$ is a weak superset of the set of outcomes to which o is preferred under $\succ_i$, the social choice rule must also choose outcome o. Intuitively, monotonicity means that an outcome o must remain the winner whenever the support for it is increased relative to a preference profile under which o was already winning. Observe that the definition imposes no constraint on the relative orderings of outcomes $o_1, o_2 \neq o$ under the two preference profiles; for example, some or all of these relative orderings could be different.

nondictatorship **Definition 9.4.7 (Nondictatorship)** *C is nondictatorial if there does not exist an agent j such that C always selects the top choice in j's preference ordering.*

Following the pattern we followed for social welfare functions, we can show that no social choice function can satisfy all three of these properties.

Theorem 9.4.8 (Muller–Satterthwaite, 1977) *If $|O| \geq 3$, any social choice function C that is weakly Pareto efficient and monotonic is dictatorial.*

Before giving the proof, we must provide a key definition.

Definition 9.4.9 (Taking O' to the top from $[\succ]$) *Let $O' \subset O$ be a finite subset of the outcomes O, and let $[\succ]$ be a preference profile. Denote the set $O \setminus O'$ as $\overline{O'}$. A second preference profile $[\succ']$ takes O' to the top from $[\succ]$ if, for all $i \in N$, $o' \succ'_i o$ for all $o' \in O'$ and $o \in \overline{O'}$ and $o'_1 \succ'_i o'_2$ if and only if $o'_1 \succ_i o'_2$.*

That is, $[\succ']$ takes O' to the top from $[\succ]$ when, under $[\succ']$:

- each outcome from O' is preferred by every agent to each outcome from $\overline{O'}$; and
- the relative preferences between pairs of outcomes in O' for every agent are the same as the corresponding relative preferences under $[\succ]$.

Observe that the relative preferences between pairs of outcomes in $\overline{O'}$ are arbitrary: they are not required to bear any relation to the corresponding relative preferences under $[\succ]$.

We can now state the proof. Intuitively, it works by constructing a social welfare function W from the given social choice function C. We show that the facts that C is weakly Pareto efficient and monotonic imply that W must satisfy PE and IIA, allowing us to apply Arrow's theorem.

> **Proof.** We will assume that C satisfies weak Pareto efficiency and monotonicity, and show that it must be dictatorial. The proof proceeds in six steps.

Step 1: *If both* $[\succ']$ *and* $[\succ'']$ *take* $O' \subset O$ *to the top from* $[\succ]$*, then* $C([\succ']) = C([\succ''])$ *and* $C([\succ']) \in O'$.

Under $[\succ']$, for all $i \in N$, $o' \succ'_i o$ for all $o' \in O'$ and all $o \in \overline{O}'$. Thus, by weak Pareto efficiency $C([\succ']) \in O'$. For every $i \in N$, every $o' \in O'$ and every $o \neq o' \in O$, $o' \succ'_i o$ if and only if $o' \succ''_i o$. Thus by monotonicity, $C([\succ']) = C([\succ''])$.

Step 2: *We define a social welfare function W from C.*

For every pair of outcomes $o_1, o_2 \in O$, construct a preference profile $[\succ^{\{o_1,o_2\}}]$ by taking $\{o_1, o_2\}$ to the top from $[\succ]$. By Step 1, $C([\succ^{\{o_1,o_2\}}])$ will be either o_1 or o_2, and will always be the same regardless of how we choose $[\succ^{\{o_1,o_2\}}]$. Now we will construct a social welfare function W from C. For each pair of outcomes $o_1, o_2 \in O$, let $o_1 \succ_W o_2$ if and only if $C([\succ^{\{o_1,o_2\}}]) = o_1$.

In order to show that W is a social welfare function, we must demonstrate that it establishes a total ordering over the outcomes. Since W is complete, it only remains to show that W is transitive. Suppose that $o_1 \succ_W o_2$ and $o_2 \succ_W o_3$; we must thus show that $o_1 \succ_W o_3$. Let $[\succ']$ be a preference profile that takes $\{o_1, o_2, o_3\}$ to the top from $[\succ]$. By Step 1, $C([\succ']) \in \{o_1, o_2, o_3\}$. We consider each possibility.

Assume for contradiction that $C([\succ']) = o_2$. Let $[\succ'']$ be a profile that takes $\{o_1, o_2\}$ to the top from $[\succ']$. By monotonicity, $C([\succ'']) = o_2$ (o_2 has weakly improved its ranking from $[\succ']$ to $[\succ'']$). Observe that $[\succ'']$ also takes $\{o_1, o_2\}$ to the top from $[\succ]$. Thus by our definition of W, $o_2 \succ_W o_1$. But we already had $o_1 \succ_W o_2$. Thus, $C([\succ']) \neq o_2$. By an analogous argument, we can show that $C([\succ']) \neq o_3$.

Thus, $C([\succ']) = o_1$. Let $[\succ'']$ be a preference profile that takes $\{o_1, o_3\}$ to the top from $[\succ']$. By monotonicity, $C([\succ'']) = o_1$. Observe that $[\succ'']$ also takes $\{o_1, o_3\}$ to the top from $[\succ]$. Thus by our definition of W, $o_1 \succ_W o_3$, and hence we have shown that W is transitive.

Step 3: *The highest-ranked outcome in* $W([\succ])$ *is always* $C([\succ])$.

We have seen that C can be used to construct a social welfare function W. It turns out that C can also be recovered from W, in the sense that the outcome given the highest ranking by $W([\succ])$ will always be $C([\succ])$. Let $C([\succ]) = o_1$, let $o_2 \in O$ be any other outcome, and let $[\succ']$ be a profile that takes $\{o_1, o_2\}$ to the top from $[\succ]$. By monotonicity, $C([\succ']) = o_1$. By the definition of W, $o_1 \succ_W o_2$. Thus, o_1 is the outcome ranked highest by W.

Step 4: *W is Pareto efficient.*

Imagine that $\forall i \in N$, $o_1 \succ o_2$. Let $[\succ']$ take $\{o_1, o_2\}$ to the top from $[\succ]$. Since C is weakly Pareto efficient, $C([\succ']) = o_1$. Thus by the definition of W from Step 2, $o_1 \succ_W o_2$, and so W is Pareto efficient.

Step 5: *W is independent of irrelevant alternatives.*

Let $[\succ^1]$ and $[\succ^2]$ be two preference profiles with the property that for all $i \in N$ and for some pair of outcomes o_1 and $o_2 \in O$, $o_1 \succ^1_i o_2$ if and only if $o_1 \succ^2_i o_2$. We must show that $o_1 \succ_{W([\succ^1])} o_2$ if and only if $o_1 \succ_{W([\succ^2])} o_2$.

Let $[\succ^{1'}]$ take $\{o_1, o_2\}$ to the top from $[\succ^1]$, and let $[\succ^{2'}]$ take $\{o_1, o_2\}$ to the top from $[\succ^2]$. From the definition of W in Step 2, $o_1 \succ_{W([\succ^{1'}])} o_2$ if and only if $C([\succ^{1'}]) = o_1$; likewise, $o_1 \succ_{W([\succ^{2'}])} o_2$ if and only if $C([\succ^{2'}]) = o_1$. Now observe that $[\succ^{1'}]$ also takes $\{o_1, o_2\}$ to the top from $[\succ^2]$, because for all $i \in N$ the relative ranking between o_1 and o_2 is the same under $[\succ^1]$ and $[\succ^2]$. Thus by Part 1, $C([\succ^{1'}]) = C([\succ^{2'}])$, and hence $o_1 \succ_{W([\succ^{1'}])} o_2$ if and only if $o_1 \succ_{W([\succ^2])} o_2$.

Step 6: *C is dictatorial.*

From Steps 4 and 5 and Theorem 9.4.4, W is dictatorial. That is, there must be some agent $i \in N$ such that, regardless of the preference profile $[\succ]$, we always have $o_1 \succ_{W([\succ'])} o_2$ if and only if $o_1 \succ_i' o_2$. Therefore, the highest-ranked outcome in $W([\succ'])$ must also be the outcome ranked highest by i. By Step 3, $C([\succ'])$ is always the outcome ranked highest in $W([\succ'])$. Thus, C is dictatorial. ∎

In effect, this theorem tells us that, perhaps contrary to intuition, social choice functions are no simpler than social welfare functions. Intuitively, the proof shows that we can repeatedly "probe" a social choice function to determine the relative social ordering between given pairs of outcomes. Because the function must be defined for all inputs, we can use this technique to construct a full social welfare ordering.

To get a feel for the theorem, consider the social choice function defined by the plurality rule.[5] Clearly, it satisfies weak Pareto efficiency and is not dictatorial. This means it must be nonmonotonic. To see why, consider the following scenario with seven voters.

$$\begin{array}{ll} 3 \text{ agents:} & a \succ b \succ c \\ 2 \text{ agents:} & b \succ c \succ a \\ 2 \text{ agents:} & c \succ b \succ a \end{array}$$

Denote these preferences as $[\succ^1]$. Under $[\succ^1]$ plurality chooses a. Now consider the situation where the final two agents increase their support for a by moving c to the bottom of their rankings as shown below; denote the new preferences as $[\succ^2]$.

$$\begin{array}{ll} 3 \text{ agents:} & a \succ b \succ c \\ 2 \text{ agents:} & b \succ c \succ a \\ 2 \text{ agents:} & b \succ a \succ c \end{array}$$

If plurality were monotonic, it would have to make the same choice under $[\succ^2]$ as under $[\succ^1]$, because for all $i \in N$, $a \succ_i^2 b$ if $a \succ_i^1 b$ and $a \succ_i^2 c$ if $a \succ_i^1 c$. However, under $[\succ^2]$ plurality chooses b. Therefore plurality is not monotonic.

---

5. Formally, we should also specify the tie-breaking rule used by plurality. However, in our example monotonicity fails even when ties never occur, so the tie-breaking rule does not matter here.

9.5 Ranking systems

We now turn to a specialization of the social choice problem that has a computational flavor, and in which some interesting progress can be made. Specifically, consider a setting in which the set of agents is *the same* as the set of outcomes—agents are asked to vote to express their opinions about each other, with the goal of determining a social ranking. Such settings have great practical importance. For example, search engines rank Web pages by considering hyperlinks from one page to another to be votes about the importance of the destination pages. Similarly, online auction sites employ *reputation systems* to provide assessments of agents' trustworthiness based on ratings from past transactions.

Let us formalize this setting, returning to our earlier assumption that agents can be indifferent between outcomes. Our setting is characterized by two assumptions. First, $N = O$: the set of agents is the same as the set of outcomes. Second, agents' preferences are such that each agent divides the other agents into a set that he likes equally, and a set that he dislikes equally (or, equivalently, has no opinion about). Formally, for each $i \in N$ the outcome set O (equivalent to N) is partitioned into two sets $O_{i,1}$ and $O_{i,2}$, with $\forall o_1 \in O_{i,1}, \forall o_2 \in O_{i,2}, o_1 \succ_i o_2$, and with $\forall o, o' \in O_{i,k}$ for $k \in \{1, 2\}$, $o \sim_i o'$. We call this the *ranking systems setting*, and call a social welfare function in this setting a *ranking rule*. Observe that a ranking rule is not required to partition the agents into two sets; it must simply return some total preordering on the agents.

ranking systems setting

ranking rule

Interestingly, Arrow's impossibility system does not hold in the ranking systems setting. The easiest way to see this is to identify a ranking rule that satisfies all of Arrow's axioms.[6]

Proposition 9.5.1 *In the ranking systems setting, approval voting satisfies IIA, PE, and nondictatorship.*

The proof is straightforward and is left as an easy exercise. Intuitively, the fact that agents partition the outcomes into only two sets is crucially important. We would be able to apply Arrow's argument if agents were able to partition the outcomes into as few as three sets. (Recall that the proof of Theorem 9.4.4 requires arbitrary preferences and $|O| \geq 3$.)

Although the possibility of circumventing Arrow's theorem is encouraging, the discussion does not end here. Due to the special nature of the ranking systems setting, there are other properties that we would like a ranking rule to satisfy.

First, consider an example in which Alice votes only for Bob, Will votes only for Liam, and Liam votes only for Vic. These votes are illustrated in Figure 9.2. Who should be ranked highest? Three of the five kids have received votes (Bob, Liam, and Vic); these three should presumably rank higher than the remaining two. But of the three, Vic is special: he is the only one whose voter (Liam) himself received a vote. Thus, intuitively, Vic should receive the highest rank. This intuition is captured by the idea of transitivity.

---

6. Note that we defined these axioms in terms of strict total orderings; nevertheless, they generalize easily to total preorderings.

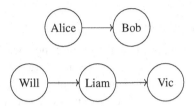

Figure 9.2 Sample preferences in a ranking system, where arrows indicate votes.

First we define *strong transitivity*. We will subsequently relax this definition; however, it is useful for what follows.

Definition 9.5.2 (Strong transitivity) *Consider a preference profile in which outcome o_2 receives at least as many votes as o_1, and it is possible to pair up all the voters for o_1 with voters from o_2 so that each voter for o_2 is weakly preferred by the ranking rule to the corresponding voter for o_1.[7] Further assume that o_2 receives more votes than o_1 and/or that there is at least one pair of voters where the ranking rule strictly prefers the voter for o_2 to the voter for o_1. Then the ranking rule satisfies* strong transitivity *if it always strictly prefers o_2 to o_1.*

strong
transitivity

Because our transitivity definition will serve as the basis for an impossibility result, we want it to be as weak as possible. One way in which this definition is quite strong is that it does not take into account the *number* of votes that a voting agent places. Consider an example in which Vic votes for almost all the kids, whereas Ray votes only for one. If Vic and Ray are ranked the same by the ranking rule, strong transitivity requires that their votes must count equally. However, we might feel that Ray has been more decisive, and therefore feel that his vote should be counted more strongly than Vic's. We can allow for such rules by weakening the notion of transitivity. The new definition is exactly the same as the old one, except that it is restricted to apply only to settings in which the voters vouch for exactly the same number of candidates.

Definition 9.5.3 (Weak transitivity) *Consider a preference profile in which outcome o_2 receives at least as many votes as o_1, and it is possible to pair up all the voters for o_1 with voters for o_2 who have both voted for exactly the same number of outcomes so that each voter for o_2 is weakly preferred by the ranking rule to the corresponding voter for o_1. Further assume that o_2 receives more votes than o_1 and/or that there is at least one pair of voters where the ranking rule strictly prefers the voter for o_2 to the voter for o_1. Then the ranking rule satisfies* weak

weak transitivity   transitivity *if it always strictly prefers o_2 to o_1.*

Recall the independence of irrelevant alternatives (IIA) property defined earlier in Definition 9.4.2, which said that the ordering of two outcomes should depend only on agents' relative preferences between these outcomes. Such an assumption is inconsistent with even our weak transitivity definitions. However, we can

---

7. The pairing must use each voter from o_2 at most once; if there are more votes for o_2 than for o_1, there will be agents who voted for o_2 who are not paired. If an agent voted for both o_1 and o_2, it is acceptable for him to be paired with himself.

broaden the scope of IIA to allow for transitive effects, and thereby still express the idea that the ranking rule should rank pairs of outcomes based only on local information.

ranked independence of irrelevant alternatives (RIIA)

Definition 9.5.4 (RIIA, informal) *A ranking rule satisfies* ranked independence of irrelevant alternatives *(RIIA) if the relative rank between pairs of outcomes is always determined according to the same rule, and this rule depends only on:*

1. *the number of votes each outcome received; and*
2. *the relative ranks of these voters.*[8]

Note that this definition prohibits the ranking rule from caring about the *identities* of the voters, which is allowed by IIA.

Despite the fact that Arrow's theorem does not apply in this setting, it turns out that another, very different impossibility result does hold.

Theorem 9.5.5 *There is no ranking system that always satisfies both weak transitivity and RIIA.*

What hope is there then for ranking systems? The obvious way forward is to consider relaxing one axiom and keeping the other. Indeed, progress can be made both by relaxing weak transitivity and by relaxing RIIA. For example, the famous PageRank algorithm (used originally as the basis of the Google search engine) can be understood as a ranking system that satisfies weak transitivity but not RIIA. Unfortunately, an axiomatic treatment of this algorithm is quite involved, so we do not provide it here.

Instead, we will consider relaxations of transitivity. First, what happens if we simply drop the weak transitivity requirement altogether? Let us add the requirements that an agent's rank can improve only when he receives more votes ("positive response") and that the agents' identities are ignored by the ranking function ("anonymity"). Then it can be shown that approval voting, which we have already considered in this setting, is the *only* possible ranking function.

Theorem 9.5.6 *Approval voting is the only ranking rule that satisfies RIIA, positive response, and anonymity.*

Finally, what if we try to modify the transitivity requirement rather than dropping it entirely? It turns out that we can also obtain a positive result here, although this comes at the expense of guaranteeing anonymity. Note that this new transitivity requirement is a different weakening of strong transitivity which does not care about the number of outcomes that agents vote for, but instead requires strict preference only when the ranking rule strictly prefers *every* paired voter for o_2 over the corresponding voter for o_1.

Definition 9.5.7 (Strong quasi-transitivity) *Consider a preference profile in which outcome o_2 receives at least as many votes as o_1, and it is possible to*

---

8. The formal definition of RIIA is more complicated than Definition 9.5.4 because it must explain precisely what is meant by depending on the relative ranks of the voters. The interested reader is invited to consult the reference cited at the end of the chapter.

forall $i \in N$ **do** $rank(i) \leftarrow 0$
repeat

> **forall** $i \in N$ **do**
>
> > **if** $|voters\_for(i)| > 0$ **then**
> >
> > > $rank(i) \leftarrow \frac{1}{n+1}[|voters\_for(i)| + \max_{j \in voters\_for(i)} rank(j)]$
> >
> > **else**
> >
> > > $rank(i) \leftarrow 0$

until $rank$ *converges*

Figure 9.3 A ranking algorithm that satisfies strong quasi-transitivity and RIIA.

pair up all the voters for o_1 with voters from o_2 so that each voter for o_2 is weakly preferred by the ranking rule to the corresponding voter for o_1. Then the ranking
strong *rule satisfies* strong quasi-transitivity *if it weakly prefers o_2 to o_1, and strictly*
quasi-transitivity *prefers o_2 to o_1 if either o_1 received no votes or each paired voter for o_2 is strictly preferred by the ranking rule to the corresponding voter for o_1.*

There exists a family of ranking algorithms that satisfy strong quasi-transitivity and RIIA. These algorithms work by assigning agents numerical ranks that depend on the number of votes they have received, and breaking ties in favor of the agent who received a vote from the highest-ranked voter. If this rule still yields a tie, it is applied recursively; when the recursion follows a cycle, the rank is a periodic rational number with period equal to the length of the cycle. One such algorithm is given in Figure 9.3. This algorithm can be proved to converge in n iterations; as each step takes $O(n^2)$ time (considering all votes for all agents), the worst-case complexity[9] of the algorithm is $O(n^3)$.

9.6   History and references

Social choice theory is covered in many textbooks on microeconomics and game theory, as well as in some specialized texts such as Feldman and Serrano [2006] and Gaertner [2006]. An excellent survey is provided in Moulin [1994].

The seminal individual publication in this area is Arrow [1970], which still remains among the best introductions to the field. The book includes Arrow's famous impossibility result (partly for which he received a 1972 Nobel Prize), though our treatment follows the elegant first proof in Geanakoplos [2005]. Plurality voting is too common and natural (it is used in 43 of the 191 countries in the United Nations for either local or national elections) to have clear origins. Borda invented his system as a fair way to elect members to the French Academy of Sciences in 1770, and first published his method in 1781 as de Borda [1781]. In 1784, Marie Jean Antoine Nicolas Caritat, aka the Marquis de Condorcet, first published his ideas regarding voting [de Condorcet, 1784]. Somewhat later,

---

9. In fact, the complexity bound on this algorithm can be somewhat improved by more careful analysis; however, the argument here suffices to show that the algorithm runs in polynomial time.

Nanson, a Briton-turned-Australian mathematician and election reformer, published his modification of the Borda count in Nanson [1882]. The Smith set was introduced in Smith [1973]. The Muller–Satterthwaite impossibility result appears in Muller and Satterthwaite [1977]; our proof follows Mas-Colell et al. [1995]. Our section on ranking systems follows Altman and Tennenholtz [2008]. Other interesting directions in ranking systems include developing practical ranking rules and/or axiomatizing such rules (e.g., Page et al. [1998], Kleinberg [1999], Borodin et al. [2005], and Altman and Tennenholtz [2005]), and exploring personalized rankings, in which the ranking function gives a potentially different answer to each agent (e.g., Altman and Tennenholtz [2007a]).

10

Protocols for Strategic Agents: Mechanism Design

As we discussed in the previous chapter, social choice theory is nonstrategic; it takes the preferences of the agents as given, and investigates ways in which they can be aggregated. But of course those preferences are usually not known. What you have, instead, is that the various agents *declare* their preferences, which they may do truthfully or not. Assuming the agents are self interested, in general they will not reveal their true preferences. Since as a designer you wish to find an optimal outcome with respect to the agents' true preferences (e.g., electing a leader that truly reflects the agents' preferences), optimizing with respect to the declared preferences will not in general achieve the objective.

10.1 Introduction

Mechanism design is a strategic version of social choice theory, which adds the assumption that agents will behave so as to maximize their individual payoffs. For example, in an election agents may not vote their true preference.

10.1.1 *Example: strategic voting*

Consider again our babysitting example. This time, in addition to Will, Liam, and Vic you must also babysit their devious new friend, Ray. Again, you invite each child to select their favorite among the three activities—going to the video arcade (a), playing basketball (b), and going for a leisurely car ride (c). As before, you announce that you will select the activity with the highest number of votes, breaking ties alphabetically. Consider the case in which the true preferences of the kids are as follows:

$$
\begin{aligned}
\text{Will:} \quad & b \succ a \succ c \\
\text{Liam:} \quad & b \succ a \succ c \\
\text{Vic:} \quad & a \succ c \succ b \\
\text{Ray:} \quad & c \succ a \succ b
\end{aligned}
$$

Will, Liam, and Vic are sweet souls who always tell you their true preferences. But little Ray, he is always figuring things out and so he goes through the following reasoning process. He prefers the most sedentary activity possible (hence his preference ordering). But he knows his friends well, an in particular he knows which activity each of them will vote for. He thus knows that if he votes for his true passion—slumping in the car for a few hours (c)—he will end up playing basketball (b). So he votes for going to the arcade (a), ensuring that this indeed is the outcome. Is there anything you can do to prevent such manipulation by little Ray?

mechanism
design

implementation
theory This is where *mechanism design*, or *implementation theory*, comes in. Mechanism design is sometimes colloquially called "inverse game theory." Our discussion of game theory in Chapter 3 was framed as follows: Given an interaction among a set of agents, how do we predict or prescribe the course of action of the various agents participating in the interaction? In mechanism design, rather than investigate a given strategic interaction, we start with certain desired behaviors on the part of agents and ask what strategic interaction among these agents might give rise to these behaviors. Roughly speaking, from the technical point of view this will translate to the following. We will assume unknown individual preferences, and ask whether we can design a game such that, no matter what the secret preferences of the agents actually are, the equilibrium of the game is guaranteed to have a certain desired property or set of properties.[1] Mechanism design is perhaps the most "computer scientific" part of game theory, since it concerns itself with designing effective protocols for distributed systems. The key difference from the traditional work in distributed systems is that in the current setting the distributed elements are not necessarily cooperative, and must be motivated to play their part. For this reason one can think of mechanism design as an exercise in "incentive engineering."

10.1.2   *Example: buying a shortest path*

Like social choice theory, the scope of mechanism design is broader than voting. auction theory The most famous application of mechanism design is *auction theory*, to which we devote Chapter 11. However, mechanism design has many other applications. Consider the transportation network depicted in Figure 10.1.

In Section 6.4.5 we considered a selfish routing problem where agents selfishly decide where to send their traffic in a network that responded to congestion in a predictable way. Here we consider a different problem. In Figure 10.1 the number next to a given edge is the cost of transporting along that edge, but these costs are the private information of the various shippers that own each edge. The task here is to find the shortest (least-cost) path from S to T; this is hard because the shippers may lie about their costs. Your one advantage is that you know that they are interested in maximizing their revenue. How can you use that knowledge to extract from them the information needed to compute the desired path?

---

1. Actually, as we shall see, technically speaking what we design is not a game but a mechanism that together with the secret utility functions defines a Bayesian game.

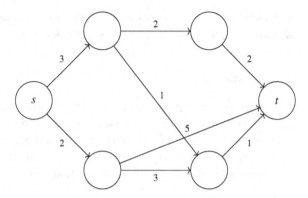

Figure 10.1 Transportation network with selfish agents.

10.2   Mechanism design with unrestricted preferences

We begin by introducing some of the broad principles of mechanism design, placing no restriction on the preferences agents can have. (We will consider such restrictions in later sections.) Because mechanism design is most often studied in settings where agents' preferences are unknown, we start by defining a Bayesian game setting, basing it on the epistemic types definition of Bayesian games that we gave in Section 6.3.1. The key difference is that the setting does not include actions for the agents, and instead defines the utility functions over the set of possible outcomes.[2]

Bayesian game setting   **Definition 10.2.1 (Bayesian game setting)** *A* Bayesian game setting *is a tuple* (N, O, Θ, p, u), *where:*

- *N is a finite set of n agents;*
- *O is a set of outcomes;*
- *$\Theta = \Theta_1 \times \cdots \times \Theta_n$ is a set of possible joint type vectors;*
- *p is a (common-prior) probability distribution on Θ; and*
- *$u = (u_1, \ldots, u_n)$, where $u_i : O \times \Theta \mapsto \mathbb{R}$ is the utility function for each player i.*

Given a Bayesian game setting, we can define a mechanism.

mechanism   **Definition 10.2.2 (Mechanism)** *A* mechanism *(for a Bayesian game setting* (N, O, Θ, p, u)*) is a pair* (A, M), *where:*

- *$A = A_1 \times \cdots \times A_n$, where A_i is the set of actions available to agent $i \in N$; and*
- *$M : A \mapsto \Pi(O)$ maps each action profile to a distribution over outcomes.*

---

2. Recall from our original discussion of utility theory in Section 3.1.2 that utility functions always map from outcomes to real values; we had previously assumed that $O = A$. We now relax this assumption, and so make explicit the utility functions' dependence on the chosen outcome.

A mechanism is *deterministic* if for every $a \in A$, there exists $o \in O$ such that $M(a)(o) = 1$; in this case we write simply $M(a) = o$.

10.2.1   *Implementation*

Together, a Bayesian game setting and a mechanism define a Bayesian game. The aim of mechanism design is to select a mechanism, given a particular Bayesian game setting, whose equilibria have desirable properties. We now define the most fundamental such property: that the outcomes that arise when the game is played are consistent with a given social choice function.

implementation in dominant strategies **Definition 10.2.3 (Implementation in dominant strategies)** *Given a Bayesian game setting* (N, O, Θ, p, u), *a mechanism* (A, M) *is an* implementation in dominant strategies *of a social choice function* C *(over* N *and* O) *if for any vector of utility functions* u, *the game has an equilibrium in dominant strategies, and in any such equilibrium* a^* *we have* $M(a^*) = C(u)$.[3]

strategy-proof A mechanism that gives rise to dominant strategies is sometimes called *strategy-proof*, because there is no need for agents to reason about each others' actions in order to maximize their utility.

In the aforementioned babysitter example, the pair consisting of "each child votes for one choice" and "the activity selected is one with the most votes, breaking ties alphabetically" is a well-formed mechanism, since it specifies the actions available to each child and the outcome depending on the choices made. Now consider the social choice function "the selected activity is that which is the top choice of the maximal number of children, breaking ties alphabetically." Clearly the mechanism defined by the babysitter does not implement this function in dominant strategies. For example, the preceding instance of it has no dominant strategy for Ray.

This suggests that the above definition can be relaxed, and can appeal to solution concepts that are weaker than dominant-strategy equilibrium. For example, one can appeal to the Bayes–Nash equilibrium.[4]

implementation in Bayes–Nash equilibrium **Definition 10.2.4 (Implementation in Bayes–Nash equilibrium)** *Given a Bayesian game setting* (N, O, Θ, p, u), *a mechanism* (A, M) *is an* implementation in Bayes–Nash equilibrium *of a social choice function* C *(over* N *and* O) *if there exists a Bayes–Nash equilibrium of the game of incomplete information*

---

3. The careful reader will notice that because we have previously defined social choice functions as deterministic, we here end up with a mechanism that selects outcomes deterministically as well. Of course, this definition can be extended to describe randomized social choice functions and mechanisms.
4. It is possible to study mechanism design in complete-information settings as well. This leads to the idea of Nash implementation, which is a sensible concept when the agents know each other's utility functions but the designer does not. This last point is crucial: if the designer did know, he could simply select the social choice directly, and we would return to the social choice setting studied in Chapter 9. We do not discuss Nash implementation further because it plays little role in the material that follows.

(N, A, Θ, p, u) *such that for every* $\theta \in \Theta$ *and every action profile* $a \in A$ *that can arise given type profile* θ *in this equilibrium, we have that* $M(a) = C(u(\cdot, \theta))$.

A classical example of Bayesian mechanism design is auction design. While we defer a lengthier discussion of auctions to Chapter 11, the basic idea is as follows. The designer wishes, for example, to ensure that the bidder with the highest valuation for a given item will win the auction, but the valuations of the agents are all private. The outcomes consist of allocating the item (in the case of a simple, single-item auction) to one of the agents, and having the agents make or receive some payments. The auction rules define the actions available to the agents (the "bidding rules"), and the mapping from action vectors to outcomes ("allocation rules" and "payment rules": who wins and who pays what as a function of the bidding). If we assume that the valuations are drawn from some known distribution, each particular auction design and particular set of agents define a Bayesian game, in which the signal of each agent is his own valuation.

Finally, there exist implementation concepts that are satisfied by a larger set of strategy profiles than implementation in dominant strategies, but that are not guaranteed to be achievable for any given social choice function and set of preferences, unlike Bayes–Nash implementation. For example, we could consider only symmetric Bayes–Nash equilibria, on the principle that strategies that depend on agent identities would be less likely to arise in practice. It turns out that symmetric Bayes–Nash equilibria always exist in symmetric Bayesian games. A second implementation notion that deserves mention is *ex post* implementation. Recall from Section 6.3.4 that an *ex post* equilibrium has the property that no agent can ever gain by changing his strategy even if he observes the other agents' types, as long as all the other agents follow the equilibrium strategies. Thus, unlike a Bayes–Nash equilibrium, an *ex post* equilibrium does not depend on the type distribution. Regardless of the implementation concept, we can require that the desired social choice function is implemented in the only equilibrium, in every equilibrium or in at least one equilibrium of the underlying game.

10.2.2 *The revelation principle*

truthfulness   One property that is often desired of mechanisms is called *truthfulness*. This property holds when agents truthfully disclose their preferences to the mechanism in equilibrium. It turns out that this property can always be achieved regardless of the social choice function implemented and of the agents' preferences.

direct mechanism   More formally, a *direct mechanism* is one in which the only action available to each agent is to announce his private information. Since in a Bayesian game an agent's private information is his type, direct mechanisms have $A_i = \Theta_i$. When an agent's set of actions is the set of all his possible types, he may lie and announce a type $\hat{\theta}_i$ that is different from his true type θ_i. A direct mechanism is said to be *truthful* (or *incentive compatible*) if, for any type vector θ, in equilibrium

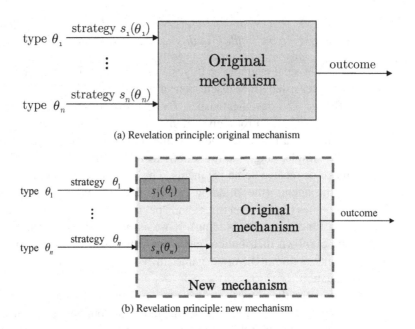

(a) Revelation principle: original mechanism

(b) Revelation principle: new mechanism

Figure 10.2 The revelation principle: how to construct a new mechanism with a truthful equilibrium, given an original mechanism with equilibrium $(s_1, \ldots, s_n)$.

of the game defined by the mechanism and under the appropriate equilibrium concept every agent i's strategy is to announce his true type, so that $\hat{\theta}_i = \theta_i$. We can thus speak about *incentive compatibility in dominant strategies* and *Bayes–Nash incentive compatibility*. Our claim that truthfulness can always be achieved implies, for example, that the social choice functions implementable by dominant-strategy truthful mechanisms are precisely those implementable by strategy-proof direct mechanisms. This means that we can, without loss of coverage, limit ourselves to a small sliver of the space of all mechanisms.

incentive compatibility in dominant strategies

Bayes–Nash incentive compatibility

revelation principle

Theorem 10.2.5 (Revelation principle) *If there exists any mechanism that implements a social choice function C in dominant strategies then there exists a direct mechanism that implements C in dominant strategies and is truthful.*

Proof. Consider an arbitrary mechanism for n agents that implements a social choice function C in dominant strategies. This mechanism is illustrated in Figure 10.2a. Let $s_1, \ldots, s_n$ denote the dominant strategies for agents $1, \ldots, n$. We will construct a new mechanism which *truthfully* implements C. Our new mechanism will ask the agents for their utility functions, use them to determine $s_1, \ldots, s_n$, the agents' dominant strategies under the original mechanism, and then choose the outcome that would have been chosen by the original mechanism for agents following the strategies $s_1, \ldots, s_n$. This new mechanism is illustrated in Figure 10.2b.

Assume that some agent i would be better off declaring a utility function u_i^* to the new mechanism rather than his true utility function u_i. This implies that i would have preferred to follow some different strategy s_i^* in the original mechanism rather than s_i, contradicting our assumption that s_i is a dominant strategy for i. (Intuitively, if i could gain by lying to the new mechanism, he could likewise gain by "lying to himself" in the original mechanism.) Thus the new mechanism is dominant-strategy truthful.   ■

In other words, any solution to a mechanism design problem can be converted into one in which agents always reveal their true preferences, if the new mechanism "lies for the agents" in just the way they would have chosen to lie to the original mechanism. The revelation principle is arguably the most basic result in mechanism design. It means that, while one might have thought *a priori* that a particular mechanism design problem calls for an arbitrarily complex strategy space, in fact one can restrict one's attention to truthful, direct mechanisms.

As we asserted earlier, the revelation principle does not apply only to implementation in dominant strategies; we have stated the theorem in this way only to keep things simple. Following exactly the same argument we can argue that, for example, a mechanism that implements a social choice function in a Bayes–Nash equilibrium can be converted into a direct, Bayes–Nash incentive compatible mechanism.

The argument we used to justify the revelation principle also applies to original mechanisms that are indirect (e.g., ascending auctions). The new, direct mechanism can take the agents' utility functions, construct their strategies for the indirect mechanism, and then simulate the indirect mechanism to determine which outcome to select. One caveat is that, even if the original indirect mechanism had a unique equilibrium, there is no guarantee that the new revelation mechanism will not have additional equilibria.

Before moving on, we finally offer some computational caveats to the revelation principle. Observe that the general effect of constructing a revelation mechanism is to push an additional computational burden onto the mechanism, as is implicit in Figure 10.2b. There are many settings in which agents' equilibrium strategies are computationally difficult to determine. When this is the case, the additional burden absorbed by the mechanism may be considerable. Furthermore, the revelation mechanism forces the agents to reveal their types completely. There may be settings in which agents are not willing to compromise their privacy to this degree. (Observe that the original mechanism may require them to reveal much less information.) Finally, even if not objectionable on privacy grounds, this full revelation can sometimes place an unreasonable burden on the communication channel. For all these reasons, in practical settings one must apply the revelation principle with caution.

10.2.3   *Impossibility of general, dominant-strategy implementation*

We now ask what social choice functions can be implemented in dominant strategies. Given the revelation principle, we can restrict our attention to truthful mechanisms. The first answer is disappointing.

Theorem 10.2.6 (Gibbard–Satterthwaite) *Consider any social choice function C of N and O. If:*

1. $|O| \geq 3$ *(there are at least three outcomes);*
2. *C is onto; that is, for every $o \in O$ there is a preference profile $[\succ]$ such that $C([\succ]) = o$ (this property is sometimes also called* citizen sovereignty*); and*
3. *C is dominant-strategy truthful,*

then C is dictatorial.

If Theorem 10.2.6 is reminiscent of the Muller–Satterthwaite theorem (Theorem 9.4.8) this is no accident, since Theorem 10.2.6 is implied by that theorem as a corollary. Note that this negative result is specific to dominant-strategy implementation. It does not hold for the weaker concepts of Nash or Bayes–Nash equilibrium implementation.

10.3 Quasilinear preferences

If we are to design a dominant-strategy truthful mechanism that is not dictatorial, we are going to have to relax some of the conditions of the Gibbard–Satterthwaite theorem. First, we relax the requirement that agents be able to express any preferences and replace it with the requirement that agents be able to express any preferences in a limited set. Second, we relax the condition that the mechanism be onto. We now introduce our limited set of preferences.

quasilinear
utility functions

quasilinear
preferences

Definition 10.3.1 (Quasilinear utility function) *Agents have* quasilinear utility functions *(or* quasilinear preferences*) in an n-player Bayesian game when the set of outcomes is $O = X \times \mathbb{R}^n$ for a finite set X, and the utility of an agent i given joint type θ is given by $u_i(o, \theta) = u_i(x, \theta) - f_i(p_i)$, where $o = (x, p)$ is an element of O, $u_i : X \times \Theta \mapsto \mathbb{R}$ is an arbitrary function and $f_i : \mathbb{R} \mapsto \mathbb{R}$ is a strictly monotonically increasing function.*

Intuitively, we split outcomes into two pieces that are linearly related. First, X represents a finite set of nonmonetary outcomes, such as the allocation of an object to one of the bidders in an auction or the selection of a candidate in an election. Second, p_i is the (possibly negative) payment made by agent i to the mechanism, such as a payment to the auctioneer.

What does it mean to assume that agents' preferences are quasilinear? First, it means that we are in a setting in which the mechanism can choose to charge or reward the agents by an arbitrary monetary amount. Second, and more restrictive, it means that an agent's degree of preference for the selection of any choice $x \in X$ is independent from his degree of preference for having to pay the mechanism some amount $p_i \in \mathbb{R}$. Thus an agent's utility for a choice cannot depend on the total amount of money that he has (e.g., an agent cannot value having a yacht more if he is rich than if he is poor). Finally, it means that agents care only about the choice selected and about their own payments: in particular, they do not care about the monetary payments made or received by other agents.

Strictly speaking, we have defined quasilinear preferences in a way that fixes the set of agents. However, we generally consider families of quasilinear problems, for any set of agents. For example, consider a voting game of the sort discussed earlier. You would want to be able to speak about a voting problem and a voting solution in a way that is not dependent on the number of agents. So in the following we assume that a quasilinear utility function is still defined when any one agent is taken away. In this case the set of nonmonetary outcomes must be updated (e.g., in an auction setting the missing agent cannot be the winner), and is denoted by O_{-i}. Similarly, the utility functions u_i and the choice function C must be updated accordingly.

10.3.1 *Risk attitudes*

There is still one part of the definition of quasilinear preferences that we have not discussed—the functions f_i. Before defining them, let us consider a question that may at first seem a bit nonsensical. Recall that we have said that p_i denotes the amount an agent i has to pay the mechanism. How much does i value a dollar? To make sense of this question, we must first note that utility is specified in its own units, rather than in units of currency, so we need to perform some kind of conversion. (Recall the discussion at the end of Section 3.1.2.) Indeed, this conversion can be understood as the purpose of f_i. However, the conversion is nontrivial because for most people the value of a dollar depends on the amount of money they start out with in the first place. (For example, if you are broke and starving then a dollar could lead to a substantial increase in your utility; if you are a millionaire, you might not bend over to pick up the same dollar if it was lying on the street.) To make the same point in another way, consider a fair lottery in which a ticket costs $\$x$ and pays off $\$2x$ half of the time. Holding your wealth constant, your willingness to participate in this lottery would probably depend on x. Most people are willing to play for sufficiently small values of x (say $\$1$), but not for larger values (say $\$10,000$). However, we have modeled agents as expected utility maximizers—how can we express the idea that an agent's willingness to participate in this lottery can depend on x, when the lottery's expected value is the same in both cases?

These two examples illustrate that we will often want the f_i functions to be
risk attitude   nonlinear. The curvature of f_i gives i's *risk attitude*, which we can understand as the way that i feels about lotteries such as the one just described.

If an agent i simply wants to maximize his expected revenue, we say the
risk neutral   agent is *risk neutral*. Such an agent has a linear value for money, as illustrated in Figure 10.3a. To see why this is so, consider Figure 10.3b. This figure describes a situation where the agent starts out with an endowment of $\$k$, and must decide whether or not to participate in a fair lottery that awards $\$k + x$ half the time, and $\$k - x$ the other half of the time. From looking at the graph, we can see that $u(k) = \frac{1}{2}u(k - x) + \frac{1}{2}u(k + x)$—the agent is indifferent between participating in the lottery and not participating. This is what we would expect, as the lottery's expected value is k, the same as the value for not participating.

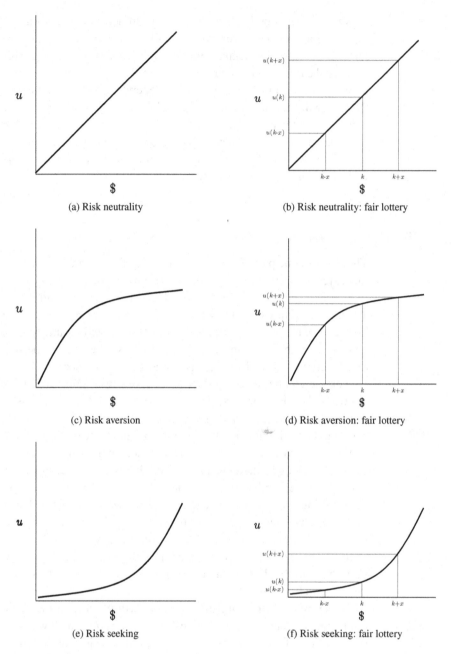

Figure 10.3 Risk attitudes: risk aversion, risk neutrality, risk seeking, and in each case, utility for the outcomes of a fair lottery.

In contrast, consider the value-for-money curve illustrated in Figure 10.3c. risk averse We call such an agent *risk averse*—he has a sublinear value for money, which means that he prefers a "sure thing" to a risky situation with the same expected value. Consider the same fair lottery described earlier from the point of view of a risk-averse agent, as illustrated in Figure 10.3d. We can see that for this agent $u(k) > \frac{1}{2}u(k-x) + \frac{1}{2}u(k+x)$—the marginal disutility of losing \$x is greater than the marginal utility of gaining \$x, given an initial endowment of k.

risk seeking   Finally, the opposite of risk aversion is *risk seeking*, illustrated in Figure 10.3e. Such an agent has a superlinear value for money, which means that the agent prefers engaging in lotteries to a sure thing with the same expected value. This is shown in Figure 10.3f. For example, an agent might prefer to spend \$1 to buy a ticket that has a $\frac{1}{1,000}$ chance of paying off \$1, 000, as compared to keeping the dollar.

     The examples above suggest that people might exhibit different risk attitudes in different regions of f_i. For example, a person could be risk seeking for very small amounts of money, risk neutral for moderate amounts and risk averse for large amounts. Nevertheless, in what follows we will assume that agents are risk neutral unless indicated otherwise. The assumption of risk neutrality is made partly for mathematical convenience, partly to avoid making an (often difficult to justify) assumption about the particular shape of agents' value-for-money curves, and partly because risk neutrality is reasonable when the amounts of money being exchanged through the mechanism are moderate. Considerable work in the literature extends results such as those presented in this chapter and the next to the case of agents with different risk attitudes.

     Even once we have assumed that agents are risk neutral, there remains one more degree of freedom in agents' utility functions: the slope of f_i. If every agent's value-for-money curve is linear and has the same slope ($\forall i \in N$, $f_i(p) = \beta p$, for

transferable utility $\beta \in \mathbb{R}_+$), then we say that the agents have *transferable utility*. This name reflects the fact that, regardless of the nonmonetary choice $x \in X$, one agent can transfer any given amount of utility to another by giving that agent an appropriate amount of money. More formally, for all $x \in X$, for any pair of agents $i, j \in N$ and for any $k \in \mathbb{R}$, i's utility is increased by exactly k and j's utility decreased by exactly k when j pays i the amount $\frac{k}{\beta}$. We will assume that this property holds for the remainder of this chapter and throughout Chapter 11, except where we indicate otherwise.

10.3.2 *Mechanism design in the quasilinear setting*

Now that we have defined the quasilinear preference model, we can talk about the design of mechanisms for agents with these preferences. As discussed earlier, we assume that agents are risk neutral and have transferable utility. For convenience, let $\beta_i = 1$, meaning that we can think of agents' utilities for different choices as being expressed in dollars. We concentrate on Bayesian games because most mechanism design is performed in such domains.

     First, we point out that since quasilinear preferences split the outcome space into two parts, we can modify our formal definition of a mechanism accordingly.

mechanism in
the quasilinear
setting

Definition 10.3.2 (Quasilinear mechanism) *A mechanism in the quasilinear setting (for a Bayesian game setting $(N, O = X \times \mathbb{R}^n, \Theta, p, u)$) is a triple $(A, \chi, \wp)$, where*

- *$A = A_1 \times \cdots \times A_n$, where A_i is the set of actions available to agent $i \in N$;*
- *$\chi : A \mapsto \Pi(X)$ maps each action profile to a distribution over choices; and*
- *$\wp : A \mapsto \mathbb{R}^n$ maps each action profile to a payment for each agent.*

choice rule

payment rule

In effect, we have split the function M into two functions χ and $\wp$, where χ is the *choice rule* and $\wp$ is the *payment rule*. We will use the notation $\wp_i$ to denote the payment function for agent i.

A direct revelation mechanism in the quasilinear setting is one in which each agent is asked to state his type.

direct
quasilinear
mechanism

Definition 10.3.3 (Direct quasilinear mechanism) *A direct quasilinear mechanism (for a Bayesian game setting $(N, O = X \times \mathbb{R}^n, \Theta, p, u)$) is a pair $(\chi, \wp)$. It defines a standard mechanism in the quasilinear setting, where for each i, $A_i = \Theta_i$.*

conditional
utility
independence

In many quasilinear mechanism design settings it is helpful to make the assumption that agents' utilities depend only on their own types, a property that we call *conditional utility independence*.[5]

conditional
utility
independence

Definition 10.3.4 (Conditional utility independence) *A Bayesian game exhibits* conditional utility independence *if for all agents $i \in N$, for all outcomes $o \in O$ and for all pairs of joint types θ and $\theta' \in \Theta$ for which $\theta_i = \theta_i'$, it holds that $u_i(o, \theta) = u_i(o, \theta')$.*

valuation

We will assume conditional utility independence for the rest of this section, and indeed for most of the rest of the chapter. When we do so, we can write an agent i's utility function as $u_i(o, \theta_i)$, since it does not depend on the other agents' types. We can also refer to an agent's *valuation* for choice $x \in X$, written $v_i(x) = u_i(x, \theta)$. v_i should be thought of as the maximum amount of money that i would be willing to pay to get the mechanism designer to implement choice x—in fact, having to pay this much would exactly make i indifferent about whether he was offered this deal or not.[6] Note that an agent's valuation depends on his type, even though we do not explicitly refer to θ_i. In the future when we discuss direct quasilinear mechanisms, we will usually mean mechanisms that ask agents to declare their valuations for each choice; of course, this alternate definition is equivalent to Definition 10.3.3.[7] Let V_i denote the set of all possible valuations for agent i. We will use the notation $\hat{v}_i \in V_i$ to denote the valuation that agent

---

5. This assumption is sometimes referred to as *privacy*. We avoid that terminology here because the assumption does not imply that agents cannot learn about others' utility functions by observing their own types.

6. Observe that here we rely upon the assumption of risk neutrality discussed earlier. Furthermore, observe that it is also meaningful to extend the concept of valuation beyond settings in which conditional utility independence holds; in such cases, we say that agents do not know their own valuations. We consider one such setting in Section 11.1.10.

7. Here we assume, as is common in the literature, that the mechanism designer knows each agent's value-for-money function f_i.

i declares to such a direct mechanism, which may be different from his true valuation v_i. We denote the vector of all agents' declared valuations as $\hat{v}$ and the set of all possible valuation vectors as V. Finally, denote the vector of declared valuations from all agents other than *i* as $\hat{v}_{-i}$.

Now we can state some properties that it is common to require of quasilinear mechanisms.

truthful   **Definition 10.3.5 (Truthfulness)** *A quasilinear mechanism is* truthful *if it is direct and* $\forall i \forall v_i$, *agent i's equilibrium strategy is to adopt the strategy* $\hat{v}_i = v_i$.

Of course, this is equivalent to the definition of truthfulness that we gave in Section 10.2.2; we have simply updated the notation for the quasilinear utility setting.

strict Pareto efficiency   **Definition 10.3.6 (Efficiency)** *A quasilinear mechanism is* strictly Pareto efficient, *or just* efficient, *if in equilibrium it selects a choice x such that*
efficiency $\forall v \forall x'$, $\sum_i v_i(x) \geq \sum_i v_i(x')$.

That is, an efficient mechanism selects the choice that maximizes the sum of agents' utilities, disregarding the monetary payments that agents are required to make. We describe this property as *economic efficiency* when there is a danger that it will be confused with other (e.g., computational) notions of efficiency. Observe that efficiency is defined in terms of agents' true valuations, not their declared valuations. This condition is also known as *social welfare maximization*.

social welfare maximization   The attentive reader might wonder about the relationship between strict Pareto efficiency as defined in Definitions 3.3.2 and 10.3.6. The underlying concept is indeed the same. The reason why we can get away with summing agents' valuations here arises from our assumption that agents' preferences are quasilinear, and hence that agents' utilities for different choices can be traded off against different payments. Recall that we observed in Section 3.3.1 that there can be many Pareto efficient outcomes because of the fact that agents' utility functions are only unique up to positive affine transformations. In a quasilinear setting, if we include the operator of the mechanism[8] as an agent who values money linearly and is indifferent between the mechanism's choices, it can be shown that all Pareto efficient outcomes involve the mechanism making the same choice and differ only in monetary allocations.

budget balance   **Definition 10.3.7 (Budget balance)** *A quasilinear mechanism is* budget balanced *when* $\forall v$, $\sum_i \wp_i(s(v)) = 0$, *where s is the equilibrium strategy profile.*

In other words, regardless of the agents' types, the mechanism collects and disburses the same amount of money from and to the agents, meaning that it makes neither a profit nor a loss.

weak budget balance   Sometimes we relax this condition and require only *weak budget balance*, meaning that $\forall v$, $\sum_i \wp_i(s(v)) \geq 0$ (i.e., the mechanism never takes a loss, but it may make a profit). Finally, we can require that either strict or weak budget

---

8. For example, this would be a seller in a single-sided auction, or a market maker in a double-sided market.

balance hold *ex ante*, which means that $\mathbb{E}_v\left[\sum_i \wp_i(s(v))\right]$ is either equal to or greater than zero. (That is, the mechanism is required to break even or make a profit only on expectation.)

Definition 10.3.8 (*Ex interim* individual rationality) *A quasilinear mechanism is* ex interim individually rational *when*

$$\forall i \forall v_i, \ \mathbb{E}_{v_{-i}|v_i}\left[v_i(\chi(s_i(v_i), s_{-i}(v_{-i}))) - \wp_i(s_i(v_i), s_{-i}(v_{-i}))\right] \geq 0,$$

where s is the equilibrium strategy profile.

This condition requires that no agent loses by participating in the mechanism. We call it *ex interim* because it holds for *every* possible valuation for agent i, but averages over the possible valuations of the other agents. This approach makes sense because it requires that, based on the information that an agent has when he chooses to participate in a mechanism, no agent would be better off choosing not to participate. Of course, we can also strengthen the condition to say that no agent *ever* loses by participation.

Definition 10.3.9 (*Ex post* individual rationality) *A quasilinear mechanism is* ex post individually rational *when* $\forall i \forall v, \ v_i(\chi(s(v))) - \wp_i(s(v)) \geq 0$, *where s is the equilibrium strategy profile.*

We can also restrict mechanisms based on their computational requirements rather than their economic properties.

Definition 10.3.10 (Tractability) *A quasilinear mechanism is* tractable *when* $\forall a \in A$, $\chi(a)$ *and* $\wp(a)$ *can be computed in polynomial time.*

Finally, in some domains there will be many possible mechanisms that satisfy the constraints we choose, meaning that we need to have some way of choosing among them. (And as we will see later, for other combinations of constraints no mechanisms exist at all.) The usual approach is to define an optimization problem that identifies the optimal outcome in the feasible set. For example, although we have defined efficiency as a constraint, it is also possible to soften the constraint and require the mechanism to achieve as much social welfare as possible. Here we define some other quantities that a mechanism designer can seek to optimize.

First, the mechanism designer can take a selfish perspective. Interestingly, this goal turns out to be quite different from the goal of maximizing social welfare. (We give an example of the differences between these approaches when we consider single-good auctions in Section 11.1.)

Definition 10.3.11 (Revenue maximization) *A quasilinear mechanism is* revenue maximizing *when, among the set of functions χ and $\wp$ that satisfy the other constraints, the mechanism selects the χ and $\wp$ that maximize* $\mathbb{E}_v\left[\sum_i \wp_i(s(v))\right]$, *where s(v) denotes the agents' equilibrium strategy profile.*

Conversely, the designer might try to collect as *little* revenue as possible, for example if the mechanism uses payments only to provide incentives, but is not intended to make money. The budget balance constraint is the best way to solve this problem, but sometimes it is impossible to satisfy. In such cases, one approach is to set weak budget balance as a constraint and then to pick the revenue minimizing mechanism, effectively softening the (strict) budget balance constraint. Here we present a *worst-case* revenue minimization objective; of course, an average-case objective is also possible.

revenue
minimization

Definition 10.3.12 (Revenue minimization) *A quasilinear mechanism is* revenue minimizing *when, among the set of functions χ and $\wp$ that satisfy the other constraints, the mechanism selects the χ and $\wp$ that minimize $\max_v \sum_i \wp_i(s(v))$ in equilibrium, where $s(v)$ denotes the agents' equilibrium strategy profile.*

The mechanism designer might be concerned with selecting a *fair* outcome. However, the notion of fairness can be tricky to formalize. For example, an outcome that fines all agents $100 and makes a choice that all agents hate equally is in some sense fair, but it does not seem desirable. Here we define so-called *maxmin fairness*, which says that the fairest outcome is the one that makes the least-happy agent the happiest. We also take an expected value over different valuation vectors, but we could instead have required a mechanism that does the best in the worst case.

maxmin fairness

Definition 10.3.13 (Maxmin fairness) *A quasilinear mechanism is* maxmin fair *when, among the set of functions χ and $\wp$ that satisfy the other constraints, the mechanism selects the χ and $\wp$ that maximize $\mathbb{E}_v[\min_{i \in N} v_i(\chi(s(v))) - \wp_i(s(v))]$, where $s(v)$ denotes the agents' equilibrium strategy profile.*

Finally, the mechanism designer might not be able to implement a social-welfare-maximizing mechanism (e.g., in order to satisfy a tractability constraint) but may want to get as close as possible. Thus, the goal could be minimizing the *price of anarchy* (see Definition 6.4.11), the worst-case ratio between optimal social welfare and the social welfare achieved by the given mechanism. Here we also consider the worst case across agent valuations.

price-of-anarchy
minimization

Definition 10.3.14 (Price-of-anarchy minimization) *A quasilinear mechanism minimizes the price of anarchy when, among the set of functions χ and $\wp$ that satisfy the other constraints, the mechanism selects the χ and $\wp$ that minimize*

$$\max_{v \in V} \frac{\max_{x \in X} \sum_{i \in N} v_i(x)}{\sum_{i \in N} v_i\left(\chi(s(v))\right)},$$

where $s(v)$ denotes the agents' equilibrium strategy profile in the worst *equilibrium of the mechanism—that is, the one in which $\sum_{i \in N} v_i(\chi(s(v)))$ is the smallest.*[9]

---

9. Note that we have to modify this definition along the lines we used in Definition 6.4.11 if $\sum_{i \in N} v_i(\chi(s(v))) = 0$ is possible.

10.4   Efficient mechanisms

Efficiency (Definition 10.3.6) is often considered to be one of the most important properties for a mechanism to satisfy in the quasilinear setting. For example, whenever an inefficient choice is selected, it is possible to find a set of side payments among the agents with the property that all agents would prefer the efficient choice in combination with the side payments to the inefficient choice. (Intuitively, the sum of agents' valuations for the efficient choice is greater than for the inefficient choice. Thus, the agents who prefer the efficient choice would still strictly prefer it even if they had to make side payments to the other agents so that each of them also strictly preferred the efficient choice.) Consequently, a great deal of research has considered the design of mechanisms that are guaranteed to select efficient choices when agents follow dominant or equilibrium strategies. In this section we survey these mechanisms.

10.4.1   *Groves mechanisms*

The most important family of efficient mechanisms are the Groves mechanisms.

Groves
mechanism

Definition 10.4.1 (Groves mechanisms) Groves mechanisms *are direct quasilinear mechanisms* $(\chi, \wp)$, *for which*

$$\chi(\hat{v}) = \arg\max_x \sum_i \hat{v}_i(x),$$

$$\wp_i(\hat{v}) = h_i(\hat{v}_{-i}) - \sum_{j \neq i} \hat{v}_j(\chi(\hat{v})).$$

In other words, Groves mechanisms are direct mechanisms in which agents can declare any valuation function $\hat{v}$ (and thus any quasilinear utility function $\hat{u}$). The mechanism then optimizes its choice assuming that the agents disclosed their true utility function. An agent is made to pay an arbitrary amount $h_i(\hat{v}_{-i})$ which does not depend on his own declaration and is paid the sum of every other agent's declared valuation for the mechanism's choice. The fact that the mechanism designer has the freedom to choose the h_i functions explains why we refer to the *family* of Groves mechanisms rather than to a single mechanism.

The remarkable property of Groves mechanisms is that they provide a dominant-strategy truthful implementation of a social-welfare-maximizing social choice function. It is easy to see that if a Groves mechanism is dominant-strategy truthful, then it must be social-welfare-maximizing: the function χ in Definition 10.4.1 performs exactly the maximization called for by Definition 10.3.6 when $\hat{v} = v$. Thus, it suffices to show the following.

Theorem 10.4.2 *Truth telling is a dominant strategy under any Groves mechanism.*

Proof. Consider a situation where every agent j other than i follows some arbitrary strategy $\hat{v}_j$. Consider agent i's problem of choosing the best strategy

$\hat{v}_i$. As a shorthand, we write $\hat{v} = (\hat{v}_{-i}, \hat{v}_i)$. The best strategy for i is one that solves

$$\max_{\hat{v}_i} \left(v_i(\chi(\hat{v})) - \wp(\hat{v}) \right).$$

Substituting in the payment function from the Groves mechanism, we have

$$\max_{\hat{v}_i} \left(v_i(\chi(\hat{v})) - h_i(\hat{v}_{-i}) + \sum_{j \neq i} \hat{v}_j(\chi(\hat{v})) \right).$$

Since $h_i(\hat{v}_{-i})$ does not depend on $\hat{v}_i$, it is sufficient to solve

$$\max_{\hat{v}_i} \left(v_i(\chi(\hat{v})) + \sum_{j \neq i} \hat{v}_j(\chi(\hat{v})) \right).$$

The only way in which the declaration $\hat{v}_i$ influences the maximization above is through the term $v_i(\chi(\hat{v}))$. If possible, i would like to pick a declaration $\hat{v}_i$ that will lead the mechanism to pick an $x \in X$ that solves

$$\max_x \left(v_i(x) + \sum_{j \neq i} \hat{v}_j(x) \right). \tag{10.1}$$

The Groves mechanism chooses an $x \in X$ as

$$\chi(\hat{v}) = \arg\max_x \left(\sum_i \hat{v}_i(x) \right) = \arg\max_x \left(\hat{v}_i(x) + \sum_{j \neq i} \hat{v}_j(x) \right).$$

Thus, agent i leads the mechanism to select the choice that he most prefers by declaring $\hat{v}_i = v_i$. Because this argument does not depend in any way on the declarations of the other agents, truth telling is a dominant strategy for agent i. ∎

Intuitively, the reason that Groves mechanisms are dominant-strategy truthful is that agents' externalities are internalized. Imagine a mechanism in which agents declared their valuations for the different choices $x \in X$ and the mechanism selected the efficient choice, but in which the mechanism did not impose any payments on agents. Clearly, agents would be able to change the mechanism's choice to another that they preferred by overstating their valuation. Under Groves mechanisms, however, an agent's utility does not depend only on the selected choice, because payments *are* imposed. Since agents are paid the (reported) utility of all the other agents under the chosen allocation, each agent becomes just as interested in maximizing the other agents' utilities as in maximizing his own. Thus, once payments are taken into account, all agents have the same interests.

Groves mechanisms illustrate a property that is generally true of dominant-strategy truthful mechanisms: an agent's payment does not depend on the amount of his own declaration. Although other dominant-strategy truthful mechanisms exist in the quasilinear setting, the next theorem shows that Groves mechanisms are the *only* mechanisms that implement an efficient allocation in dominant strategies among agents with arbitrary quasilinear utilities.

Theorem 10.4.3 (Green–Laffont) *An efficient social choice function $C:$ $\mathbb{R}^{Xn} \mapsto X \times \mathbb{R}^n$ can be implemented in dominant strategies for agents with unrestricted quasilinear utilities only if $\wp_i(v) = h(v_{-i}) - \sum_{j \neq i} v_j(\chi(v))$.*

Proof. From the revelation principle, we can assume that C is *truthfully implementable in dominant strategies*. Thus, from the definition of efficiency, the choice must be selected as

$$x = \arg\max_x \sum_i v_i(x)$$

We can write the payment function as

$$\wp_i(v) = h(v_i, v_{-i}) - \sum_{j \neq i} v_j(\chi(v)).$$

Observe that we can do this without loss of generality because h can be an arbitrary function that cancels out the second term. Now for contradiction, assume that there exist some v_i and v_i' such that $h(v_i, v_{-i}) \neq h(v_i', v_{-i})$.

Case 1: $\chi(v_i, v_{-i}) = \chi(v_i', v_{-i})$. Since C is truthfully implementable in dominant strategies, an agent i whose true valuation was v_i would be better off declaring v_i than v_i':

$$v_i(\chi(v_i, v_{-i})) - \wp_i(v_i, v_{-i}) \geq v_i(\chi(v_i', v_{-i})) - \wp_i(v_i', v_{-i}),$$
$$\wp_i(v_i, v_{-i}) \leq \wp_i(v_i', v_{-i}).$$

In the same way, an agent i whose true valuation was v_i' would be better off declaring v_i' than v_i:

$$v_i'(\chi(v_i', v_{-i})) - \wp_i(v_i', v_{-i}) \geq v_i'(\chi(v_i, v_{-i})) - \wp_i(v_i, v_{-i}),$$
$$\wp_i(v_i', v_{-i}) \leq \wp_i(v_i, v_{-i}).$$

Thus, we must have

$$\wp_i(v_i, v_{-i}) = \wp_i(v_i', v_{-i}),$$
$$h(v_i, v_{-i}) - \sum_{j \neq i} v_j(\chi(v_i, v_{-i})) = h(v_i', v_{-i}) - \sum_{j \neq i} v_j(\chi(v_i', v_{-i})).$$

We are currently considering the case where $\chi(v_i, v_{-i}) = \chi(v_i', v_{-i})$. Thus we can write

$$h(v_i, v_{-i}) - \sum_{j \neq i} v_j(\chi(v_i, v_{-i})) = h(v_i', v_{-i}) - \sum_{j \neq i} v_j(\chi(v_i, v_{-i})),$$
$$h(v_i, v_{-i}) = h(v_i', v_{-i}).$$

This is a contradiction.

Case 2: $\chi(v_i, v_{-i}) \neq \chi(v'_i, v_{-i})$. Without loss of generality, let $h(v_i, v_{-i}) < h(v'_i, v_{-i})$. Since this inequality is strict, there must exist some $\epsilon \in \mathbb{R}_+$ such that $h(v_i, v_{-i}) < h(v'_i, v_{-i}) - \epsilon$.

Our mechanism must work for every v. Consider a case where i's valuation is

$$
v''_i(x) = \begin{cases} -\sum_{j \neq i} v_j(\chi(v_i, v_{-i})) & x = \chi(v_i, v_{-i}); \\ -\sum_{j \neq i} v_j(\chi(v'_i, v_{-i})) + \epsilon & x = \chi(v'_i, v_{-i}); \\ -\sum_{j \neq i} v_j(x) - \epsilon & \text{for any other } x. \end{cases}
$$

Note that agent i still declares his valuations as real numbers; they just happen to satisfy the constraints given above. Also note that the ϵ used here is the same $\epsilon \in \mathbb{R}_+$ mentioned earlier. From the fact that C is truthfully implementable in dominant strategies, an agent i whose true valuation was v''_i would be better off declaring v''_i than v_i:

$$
v''_i(\chi(v''_i, v_{-i})) - \wp_i(v''_i, v_{-i}) \geq v''_i(\chi(v_i, v_{-i})) - \wp_i(v_i, v_{-i}). \tag{10.2}
$$

Because our mechanism is efficient, it must pick the choice that solves

$$
f = \max_x \left(v''_i(x) + \sum_j v_j(x) \right).
$$

Picking $x = \chi(v'_i, v_{-i})$ gives $f = \epsilon$; picking $x = \chi(v_i, v_{-i})$ gives $f = 0$, and any other x gives $f = -\epsilon$. Therefore, we can conclude that

$$
\chi(v''_i, v_{-i}) = \chi(v'_i, v_{-i}). \tag{10.3}
$$

Substituting Equation (10.3) into Equation (10.2), we get

$$
v''_i(\chi(v'_i, v_{-i})) - \wp_i(v''_i, v_{-i}) \geq v''_i(\chi(v_i, v_{-i})) - \wp_i(v_i, v_{-i}). \tag{10.4}
$$

Expand Equation (10.4):

$$
\left(-\sum_{j \neq i} v_j(\chi(v'_i, v_{-i})) + \epsilon \right) - \left(h(v''_i, v_{-i}) - \sum_{j \neq i} v_j(\chi(v''_i, v_{-i})) \right)
$$
$$
\geq \left(-\sum_{j \neq i} v_j(\chi(v_i, v_{-i})) \right) - \left(h(v_i, v_{-i}) - \sum_{j \neq i} v_j(\chi(v_i, v_{-i})) \right).
$$
$$\tag{10.5}$$

We can use Equation (10.3) to replace $\chi(v''_i, v_{-i})$ by $\chi(v'_i, v_{-i})$ on the left-hand side of Equation (10.5). The sums then cancel out, and the inequality simplifies to

$$
h(v_i, v_{-i}) \geq h(v''_i, v_{-i}) - \epsilon. \tag{10.6}
$$

Since $\chi(v''_i, v_{-i}) = \chi(v'_i, v_{-i})$, by the argument from Case 1 we can show that

$$
h(v''_i, v_{-i}) = h(v'_i, v_{-i}). \tag{10.7}
$$

Substituting Equation (10.7) into Equation (10.6), we get

$$h(v_i, v_{-i}) \geq h(v_i', v_{-i}) - \epsilon.$$

This contradicts our assumption that $h(\chi(v_i, v_{-i})) < h(\chi(v_i', v_{-i})) - \epsilon$. We have thus shown that there cannot exist v_i, v_i' such that $h(v_i, v_{-i}) \neq h(v_i', v_{-i})$.   ∎

Although we do not give the proof here, it has also been shown that Groves mechanisms are unique among Bayes–Nash incentive compatible efficient mechanisms, in a weaker sense. Specifically, any Bayes–Nash incentive compatible efficient mechanism corresponds to a Groves mechanism in the sense that each agent makes the same *ex interim* expected payments and hence has the same *ex interim* expected utility under both mechanisms.

10.4.2   *The VCG mechanism*

So far, we have said nothing about how to set the function h_i in a Groves mechanism's payment function. Here we will discuss the most popular answer, which is called the Clarke tax. In the subsequent sections we will discuss some of its properties, but first we define it.

Clarke tax   **Definition 10.4.4 (Clarke tax)** *The* Clarke tax *sets the h_i term in a Groves mechanism as*

$$h_i(\hat{v}_{-i}) = \sum_{j \neq i} \hat{v}_j \left(\chi(\hat{v}_{-i}) \right),$$

where χ is the Groves mechanism allocation function.

The resulting Groves mechanism goes by many names. We will see in Chapter 11 that the Vickrey auction (invented in 1961) is a special case; thus, in resource allocation settings the mechanism is sometimes known as the *generalized Vickrey auction*. Second, the mechanism is also known as the *pivot mechanism*; we will explain the rationale behind this name in a moment. From now on, though, we will refer to it as the *Vickrey–Clarke–Groves mechanism* (VCG), naming its contributors in chronological order of their contributions. We restate the full mechanism here.

Vickrey–
Clarke–Groves
(VCG)
mechanism

Definition 10.4.5 (Vickrey–Clarke–Groves (VCG) mechanism)   *The* VCG mechanism *is a direct quasilinear mechanism $(\chi, \wp)$, where*

$$\chi(\hat{v}) = \arg\max_x \sum_i \hat{v}_i(x),$$

$$\wp_i(\hat{v}) = \sum_{j \neq i} \hat{v}_j \left(\chi(\hat{v}_{-i}) \right) - \sum_{j \neq i} \hat{v}_j(\chi(\hat{v})).$$

First, note that because the Clarke tax does not depend on an agent i's own declaration $\hat{v}_i$, our previous arguments that Groves mechanisms are dominant-strategy truthful and efficient, carry over immediately to the VCG mechanism. Now, we try to provide some intuition about the VCG payment rule. Assume that all agents follow their dominant strategies and declare their valuations truthfully. The second sum in the VCG payment rule pays each agent i the sum of every other agent $j \neq i$'s utility for the mechanism's choice. The first sum charges each agent i the sum of every other agent's utility for the choice that *would have been made* had i not participated in the mechanism. Thus, each agent is made to pay his *social cost*—the aggregate impact that his participation has on other agents' utilities.

What can we say about the amounts of different agents' payments to the mechanism? If some agent i does not change the mechanism's choice by his participation—that is, if $\chi(v) = \chi(v_{-i})$—then the two sums in the VCG payment function will cancel out. The social cost of i's participation is zero, and so he has to pay nothing. In order for an agent i to be made to pay a nonzero amount, he must be *pivotal* in the sense that the mechanism's choice $\chi(v)$ is different from its choice without i, $\chi(v_{-i})$. This is why VCG is sometimes called the pivot mechanism—only pivotal agents are made to pay. Of course, it is possible that some agents will *improve* other agents' utilities by participating; such agents will be made to pay a negative amount, or in other words will be paid by the mechanism.

Let us see an example of how the VCG mechanism works. Recall that Section 10.1.2 discussed the problem of buying a shortest path in a transportation network. We will now reconsider that example, and determine what route and what payments the VCG mechanism would select. For convenience, we reproduce Figure 10.1 as Figure 10.4, and label the nodes so that we have names to refer to the agents (the edges).

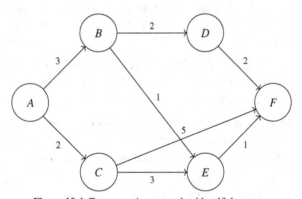

Figure 10.4 Transportation network with selfish agents.

Note that in this example, the numbers labeling the edges in the graph denote agents' costs rather than utilities; thus, an agent's utility is $-c$ if a route involving his edge (having cost c) is selected, and zero otherwise. The arg max in χ will amount to cost minimization. Thus, $\chi(v)$ will return the shortest path in the graph, which is $ABEF$.

How much will agents have to pay? First, let us consider the agent AC. The shortest path taking his declaration into account has a length of 5 and imposes a cost of -5 on agents other than him (because it does not involve him). Likewise, the shortest path without AC's declaration also has a length of 5. Thus, his payment is $p_{AC} = (-5) - (-5) = 0$. This is what we expect, since AC is not pivotal. Clearly, by the same argument BD, CE, CF, and DF will all be made to pay zero.

Now let us consider the pivotal agents. The shortest path taking AB's declaration into account has a length of 5, and imposes a cost of 2 on other agents. The shortest path without AB is $ACEF$, which has a cost of 6. Thus $p_{AB} = (-6) - (-2) = -4$: AB is paid 4 for his participation. Arguing similarly, you can verify that $p_{BE} = (-6) - (-4) = -2$, and $p_{EF} = (-7) - (-4) = -3$. Note that although EF had the same cost as BE, they are paid different amounts for the use of their edges. This occurs because EF has more *market power*: for the other agents, the situation without EF is worse than the situation without BE.

10.4.3   *VCG and individual rationality*

We have seen that Groves mechanisms are dominant-strategy truthful and efficient. We have also seen that no other mechanism has both of these properties in general quasilinear settings. Thus, we might be a bit worried that we have not been able to guarantee either individual rationality or budget balance, two properties that are quite important in practice. (Recall that individual rationality means that no agent would prefer not to participate in the mechanism; budget balance means that the mechanism does not lose money.) We will consider budget balance in Section 10.4.6; here we investigate individual rationality.

As it turns out, our worry is well founded: even with the freedom to set h_i, we cannot find a mechanism that guarantees us individual rationality in an unrestricted quasilinear setting. However, we are often able to guarantee the strongest variety of individual rationality when the setting satisfies certain mild restrictions.

choice-set monotonicity **Definition 10.4.6 (Choice-set monotonicity)** *An environment exhibits* choice-set monotonicity *if* $\forall i$, $X_{-i} \subseteq X$ *(removing any agent weakly decreases—that is, never increases—the mechanism's set of possible choices X).*

no negative externalities **Definition 10.4.7 (No negative externalities)** *An environment exhibits* no negative externalities *if* $\forall i \forall x \in X_{-i}$, $v_i(x) \geq 0$ *(every agent has zero or positive utility for any choice that can be made without his participation).*

These assumptions are often quite reasonable, as we illustrate with two examples. First, consider running VCG to decide whether or not to undertake a public project such as building a road. In this case, the set of choices is independent of the number of agents, satisfying choice-set monotonicity. No agent negatively values the project, though some might value the situation in which the project is not undertaken more highly than the situation in which it is.

Second, consider a market consisting of a set of agents interested in buying a single unit of a good such as a share of stock and another set of agents interested in selling a single unit of this good. The choices in this environment are sets of buyer–seller pairings. (Prices are imposed through the payment function.) If a new agent is introduced into the market, no previously existing pairings become infeasible, but new ones become possible; thus choice-set monotonicity is satisfied. Because agents have zero utility both for choices that involve trades between other agents and no trades at all, there are no negative externalities.

Under these restrictions, it turns out that the VCG mechanism ensures *ex post* individual rationality.

Theorem 10.4.8 *The VCG mechanism is* ex post *individually rational when the choice-set monotonicity and no negative externalities properties hold.*

Proof. All agents truthfully declare their valuations in equilibrium. Then we can write agent i's utility as

$$u_i = v_i(\chi(v)) - \left(\sum_{j \neq i} v_j(\chi(v_{-i})) - \sum_{j \neq i} v_j(\chi(v)) \right)$$

$$= \sum_j v_j(\chi(v)) - \sum_{j \neq i} v_j(\chi(v_{-i})). \quad (10.8)$$

We know that $\chi(v)$ is the choice that maximizes social welfare, and that this optimization could have picked $\chi(v_{-i})$ instead (by choice-set monotonicity). Thus,

$$\sum_j v_j(\chi(v)) \geq \sum_j v_j(\chi(v_{-i})).$$

Furthermore, from no negative externalities,

$$v_i(\chi(v_{-i})) \geq 0.$$

Therefore,

$$\sum_i v_i(\chi(v)) \geq \sum_{j \neq i} v_j(\chi(v_{-i})),$$

and thus Equation (10.8) is nonnegative. ∎

10.4.4 *VCG and weak budget balance*

What about weak budget balance, the requirement that the mechanism will not lose money? Our two previous conditions, choice-set monotonicity and no negative externalities, are not sufficient to guarantee weak budget balance: for example, the "buying the shortest path" example given earlier satisfied these two conditions, but we saw that the VCG mechanism paid out money and did not collect any. Thus, we will have to explore further restrictions to the quasilinear setting.

no single-agent effect **Definition 10.4.9 (No single-agent effect)** *An environment exhibits* no single-agent effect *if* $\forall i$, $\forall v_{-i}$, $\forall x \in \arg\max_y \sum_j v_j(y)$ *there exists a choice x' that is feasible without i and that has* $\sum_{j \neq i} v_j(x') \geq \sum_{j \neq i} v_j(x)$.

In other words, removing any agent does not worsen the total value of the best solution to the others, regardless of their valuations. For example, this property is satisfied in a single-sided auction—dropping an agent just reduces the amount of competition in the auction, making the others better off.

Theorem 10.4.10 *The VCG mechanism is weakly budget balanced when the no single-agent effect property holds.*

> **Proof.** As before, we start by assuming truth telling in equilibrium. We must show that the sum of transfers from agents to the center is greater than or equal to zero.
>
> $$\sum_i \wp_i(v) = \sum_i \left(\sum_{j \neq i} v_j(\chi(v_{-i})) - \sum_{j \neq i} v_j(\chi(v)) \right)$$
>
> From the no single-agent effect condition we have that
>
> $$\forall i \ \sum_{j \neq i} v_j(\chi(v_{-i})) \geq \sum_{j \neq i} v_j(\chi(v)).$$
>
> Thus the result follows directly.   ∎

Indeed, we can say something more about VCG's revenue properties: restricting ourselves to settings in which VCG is *ex post* individually rational as discussed earlier, and comparing to all other efficient and *ex interim* IR mechanisms, VCG turns out to collect the maximal amount of revenue from the agents. This is somewhat surprising, since this result does not require dominant strategies, and hence compares VCG to all Bayes–Nash mechanisms. A useful corollary of this result is that VCG is as budget balanced as any efficient mechanism can be: it satisfies weak budget balance in every case where *any* dominant strategy, efficient and *ex interim* IR mechanism would be able to do so.

10.4.5 *Drawbacks of VCG*

The VCG mechanism is one of the most powerful positive results in mechanism design: it gives us a general way of constructing dominant-strategy truthful mechanisms to implement social-welfare-maximizing social choice functions in quasilinear settings. We have seen that no fundamentally different mechanism could do the same job. And VCG gives us even more: under the right conditions it further guarantees *ex post* individual rationality and weak budget balance. Thus, it is not surprising that this mechanism has been enormously influential and continues to be widely studied.

However, despite these attractive properties, VCG also has some undesirable characteristics. In this section, we survey six of them. Before we go on, however, we offer a caveat: although there exist mechanisms that circumvent each of the drawbacks we discuss, none of the drawbacks are *unique* to VCG, or even to Groves mechanisms. Indeed, in some cases the problems are known to crop up in extremely broad classes of mechanisms; we cite some arguments to this effect at the end of the chapter.

1. Agents must fully disclose private information

VCG requires agents to fully reveal their private information (e.g., in the transportation network example, every agent has to tell the mechanism his costs

Agent	U (build road)	U (do not build road)	Payment
1	200	0	150
2	100	0	50
3	0	250	0

Table 10.1 Valuations for agents in the road-building referendum example.

Agent	U (build road)	U (do not build road)	Payment
1	250	0	100
2	150	0	0
3	0	250	0

Table 10.2 Agents in the road-building referendum can gain by colluding.

exactly). In some real-world domains, this private information may have value to agents that extends beyond the current interaction—for example, the agents may know that they will compete with each other again in the future. In such settings, it is often preferable to elicit only as much information from agents as is required to determine the social welfare maximizing choice and compute the VCG payments. We discuss this issue further when we come to ascending auctions in Chapter 11.

2. Susceptibility to collusion

Consider a referendum setting in which three agents use the VCG mechanism to decide between two choices. For example, this mechanism could be useful in the road-building referendum setting discussed earlier. Table 10.1 shows a set of valuations and the VCG payments that each agent would be required to make.

We know from Theorem 10.4.2 that no agent can gain by changing his declaration. However, the same cannot be said about groups of agents. It turns out that groups of colluding agents can achieve higher utility by coordinating their declarations rather than honestly reporting their valuations. For example, Table 10.2 shows that agents 1 and 2 can reduce both of their payments without changing the mechanism's decision by both increasing their declared valuations by $50.

3. VCG is not frugal

Consider again the transportation network example that we worked through in Section 10.4.2. We saw that the shortest path has a length of 5, the second shortest disjoint path has a length of 7, and VCG ends up paying 9. Can we give a bound on how much more than the agents' costs VCG can pay? Loosely speaking, mechanisms whose payments are small by such a measure are called *frugal*.

frugal   Before deciding whether VCG is frugal, we must determine what kind of
mechanism bound to look for. We might want VCG to pay an amount similar to the agents'

Agent	U (build road)	U (do not build road)	Payment
1	0	90	0
2	100	0	90

Table 10.3 Valuations for agents in the road-building referendum example.

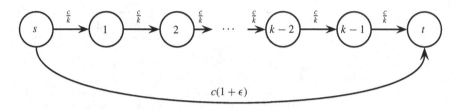

Figure 10.5 A transportation network example for which VCG's payments are not even close to the cost of the second disjoint path.

true costs. However, even in the simplest possible network it is easy to see that this is not possible. Consider a graph where there are only two paths, each owned by a single agent. In this case VCG selects the shortest path and pays the cost of the longer path, no matter how much longer it is.

It might seem more promising to hope that VCG's payments would be at least close to the cost of the *second* shortest disjoint path. Indeed, in our two-agent example this is always exactly what VCG pays. However, now consider a different graph that has two paths as illustrated in Figure 10.5. The top path involves k agents, each of whom has a cost of $\frac{c}{k}$; thus, the path has a total cost of c. The lower path involves a single agent with a cost of $c(1 + \epsilon)$. VCG would select the path with cost c, and pay each of the k agents $c(1 + \epsilon) - (k - 1)\frac{c}{k}$. Hence VCG's total payment would be $c(1 + k\epsilon)$. For fixed ϵ, this means that VCG's payment is $\Theta(k)$ times the cost of the second shortest disjoint path. Thus VCG is said not to be a frugal mechanism.

4. Dropping bidders can increase revenue

revenue monotonicity   Now we will consider *revenue monotonicity*: the property that a mechanism's revenue always weakly increases as agents are added to the mechanism. Although it may seem intuitive that having more agents should never mean less revenue, in fact VCG does not satisfy this property. To see why, let us return to the road-building example.

Consider the new valuations given in Table 10.3. Observe that the social-welfare-maximizing choice is to build the road. Agent 2 is pivotal and so would be made to pay 90, his social cost. Now see what happens when we add a third agent, as shown in Table 10.4. Again, VCG would decide that the road should be built. However, since in this second case the choice does not change when *either* winning agent is dropped, neither of them is made to pay anything, and

Agent	U (build road)	U (do not build road)	Payment
1	0	90	0
2	100	0	0
3	100	0	0

Table 10.4 Adding agent 3 causes VCG to select the same choice but to collect zero revenue.

so the mechanism collects zero revenue. Observe that the road-building referendum problem satisfies the "no single-agent effect" property; thus revenue monotonicity can fail even when the mechanism is guaranteed to be weakly budget balanced.

The fact that VCG is not revenue monotonic can also be understood as a strategic opportunity for agent 2, in the setting where agent 3 does not exist. Specifically, agent 2 can reduce his payment to zero if he is able to participate in the mechanism under multiple identities, submitting valuations both as himself and as agent 3. (This assumption might be reasonable, for example, if the mechanism is executed over the Internet.) Note that this strategy is not without its risks, however: for example, if agent 1's valuation were 150, both of agent 2's identities would be pivotal and so agent 2 would end up paying more than his true valuation.

5. Cannot return all revenue to the agents

In a setting such as this road-building example, we may want to use VCG to induce agents to report their valuations honestly, but may not want to make a profit by collecting money from the agents. In our example this might be true if the referendum was held by a government interested only in maximizing social welfare. Thus, we would want to find some way of returning the mechanism's profits back to the agents—that is, we would want a (strictly) budget-balanced mechanism rather than a weakly budget-balanced one. This turns out to be surprisingly hard to achieve, even when the "no single-agent effect" property holds, because the possibility of receiving a rebate after the mechanism has been run changes the agents' incentives. In fact, even if profits are given to a charity that the agents care about, or spent in a way that benefits the local economy and hence benefits the agents, the VCG mechanism can be undermined. This having been said, it *is* possible to return at least some of the revenues to the agents, although this must be done carefully. We give pointers to the relevant literature at the end of the chapter.

6. Computational intractability

Finally, even when there are no problems *in principle* with using VCG, there can still be practical obstacles. Perhaps the biggest such problem is that efficient mechanisms can require unreasonable amounts of computation: evaluating the arg max can require solving an NP-hard problem in many practical domains.

Thus, VCG can fail to satisfy the *tractability* property (Definition 10.3.10). This problem is not just theoretical: we present important examples in which VCG is intractable in Sections 11.2.3 and 11.3.2. In Section 10.5 below, we consider some alternatives to VCG for use in such settings.

10.4.6 *Budget balance and efficiency*

In Section 10.4.4 we identified a realistic case in which the VCG mechanism is weakly budget balanced. However, we also noted that there exist other important and practical settings in which the no single-agent effect property does not hold. simple exchange   For example, define a *simple exchange* as an environment consisting of buyers and sellers with quasilinear utility functions, all interested in trading a single identical unit of some good. The no single-agent effect property is not satisfied in a simple exchange because dropping a seller could make some buyer worse off and vice versa. Can we find some other argument to show that VCG will remain budget balanced in this important setting?

It turns out that neither VCG nor any other Groves mechanism is budget balanced in the simple exchange setting. (Recall Theorem 10.4.3, which showed that only Groves mechanisms are both dominant-strategy incentive-compatible and efficient.)

Theorem 10.4.11 (Green–Laffont; Hurwicz) *No dominant-strategy incentive-compatible mechanism is always both efficient and weakly budget balanced, even if agents are restricted to the simple exchange setting.*

Furthermore, another seminal result showed that a similar problem arises in the broader class of Bayes–Nash incentive-compatible mechanisms (which, recall, includes the class of dominant-strategy incentive-compatible mechanisms) if we also require *ex interim* individual rationality and allow general quasilinear utility functions.

Theorem 10.4.12 (Myerson–Satterthwaite) *No   Bayes–Nash   incentive-compatible mechanism is always simultaneously efficient, weakly budget balanced, and* ex interim *individually rational, even if agents are restricted to quasilinear utility functions.*

On the other hand, it turns out that it *is* possible to design a Bayes–Nash incentive compatible mechanism that achieves any two of these three properties.

10.4.7 *The AGV mechanism*

Of particular interest is the AGV mechanism, which trades away *ex interim* individual rationality and dominant strategies in exchange for budget balance and *ex ante* individual rationality.

Arrow;
d'Aspremont–
Gérard-Varet
(AGV)
mechanism

Definition 10.4.13 (Arrow; d'Aspremont–Gérard-Varet (AGV) mechanism)
The Arrow; d'Aspremont–Gérard-Varet mechanism *(AGV) is a direct quasilinear mechanism* $(\chi, \wp)$, *where*

$$\chi(\hat{v}) = \arg\max_x \sum_i \hat{v}_i(x),$$

$$\wp_i(\hat{v}) = \left(\frac{1}{n-1} \sum_{j \neq i} ESW_{-j}(\hat{v}_j)\right) - ESW_{-i}(\hat{v}_i),$$

$$ESW_{-i}(\hat{v}_i) = \mathbb{E}_{v_{-i}}\left[\sum_{j \neq i} v_j\left(\chi(\hat{v}_i, v_{-i})\right)\right].$$

ESW (standing for "expected social welfare") is an intermediate term that is used to make the definition of $\wp$ more concise. Observe that AGV's allocation rule is the same as under Groves mechanisms. Although we will not prove this or any of the other properties we mention here, AGV is incentive compatible, from which we can conclude that it is also efficient. Again like Groves mechanisms, each agent i is given a payment reflecting the other agents' valuations for the choice selected given his declaration. While in Groves mechanisms this calculation used $-i$'s declared valuations, however, AGV computes $-i$'s *ex ante* expected social welfare given i's declaration. The rest of the payment is computed very differently than it is under VCG: each agent is charged a $\frac{1}{n-1}$ share of the payments made to each of the other agents. This guarantees that the mechanism is budget balanced (i.e., that it always collects from the agents exactly the total amount that it pays back to them). Two sacrifices are made in exchange for this property: AGV is truthful only in Bayes–Nash equilibrium rather than in dominant strategies and is only *ex ante* individually rational.

The AGV mechanism illustrates two senses in which we can discover new, useful mechanisms by relaxing properties that we had previously insisted on. First, our move from dominant-strategy incentive compatibility to Bayes–Nash incentive compatibility allowed us to circumvent Theorem 10.4.3, which told us that efficiency can be achieved under dominant strategies only by Groves mechanisms. (AGV is also efficient, but is not a Groves mechanism.) Second, moving from *ex interim* to *ex ante* individual rationality is sufficient to get around the negative result from Theorem 10.4.12, that we cannot simultaneously achieve weak budget balance, efficiency, and *ex interim* individual rationality, even under Bayes–Nash equilibrium.

10.5 Beyond efficiency

Throughout our consideration of the quasilinear setting in this chapter we have so far focused on efficient mechanisms. As we discussed in Section 10.4.5, efficient choice rules can require unreasonable amounts of computation, and hence both Groves mechanisms and AGV can fail to satisfy the *tractability* property. In this section we consider two ways of addressing this issue. The first is to explore dominant-strategy mechanisms that implement different social choice functions. We have already seen that the quasilinear preference setting is considerably more amenable to dominant strategy implementation than the unrestricted preferences setting. However, there are still limits—what are they? Second, we will examine an alternate way of building mechanisms, by using a Groves payment rule with an alternate choice function, and leveraging agents' computational limitations in order to achieve the implementation.

10.5.1   *What else can be implemented in dominant strategies?*

Here we give some characterizations of the social choice functions that can be implemented in dominant strategies in the quasilinear setting and of how payments must be constrained in order to enable such implementations. As always, the revelation principle allows us to restrict ourselves to truthful mechanisms without loss of generality. We also restrict ourselves to *deterministic* mechanisms: this restriction does turn out to be substantive.

Let $\mathcal{X}_i(\hat{v}_{-i}) \subseteq X$ denote the set of choices that can be selected by the choice rule χ given the declaration $\hat{v}_{-i}$ by the agents other than i (i.e., the range of $\chi(\cdot, \hat{v}_{-i})$). Now we can state conditions that are both necessary and sufficient for dominant-strategy truthfulness that are both intuitive and useful.

Theorem 10.5.1 *A direct, deterministic mechanism is dominant-strategy incentive-compatible if and only if, for every $i \in N$ and every $\hat{v}_{-i} \in V_{-i}$:*

1. *the payment function $\wp_i(\hat{v})$ can be written as $\wp_i(\hat{v}_{-i}, \chi(\hat{v}))$; and*
2. *for every $\hat{v}_i \in V_i$, $\chi(\hat{v}_i, \hat{v}_{-i}) \in \arg\max_{x \in \mathcal{X}_i(\hat{v}_{-i})}(\hat{v}_i(x) - \wp_i(\hat{v}_{-i}, x))$.*

The first condition says that an agent's payment can only depend on other agents' declarations and the selected choice; it *cannot* depend otherwise on the agent's own declaration. The second condition says that, taking the other agent's declarations and the payment function into account, from every player's point of view the mechanism selects the most preferable choice. This result is not very difficult to prove; the interested reader is encouraged to try.

As the above characterization suggests, there is a tight connection between the choice rules and payment rules of dominant-strategy truthful mechanisms. In fact, under reasonable assumptions about the valuation space, once a choice rule is chosen, all possible payment rules differ only in their choice of a function $h_i(\hat{v}_{-i})$ that is added to the rest of the payment. We already saw an example of this with the Groves family of mechanisms: these mechanisms share the same choice rule, and their payment rules differ only in a constant $h_i(\hat{v}_{-i})$.

Given this strong link between choice rules and payment rules, it is interesting to characterize a set of *choice rules* that can be implemented in dominant strategies, without reference to payments. Here we will consider such a characterization, though in general it turns out only to offer a *necessary* condition for dominant-strategy truthfulness.

weak
monotonicity
(WMON)

Definition 10.5.2 (WMON) *A social choice function* C *satisfies* weak monotonicity *(WMON) if for all* $i \in N$ *and all* $v_{-i} \in V_{-i}$, $C(v_i, v_{-i}) \neq C(v_i', v_{-i})$ *implies that* $v_i(C(v_i, v_{-i})) - v_i(C(v_i', v_{-i})) \geq v_i'(C(v_i, v_{-i})) - v_i'(C(v_i', v_{-i}))$.

In words, WMON says that any time the choice function's decision can be altered by a single agent changing his declaration, it must be the case that this change expressed a relative increase in preference for the new choice over the old choice.

Theorem 10.5.3 *All social choice functions implementable by deterministic dominant-strategy incentive-compatible mechanisms in quasilinear settings satisfy WMON. Furthermore, let* C *be an arbitrary social choice function* C : $V_1 \times \cdots \times V_n \mapsto X$ *satisfying WMON and having the property that* $\forall i \in N$, V_i *is a convex set. Then* C *can be implemented in dominant strategies.*

Although Theorem 10.5.3 does not provide a full characterization of those social choice functions that can be implemented in dominant strategies, it gets pretty close—the convexity restriction is often acceptable. A bigger problem is that WMON is a *local* characterization, speaking about how the mechanism treats each agent individually. It would be desirable to have a global characterization that gave the social choice function directly. This also turns out to be possible.

affine maximizer

Definition 10.5.4 (Affine maximizer) *A social choice function is an* affine maximizer *if it has the form*

$$\arg\max_{x \in X} \left(\gamma_x + \sum_{i \in N} w_i v_i(x) \right),$$

where each γ_x *is an arbitrary constant (may be* $-\infty$*) and each* $w_i \in \mathbb{R}_+$.

In the case of general quasilinear preferences (i.e., when each agent can have any valuation for each choice $x \in X$) and where the choice function selects from more than two alternatives, affine maximizers turn out to be the only social choice functions implementable in dominant strategies.

Theorem 10.5.5 (Roberts) *If there are at least three choices that a social choice function will select given some input, and if agents have general quasilinear preferences, then the set of (deterministic) social choice functions implementable in dominant strategies is precisely the set of affine maximizers.*

Note that efficiency is an affine-maximizing social choice function for which $\forall x \in X, \gamma_x = 0$ and $\forall i \in N, w_i = 1$. Indeed, affine maximizing mechanisms can be seen as weighted Groves mechanisms—they transform both the choices and the agents' valuations by applying linear weights, and then effectively run a Groves mechanism in the transformed space. Thus, Theorem 10.5.5 says that we cannot stray very far from Groves mechanisms even if we are willing to give up on efficiency.

Is this the end of the story on dominant-strategy implementation in quasilinear settings? Not quite. It turns out that the assumption that agents have general quasilinear preferences is a strong one, and does not hold in many domains of

single-parameter
valuation

interest. As another extreme, we can consider *single-parameter valuations*: each agent i partitions the set of choices X into a set $X_{i,wins}$ in which i "wins" and receives some constant payoff v_i that does not depend on the choice $x \in X_{i,wins}$, and a set of choices $X_{i,loses} = X \setminus X_{i,wins}$ in which i "loses" and receives zero payoff.[10] Importantly, the sets $X_{i,wins}$ and $X_{i,loses}$ are assumed to be common knowledge, and so the agent's private information can be summarized by a single parameter, v_i. Such settings are quite practical: we will see several that satisfy these conditions in Chapter 11. Single-parameter settings are interesting because for such preferences, it *is* possible to go well beyond affine maximizers. In fact, additional characterizations exist describing the social choice functions that can be implemented in this and other restricted-preference settings. We will not describe them here, instead referring interested readers to the works cited at the end of the chapter. However, we do present a dominant-strategy incentive-compatible, non-affine-maximizing mechanism for a single-parameter setting in Section 11.3.5.

10.5.2   *Tractable Groves mechanisms*

Now we consider a general approach that attempts to implement tractable, inefficient social choice functions by sticking with Groves mechanisms, but replacing the (possibly exponential-time) computation of the arg max with some other polynomial-time algorithm. The very clever idea here is not to build mechanisms that are impossible to manipulate (indeed, in many cases it can be shown that this cannot be done), but rather to build mechanisms that agents will be unable to manipulate in practice, given their computational limitations.

First, we define the class of mechanisms being considered.

Groves-based
mechanism

Definition 10.5.6 (Groves-based mechanisms) Groves-based mechanisms *are direct quasilinear mechanisms* $(\chi, \wp)$, *for which:*

> $\chi(\hat{v})$ *is an arbitrary function mapping type declarations to choices; and*

$$\wp_i(\hat{v}) = h_i(\hat{v}_{-i}) - \sum_{j \neq i} \hat{v}_j(\chi(\hat{v})).$$

That is, a mechanism is Groves based if it uses the Groves payment function, regardless of what allocation function it uses. (Contrast Definition 10.5.6 with Definition 10.4.1, which defined a Groves mechanism.)

Most interesting Groves-based mechanisms are not dominant-strategy truthful. For example, consider a property sometimes called *reasonableness*: if there exists some good g that only one agent i values above zero, g should be allocated to i. It can be shown that the only dominant-strategy truthful Groves-based mechanisms that satisfy reasonableness are the Groves mechanisms themselves. This rules out the use of most greedy algorithms as candidates for the allocation function χ in truthful Groves-based mechanisms, as most such algorithms would select "reasonable" allocations.

---

10. The assumption that this second payoff is zero can be understood as a normalization and does not change the set of social choice functions that can be implemented.

If tractable Groves-based mechanisms lose the all-important property of dominant-strategy truthfulness, why are they still interesting? The proof of Theorem 10.4.2 essentially argued that the Groves payment function aligns agents' utilities, making all agents prefer the optimal allocation. Groves-based mechanisms still have this property, but may not select the optimal allocation. We can conclude that the only way an agent can gain by lying to a Groves-based mechanism is to *help it* by causing it to select a more efficient allocation.

We now come to the idea of a second-chance mechanism. Intuitively, since lies by agents can only help the mechanism, the mechanism can simply ask the agents how they intend to lie and select an choice that would be picked because of such a lie if it turns out to be better than what the mechanism would have picked otherwise.

second-chance
mechanism

Definition 10.5.7 (Second-chance mechanisms) *Given a Groves-based mechanism $(\chi, \wp)$, a second-chance mechanism works as follows:*

appeal function

1. *Each agent i is asked to submit a valuation declaration $\hat{v}_i \in V_i$ and an appeal function $l : V \mapsto V$.*
2. *The mechanism computes $\chi(\widehat{v})$, and also $\chi(l_i(\hat{v}))$ for all $i \in N$. From the set of choices thus identified, the mechanism keeps one that maximizes the sum of agents' declared valuations.*
3. *The mechanism charges each agent i $\wp(\hat{v})$.*

Intuitively, an appeal function maps agents' valuations to valuations that they might instead have chosen to report by lying. It is important that the appeal functions be computationally bounded (e.g., their execution could be time limited). Otherwise, these functions can solve the social welfare maximization problem and then select an input that would cause χ to select this choice. When appeal functions are computationally restricted, we cannot in general say that second-chance mechanisms are truthful. However, they are *feasibly truthful*, because an agent can use the appeal function to try out any lie that he believes might help him. Thus in a second-chance mechanism, a computationally limited agent can do no better than to declare his true valuation along with the best appeal function he is able to construct.

feasibly truthful

10.6 Computational applications of mechanism design

We now survey some applications of mechanism design, to give a sense of some more recent work that has drawn on the theory we have described so far. However, we must offer two caveats. First, we speak here only about *computational* applications of mechanism design, by which we mean mechanisms that contain an interesting computational ingredient and/or mechanisms applied to computational domains (e.g., computer networks). Thus we skip over some highly influential applications from economics, such as theories of taxation, government regulation, and corporate finance. Second, without a doubt the most significant application of mechanism design—computational or not—is the design and analysis of auctions. Because there is so much to say about this application, we defer its discussion to Chapter 11.

algorithmic mechanism design

Some of the mechanism design applications we discuss in this section are examples of so-called *algorithmic mechanism design*. This term describes settings in which a center wants to solve an optimization problem, but the inputs to this problem are the private information of self-interested agents. The center must thus design a mechanism that solves the optimization while inducing the agents to reveal their information truthfully. Observe that this setting does not really describe a different problem from classical mechanism design, though it does adopt a different perspective. It also tends to describe work that has a somewhat different flavor, often emphasizing approximation algorithms and worst-case analysis.

10.6.1 *Task scheduling*

One problem that has been well studied in the context of algorithmic mechanism design is that of task scheduling. Consider n agents who can perform tasks and a set T tasks that must be allocated. Each agent i's type t_i is a vector, giving the minimum amount of time $t_{i,j}$ in which i can perform each task j. The center's goal is to minimize the completion time of the last task, called the *makespan*. A choice x by the mechanism is an allocation of each task to some agent; agents must perform the tasks they are assigned. Let $x(i, j)$ equal 1 if an agent i is assigned task j, and zero otherwise. Note that some agents may be given more than one task and some may not be given a task at all. The mechanism is able to *verify* the agents' work, observing the true amount of time it took an agent to complete his tasks. We write the true amount of time i spent on task j as $\tilde{t}_{i,j}$; of course $\tilde{t}_{i,j}$ must always be greater than or equal to $t_{i,j}$. An agent i's valuation for a choice x by the mechanism is $-\sum_{j \in T} x(i, j)\tilde{t}_{i,j}$, the sum of the true amounts of time he spends on his assigned tasks. Of course, an agent i can lie about the amount of time it will take him to perform a task. We denote the tuple of all agents' declarations as $\hat{t}$.

The task scheduling problem cannot be solved with a Groves mechanism. While such a mechanism would indeed provide agents with dominant strategies for truthfully revealing their types, it would choose the wrong allocation, maximizing the sum of agents' welfare rather than minimizing the makespan. Indeed, note that makespan is like a worst-case version of social welfare: it measures the unhappiness of the unhappiest agent, and ignores the other agents completely. Another family of mechanisms does work for solving the task allocation scheduling problem. These mechanisms can be understood as generalizing Groves mechanisms to objective functions other than social welfare.

compensation and penalty mechanism

Definition 10.6.1 (Compensation and penalty mechanisms) Compensation and penalty mechanisms *are quasilinear mechanisms* $(\chi, \wp)$, *for which*

$$\chi(\hat{t}) = \arg\min_x \left(\max_{i \in N} \sum_{j \in T} x(i, j)\hat{t}_{i,j} \right)$$

$$\wp_i(\hat{t}) = h_i(\hat{t}_{-i}) - \sum_{j \in T} x(i, j)\tilde{t}_{i,j} + \max\left\{ \sum_{j \in T} x(i, j)\tilde{t}_{i,j}, \max_{i' \neq i \in N} \sum_{j \in T} x(i', j)\hat{t}_{i',j} \right\}.$$

Thus, the mechanism selects the choice that minimizes makespan, given the agents' declarations. What types should agents declare? Should agents solve tasks as quickly as possible, or can they increase their utilities by taking longer? An answer is given by the following theorem.

Theorem 10.6.2 *Compensation and penalty mechanisms are dominant-strategy incentive compatible: agents choose to complete their tasks as quickly as possible ($\tilde{t}_{i,j} = t_{i,j}$) and to report these completion times truthfully ($\hat{t}_{i,j} = t_{i,j}$).*

Proof. The first term in the payment function $\wp_i$, $h_i(\hat{t}_{-i})$, does not depend on i's declaration. Thus it does not affect i's incentives, and so we can disregard it.

The rest of $\wp_i$ consists of two terms. The second term is a payment to agent i equal to his true cost for his assigned tasks. This payment exactly *compensates i* for any tasks he was assigned, making him indifferent between all task assignments regardless of how long he spent completing his tasks.

The third term of $\wp_i$ is a *penalty* to i in the amount of the mechanism's objective function, except that i's actual task completion time is used instead of his declared time. The strategic problem for i is thus to choose the $\tilde{t}$ and $\hat{t}$ that will lead the mechanism to select the x that makes this penalty as small as possible. By choosing $\tilde{t}_{i,j} > t_{i,j}$, i does not influence x (this depends only on $\hat{t}_{i,j}$) and can only increase his penalty. $\tilde{t}_{i,j} < t_{i,j}$ is impossible, and so it is a dominant strategy for i to choose $\tilde{t}_{i,j} = t_{i,j}$. If i declares $\hat{t}_{i,j} > t_{i,j}$, then he can only increase the makespan and hence his penalty, by making the mechanism allocate tasks suboptimally to the other agents. If i declares $\hat{t}_{i,j} < t_{i,j}$, he can reduce the makespan; however, he cannot reduce his penalty since it depends on $\tilde{t}_{i,j}$ rather than $\hat{t}_{i,j}$. In this case he still can *increase* his penalty by causing the mechanism to allocate tasks suboptimally. Thus, i's dominant strategy is to declare $\hat{t}_{i,j} = t_{i,j}$. ∎

Observe that it is important that the mechanism can verify the amount of time an agent took to complete the task. If this were not the case, the agent could under-report his completion time, driving down the makespan and hence reducing his own penalty. Note also that these mechanisms really do generalize Groves mechanisms: if we replace the mechanism's objective function (in χ and the third term of $\wp$) with social welfare, we recover Groves.

While compensation and penalty mechanisms are truthful even with $h_i = 0$, they are not individually rational, as an agent's utility is always simply the negative of his penalty, and this penalty is always positive. However, we can regain individual rationality in the same way as we did in moving from Groves mechanisms to VCG. Specifically, we can set h_i to be the mechanism's objective function when i does not participate,

$$h_i(\hat{t}_{-i}) = \min_x \left(\max_{i' \neq i \in N} \sum_{j \in T} x(i', j) \hat{t}_{i',j} \right).$$

Now h_i will always be greater than or equal to i's penalty, because the makespan is guaranteed to weakly increase if we omit i. This ensures that i never loses by participating in the mechanism.

As we indicated at the beginning of the section, work on algorithmic mechanism design often focuses on the use of approximation algorithms. Such an approach is sensible for the task scheduling problem because finding the makespan-minimizing allocation ($\chi(\hat{t})$ in compensation and penalty mechanisms) is an NP-hard problem, whereas approximation algorithms can run in polynomial time. Although we do not go into the details here, there is a whole constellation of results about what approximation bounds are achievable by which variety of dominant-strategy approximation-algorithm-based mechanism, under what assumptions (e.g., verification possible or not; restrictions on valuations). For example, in the case without verification no deterministic mechanism based on an approximation algorithm can achieve better than a 2-approximation; this bound is tight for the 2 agent case. On the other hand, randomized mechanisms can do better, achieving a 1.75-approximation. More details are available in the paper cited at the end of the chapter.

10.6.2   *Bandwidth allocation in computer networks*

When designing a computer network, the network operator wants to ensure that the most important network traffic gets the best performance and/or that some fairness criterion is satisfied. However, optimizing traffic in this way is difficult because the network operator does not know the users' tasks or how important they are. Thus, the network operator faces a mechanism design problem. Although much more elaborate settings have been studied (see the notes at the end of the chapter), in this section we will consider the problem of allocating the capacity of a single link in a network. The reason that this problem is still tricky is that the bandwidth of a link is not allocated all-or-nothing to a single buyer, as it was in our example at the beginning of the chapter (Section 10.1.2). Instead, the link has a real-valued capacity that can be divided arbitrarily between the agents. Thus, even this simple problem considers a choice space X that is uncountably infinite, and valuation functions that can be understood as continuous "demand curves."

Formally, consider a domain with N users who want to use a network resource with capacity $C \in \mathbb{R}_+$. Each user has a valuation function $v_i : \mathbb{R}_+ \mapsto \mathbb{R}$ expressing his happiness for being allocated any nonnegative amount of capacity d_i. We will assume throughout that this function v_i is concave, strictly increasing, and continuous.[11]

proportional
allocation
mechanism

We will begin by considering a particular mechanism that has been widely studied: the *proportional allocation mechanism*. This is a quasilinear mechanism in which agents are invited to specify a single value $w_i \in \mathbb{R}_+$. The mechanism interprets each value w_i as the payment that user i offers to make to the network. In order to determine the amount of capacity that each user will receive, we start

---

11. Furthermore, it is necessary to make some differentiability assumptions about the valuation functions; for details see the references cited at the end of the chapter.

from the assumption that each user must be charged for his use of the resource
at the same rate, μ. Assuming that the network operator wants to allocate all
capacity, we can then calculate this rate uniquely as $\mu = \frac{\sum_i w_i}{C}$, implying that
each agent i receives the allocation $d_i = \frac{w_i}{\mu}$.

Unlike most of the mechanisms discussed in this chapter, the proportional
allocation mechanism is not direct. However, this is one of its attractive qualities.
Even under our assumptions of concavity, continuity, and monotonicity, an agent's
valuation function can be arbitrarily complex. In a real network system, it would
defeat the purpose of an allocation mechanism to allow agents to communicate
a great deal of information about their valuation functions—the whole idea is
to allocate bandwidth efficiently. Since the proportional allocation mechanism
requires each agent to declare only a single real number, its proponents have
argued that it is practical and have even gone so far as to describe ways that it
could be added to existing (e.g., TCP/IP) network architectures.

A more serious concern is that the proportional allocation mechanism appears
strategically complex, since agents can affect their payments (rather than just their
allocations) by changing their declarations. Nevertheless, there are a number of
interesting things that we can say about the mechanism. First, let us set aside our
usual game-theoretic assumption that agents play best responses to each other and
price taker   to the rules of the mechanism. Instead, let us assume that agents are *price takers*:
that they consider the rate μ to be fixed and that they select the best declarations
w_i given μ. (In fact, an agent's declaration w_i is used in the calculation of μ;
thus, we assume that agents disregard this connection.) Given this assumption,
it is interesting to ask whether allocations chosen by our mechanism constitute a
competitive   *competitive equilibrium* (Definition 2.3.4). Formally, a declaration profile w and
equilibrium   rate μ constitute a competitive equilibrium if each w_i maximizes i's quasilinear
valuation $v_i(\frac{w_i}{\mu}) - w_i$, and if $\mu = \frac{\sum_i w_i}{C}$. It is possible to prove the following result.

Theorem 10.6.3 *Given n agents with valuation functions $(v_1, \ldots, v_n)$ and a
resource with capacity $C > 0$, there exists a competitive equilibrium (w, μ) of the
proportional allocation mechanism. Furthermore, the allocation is efficient: the
choices $d_i = \frac{w_i}{\mu}$ maximize the social welfare $\sum_i v_i(d_i) - w_i$ subject to capacity
constraints.*

Thus, given price-taking agents, full efficiency can be achieved by the proportional allocation mechanism, even though it only elicits a single scalar value
from each agent.

Now, let us return to the more standard game-theoretic setting, in which
agents take into account their abilities to affect μ through their own declarations.
Thus, our solution concept shifts from the competitive equilibrium to the Nash
equilibrium. It is possible to show that a Nash equilibrium exists[12] and that it is
unique. How does this Nash equilibrium compare to the competitive equilibrium
described in Theorem 10.6.3? The natural way to formalize this question is to
ask what fraction of the social welfare achieved in the competitive equilibrium
is also achieved in the Nash equilibrium. When we ask how small this fraction

---

12. In settings with continuous action spaces, the existence of Nash equilibrium is not guaranteed.

becomes in the worst case, we arrive precisely at the notion of minimizing the price of anarchy (see Definition 10.3.14; recall also our previous use of the price of anarchy in the context of "selfish routing" in Section 6.4.5).

price of anarchy

Theorem 10.6.4 *Let $n \geq 2$, let d^{CE} be an allocation profile achievable in competitive equilibrium and let d^{NE} be the unique allocation profile achievable in Nash equilibrium. Then any profile of valuation functions v for which $\forall i,\ v_i(0) \geq 0$ satisfies*

$$\sum_i v_i(d_i^{NE}) \geq \frac{3}{4} \sum_i v_i(d_i^{CE}).$$

In other words, the price of anarchy is $\frac{4}{3}$; in the worst case, the Nash equilibrium achieves 25% less efficiency than the competitive equilibrium. While it is always disappointing not to achieve full efficiency, this result should be understood as good news. Even in the worst case, strategic behavior by agents will only cause a small reduction in social welfare.

So far, we have analyzed a given mechanism rather than showing that this mechanism optimizes some objective function. However, the proportional allocation mechanism can indeed be justified in this way. Specifically, it achieves minimal price of anarchy, as compared to a broad family of mechanisms in which agents' declarations are a single scalar and the mechanism charges all users the same rate. We do not state this result formally here, as the precise definition of the family of mechanisms is quite technical; instead, we refer the reader to the references cited at the end of the chapter. We also note that when the setting is relaxed so that users still submit only a single scalar but the mechanism is allowed to charge different users at different rates, a VCG-like mechanism can be used to achieve full efficiency.

10.6.3 *Multicast cost sharing*

Consider the problem of streaming media (e.g., a television broadcast) over a digital network. If this information is sent naively (e.g., using the TCP/IP protocol), then each user establishes a separate connection with the server and the same information may be sent many times over the same network links. This approach can easily overwhelm a link's capacity. A more sensible alternative is *multicast routing*, in which information is sent only once across each link, and it is replicated onto multiple outgoing links where necessary. Besides saving bandwidth, this approach can also make more economic sense. For example, individual users sharing a satellite link might not be willing to pay the full cost of receiving a high-bandwidth video stream, but could be willing to split the cost among themselves. Such a system faces the problem of *multicast cost sharing*: given a set of users with different values for receiving the transmission and a network with costly links, who should receive the transmission and how much should they each pay? This is a mechanism design problem.

multicast routing

multicast cost sharing

Formally, consider an undirected graph with nodes N (a set of agents) and links L. Each link $l \in L$ has a cost $c(l) \geq 0$. One of the agents, $\alpha_0 \in N$ is the

```
S ← N*                          // assume that every agent will receive the transmission
repeat
    Find the multicast routing tree T(S)
    Compute payments pᵢ such that each agent i pays an equal share of the
    cost for every link in T({i})
    foreach i ∈ S do
        if v̂ᵢ < pᵢ then  S ← S \ {i}          // i is dropped from S
until no agents were dropped from S
```

Figure 10.6 An algorithm for computing the allocation and payments for the Shapley value mechanism.

source of the transmission; there is also a set of agents $N^* \subseteq N$ who are interested in receiving it. Each $i \in N^*$ values the transmission at $v_i > 0$.

Our goal is to find a cost-sharing mechanism, a direct quasilinear mechanism $(\chi, \wp)$ that receives declarations $\hat{v}_i$ of each agent i's utility and determines which agents will receive the transmission and how much they will pay. The function χ determines a set of users $S \subseteq N^*$ who will receive the transmission. In order to

multicast   do so, we must find a *multicast routing tree* $T(S) \subseteq L$ rooted at α_0 that spans S.
routing tree   We make a monotonicity assumption about the algorithm used to find $T(S)$,

$$S_1 \subseteq S_2 \Rightarrow T(S_1) \subseteq T(S_2).$$

The mechanism also includes a payment function $\wp$ that ensures that the agents in S share the costs of the links in $T(S)$. We denote the payment collected from $i \in S$ as p_i. We assume that the mechanism is computed by trusted hardware in the network (e.g., the routers); however, we will be concerned with communication complexity, and hence will look for ways that this computation can be distributed throughout the system.

Ideally, we would like a cost-sharing mechanism to be dominant-strategy incentive compatible, budget balanced, and efficient. However, we have already seen (Theorem 10.4.11) that such mechanisms do not exist. Thus, we will consider mechanisms that achieve two of these properties and relax the third.

Truthful and budget balanced: The Shapley value mechanism

First, we will describe a dominant-strategy truthful mechanism that achieves budget balance at the expense of efficiency. Intuitively, this mechanism is built around the idea that the cost of a link should be divided equally among the agents that use it. Its name comes from the fact that this objective can be seen as a special case of the Shapley value from coalitional game theory (see Section 12.2.1). We describe a centralized version of the mechanism in Figure 10.6.

To see why this algorithm leads to a dominant-strategy truthful mechanism, observe that the payments are "cross-monotonic." This means that each agent's payment can only increase when another agent is dropped, and hence that an agent's incentives are not affected by the order in which agents are dropped by the algorithm. That is, if the payment that the mechanism would charge an agent i given a set of other agents S exceeds i's utility, then i's payment is guaranteed to exceed his utility for all subsets of the other agents $S' \subset S$. Since we only drop agents when their proposed payments exceed their utilities, the order in which we drop them is unimportant. Because we can drop agents "greedily" (i.e., without having to consider the possibility of reinstating them) the algorithm runs in polynomial time.

This algorithm can be run in a network by having all agents send their utilities to some node (e.g., the source α_0) and then running the algorithm there. However, although the algorithm is computationally tractable, this centralized approach requires an unreasonable amount of *communication* as the network becomes large. Thus, we would prefer a distributed solution. Unfortunately, no distributed algorithm can compute the same allocation and payments using asymptotically less communication than the centralized solution.

Theorem 10.6.5 *Any (deterministic or randomized) distributed algorithm that computes the same allocation and payments as the Shapley value algorithm must send $\Omega(|N^*|)$ bits over linearly many links in the worst case.*

Truthful and efficient: The VCG mechanism

Now we consider relaxing the budget balance requirement and instead insisting on efficiency. Unsurprisingly (consider Theorem 10.4.3) we must obtain a Groves mechanism in this case; VCG is the obvious choice. VCG can be easily used as a cost-sharing mechanism in the centralized case. Like the Shapley value mechanism, it requires only polynomial computation and hence is tractable. However, it has an interesting and important advantage over the Shapley value mechanism: it can also be made to work efficiently as a distributed algorithm.

Theorem 10.6.6 *A distributed algorithm can compute the same allocation and payments as VCG by sending exactly two values across each link.*

> **Proof.** The algorithm in Figure 10.7 computes VCG payments and allocations. Let l_i be the link connecting node i to its parent. Every nonroot node i sends and receives a single real-valued message over l_i. ∎

This algorithm can be understood as passing messages from one node in the tree to the next. Observe that the first for loop proceeds "bottom up" (i.e., computing m for all children of a node i before computing m for node i itself), while the second for loop proceeds "top down" (i.e., computing s for a node i before computing s for any of i's children). Thus we can see the m's as messages that are passed up the tree, starting at the leaves, and the s's as messages that are passed back down, starting at the root.

Let us consider applying this algorithm to a sample multicast spanning tree (Figure 10.8a). In the upward pass (Figure 10.8b), every node i computes m_i, the marginal value connecting i to the network, given that its parent is connected. This is the maximum amount the agents on the subtree rooted at i would be willing to pay to join the multicast. In the downward pass (Figure 10.8c), every node i computes s_j for each child node j. s_j is the actual total surplus generated by connecting j to the multicast tree. If m_j or s_j is negative, the efficient allocation

```
// Upward pass
foreach node i, bottom up do
    m_i ← v̂_i − c(l_i)
    foreach node j ∈ children of i do
        m_i ← m_i + max(m_j, 0)

// Downward pass
S ← ∅
s_root ← m_root
foreach node i, top down do
    if s_i ≥ 0 then
        S ← S ∪ {i}
        p_i ← max(v̂_i − s_i, 0)
    foreach node j ∈ children of i do
        s_j ← min(s_i, m_j)
```

Figure 10.7 A distributed algorithm for computing the efficient allocation and VCG payments for multicast cost sharing.

does not connect j. Thus, s_j can also be seen as the maximum amount by which agents in the subtree rooted at s_j could reduce their joint value declaration while remaining connected. Each connected node j is charged $\max(\hat{v}_j - s_j, 0)$, meaning that his surplus is equal to the amount he could have under-reported his value without being disconnected. These payments are illustrated in Figure 10.8d.

10.6.4 *Two-sided matching*

So far in this chapter we have concentrated on mechanism design in quasilinear settings, meaning that we have assumed that money can be transferred between agents and the mechanism. However, there exist many interesting settings where such transfers are impossible, for example, because of legal restrictions. Examples of such problems include kidney exchanges, college admissions, and the assignment of medical interns to hospitals. *Two-sided matching* is a widely studied model that can be used to describe such domains. Under this model, each agent belongs to one of two groups. Members of each group are matched up, based on their declared preferences over their candidate partners. The mechanism design problem is to induce agents to disclose these preferences in a way that allows a desirable matching to be chosen, despite the restriction that payments cannot be imposed.

two-sided matching

We will use the running example of a cohort of graduate students who must align with thesis advisors. Each student has a preference ordering over advisors (depending on their research interests, personalities, etc.), and likewise each potential advisor has a preference ordering over students. In this setting a social choice function is a decision about which students should be assigned to which advisors, given their preferences; as always, the mechanism design concern is how to implement a desired social choice function.

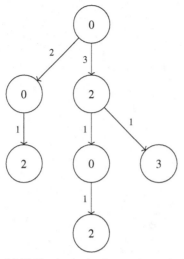

(a) Multicast spanning tree: nodes are labeled with values, edges are labeled with costs.

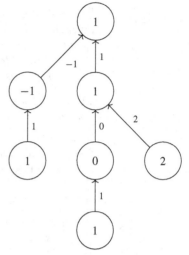

(b) Upward pass: each node i computes m_i and passes it to its parent.

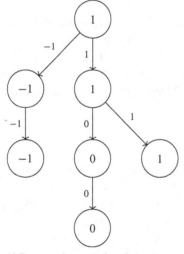

(c) Downward pass: each node i computes s_j for child j and passes it down.

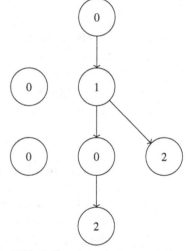

(d) Final allocation: only connected edges are shown; nodes are labeled with payments.

Figure 10.8 An example run of the algorithm from Figure 10.7.

We now define the setting more formally. Let A be a set of advisors and let S be a set of graduate students. We do not assume that $|A| = |S|$; thus, some students and/or advisors may remain unpaired. We assume that each student can have at most one advisor and each advisor will take at most one new student.[13] Each student i has a preference ordering $\succ_i$ over the advisors, and each advisor j has a preference ordering $\succ_j$ over the students. We write $a \succ_s a'$ to mean that student

---

13. Many but not all of the results in this section can also be extended to the case where advisors can take multiple students.

unacceptable matching

s prefers advisor a to advisor a', and $\varnothing \succ_s a$ to mean that s prefers not finding a supervisor to aligning with advisor a. In the latter case we say that advisor a is *unacceptable* to student s. Similarly, we write $s \succ_a s'$ and $\varnothing \succ_a s$. Note that we have assumed that all preferences are strict,[14] but that each agent can identify a set of partners with whom he would prefer not to be matched, effectively expressing a tie among unacceptable partners. In what follows, we adopt the convention that all advisors are female and all students are male. The resulting problem of finding good male–female pairings pays homage to the problem introduced in the two-sided matching literature a half-century ago, so-called *stable marriage*.

stable marriage

matching

Definition 10.6.7 (Matching) *A matching* $\mu : A \cup S \to A \cup S \cup \{\varnothing\}$ *is an assignment of advisors to students such that each advisor is assigned to at most one student and vice versa. More formally,* $\mu(s) = a$ *if and only if* $\mu(a) = s$. *Furthermore,* $\forall s \in S$, *either* $\exists a \in A$, $\mu(s) = a$ *or* $\mu(s) = \varnothing$ *(the student is unpaired), and likewise* $\forall a \in A$, *either* $\exists s \in S$, $\mu(a) = s$ *or* $\mu(a) = \varnothing$.

Note that it is always possible that some student s has the same match under two different matchings μ and μ', that is $\mu(s) = \mu'(s)$. In this case, s must be indifferent between matchings μ and μ'. A similar argument is true for advisors. Therefore, we use the operator $\succeq$ as well as $\succ$ when describing an agent's preference relation over matchings. More formally, $\mu(s) \succeq_s \mu'(s)$ means that either $\mu(s) \succ_s \mu'(s)$ or $\mu(s) = \mu'(s)$. Similarly, $\mu(a) \succeq_a \mu'(a)$ means that either $\mu(a) \succ_a \mu'(a)$ or $\mu(a) = \mu'(a)$.

Clearly, there are many possible matchings. The key question is which matching should be chosen, given students' and advisors' preference orderings. In other words, what properties does a desirable matching have? We identify two.

individually rational matching

Definition 10.6.8 (Individual rationality) *A matching* μ *is* individually rational *if no agent i prefers to remain unmatched than to be matched to* $\mu(i)$.

unblocked matching

Definition 10.6.9 (Unblocked) *A matching* μ *is* unblocked *if there exists no pair* (s, a) *such that* $\mu(s) \neq a$, *but* $a \succ_s \mu(s)$ *and* $s \succ_a \mu(a)$.

Intuitively, a matching is individually rational if no agent is matched with an unacceptable partner; a matching is unblocked if there exists no pair that would prefer to be matched with each other than with their respective partners. Putting these two definitions together, we obtain the concept of a stable matching.

stable matching

Definition 10.6.10 (Stable matching) *A matching* μ *is* stable *if and only if it is individually rational and unblocked.*

It turns out that in the setting we have defined above, no matter how many students and advisors there are and what preferences they have, there always exists at least one stable matching.

---

14. Our assumption that preferences are strict is restrictive; some of the results presented in this section no longer hold if it is relaxed.

Step 1: *each student applies to his most preferred advisor.*
repeat
> **Step 2:** *each advisor keeps her most preferred acceptable application (if any) and rejects the rest (if any).*
> **Step 3:** *each student who was rejected at the previous step applies to his next acceptable choice.*

until *no student applied in the last step*

Figure 10.9 Deferred acceptance algorithm, student-application version.

Theorem 10.6.11 (Gale and Shapley, 1962) *A stable matching always exists.*

> **Proof.** The proof is obtained by giving a procedure that produces a stable matching given any set of student and advisor preferences. Here, we describe the so-called "student-application" version of the algorithm (Figure 10.9). There is an analogous algorithm in which advisors apply to students. This algorithm must stop in at most a quadratic number of steps, since no student ever applies more than once to any advisor. The outcome is a matching, since at any step each student is paired with at most one advisor and vice versa. The matching is individually rational, since no student or advisor is ever matched to an unacceptable agent.
>
> It only remains to show that the matching is unblocked. Let μ be the matching produced by the algorithm. Assume for contradiction that μ is blocked by some student s and advisor a. Since s prefers a to his own match at μ, a must be acceptable to s, and so he must have applied to her before having applied to his match. Since s is not matched to a in μ, he must have been rejected by her in favor of someone she liked better. Therefore, (s, a) does not block μ, a contradiction. ∎

Thus, there always exists at least one stable matching. However, these matchings are not necessarily unique—given a set of student and advisor preferences, there may exist many stable matchings. Let us now consider how different matchings can be compared.

student-optimal matching **Definition 10.6.12** *A stable matching μ is* student optimal *if every student likes it at least as well as any other stable matching; that is, $\forall s \in S$ and for every other stable matching μ', $\mu(s) \succeq_s \mu'(s)$.*

advisor-optimal matching Along the same lines, we can define *advisor-optimal* matching. Now we can draw the following conclusions about stable matchings.

Theorem 10.6.13 *There exists exactly one student-optimal stable matching and one advisor-optimal stable matching. The matching produced by the student-application version of the deferred application algorithm is the student-optimal stable matching, and the matching produced by the advisor-application version of the deferred application algorithm is the advisor-optimal stable matching.*

Next, it turns out that any stable matching that is better for all the students is worse for all the advisors and vice versa.

Theorem 10.6.14 *If μ and μ' are stable matchings, $\forall s \in S$, $\mu(s) \succeq_s \mu'(s)$ if and only if $\forall a \in A$, $\mu'(a) \succeq_a \mu(a)$.*

achievable
match
 Say that an advisor a is *achievable* for student s, and vice versa, if there is a stable matching μ that matches a to s. Then we can state one implication of the above theorem: that the student-optimal stable matching is the worst stable matching from each advisor's point of view, and vice versa.

Corollary 10.6.15 *The student-optimal stable matching matches each advisor with her least preferred achievable student, and the advisor-optimal stable matching matches each student with her least preferred achievable advisor.*

Now let us move to the mechanism design question. If agents' preferences are private information, can we find a mechanism that ensures that a stable matching will be achieved? As is common in the matching literature, we restrict our attention to settings in which neither the agents' equilibrium strategies nor the mechanism itself are allowed to depend on the distribution over agents' preferences. Thus, we must rely on either dominant-strategy or *ex post* equilibrium implementation. Unfortunately, it turns out that stable matchings cannot be implemented under either equilibrium concept.

Theorem 10.6.16 *No mechanism implements stable matching in dominant strategies.*

> **Proof.** By the revelation principle, if such a mechanism exists, then there also exists a direct truthful mechanism that selects matchings that are stable with respect to the declared preference orderings. Consider a setting with two students, s_1 and s_2, and two advisors, a_1, and a_2. Imagine that s_1, s_2 and a_1 declare the following preference orderings: $a_1 \widehat{\succ}_{s_1} a_2$, $a_2 \widehat{\succ}_{s_2} a_1$, and $s_2 \widehat{\succ}_{a_1} s_1$. Assume that a_2's true preference ordering is the following: $s_1 \succ_{a_2} s_2$. If a_2 declares the truth, then (1) the setting will have two stable matchings, μ and μ', given by $\mu(s_i) = a_i$ for $i \in \{1, 2\}$, and $\mu'(s_i) = a_j$ for $i, j \in \{1, 2\}$, $j \neq i$, and (2) any stable matching mechanism must choose one of μ or μ'. Suppose the mechanism chooses μ. Observe that if a_2 declares that her only acceptable student is s_1, then μ' is the only stable matching with respect to the stated preferences and the mechanism must select μ'—which a_2 prefers to μ. Similarly, we can show that if the mechanism chooses μ' when the above preference orderings are stated, then in a setting where $a_2 \succ_{s_2} a_1$ is s_2's true preference ordering, s_2 benefits by misreporting his preference ordering. Therefore, declaring the truth is not a dominant strategy for every agent. ∎

Furthermore, it does not help to move to the *ex post* equilibrium concept, as can be proved along the same lines as Theorem 10.6.16.

Theorem 10.6.17 *No mechanism implements stable matching in* ex post *equilibrium.*

All is not hopeless, however—it turns out that we can obtain a positive mechanism design result for stable two-sided matching. The key is to relax our assumption that *all* agents are strategic. In our setting we will assume that advisors can be compelled to behave honestly. Under this assumption, it is enough to prove the following result.

Theorem 10.6.18 *Under the direct mechanism associated with the student-application version of the deferred acceptance algorithm, it is a dominant strategy for each student to declare his true preferences.*

Proof. This proof proceeds by contradiction. Suppose that the claim is not true and, without loss of generality, say that it is not a dominant strategy for student s_1 to state his true preference ordering. Then, there is a preference profile $[\hat{\succ}] = (\succ_{s_1}, \hat{\succ}_{s_2}, \ldots, \hat{\succ}_{s_{|S|}}, \hat{\succ}_{a_1}, \ldots, \hat{\succ}_{a_{|A|}})$ such that s_1 benefits from reporting $\succ'_{s_1} \neq \succ_{s_1}$. Let μ be the stable matching obtained by applying the student application version of the deferred acceptance algorithm to $[\hat{\succ}]$. By Theorem 10.6.13, μ is student optimal with respect to $[\hat{\succ}]$. Let μ' be the stable matching obtained by applying the same algorithm to $[\hat{\succ}'] = (\succ'_{s_1}, \hat{\succ}_{s_2}, \ldots, \hat{\succ}_{s_{|S|}}, \hat{\succ}_{a_1}, \ldots, \hat{\succ}_{a_{|A|}})$. Note that except for s_1, all the other students and advisors declare the same preference ordering under $[\hat{\succ}]$ and $[\hat{\succ}']$.

Let $R = \{s_1\} \cup \{s : \mu'(s)\hat{\succ}_s\mu(s)\}$ denote the set of students who strictly prefer μ' to μ (with respect to their declared preferences $[\hat{\succ}]$). Note that we have included s_1 in R because, by assumption, $\mu'(s_1) \succ_{s_1} \mu(s_1)$. Let $T = \{a : \mu'(a) \in R\}$ denote the set of advisors who are matched with some student from R under μ'. In what follows we first show (Part 1) that any advisor $a \in T$ is matched with an (always different) student from R under μ; that is, $\{a : \mu'(a) \in R\} = \{a : \mu(a) \in R\} = T$. Then (Part 2) we show that there exist some $a_\ell \in T$ and $s_r \notin R$ such that (s_r, a_ℓ) blocks μ' at $[\hat{\succ}']$ and therefore μ' is not stable with respect to $[\hat{\succ}']$. This contradicts our assumption that μ' is a stable matching with respect to $[\hat{\succ}']$.

Part 1: For any $s \in R$, let $a = \mu'(s)$. Stability of μ with respect to $[\hat{\succ}]$ requires that advisor a be matched to some student under μ (rather than being unpaired), as otherwise (s, a) would block μ at $[\hat{\succ}]$; let $s' = \mu(a)$. If $s' = s_1$, then since s_1 prefers his match under μ' to his match under μ, $s' \in R$. Otherwise, since (with respect to his preferences declared in $[\hat{\succ}]$) s strictly prefers $\mu'(s)$ to $\mu(s)$, stability of μ with respect to $[\hat{\succ}]$ implies that $s'\hat{\succ}_a s$. Since we defined $s = \mu'(a)$, thus $s'\hat{\succ}_a\mu'(a)$. Then, stability of μ' with respect to $[\hat{\succ}']$ implies that $\mu'(s')\hat{\succ}_{s'}a$. Since we defined $a = \mu(s')$, thus $\mu'(s')\hat{\succ}_{s'}\mu(s')$ and therefore $s' \in R$. As a result, we can write $T = \{a : \mu'(a) \in R\} = \{a : \mu(a) \in R\}$.

Part 2: Since every student $s \in R$ prefers $\mu'(s)$ to $\mu(s)$, stability of μ with respect to $[\hat{\succ}]$ implies that $\forall a \in T$, $\mu(a)\hat{\succ}_a\mu'(a)$. Therefore, during the execution of the student-application algorithm on $[\hat{\succ}]$, each student $s \in R$ will apply to $\mu'(s)$ and will get rejected by $\mu'(s)$ at some iteration. In other words, each $a \in T$ rejects $\mu'(a) \in R$ at some iteration. Let s_ℓ be (weakly) the last student in R who applies to an advisor during the execution of the student-application algorithm. This application is sent to $\mu(s_\ell) \in T$; let $\mu(s_\ell) = a_\ell$. By construction, a_ℓ must have rejected $\mu'(a_\ell)$ at some strictly earlier iteration of the algorithm. Thus, when s_ℓ applies to a_ℓ, a_ℓ must reject an application from some $s_r \notin R$ such that $s_r\hat{\succ}_{a_\ell}\mu'(a_\ell)$ (fact 1). Note that $s_r \neq s_1$, since

$s_r \notin R$ and $s_1 \in R$. Since s_r applies to a_ℓ before he finally gets matched to $\mu(s_r)$, we have that $a_\ell \widehat{\succ}_{s_r} \mu(s_r)$. Furthermore, since $s_r \notin R$, we also have that $\mu(s_r) \widehat{\succeq}_{s_r} \mu'(s_r)$. Therefore $a_\ell \widehat{\succ}_{s_r} \mu'(s_r)$ (fact 2). Thus, from (fact 1) and (fact 2), (s_r, a_ℓ) blocks μ' at $[\widehat{\succ}']$ and μ' is not stable with respect to $[\widehat{\succ}']$, yielding our contradiction. ∎

Of course, it is similarly possible to achieve a direct mechanism under which truth telling is a dominant strategy for advisors by using the advisor-application version of the deferred acceptance algorithm.

10.7 Constrained mechanism design

So far we have assumed that the mechanism designer is free to design any mechanism, but this assumption is violated in many applications—the ones discussed in this section, and many others. In particular, often one starts with given strategy spaces for each of the agents, with limited or no ability to change those. Examples abound:

- A city official who wishes to improve the traffic flow in the city cannot redesign cars or build new roads;
- A UN mediator who wishes to incent two countries fighting over a scarce resource to cease hostilities cannot change their military capabilities or the amount or value of the resource;
- A computer network operator who wishes to route traffic a certain way cannot change the network topology or the underlying routing algorithm.

Many other examples exist, and in fact such constraints can be thought of as the norm rather than the exception. How can such would-be mechanism designers intervene to influence the course of events?

social law   In Chapter 2 we already encountered this problem. Specifically, in Section 2.4 we saw how imposing *social laws*—that is, restricting the options available to agents—can be beneficial to all agents. Social laws played an important coordinating role (as in "drive on the right side of the road") and, furthermore, in some cases prevented the narrow self interests of the agents from hurting them (e.g., allowing cooperation in the Prisoners' Dilemma game). However, in that discussion we made the important assumption that once a social law was imposed (or agreed upon, depending on the interpretation), the agents could be assumed to follow it.

Here we relax this assumption, and we do so in three ways. First, we view the players as having the option of entering into a contract among themselves. Once they do—and only then—the center can impose arbitrary fines on law breakers, if he is aware of such deviations. The question in this case is which contracts the agents can be expected to enter, and how the work of the center can be minimized or even eliminated. Second, we consider the case in which the center can simply bribe the players to play a certain way (or, in more neutral language, offer positive incentives for certain actions). The question in this case is how the center can bias the outcome toward the desired one while minimizing his cost. Finally, we consider a center who offers to play on behalf of the agents, who in turn are free to accept or reject the offer. We look at each setting in turn.

10.7.1 *Contracts*

Consider any given game G, and a center who can do the following.

1. Propose a contract before G is played. This contract specifies a particular outcome, that is, a unique action for each agent,[15] and a penalty for deviating from it.
2. Collect signatures on the contract and make it common knowledge who signed.
3. Monitor the players' actions during the execution of G.
4. If the contract was signed by all agents, fine anyone who deviated from it as specified by the contract.

Here we assume that players still have the freedom to choose whether or not to honor the agreement; the challenge is to design a mechanism such that, in equilibrium, they will do so.

The technical results in this line of work will refer to games of complete information, but for intuition consider the example of an online marketplace such as eBay. (We discuss auctions in detail in Chapter 11, but those details are not needed here.) Consider the entire game being played, including the decision after the close of the auction by the seller of whether to deliver the good and by the buyer of whether to send payment. Straightforward analysis shows that that the equilibrium is for neither to keep his promise, and the experience with fraud in online auctions demonstrates that the problem is not merely theoretical. It would be in an online auction site's interest to find a way to bind its customers to their promises.

The first question one may ask is what the achievable outcomes are. What outcomes may the center suggest, with associated penalties, that the agents will accept? However, once the problem is couched in a formal setting, it is not hard to show a folk theorem of sorts: any outcome will be accepted when accompanied by appropriate fines, so long as the payoffs of each agent in that outcome are better than that player's payoffs in *some* equilibrium of the original game.

Although the center can achieve almost any outcome, it would seem to require great effort: suggesting an outcome, collecting signatures, observing the game, and enforcing the contracts. If this procedure happens not just for one game, but for hundreds or thousands per day, the center may wish to find a way to avoid this burden while still achieving the same effect.

However, one can often achieve the same effects with much less effort on the part of the center. We continue to assume that the center still needs to propose a contract. We also simply assume that it does not monitor the game. Nor does it participate in the signing phase; the agents do that among themselves

---

15. In the parlance of Section 2.4, a *convention*.

using a broadcast channel. While we might imagine that the players could simply broadcast their signatures, this protocol allows a single player to learn the others' signatures and threaten them with fines. Nonetheless, one can construct a more complicated protocol—using a second stage of contracts—that does not require the center's participation. The only phase in which the center's protocol requires it to get involved under some conditions is the enforcement stage. However, here too one can minimize the effort required in actuality. This is done by devising contracts that, *in equilibrium*, at this stage too the center sits idle. Among other things, one can show that if the game play is *verifiable* (if the center can discover after the fact whether players obeyed the contract), then anything achievable by a fully engaged center is also achievable by a center that in equilibrium always sits idle.

10.7.2 *Bribes*

Consider the following simple congestion setting, similar to the one discussed in Section 10.1.2. Assume that there are two agents, 1 and 2, who must select among two service providers. One of the service providers, f, is a fast one, while the other, s, is a slower one. We capture this by having an agent obtain a payoff of 6 when he is the only one who uses f, and a payoff of 4 when he is the only one who uses s. If both agents select the same service provider then the speeds they each obtain decrease by a factor of 2, leading to half the payoff. Thus, if both agents use f then each of them obtains a payoff of 3, while if both use s then each obtains 2. Written in normal form, this game is described as follows.

$$
M = \quad
\begin{array}{c|c|c|}
 & f & s \\
\hline
f & 3,3 & 6,4 \\
\hline
s & 4,6 & 2,2 \\
\hline
\end{array}
$$

Assume that the mechanism designer wishes to prevent the agents from using the same service provider (leading to low payoffs for both) and further wants to obtain a mechanism in which each agent has a dominant strategy. Then it can do as follows: it can promise to pay agent 1 a value of 10 if both agents will use f, and promise to pay agent 2 a value of 10 if both agents will use s. These promises transform M to the following game.

$$
M' = \quad
\begin{array}{c|c|c|}
 & f & s \\
\hline
f & 13,3 & 6,4 \\
\hline
s & 4,6 & 2,12 \\
\hline
\end{array}
$$

Notice that in M', strategy f is dominant for agent 1, and strategy s is dominant for agent 2. As a result the only rational strategy profile is the one in which agent 1 chooses f and agent 2 chooses s. Hence, the mechanism designer implements one of the desired outcomes. Moreover, given that the strategy profile (f, s) is selected, the mechanism will have to pay nothing. It has just implemented, *in dominant strategies*, a desired behavior (which had previously been obtained in one of the game's Nash equilibria) at zero cost, relying only on its creditability, without modifying the rules of interactions or enforcing any type of behavior! In this case we say that the desired behavior has a 0-implementation. More generally, an outcome has a *k-implementation* if it can be implemented in dominant strategies using such payments with a cost in equilibrium of at most k. This definition can be used to prove the following result.

k-implementation

Theorem 10.7.1 *An outcome is 0-implementable iff it is a Nash equilibrium.*

10.7.3   *Mediators*

We have so far considered a center who can enforce contracts, and one who can offer monetary incentives. We now consider a more active center, one who can play on behalf of agents.

Consider the ever-recurring example of the Prisoners' Dilemma game.

	C	D
C	4, 4	0, 6
D	6, 0	1, 1

As you know, the strategy profile (D, D) is a Nash equilibrium, and even an equilibrium in weakly dominant strategies. However, it is not what is called a *strong equilibrium*, that is, a strategy profile that is stable against group deviations. If *both* players deviate to (C, C), the payoff of each one of them will increase.

strong equilibrium

Now consider a reliable mediator who offers the agents the following protocol. If both agents agree to use the mediator's services then he will perform the action C (cooperate) on behalf of both agents. However, if only one agent agrees to use his services then he will perform the action D (defect) on behalf of that agent. We assume that when accepting the mediator's offer the agent is committed to using the mediator and forgoes the option of acting on his own; however, he is free to reject the offer, in which case he is free to use any strategy. This induces the following game between the agents.

Mediator	C	D
4, 4	6, 0	1, 1
0, 6	4, 4	0, 6
1, 1	6, 0	1, 1

The mediated game has a most desirable property: it is a strong equilibrium for the two agents to use the mediator's services, guaranteeing each a payoff of 4. No coalition (i.e., either of the two agents alone, or the pair) can deviate and achieve for all coalition members a payoff greater than 4.

This example turns out to be more than a happy coincidence. While strong equilibria are rare in general, adding mediators make them less rare. For example, *balanced game*   adding a mediator to any *balanced symmetric game* yields a strong equilibrium with optimal surplus.[16] Also, if we consider only deviations by coalitions of size at most k (a so-called k-strong equilibrium), we have the following. For any symmetric game with n agents, if $k!$ divides n then there exists a k-strong mediated equilibrium, leading to optimal surplus.[17] However, if $k!$ does not divide n, then it can be shown that the game may or may not possess a k-strong equilibrium.

10.8   History and references

Mechanism design is covered to varying degrees in modern game theory text-books, but even better are the microeconomic textbook of Mas-Colell et al. [1995] and books on auction theory such as Krishna [2002]. Good overviews from a computer science perspective are given in the introductory chapters of Parkes [2001] and in Nisan [2007]. Specific publications that underlie some of the results covered in this chapter are as follows.

The foundational idea of mechanisms as communication systems that select outcomes based on messages from agents is due to Hurwicz [1960], who also elaborated the theory to include the idea that mechanisms should be "incentive compatible" [1972]. The revelation principle was first articulated by Gibbard [1973] and was developed in the greatest generality by [Myerson, 1979, 1982, 1986]. In 2007, Hurwicz and Myerson shared a Nobel Prize (along with Maskin, whose work we do not discuss in this book), "for having laid the foundations

---

16. Full discussion of balanced games is beyond the scope of this discussion. However, we remark that a game in strategic form is called balanced if its associated core is nonempty. The core of a game is defined in the context of coalitional games in Chapter 12.

17. As an anecdote, we note that the Israeli parliament consists of $120 = 5!$ members. Hence, every anonymous game played by this parliament possesses an optimal surplus symmetric 5-strong equilibrium. While no Parliament member is able to give the right of voting to a mediator, this right of voting could be replaced in real life by a commitment to follow the mediator's algorithm.

of mechanism design theory." Theorem 10.2.6 is due to both Satterthwaite and Gibbard, in two separate publications [Gibbard, 1973; Satterthwaite, 1975].

The VCG mechanism was anticipated by Vickrey [1961], who outlined an extension of the second-price auction to multiple identical goods. Groves [1973] explicitly considered the general family of truthful mechanisms applying to multiple distinct goods (though the result had appeared already in his 1969 Ph.D. dissertation). Clarke [1971] proposed his tax for use with public goods (i.e., goods such as roads and national defense that are paid for by all regardless of personal use). Theorem 10.4.3 is due to Green and Laffont [1977]; Theorem 10.4.11 is due to that paper as well as to the earlier Hurwicz [1975]. The fact that Groves mechanisms are payoff equivalent to all other Bayes–Nash incentive-compatible efficient mechanisms was shown by Krishna and Perry [1998] and Williams [1999]; the former reference also gave the results that VCG is *ex interim* individually rational and that VCG collects the maximal amount of revenue among all *ex interim* individually-rational Groves mechanisms. Recent work shows that some "VCG drawbacks" are also problems with broad classes of mechanisms; for example, this has been shown for nonfrugality [Archer and Tardos, 2002; Elkind et al., 2004] and for revenue monotonicity [Rastegari et al., 2007]. The problem of participating in Groves mechanisms under multiple identities (specifically in the case of combinatorial auctions, which are described in Section 11.3) was investigated by Yokoo [2006]. Although it is not generally possible to return *all* VCG revenue to the agents, recent research has investigated VCG-like mechanisms that collect as little revenue from the agents as possible and thus minimize the extent to which they violate (strong) budget balance [Porter et al., 2004b; Cavallo, 2006; Guo and Conitzer, 2007]. Interestingly, the first of these papers came to the problem through a desire to achieve *fair* outcomes. The Myerson–Satterthwaite theorem (10.4.12) appears in Myerson and Satterthwaite [1983]. The AGV mechanism is due (independently) to Arrow [1977] and d'Aspremont and Gérard-Varet [1979].

The section on implementation in dominant strategies follows Nisan [2007]; Theorem 10.5.5 is due to Roberts [1979]. Second-chance mechanisms are due to Nisan and Ronen [2007]. (One difference: we preferred the term *Groves-based mechanisms* to Nisan and Ronen's *VCG-based mechanisms.*)

Our section on task scheduling reports results due to Nisan and Ronen [2001]; this work also introduced the term *algorithmic mechanism design*. Our section on bandwidth allocation in computer networks follows Johari [2007], which in turn draws on Johari and Tsitsiklis [2004]; the proportional allocation mechanism is due to Kelly [1997], and the VCG-like mechanism is described in Johari and Tsitsiklis [2005]. Our section on multicast cost sharing follows Feigenbaum et al. [2007], which draws especially on Feigenbaum et al. [2001; 2003]. Our discussion of mechanisms for two-sided matching draws on Roth and Sotomayor [1990], Schummer and Vohra [2007] and Gale and Shapley [1962]. The first algorithm for finding stable matchings was developed by Stalnaker [1953], and was used to match medical interns to hospitals. The stable matching problem was formalized by Gale and Shapley [1962], who also introduced the deferred acceptance algorithm. Theorems 10.6.13 and 10.6.14 follow Knuth [1976];

Theorems 10.6.16 and 10.6.17 are due to Roth [1984]; and Theorem 10.6.18 draws partly on Schummer and Vohra [2007] and subsequent unpublished correspondence between Baharak Rastegari and Rakesh Vohra. A more general version of Theorem 10.6.18 appeared in Roth and Sotomayor [1990] and Dubins and Freedman [1981].

The notion of social laws and conventions are introduced in Shoham and Tennenholtz [1995]. The use of contracts to influence the outcome of a game is discussed in McGrew and Shoham [2004]. The use of monetary incentives to influence the outcome of a game, or k-implementation, is introduced in Monderer and Tennenholtz [2003]. Mediators and their connections to strong equilibria are discussed in Monderer and Tennenholtz [2006].

11

Protocols for Multiagent
Resource Allocation: Auctions

In this chapter we consider the problem of allocating (discrete) resources among selfish agents in a multiagent system. Auctions—an interesting and important application of mechanism design—turn out to provide a general solution to this problem. We describe various different flavors of auctions, including single-good, multiunit, and combinatorial auctions. In each case, we survey some of the key theoretical, practical, and computational insights from the literature.

The auction setting is important for two reasons. First, auctions are widely used in real life, in consumer, corporate, as well as government settings. Millions of people use auctions daily on Internet consumer Web sites to trade goods. More complex types of auctions have been used by governments around the world to sell important public resources such as access to electromagnetic spectrum. Indeed, all financial markets constitute a type of auction (one of the family of so-called *double auctions*). Auctions are also often used in computational settings, to efficiently allocate bandwidth and processing power to applications and users.

The second—and more fundamental—reason to care about auctions is that they provide a general theoretical framework for understanding resource allocation problems among self-interested agents. Formally speaking, an auction is any protocol that allows agents to indicate their interest in one or more resources and that uses these indications of interest to determine both an allocation of resources and a set of payments by the agents. Thus, auctions are important for a wide range of computational settings (e.g., the sharing of computational power in a grid computer on a network) that would not normally be thought of as auctions and that might not even use money as the basis of payments.

11.1 Single-good auctions

It is important to realize that the most familiar type of auction—the ascending-bid, English auction—is a drop in the ocean of auction types. Indeed, since auctions are simply mechanisms (see Chapter 10) for allocating goods, there is an infinite number of auction types. In the most familiar types of auctions there is one good for sale, one seller, and multiple potential buyers. Each buyer has his own valuation for the good, and each wishes to purchase it at the lowest possible price. These auctions are called *single-sided*, because there are multiple agents

single-sided
auction

on only one side of the market. Our task is to design a protocol for this auction that satisfies certain desirable global criteria. For example, we might want an auction protocol that maximizes the expected revenue of the seller. Or, we might want an auction that is economically efficient; that is, one that guarantees that the potential buyer with the highest valuation ends up with the good.

Given the popularity of auctions, on the one hand, and the diversity of auction mechanisms, on the other, it is not surprising that the literature on the topic is vast. In this section we provide a taste for this literature, concentrating on auctions for selling a single good. We explore richer settings later in the chapter.

11.1.1   *Canonical auction families*

To give a feel for the broad space of single-good auctions, we start by describing some of the most famous families: English, Japanese, Dutch, and sealed-bid auctions. We end the section by presenting a unifying view of auctions as structured negotiations.

English auctions

English auction   The *English auction* is perhaps the best-known family of auctions, since in one form or another such auctions are used in the venerable, old-guard auction houses, as well as most of the online consumer auction sites. The auctioneer sets a starting price for the good, and agents then have the option to announce successive bids, each of which must be higher than the previous bid (usually by some minimum increment set by the auctioneer). The rules for when the auction closes vary; in some instances the auction ends at a fixed time, in others it ends after a fixed period during which no new bids are made, in others at the latest of the two, and in still other instances at the earliest of the two. The final bidder, who by definition is the agent with the highest bid, must purchase the good for the amount of his final bid.

Japanese auctions

Japanese auction   The *Japanese auction*[1] is similar to the English auction in that it is an ascending-bid auction but is different otherwise. Here the auctioneer sets a starting price for the good, and each agent must choose whether or not to be "in," that is, whether he is willing to purchase the good at that price. The auctioneer then calls out successively increasing prices in a regular fashion,[2] and after each call each agent must announce whether he is still in. When an agent drops out it is irrevocable, and he cannot reenter the auction. The auction ends when there is exactly one agent left in; the agent must then purchase the good for the current price.

Dutch auctions

Dutch auction   In a *Dutch auction* the auctioneer begins by announcing a high price and then proceeds to announce successively lower prices in a regular fashion. In practice,

---

1. Unlike the terms *English* and *Dutch*, the term *Japanese* is not used universally; however, it is commonly used, and there is no competing name for this family of auctions.
2. In the theoretical analyses of this auction the assumption is usually that the prices rise continuously.

the descending prices are indicated by a clock that all of the agents can see. The auction ends when the first agent signals the auctioneer by pressing a buzzer and stopping the clock; the signaling agent must then purchase the good for the displayed price. This auction gets its name from the fact that it is used in the Amsterdam flower market; in practice, it is most often used in settings where goods must be sold quickly.

Sealed-bid auctions

open-outcry auction

sealed-bid auction

first-price auction

second-price auction

kth-price auction

All the auctions discussed so far are considered *open-outcry* auctions, in that all the bidding is done by calling out the bids in public (however, as we will discuss shortly, in the case of the Dutch auction this is something of an optical illusion). The family of *sealed-bid auctions*, probably the best known after English auctions, is different. In this case, each agent submits to the auctioneer a secret, "sealed" bid for the good that is not accessible to any of the other agents. The agent with the highest bid must purchase the good, but the price at which he does so depends on the type of sealed-bid auction. In a first-price sealed-bid auction (or simply *first-price auction*) the winning agent pays an amount equal to his own bid. In a *second-price auction* he pays an amount equal to the next highest bid (i.e., the highest rejected bid). The second-price auction is also called the *Vickrey auction*. In general, in a *kth-price auction* the winning agent purchases the good for a price equal to the kth highest bid.

Auctions as structured negotiations

While it is useful to have reviewed the best-known auction types, this list is far from exhaustive. For example, consider the following auction, consisting of a sequence of sealed bids. In the first round the lowest bidder drops out; his bid is announced and becomes the minimum bid in the next round for the remaining bidders. This process continues until only one bidder remains; this bidder wins and pays the minimum bid in the final round. This auction, called the *elimination auction*, is different from the auctions described earlier, and yet makes perfect sense. Or consider a procurement reverse auction, in which an initial sealed-bid auction is conducted among the interested suppliers, and then a reverse English auction is conducted among the three cheapest suppliers (the "finalists") to determine the ultimate supplier. This two-phase auction is not uncommon in industry.

elimination auction

Indeed, a taxonomical perspective obscures the elements common to all auctions, and thus the infinite nature of the space. What is an auction? At heart it is simply a structured framework for negotiation. Each such negotiation has certain rules, which can be broken down into three categories.

1. *Bidding rules:* How are offers made (by whom, when, what can their content be)?
2. *Clearing rules:* When do trades occur, or what are those trades (who gets which goods, and what money changes hands) as a function of the bidding?
3. *Information rules:* Who knows what when about the state of negotiation?

The different auctions we have discussed make different choices along these three axes, but it is clear that other rules can be instituted. Indeed, when viewed this way, it becomes clear that what seem like three radically different commerce mechanisms—the hushed purchase of a Matisse at a high-end auction house in London, the mundane purchase of groceries at the local supermarket, and the one-on-one horse trading in a Middle Eastern *souk*—are simply auctions that make different choices along these three dimensions.

11.1.2 *Auctions as Bayesian mechanisms*

We now move to a more formal investigation of single-good auctions. Our starting point is the observation that choosing an auction that has various desired properties is a mechanism design problem. Ordinarily we assume that agents' utility functions in an auction setting are quasilinear. To define an auction as a quasilinear mechanism (see Definition 10.3.2) we must identify the following elements:

- set of agents N;
- set of outcomes $O = X \times \mathbb{R}^n$;
- set of actions A_i available to each agent $i \in N$;
- choice function χ that selects one of the outcomes given the agents' actions; and
- payment function $\wp$ that determines what each agent must pay given all agents' actions.

In an auction, the possible outcomes O consist of all possible ways to allocate the good—the set of choices X—and all possible ways of charging the agents. The agents' actions will vary in different auction types. In a sealed-bid auction, each set A_i is an interval from $\mathbb{R}$ (i.e., an agent's action is the declaration of a bid amount between some minimum and maximum value). A Japanese auction is an extensive-form game with chance nodes (see Section 5.2), and so in this case the action space is the space of all policies the agent could follow (i.e., all different ways of acting conditioned on different observed histories). As in all mechanism design problems, the choice and payment functions χ and $\wp$ depend on the objective of the auction, such as achieving an efficient allocation or maximizing revenue.

A Bayesian game with quasilinear preferences includes two more ingredients that we need to specify: the common prior and the agents' utility functions. We will say more about the common prior—the distribution from which the agents' types are drawn—later; here, just note that the definition of an auction as a Bayesian game is incomplete without it. Considering the agents' utility functions, first note that the quasilinearity assumption (see Definition 10.3.1) allows us to write $u_i(o, \theta_i) = u_i(x, \theta_i) - f_i(p_i)$. The function f_i indicates the agent's risk attitude, as discussed in Section 10.3.1. Unless we indicate otherwise, we will commonly assume risk neutrality.

We are left with the task of describing the agents' valuations: their utilities for different allocations of the goods $x \in X$. Auction theory distinguishes between a number of different settings here. One of the best-known and most extensively studied is the *independent private value* (IPV) setting. In this setting all agents' valuations are drawn independently from the same (commonly known) distribution, and an agent's type (or "signal") consists only of his own valuation, giving him no information about the valuations of the others. An example where the IPV setting is appropriate is in auctions consisting of bidders with personal tastes who aim to buy a piece of art purely for their own enjoyment. In most of this section we will assume that agents have independent private values, though we will explore an alternative, the common-value assumption, in Section 11.1.10.

independent
private value
(IPV)

11.1.3 *Second-price, Japanese, and English auctions*

Let us now consider whether the second-price sealed-bid auction, which is a direct mechanism, is truthful (i.e., whether it provides incentive for the agents to bid their true values). The following, very conceptually straightforward proof shows that in the IPV case it is.

Theorem 11.1.1 *In a second-price auction where bidders have independent private values, truth telling is a dominant strategy.*

The second-price auction is a special case of the VCG mechanism, and hence of the Groves mechanism. Thus, Theorem 11.1.1 follows directly from Theorem 10.4.2. However, a proof of this narrower claim is considerably more intuitive than the general argument.

> **Proof.** Assume that all bidders other than i bid in some arbitrary way, and consider i's best response. First, consider the case where i's valuation is larger than the highest of the other bidders' bids. In this case i would win and would pay the next-highest bid amount, as illustrated in Figure 11.1a. Could i be better off by bidding dishonestly in this case? If he bid higher, he would still win and would still pay the same amount, as illustrated in Figure 11.1b. If he bid lower, he would either still win and still pay the same amount (Figure 11.1c) or lose and pay zero (Figure 11.1d).[3] Since i gets nonnegative utility for receiving the good at a price less than or equal to his valuation, i cannot gain, and would sometimes lose by bidding dishonestly in this case. Now consider the other case, where i's valuation is less than at least one other bidder's bid. In this case i would lose and pay zero (Figure 11.1e). If he bid less, he would still lose and pay zero (Figure 11.1f). If he bid more, either he would still lose and pay zero (Figure 11.1g) or he would win and pay more than his valuation (Figure 11.1h), achieving negative utility. Thus again, i cannot gain, and would sometimes lose by bidding dishonestly in this case. ∎

---

3. Figure 11.1d is oversimplified: the winner will not always pay i's bid in this case. (Do you see why?)

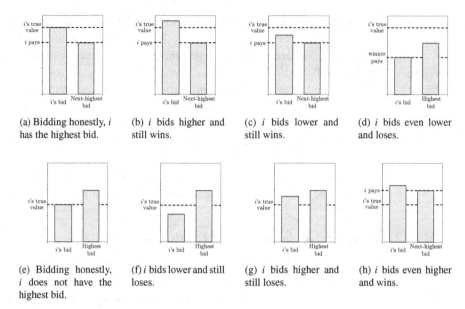

(a) Bidding honestly, *i* has the highest bid.

(b) *i* bids higher and still wins.

(c) *i* bids lower and still wins.

(d) *i* bids even lower and loses.

(e) Bidding honestly, *i* does not have the highest bid.

(f) *i* bids lower and still loses.

(g) *i* bids higher and still loses.

(h) *i* bids even higher and wins.

Figure 11.1 A case analysis to show that honest bidding is a dominant strategy in a second-price auction with independent private values.

Notice that this proof does not depend on the agents' risk attitudes. Thus, an agent's dominant strategy in a second-price auction is the same regardless of whether the agent is risk neutral, risk averse or risk seeking.

In the IPV case, we can identify strong relationships between the second-price auction and Japanese and English auctions. Consider first the comparison between second-price and Japanese auctions. In both cases the bidder must select a number (in the sealed-bid case the number is the one written down, and in the Japanese case it is the price at which the agent will drop out); the bidder with highest amount wins, and pays the amount selected by the second-highest bidder. The difference between the auctions is that information about other agents' bid amounts is disclosed in the Japanese auction. In the sealed-bid auction an agent's bid amount must be selected without knowing anything about the amounts selected by others, whereas in the Japanese auction the amount can be updated based on the prices at which lower bidders are observed to drop out. In general, this difference can be important (see Section 11.1.10); however, it makes no difference in the IPV case. Thus, Japanese auctions are also dominant-strategy truthful when agents have independent private values.

Obviously, the Japanese and English auctions are closely related. Thus, it is not surprising to find that second-price and English auctions are also similar. One connection can be seen through *proxy bidding*, a service offered on some online auction sites such as eBay. Under proxy bidding, a bidder tells the system the maximum amount he is willing to pay. The user can then leave the site, and the system bids as the bidder's proxy: every time the bidder is outbid, the system will respond with a bid one increment higher, until the bidder's maximum is

proxy bidding

reached. It is easy to see that if all bidders use the proxy service and update it only once, what occurs will be identical to a second-price auction (excepting that the winner's payment may be one bid increment higher).

The main complication with English auctions is that bidders can place so-called *jump bids*: bids that are greater than the previous high bid by more than the minimum increment. Although it seems relatively innocuous, this feature complicates analysis of such auctions. Indeed, when an ascending auction is analyzed it is almost always the Japanese variant, not the English.

11.1.4 *First-price and Dutch auctions*

Let us now consider first-price auctions. The first observation we can make is that the Dutch auction and the first-price auction, while quite different in appearance, are actually the same auction (in the technical jargon, they are *strategically equivalent*). In both auctions each agent must select an amount without knowing about the other agents' selections; the agent with the highest amount wins the auction, and must purchase the good for that amount. Strategic equivalence is a very strong property: it says the auctions are exactly the same no matter what risk attitudes the agents have, and no matter what valuation model describes their utility functions. This being the case, it is interesting to ask why both auction types are held in practice. One answer is that they make a trade-off between time complexity and communication complexity. First-price auctions require each bidder to send a message to the auctioneer, which could be unwieldy with a large number of bidders. Dutch auctions require only a single bit of information to be communicated to the auctioneer, but requires the auctioneer to broadcast prices.

Of course, all this talk of equivalence does not help us to understand anything about how an agent should actually *bid* in a first-price or Dutch auction. Unfortunately, unlike the case of second-price auctions, here we do not have the luxury of dominant strategies, and must thus resort to Bayes–Nash equilibrium analysis. Let us assume that agents have independent private valuations. Furthermore, in a first-price auction, an agent's risk attitude also matters. For example, a risk-averse agent would be willing to sacrifice some expected utility (by increasing his bid over what a risk-neutral agent would bid), in order to increase his probability of winning the auction. Let us assume that agents are risk neutral and that their valuations are drawn uniformly from some interval, say $[0, 1]$. Let s_i denote the bid of player i, and v_i denote his true valuation. Thus if player i wins, his payoff is $u_i = v_i - s_i$; if he loses, it is $u_i = 0$. Now we prove in the case of two agents that there is an equilibrium in which each player bids half of his true valuation. (This also happens to be the *unique* symmetric equilibrium, but we do not demonstrate that here.)

Proposition 11.1.2 *In a first-price auction with two risk-neutral bidders whose valuations are drawn independently and uniformly at random from the interval $[0, 1]$, $(\frac{1}{2}v_1, \frac{1}{2}v_2)$ is a Bayes–Nash equilibrium strategy profile.*

Proof. Assume that bidder 2 bids $\frac{1}{2}v_2$. From the fact that v_2 was drawn from a uniform distribution, all values of v_2 between 0 and 1 are equally likely. Now consider bidder 1's expected utility, in order to write an expression for his best response.

$$E[u_1] = \int_0^1 u_1 dv_2 \qquad (11.1)$$

The integral in Equation (11.1) can be broken up into two smaller integrals that describe cases in which player 1 does and does not win the auction.

$$E[u_1] = \int_0^{2s_1} u_1 dv_2 + \int_{2s_1}^1 u_1 dv_2$$

We can now substitute in values for u_1. In the first case, because 2 bids $\frac{1}{2}v_2$, 1 wins when $v_2 < 2s_1$ and gains utility $v_1 - s_1$. In the second case 1 loses and gains utility 0. Observe that we can ignore the case where the agents tie, because this occurs with probability zero.

$$\begin{aligned}
E[u_1] &= \int_0^{2s_1} (v_1 - s_1) dv_2 + 0 \\
&= (v_1 - s_1)v_2 \Big|_0^{2s_1} \\
&= 2v_1 s_1 - 2s_1^2 \qquad (11.2)
\end{aligned}$$

We can find bidder 1's best response to bidder 2's strategy by taking the derivative of Equation (11.2) and setting it equal to zero.

$$\frac{\partial}{\partial s_1}(2v_1 s_1 - 2s_1^2) = 0$$

$$2v_1 - 4s_1 = 0$$

$$s_1 = \frac{1}{2}v_1$$

Thus when player 2 is bidding half her valuation, player 1's best strategy is to bid half his valuation. The calculation of the optimal bid for player 2 is analogous, given the symmetry of the game and the equilibrium. ∎

This proposition was quite narrow: it spoke about the case of only two bidders, and considered valuations that were drawn uniformly at random from a particular interval of the real numbers. Nevertheless, this is already enough for us to be able to observe that first-price auctions are not incentive compatible (and hence, unsurprisingly, are not equivalent to second-price auctions).

Somewhat more generally, we have the following theorem.

Theorem 11.1.3 *In a first-price sealed-bid auction with n risk-neutral agents whose valuations are independently drawn from a uniform distribution on the same bounded interval of the real numbers, the unique symmetric equilibrium is given by the strategy profile $(\frac{n-1}{n}v_1, \ldots, \frac{n-1}{n}v_n)$.*

In other words, the unique equilibrium of the auction occurs when each player bids $\frac{n-1}{n}$ of his valuation. This theorem can be proved using an argument similar to that used in Proposition 11.1.2, although the calculus gets a bit more involved (for one thing, we must reason about the fact that each of several opposing agents may place the high bid). However, there is a broader problem: that proof only showed how to *verify* an equilibrium strategy. How do we identify one in the first place? Although it is also possible to do this from first principles (at least for straightforward auctions such as first-price), we will explain a simpler technique in the next section.

11.1.5 *Revenue equivalence*

Of the large (in fact, infinite) space of auctions, which one should an auctioneer choose? To a certain degree, the choice does not matter, a result formalized by the following important theorem.[4]

Theorem 11.1.4 (Revenue equivalence theorem) *Assume that each of n risk-neutral agents has an independent private valuation for a single good at auction, drawn from a common cumulative distribution $F(v)$ that is strictly increasing and atomless on $[\underline{v}, \bar{v}]$. Then any efficient[5] auction mechanism in which any agent with valuation $\underline{v}$ has an expected utility of zero yields the same expected revenue, and hence results in any bidder with valuation v_i making the same expected payment.*

> **Proof.** Consider any mechanism (direct or indirect) for allocating the good. Let $u_i(v_i)$ be i's expected utility given true valuation v_i, assuming that all agents including i follow their equilibrium strategies. Let $P_i(v_i)$ be i's probability of being awarded the good given (i) that his true type is v_i; (ii) that he follows the equilibrium strategy for an agent with type v_i; and (iii) that all other agents follow their equilibrium strategies.
>
> $$u_i(v_i) = v_i P_i(v_i) - E[\text{payment by type } v_i \text{ of player } i] \qquad (11.3)$$
>
> From the definition of equilibrium, for any other valuation $\hat{v}_i$ that i could have,
>
> $$u_i(v_i) \geq u_i(\hat{v}_i) + (v_i - \hat{v}_i)P_i(\hat{v}_i). \qquad (11.4)$$
>
> To understand Equation (11.4), observe that if i followed the equilibrium strategy for a player with valuation $\hat{v}_i$ rather than for a player with his (true) valuation v_i, i would make all the same payments and would win the good with the same probability as an agent with valuation $\hat{v}_i$. However, whenever he wins the good, i values it $(v_i - \hat{v}_i)$ more than an agent of type $\hat{v}_i$ does. The inequality must hold because in equilibrium this deviation must

---

4. What is stated, in fact, is the revenue equivalence theorem for the private-value, single-good case. Similar theorems hold for other—though not all—cases.
5. Here we make use of the definition of economic efficiency given in Definition 10.3.6. Equivalently, we could require that the auction has a symmetric and increasing equilibrium and always allocates the good to an agent who placed the highest bid.

be unprofitable. Consider $\hat{v}_i = v_i + dv_i$, by substituting this expression into Equation (11.4):

$$u_i(v_i) \geq u_i(v_i + dv_i) + dv_i P_i(v_i + dv_i). \tag{11.5}$$

Likewise, considering the possibility that i's true type could be $v_i + dv_i$,

$$u_i(v_i + dv_i) \geq u_i(v_i) + dv_i P_i(v_i). \tag{11.6}$$

Combining Equations (11.5) and (11.6), we have

$$P_i(v_i + dv_i) \geq \frac{u_i(v_i + dv_i) - u_i(v_i)}{dv_i} \geq P_i(v_i). \tag{11.7}$$

Taking the limit as $dv_i \rightarrow 0$ gives

$$\frac{du_i}{dv_i} = P_i(v_i). \tag{11.8}$$

Integrating up,

$$u_i(v_i) = u_i(\underline{v}) + \int_{x=\underline{v}}^{v_i} P_i(x)dx. \tag{11.9}$$

Now consider any two efficient auction mechanisms in which the expected payment of an agent with valuation $\underline{v}$ is zero. A bidder with valuation $\underline{v}$ will never win (since the distribution is atomless), so his expected utility $u_i(\underline{v}) = 0$. Because both mechanisms are efficient, every agent i always has the same $P_i(v_i)$ (his probability of winning given his type v_i) under the two mechanisms. Since the right-hand side of Equation (11.9) involves only $P_i(v_i)$ and $u_i(\underline{v})$, each agent i must therefore have the same expected utility u_i in both mechanisms. From Equation (11.3), this means that a player of any given type v_i must make the same expected payment in both mechanisms. Thus, i's *ex ante* expected payment is also the same in both mechanisms. Since this is true for all i, the auctioneer's expected revenue is also the same in both mechanisms.   ∎

Thus, when bidders are risk neutral and have independent private valuations, all the auctions we have spoken about so far—English, Japanese, Dutch, and all sealed-bid auction protocols—are revenue equivalent. The revenue equivalence theorem is useful beyond telling the auctioneer that it does not much matter which auction she holds, however. It is also a powerful analytic tool. In particular, we can make use of this theorem to identify equilibrium bidding strategies for auctions that meet the theorem's conditions.

For example, let us consider again the n-bidder first-price auction discussed in Theorem 11.1.3. Does this auction satisfy the conditions of the revenue equivalence theorem? The second condition is easy to verify; the first is harder, because it speaks about the outcomes of the auction under the equilibrium bidding strategies. For now, let us assume that the first condition is satisfied as well.

The revenue equivalence theorem only helps us, of course, if we use it to compare the revenue from a first-price auction with that of another auction that we already understand. The second-price auction serves nicely in this latter role: we already know its equilibrium strategy, and it meets the conditions of the theorem. We know from the proof that a bidder of the same type will make the same expected payment in both auctions. In both of the auctions we are considering, a bidder's payment is zero unless he wins. Thus a bidder's expected payment conditional on being the winner of a first-price auction must be the same as his expected payment conditional on being the winner of a second-price auction. Since the first-price auction is efficient, we can observe that under the symmetric equilibrium agents will bid this amount all the time: if the agent is the high bidder then he will make the right expected payment, and if he is not, his bid amount will not matter.

We must now find an expression for the expected value of the second-highest valuation, given that bidder i has the highest valuation. It is helpful to know order statistic the formula for the kth *order statistic*, in this case of draws from the uniform distribution. The kth order statistic of a distribution is a formula for the expected value of the kth-largest of n draws. For n IID draws from $[0, v_{max}]$, the kth order statistic is

$$\frac{n + 1 - k}{n + 1} v_{max}. \tag{11.10}$$

If bidder i's valuation v_i is the highest, then there are $n - 1$ other valuations drawn from the uniform distribution on $[0, v_i]$. Thus, the expected value of the second-highest valuation is the first-order statistic of $n - 1$ draws from $[0, v_i]$. Substituting into Equation (11.10), we have $\frac{(n-1)+1-(1)}{(n-1)+1}(v_i) = \frac{n-1}{n} v_i$. This confirms the equilibrium strategy from Theorem 11.1.3. It also gives us a suspicion (that turns out to be correct) about the equilibrium strategy for first-price auctions under valuation distributions other than uniform: each bidder bids the expectation of the second-highest valuation, conditioned on the assumption that his own valuation is the highest.

A caveat must be given about the revenue equivalence theorem: this result makes an "if" statement, not an "if and only if" statement. That is, while it is true that all auctions satisfying the theorem's conditions must yield the same expected revenue, it is *not* true that all strategies yielding that expected revenue constitute equilibria. Thus, after using the revenue equivalence theorem to identify a strategy profile that one believes to be an equilibrium, one must then prove that this strategy profile is indeed an equilibrium. This should be done in the standard way, by assuming that all but one of the agents play according to the equilibrium and show that the equilibrium strategy is a best response for the remaining agent.

Finally, recall that we assumed above that the first-price auction allocates the good to the bidder with the highest valuation. The reason it was reasonable to do this (although we could instead have proved that the auction has a symmetric, increasing equilibrium) is that we have to check the strategy profile derived using the revenue equivalence theorem anyway. Given the equilibrium strategy, it is

Risk-neutral, IPV		=		=		=		=	
Risk-averse, IPV	Jap	=	Eng	=	2nd	<	1st	=	Dutch
Risk-seeking, IPV		=		=		>		=	

Table 11.1 Relationships between revenues of various single-good auction protocols.

easy to confirm that the bidder with the highest valuation will indeed win the good.

11.1.6 *Risk attitudes*

One of the key assumptions of the revenue equivalence theorem is that agents are risk neutral. It turns out that many of the auctions we have been discussing cease to be revenue-equivalent when agents' risk attitudes change. Recall from Section 10.3.1 that an agent's risk attitude can be understood as describing his preference between a sure payment and a gamble with the same expected value. (Risk-averse agents prefer the sure thing; risk-neutral agents are indifferent; risk-seeking agents prefer to gamble.)

To illustrate how revenue equivalence breaks down when agents are not risk-neutral, consider an auction environment involving n bidders with IPV valuations drawn uniformly from $[0, 1]$. Bidder i, having valuation v_i, must decide whether he would prefer to engage in a first-price auction or a second-price auction. Regardless of which auction he chooses (presuming that he, along with the other bidders, follows the chosen auction's equilibrium strategy), i knows that he will gain positive utility only if he has the highest utility. In the case of the first-price auction, i will always gain $\frac{1}{n}v_i$ when he has the highest valuation. In the case of having the highest valuation in a second-price auction i's *expected* gain will be $\frac{1}{n}v_i$, but because he will pay the second-highest actual bid, the amount of i's gain will vary based on the other bidders' valuations. Thus, in choosing between the first-price and second-price auctions and conditioning on the belief that he will have the highest valuation, i is presented with the choice between a sure payment and a risky payment with the same expected value. If i is risk averse, he will value the sure payment more highly than the risky payment, and hence will bid more aggressively in the first-price auction, causing it to yield the auctioneer a higher revenue than the second-price auction. (Note that it is i's behavior in the *first-price* auction that will change: the second-price auction has the same dominant strategy regardless of i's risk attitude.) If i is risk seeking he will bid *less* aggressively in the first-price auction, and the auctioneer will derive greater profit from holding a second-price auction.

The strategic equivalence of Dutch and first-price auctions continues to hold under different risk attitudes; likewise, the (weaker) equivalence of Japanese, English, and second-price auctions continues to hold as long as bidders have IPV valuations. These conclusions are summarized in Table 11.1.

A similar dynamic holds if the bidders are all risk neutral, but the *seller* is either risk averse or risk seeking. The variations in bidders' payments are greater in second-price auctions than they are in first-price auctions, because the former depends on the two highest draws from the valuation distribution, while the latter depends on only the highest draw. However, these payments have the same expectation in both auctions. Thus, a risk-averse seller would prefer to hold a first-price auction, while a risk-seeking seller would prefer to hold a second-price auction.

11.1.7 *Auction variations*

In this section we consider three variations on our auction model. First, we consider reverse auctions, in which one buyer accepts bids from a set of sellers. Second, we discuss the effect of entry costs on equilibrium strategies. Finally, we consider auctions with uncertain numbers of bidders.

Reverse auctions

So far, we have considered auctions in which there is one seller and a set of buyers. What about the opposite: an environment in which there is one buyer and a set of sellers? This is what occurs when a buyer engages in a *request for quote* (RFQ). Broadly, this is called a *reverse auction*, because in its open-outcry variety this scenario involves prices that descend rather than ascending.

request for quote

reverse auction

It turns out that everything that we have said about auctions also applies to reverse auctions. Reverse auctions are simply auctions in which we substitute the word "seller" for "buyer" and vice versa and furthermore, negate all numbers indicating prices or bid amounts. Because of this equivalence we will not discuss reverse auctions any further; note, however, that our concentration on (nonreverse) auctions is without loss of generality.

Auctions with entry costs

A second auction variation *does* complicate things, though we will not analyze it here. This is the introduction of an *entry cost* to an auction. Imagine that a first-price auction cost $1 to attend. How should bidders decide whether or not to attend, and then how should they decide to bid given that they're no longer sure how many other bidders will have chosen to attend? This is a realistic way of augmenting our auction model: for example, it can be used to model the cost of researching an auction, driving (or navigating a Web browser) to it, and spending the time to bid. However, it can make equilibrium analysis much more complex.

entry cost

Things are straightforward for second-price (or, for IPV valuations, Japanese and English) auctions. To decide whether to participate, bidders must evaluate their expected gain from participation. This means that the equilibrium strategy in these auctions now *does* depend on the distribution of other agents' valuations and on the number of these agents. The good news is that, once they have decided to bid, it remains an equilibrium for bidders to bid truthfully.

In first-price auctions (and, generally, other auctions that do not have a dominant-strategy equilibrium) auctions with entry costs are harder—though certainly not impossible—to analyze. Again, bidders must make a trade-off between their expected gain from participating in the auction and the cost of doing so. The complication here is that, since he is uncertain about other agents' valuations, a given bidder will thus also be uncertain about the number of agents who will decide that participating in the auction is in their interest. Since an agent's equilibrium strategy given that he has chosen to participate depends on the number of other participating agents, this makes that equilibrium strategy more complicated to compute. And that, in turn, makes it more difficult to determine the agent's expected gain from participating in the first place.

Auctions with uncertain numbers of bidders

Our standard model of auctions has presumed that the number of bidders is common knowledge. However, it may be the case that bidders are uncertain about the number of competitors they face, especially in a sealed-bid auction or in an auction held over the internet. The preceding discussion of entry costs gave another example of how this could occur. Thus, it is natural to elaborate our model to allow for the possibility that bidders might be uncertain about the number of agents participating in the auction.

It turns out that modeling this scenario is not as straightforward as it might appear. In particular, one must be careful about the fact that bidders will be able to update their *ex ante* beliefs about the total number of participants by conditioning on the fact of their own selection, and thus may lead to a situation in which bidders' beliefs about the number of participants may be asymmetric. (This can be especially difficult when the model does not place an upper bound on the number of agents who can participate in an auction.) We will not discuss these modeling issues here; interested readers should consult the notes at the end of the chapter. Instead, simply assume that the bidders hold symmetric beliefs, each believing that the probability that the auction will involve j bidders is $p(j)$.

Because the dominant strategy for bidding in second-price auctions does not depend on the number of bidders in the auction, it still holds in this environment. The same is not true of first-price auctions, however. Let $F(v)$ be a cumulative probability density function indicating the probability that a bidder's valuation is greater than or equal to v, and let $b^e(v_i, j)$ be the equilibrium bid amount in a (classical) first-price auction with j bidders, for a bidder with valuation j. Then the symmetric equilibrium of a first-price auction with an uncertain number of bidders is

$$b(v_i) = \sum_{j=2}^{\infty} \frac{F^{j-1}(v_i)p(j)}{\sum_{k=2}^{\infty} F^{k-1}(v_i)p(k)} \, b^e(v_i, j).$$

Interestingly, because the proof of the revenue equivalence theorem does not depend on the number of agents, that theorem applies directly to this environment. Thus, in this stochastic environment the seller's revenue is the same when she runs a first-price and a second-price auction. The revenue equivalence theorem can thus be used to derive the strategy above.

11.1.8 *"Optimal" (revenue-maximizing) auctions*

So far in our theoretical analysis we have considered only those auctions in which the good is allocated to the high bidder and the seller imposes no reserve price. These assumptions make sense, especially when the seller wants to ensure *economic efficiency*—that is, that the bidder who values the good most gets it. However, we might instead believe that the seller does not care who gets the good, but rather seeks to maximize her expected revenue. In order to do so, she may be willing to risk failing to sell the good even when there is an interested buyer, and furthermore might be willing sometimes to sell to a buyer who did not make the highest bid, in order to encourage high bidders to bid more aggressively. Mechanisms that are designed to maximize the seller's expected revenue are

optimal auction   known as *optimal auctions*.

Consider an IPV setting where bidders are risk neutral and each bidder i's valuation is drawn from some strictly increasing cumulative density function $F_i(v)$, having probability density function $f_i(v)$. Note that we allow for the possibility that $F_i \neq F_j$: bidders' valuations can come from different distributions. Such

asymmetric   interactions are called *asymmetric auctions*. We do assume that the seller knows

auction   the distribution from which each individual bidder's valuation is drawn and hence is able to distinguish strong bidders from weak bidders.

virtual valuation   Define bidder i's *virtual valuation* as

$$\psi_i(v_i) = v_i - \frac{1 - F_i(v_i)}{f_i(v_i)},$$

and assume that the valuation distribution is such that each ψ_i is increasing in v_i. Also define an agent-specific reserve price r_i^* as the value for which $\psi_i(r_i^*) = 0$. The optimal (single-good) auction is a sealed-bid auction in which every agent is asked to declare his true valuation. These declarations are used to compute a virtual (declared) valuation for each agent. The good is sold to the agent i whose virtual valuation $\psi_i(\hat{v}_i)$ is the highest, as long as this value is positive (i.e., the agent's declared valuation v_i exceeds his reserve price r_i^*). If every agent's virtual valuation is negative, the seller keeps the good and achieves a revenue of zero. If the good is sold, the winning agent i is charged the smallest valuation that he could have declared while still remaining the winner: $\inf\{v_i^* : \psi_i(v_i^*) \geq 0 \text{ and } \forall j \neq i, \ \psi_i(v_i^*) \geq \psi_j(\hat{v}_j)\}$.

How would bidders behave in this auction? Note that it can be understood as a second-price auction with a reserve price, held in virtual valuation space rather than in the space of actual valuations. However, since neither the reserve prices nor the transformation between actual and virtual valuations depends on the agent's declaration, the proof that a second-price auction is dominant-strategy truthful applies here as well, and hence the optimal auction remains strategy-proof.

We began this discussion by introducing a new assumption: that different bidders' valuations could be drawn from different distributions. What happens when this does not occur, and instead all bidders' valuations come from the same distribution? In this case, the optimal auction has a simpler interpretation: it is simply a second-price auction (without virtual valuations) in which the seller sets a reserve price r^* at the value that satisfies $r^* - \frac{1-F_i(r^*)}{f_i(r^*)} = 0$. For this reason, it is common to hear the claim that optimal auctions correspond to setting reserve prices optimally. It is important to recognize that this claim holds only in the case of *symmetric* IPV valuations. In the asymmetric case, the virtual valuations can be understood as artificially increasing the amount of weak bidders' bids in order to make them more competitive. This sacrifices efficiency, but more than makes up for it on expectation by forcing bidders with higher expected valuations to bid more aggressively.

Although optimal auctions are interesting from a theoretical point of view, they are rarely to never used in practice. The problem is that they are not *detail* *free*: they require the seller to incorporate information about the bidders' valuation distributions into the mechanism. Such auctions are often considered impractical; famously, the *Wilson doctrine* urges auction designers to consider only detail free mechanisms. With this criticism in mind, it is interesting to ask the following question. In a symmetric IPV setting, is it better for the auctioneer to set an optimal reserve price (causing the auction to depend on bidders' valuation distribution) or to attract one additional bidder to the auction? Interestingly, the auctioneer is better off in the latter case. Intuitively, an extra bidder is similar to a reserve price in the sense that his addition to the auction increases competition among the other bidders, but differs because he can also buy the good himself. This suggests that trying to attract as many bidders as possible (by, among other things, running an auction protocol with which bidders are comfortable) may be more important than trying to figure out the bidders' valuation distributions in order to run an optimal auction.

detail-free auction (margin note)

Wilson doctrine (margin note)

11.1.9 *Collusion*

Since we have seen that an auctioneer can increase her expected revenue by increasing competition among bidders, it is not surprising that bidders, conversely, can reduce their expected payments to the auctioneer by reducing competition among themselves. Such cooperation between bidders is called *collusion*. Collusion is usually illegal; interestingly, however, it is also notoriously difficult for agents to pull off. The reason is conceptually similar to the situation faced by agents playing the Prisoner's Dilemma (see Section 3.4.3): while a given agent is better off if everyone cooperates than if everyone behaves selfishly, he is *even* better off if everyone else cooperates and he behaves selfishly himself. An interesting question to ask about collusion, therefore, is which collusive protocols have the property that agents will gain by colluding while being unable to gain further by deviating from the protocol.

collusion (margin note)

Second-price auctions

First, consider a protocol for collusion in second-price (or Japanese/English) auctions. We assume that a set of two or more colluding agents is chosen exogenously; this set of agents is called a *cartel* or a *bidding ring*. Assume that the agents are risk neutral and have IPV valuations. It is sometimes necessary (as it is in this case) to assume the existence of an agent who is not interested in the good being auctioned, but who serves to run the bidding ring. This agent does not behave strategically, and hence could be a simple computer program. We will refer to this agent as the *ring center*. Observe that there may be agents

cartel (margin note)

bidding ring (margin note)

ring center (margin note)

who participate in the main auction and do not participate in the cartel; there may even be multiple cartels. The protocol follows.

1. Each agent in the cartel submits a bid to the ring center.
2. The ring center identifies the maximum bid that he received, $\hat{v}_1^r$; he submits this bid in the main auction and drops the other bids. Denote the highest dropped bid as $\hat{v}_2^r$.
3. If the ring center's bid wins in the main auction (at the second-highest price in that auction, $\hat{v}_2$), the ring center awards the good to the bidder who placed the maximum bid in the cartel and requires that bidder to pay $\max(\hat{v}_2, \hat{v}_2^r)$.
4. The ring center gives every agent who participated in the bidding ring a payment of k, regardless of the amount of that agent's bid and regardless of whether or not the cartel's bid won the good in the main auction.

How should agents bid if they are faced with this bidding ring protocol? First of all, consider the case where $k = 0$. Here it is easy to see that this protocol is strategically equivalent to a second-price auction in a world where the bidder's cartel does not exist. The high bidder always wins, and always pays the globally second-highest price (the max of the second-highest prices in the cartel and in the main auction). Thus the auction is dominant-strategy truthful, and agents have no incentive to cheat each other in the bidding ring's "preauction." At the same time, however, agents also do not gain by participating in the bidding ring: they would be just as happy if the cartel disbanded and they had to bid directly in the main auction.

Although for $k = 0$ the situation with and without the bidding ring is equivalent from the bidders' point of view, it is different from the point of view of the ring center. In particular, with positive probability $\hat{v}_2^r$ will be the globally second-highest valuation, and hence the ring center will make a profit. (He will pay $\hat{v}_2$ for the good in the main auction, and will be paid $\hat{v}_2^r > \hat{v}_2$ for it by the winning bidder.) Let $c > 0$ denote the ring center's expected profit. If there are n_r agents in the bidding ring, the ring center could pay each agent up to $k = \frac{c}{n_r}$ and still budget balance on expectation. For values of k smaller than this amount but greater than zero, the ring center will profit on expectation while still giving agents a strict preference for participation in the bidding ring.

How are agents able to gain in this setting—doesn't the revenue equivalence theorem say that their gains should be the same in all efficient auctions? Observe that the agents' expected payments are in fact unchanged, although not all of this amount goes to the auctioneer. What does change is the unconditional payment that every agent receives from the ring center. The second condition of the revenue-equivalence theorem states that a bidder with the lowest possible valuation must receive zero expected utility. This condition is violated under our bidding ring protocol, in which such an agent has an expected utility of k.

First-price auctions

The construction of bidding ring protocols is much more difficult in the first-price auction setting. This is for a number of reasons. First, in order to make a lower expected payment, the winner must actually place a lower bid. In a second-price auction, a winner can instead persuade the second-highest bidder to leave the auction and make the same bid he would have made anyway. This difference matters because in the second-price auction the second-highest bidder has no incentive to renege on his offer to drop out of the auction; by doing so, he can only make the winner pay more. In the first-price auction, the second-highest bidder could trick the highest bidder into bidding lower by offering to drop out, and then could still win the good at less than his valuation. Some sort of enforcement mechanism is therefore required for punishing cheaters. Another problem with bidding rings for first-price auctions concerns how we model what noncolluding bidders know about the presence of a bidding ring in their auction. In the second-price auction we were able to gloss over this point: the noncolluding agents did not care whether other agents might have been colluding, because their dominant strategy was independent of the number of agents or their valuation distributions. (Observe that in our previous protocol, if the cumulative density function of bidders' valuation distribution was F, the ring center could be understood as an agent with a valuation drawn from a distribution with CDF F^{n_r}.) In a first-price auction, the number of bidders and their valuation distributions matter to bidders' equilibrium strategies. If we assume that bidders know the true number of bidders, then a collusive protocol in which bidders are dropped does not make much sense. (The strategies of other bidders in the main auction would be unaffected.) If we assume that noncolluding bidders follow the equilibrium strategy based on the number of bidders who actually bid in the main auction, bidder-dropping collusion does make sense, but the noncolluding bidders no longer follow an equilibrium strategy. (They would gain on expectation if they bid more aggressively.)

For the most part, the literature on collusion has sidestepped this problem by considering first-price auctions only under the assumption that all n bidders belong to the cartel. In this setting, two kinds of bidding ring protocols have been proposed.

The first assumes that the same bidders will have repeated opportunities to collude. Under this protocol all bidders except one are dropped, and this bidder bids zero (or the reserve price) in the main auction. Clearly, other bidders could gain by cheating and also placing bids in the main auction; however, they are dissuaded from doing so by the threat that if they cheat, the cartel will be disbanded and they will lose the opportunity to collude in the future. Under appropriate assumptions about agents' discount rates (their valuations for profits in the future), their number, their valuation distribution, and so on, it can be shown that it constitutes an equilibrium for agents to follow this protocol. A variation on the protocol, which works almost regardless of the values of these variables, has the other agents forever punish any agent who cheats, following a grim trigger strategy (see Section 6.1.2).

The second protocol works in the case of a single, unrepeated, first-price auction. It is similar to the protocol introduced in the previous section.

1. Each agent in the cartel submits a bid to the ring center.
2. The ring center identifies the maximum bid that he received, $\hat{v}_1$. The bidder who placed this bid must pay the full amount of his bid to the ring center.

3. The ring center bids in the main auction at 0. Note that the bidding ring always wins in the main auction as there are no other bidders.
4. The ring center gives the good to the bidder who placed the winning bid in the preauction.
5. The ring center pays every bidder other than the winner $\frac{1}{n-1}\hat{v}_1$.

Observe that this protocol can be understood as holding a first-price auction for the right to bid the reserve price in the main auction, with the profits of this preauction split evenly among the losing bidders. (We here assume a reserve price of zero; the protocol can easily be extended to work for other reserve prices.) Let $b^{n+1}(v_i)$ denote the amount that bidder i would bid in the (standard) equilibrium of a first-price auction with a total of $n + 1$ bidders. The symmetric equilibrium of the bidding ring preauction is for each bidder i to bid

$$\hat{v}_i = \frac{n - 1}{n}b^{n+1}(v_i).$$

Demonstrating this fact is not trivial; details can be found in the paper cited at the end of the chapter. Here we point out only the following. First, the $\frac{n-1}{n}$ factor has nothing to with the equilibrium bid amount for first-price auctions with a uniform valuation distribution; indeed, the result holds for any valuation distribution. Rather, it can be interpreted as meaning that each bidder offers to pay everyone else $\frac{1}{n}b^{n+1}(v_i)$, and thereby also to gain utility of $\frac{1}{n}b^{n+1}(v_i)$ for himself. Second, although the equilibrium strategy depends on b^{n+1}, there are really only n bidders. Finally, observe that this mechanism is budget balanced (i.e., not just on expectation).

11.1.10 *Interdependent values*

So far, we have only considered the independent private values (IPV) setting. As we discussed earlier, this setting is reasonable for domains in which the agents' valuations are unrelated to each other, depending only on their own signals—for example, because an agent is buying a good for his own personal use. In this section, we discuss different models, in which agents' valuations depend on both their own signals and other agents' signals.

Common values

common value   First of all, we discuss the *common value* (CV) setting, in which all agents value the good at exactly the same amount. The twist is that the agents do not know this amount, though they have (common) prior beliefs about its distribution. Each agent has a private signal about the value, which allows him to condition his prior beliefs to arrive at a posterior distribution over the good's value.[6]

---

6. In fact, most of what we say in this section also applies to a much more general valuation model in which each bidder may value the good differently. Specifically, in this model each bidder receives a signal drawn independently from some distribution, and bidder i's valuation for the good is some arbitrary function of all of the bidders' signals, subject to a symmetry condition that states that i's valuation does not depend on *which* other agents received which signals. We focus here on the common value model to simplify the exposition.

For example, consider the problem of buying the rights to drill for oil in a particular oil field. The field contains some (uncertain but fixed) amount of oil, the cost of extraction is about the same no matter who buys the contract, and the value of the oil will be determined by the price of oil when it is extracted. Given publicly available information about these issues, all oil drilling companies have the same prior distribution over the value of the drilling rights. The difference between agents is that each has different geologists who estimate the amount of oil and how easy it will be to extract, and different financial analysts who estimate the way oil markets will perform in the future. These signals cause agents to arrive at different posterior distributions over the value of the drilling rights based on which, each agent i can determine an expected value v_i. How can this value v_i be interpreted? One way of understanding it is to note that if a single agent i was selected at random and offered a take-it-or-leave-it offer to buy the drilling contract for price p, he would achieve positive expected utility by accepting the offer if and only if $p < v_i$.

Now consider what would happen if these drilling rights were sold in a second-price auction among k risk-neutral agents. One might expect that each bidder i ought to bid v_i. However, it turns out that bidders would achieve negative expected utility by following this strategy.[7] How can this be—didn't we previously claim that i would be happy to pay any amount up to v_i for the rights? The catch is that, since the value of the good to each bidder is the same, each bidder cares as much about *other* bidders' signals as he does about his own. When he finds out that he won the second-price auction, the winning bidder also learns that he had the most optimistic signal. This information causes him to downgrade his expectation about the value of the drilling rights, which can make him conclude that he paid too much! This phenomenon is called the *winner's*

winner's curse   *curse.*

Of course, the winner's curse does not mean that in the CV setting the winner of a second-price auction always pays too much. Instead, it goes to show that truth telling is no longer a dominant strategy (or, indeed, an equilibrium strategy) of the second-price auction in this setting. There is still an equilibrium strategy that bidders can follow in order to achieve positive expected utility from participating in the auction; this simply requires the bidders to consider how they would update their beliefs on finding that they were the high bidder. The symmetric equilibrium of a second-price auction in this setting is for each bidder i to bid the amount $b(v_i)$ at which, if the second-highest bidder also happened to have bid $b(v_i)$, i would achieve zero expected gain for the good, conditioned on the two highest signals both being v_i.[8] We do not prove this result—or even state it more formally—as doing so would require the introduction of considerable notation.

What about auctions other than second-price in the CV setting? Let us consider Japanese auctions, recalling from Section 11.1.3 that the this auction can be used

---

7. As it turns out, we can make this statement only because we assumed that $k > 2$. For the case of exactly two bidders, bidding v_i is the right thing to do.
8. We do not need to discuss how ties are broken since i achieves zero expected utility whether he wins or loses the good.

as a model of the English auction for theoretical analysis. Here the winner of the auction has the opportunity to learn more about his opponents' signals, by observing the time steps at which each of them drops out of the auction. The winner will thus have the opportunity to condition his strategy on each of his opponents' signals, unless all of his opponents drop out at the same time. Let us assume that the sequence of prices that will be called out by the auctioneer is known: the tth price will be p_t. The symmetric equilibrium of a Japanese auction in the CV setting is as follows. At each time step t, each agent i computes the expected utility of winning the good v_{i,t_i}, given what he has learned about the signals of opponents who dropped out in previous time steps, and assuming that all remaining opponents drop out at the current time step. (Bidders can determine the signals of opponents who dropped out, at least approximately, by inverting the equilibrium strategy to determine what opponents' signals must have been in order for them to have dropped out when they did.) If $v_{i,t_i} > p_{t+1}$, then if all remaining agents actually did drop out at time t and made i the winner at time $t + 1$, i would gain on expectation. Thus, i remains in the auction at time t if $v_{i,t_i} > p_{t+1}$, and drops out otherwise.

Observe that the stated equilibrium strategy is different from the strategy given above for second-price auctions: thus, while second-price and Japanese auctions are strategically equivalent in the IPV case, this equivalence does not hold in CV domains.

Affiliated values and revenue comparisons

affiliated values
The common value model is generalized by another valuation model called *affiliated values*, which permits correlations between bidders' signals. For example, this latter model can describe cases where a bidder's valuation is divided into a private-value component (e.g., the bidder's inherent value for the good) and a common-value component (e.g., the bidder's private, noisy signal about the good's resale value). Technically, we say that agents have affiliated values when a high value of one agent's signal increases the probability that other agents will have high signals as well. A thorough treatment is beyond the scope of this book; however, we make two observations here.

First, in affiliated values settings generally—and thus in common-value settings as a special case—Japanese (and English) auctions lead to higher expected prices than sealed-bid second-price auctions. Even lower is the expected revenue from first-price sealed-bid auctions. The intuition here is that the winner's gain depends on the privacy of his information. The more the price paid depends on others' information (rather than on expectations of others' information), the more closely this price is related to the winner's information, since valuations are affiliated. As the winner loses the privacy of his information, he can extract a smaller "information rent," and so must pay more to the seller.

linkage principle
Second, this argument leads to a powerful result known as the *linkage principle*. If the seller has access to any private source of information that she knows is affiliated with the bidders' valuations, she is better off precommitting to reveal it honestly. Consider the example of an auction of used cars, where the quality of each car is a random variable about which the seller, and each bidder, receives

some information. The linkage principle states that the seller is better off committing to declare everything she knows about each car's defects before the auctions, even though this will sometimes lower the price at which she will be able to sell an individual car. The reason the seller gains by this disclosure is that making her information public also reveals information about the winner's signal and hence reduces his ability to charge information rent. Note that the seller's "commitment power" is crucial to this argument. Bidders are only affected in the desired way if the seller is able to convince them that she will always tell the truth, for example, by agreeing to subject herself to an audit by a trusted third party.

11.2 Multiunit auctions

We have so far considered the problem of selling a single good to one winning bidder. In practice there will often be more than one good to allocate, and different goods may end up going to different bidders. Here we consider *multiunit auctions*, in which there is still only one *kind* of good available, but there are now multiple identical copies of that good. (Think of new cars, tickets to a movie, MP3 downloads, or shares of stock in the same company.) Although this setting seems like only a small step beyond the single-item case we considered earlier, it turns out that there is still a lot to be said about it.

multiunit auctions

11.2.1 *Canonical auction families*

In Section 11.1.1 we surveyed some canonical single-good auction families. Here we review the same auctions, explaining how each can be extended to the multiunit case.

Sealed-bid auctions

Overall, sealed-bid auctions in multiunit settings differ from their single-unit cousins in several ways. First, consider payment rules. If there are three items for sale, and each of the top three bids requests a single unit, then each bid will win one good. In general, these bids will offer different amounts; the question is what each bidder should pay. In the pay-your-bid scheme (the so-called *discriminatory pricing rule*) each of the three top bidders pays a different amount, namely, his own bid. This rule therefore generalizes the first-price auction. Under the *uniform pricing rule* all winners pay the same amount; this is usually either the highest among the losing bids or the lowest among the winning bids.

discriminatory pricing rule

uniform pricing rule

Second, instead of placing a single bid, bidders generally have to provide a price offer for every number of units. If a bidder simply names one number of units and is unwilling to accept any fewer, we say he has placed an *all-or-nothing bid*. If he names one number of units but will accept any smaller number at the same price-per-unit we call the bid *divisible*. We investigate some richer ways for bidders to specify multiunit valuations towards the end of Section 11.2.3.

all-or-nothing bid

divisible bid

Finally, tie-breaking can be tricky when bidders place all-or-nothing bids. For example, consider an auction for 10 units in which the highest bids are as follows, all of them all-or-nothing: 5 units for $20/unit, 3 units for $15/unit, 5 units for $15/unit, and 1 unit for $15/unit. Presumably, the first bid should be satisfied, as well as two of the remaining three—but which? Here one sees different tie-breaking rules—by quantity (larger bids win over smaller ones), by time (earlier bids win over later bids), and combinations thereof.

English auctions

When moving to the multiunit case, designers of English auctions face all of the problems discussed above. However, since bidders can revise their offers from one round to the next, multiunit English auctions rarely ask bidders to specify more than one number of units along with their price offer. Auction designers still face the choice of whether to treat bids as all-or-nothing or divisible. Another subtlety arises when you consider minimum increments. Consider the following example, in which there is a total of 10 units available, and two bids: one for 5 units at $1/unit, and one for 5 units at $4/unit. What is the lowest acceptable next bid? Intuitively, it depends on the quantity—a bid for 3 units at $2/unit can be satisfied, but a bid for 7 units at $2/unit cannot. This problem is avoided if the latter bid is divisible, and hence can be partially satisfied.

Japanese auctions

Japanese auctions can be extended to the multiunit case in a similar way. Now after each price increase each agent calls out a number rather than the simple in/out declaration, signifying the number of units he is willing to buy at the current price. A common restriction is that the number must decrease over time; the agent cannot ask to buy more at a high price than he did at a lower price. The auction is over when the supply equals or exceeds the demand. Different implementations of this auction variety differ in what happens if supply exceeds demand: all bidders can pay the last price at which demand exceeded supply, with some of the dropped bidders reinserted according to one of the tie-breaking schemes above; goods can go unsold; one or more bidders can be offered partial satisfaction of their bids at the previous price; and so on.

Dutch auctions

In multiunit Dutch auctions, the seller calls out descending per unit prices, and agents must augment their signals with the quantity they wish to buy. If that is not the entire available quantity, the auction continues. Here there are several options—the price can continue to descend from the current level, can be reset to a set percentage above the current price, or can be reset to the original high price.

11.2.2   *Single-unit demand*

Let us now investigate multiunit auctions more formally, starting with a very simple model. Specifically, consider a setting with k identical goods for sale and risk-neutral bidders who want only 1 unit each and have independent private

Bidder	Bid amount
1	$25
2	$20
3	$15
4	$8

Table 11.2 Example valuations in a single-unit demand multiunit auction.

values for these single units. Observe that restricting ourselves to this setting gets us around some of the tricky points above such as complex tie breaking.

We saw in Section 11.1.3 that the VCG mechanism can be applied to provide useful insight into auction problems, yielding the second-price auction in the single-good case. What sort of auction does VCG correspond to in our simple multiunit setting? (You may want to think about this before reading on.) As before, since we will simply apply VCG, the auction will be efficient and dominant-strategy truthful; since the market is one-sided it will also satisfy *ex post* individual rationality and weak budget balance. The auction mechanism is to sell the units to the k highest bidders for the same price, and to set this price at the amount offered by the highest losing bid. Thus, instead of a second-price auction we have a $k + 1^{\text{st}}$-price auction.

One immediate observation that we can make about this auction mechanism is that a seller will not necessarily achieve higher profits by selling more units. For example, consider the valuations in Table 11.2.

If the seller were to offer only a single unit using VCG, he would receive revenue of $20. If he offered two units, he would receive $30: less than before on a per unit basis, but still more revenue overall. However, if the seller offered three units he would achieve total revenue of only $24, and if he offered four units he would get no revenue at all. What is going on? The answer points to something fundamental about markets. A dominant-strategy, efficient mechanism can use nothing but losing bidders' bids to set prices, and as the seller offers more and more units, there will necessarily be a weaker and weaker pool of losing bidders to draw upon. Thus the per unit price will weakly fall as the seller offers additional units for sale, and depending on the bidders' valuations, his total revenue can fall as well. What can be done to fix this problem? As we saw for the single-good case in Section 11.1.8, the seller's revenue can be increased on expectation by permitting inefficient allocations, for example, by using knowledge of the valuation distribution to set reserve prices. In the preceding example, the seller's revenue would have been maximized if he had been lucky enough to set a $15 reserve price. (To see how the auction behaves in this case, think of the reserve price simply as k additional bids placed in the auction by the seller.) However, these tactics only go so far. In the end, the law of supply and demand holds—as the supply goes up, the price goes down, since competition between bidders is reduced. We will return to the importance of this idea for multiunit auctions in Section 11.2.4.

The $k + 1^{st}$-price auction can be contrasted with another popular payment rule, used for example in a multiunit English auction variant by the online auction site eBay. In this auction bidders are charged the lowest winning bid rather than the highest losing bid.[9] This has the advantage that winning bidders always pay a nonzero amount, even when there are fewer bids than there are units for sale. In essence, this makes the bidders' strategic problem somewhat harder (the lowest winning bidder is able to improve his utility by bidding dishonestly, and so overall, bidders no longer have dominant strategies) in exchange for making the seller's strategic problem somewhat easier. (While the seller can still lower his revenue by selling too many units, he does not have to worry about the possibility of giving them away for nothing.)

Despite such arguments for and against different mechanisms, as in the single-good case, in some sense it does not matter what auction the seller chooses. This is because the revenue equivalence theorem for that case (Theorem 11.1.4) can be extended to cover multiunit auctions.[10] The proof is similar, so we omit it.

Theorem 11.2.1 (Revenue equivalence theorem, multiunit version) *Assume that each of n risk-neutral agents has an independent private valuation for a single unit of k identical goods at auction, drawn from a common cumulative distribution $F(v)$ that is strictly increasing and atomless on $[\underline{v}, \bar{v}]$. Then any efficient auction mechanism in which any agent with valuation $\underline{v}$ has an expected utility of zero yields the same expected revenue, and hence results in any bidder with valuation v_i making the same expected payment.*

Thus all of the payment rules suggested in the previous paragraph must yield the same expected revenue to the seller. Of course, this result holds only if we believe that bidders are correctly described by the theorem's assumptions (e.g., they are risk neutral) and that they will play equilibrium strategies. The fact that auction houses like eBay opt for non-dominant-strategy mechanisms suggests that these beliefs may not always be reasonable in practice.

sequential auction
We can also use this revenue equivalence result to analyze another setting: repeated single-good auction mechanisms, or so-called *sequential auctions*. For example, imagine a car dealer auctioning off a dozen new cars to a fixed set of bidders through a sequence of second-price auctions. With a bit of effort, it can be shown that for such an auction there is a symmetric equilibrium in which bidders' bids increase from one auction to the next, and in a given auction bidders with higher valuations place higher bids. (To offer some intuition for the first of these claims, bidders still have a dominant strategy to bid truthfully in the final auction. In previous auctions, bidders have positive expected utility after losing the auction, because they can participate in future rounds. As the number of future rounds decreases, so does this expected utility; hence in equilibrium bids rise.) We

---

9. Confusingly, this multiunit English auction variant is sometimes called a *Dutch auction*. This is a practice to be discouraged; the correct use of the term is in connection with the descending open-outcry auction.

10. As before, we state a more restricted version of this revenue equivalence theorem than necessary. For example, revenue equivalence holds for *all* pairs of auctions that share the same allocation rule (not just for efficient auctions) and does not require our assumption of single-unit demand.

can therefore conclude that the auction is efficient, and thus by Theorem 11.2.1 each bidder makes the same expected payment as under VCG. Thus there exists a symmetric equilibrium in which bidders bid honestly in the final auction k, and in each auction $j < k$, each bidder i bids the expected value of the kth-highest of the other bidders' valuations, conditional on the assumption that his valuation v_i lies between the jth-highest and the $j + 1^{\text{st}}$-highest valuations. This makes sense because in each auction the bidder who is correct in making this assumption will be the bidder who places the second-highest bid and sets the price for the winner. Thus, the winner of each auction will pay an unbiased estimate of the overall $k + 1^{\text{st}}$-highest valuation, resulting in an auction that achieves the same expected revenue as VCG.

Very similar reasoning can be used to show that a symmetric equilibrium for k sequential *first-price* auctions is for each bidder i in each auction $j \leq k$ to bid the expected value of the kth-highest of the other bidders' valuations, conditional on the assumption that his valuation v_i lies between the $j - 1^{\text{st}}$-highest and the jth-highest valuations. Thus, each bidder conditions on the assumption that he is the highest bidder remaining; the bidder who is correct in making this assumption wins, and hence pays an amount equal to the expected value of the overall $k + 1^{\text{st}}$-highest valuation.

11.2.3   *Beyond single-unit demand*

Now let us investigate how things change when we relax the restriction that each bidder is only interested in a single unit of the good.

VCG for general multiunit auctions

How does VCG behave in this more general setting? We no longer have something as simple as the $k + 1^{\text{st}}$-price auction we encountered in Section 11.2.2. Instead, we can say that all winning bidders who won the same number of units will pay the same amount as each other. This makes sense because the change in social welfare that results from dropping any one of these bidders will be the same. Bidders who win different numbers of units will not necessarily pay the same per unit prices. We can say, however, that bidders who win larger numbers of units will pay at least as much (in total, though not necessarily per unit) as bidders who won smaller numbers of units, as their impact on social welfare will always be at least as great.

VCG can also help us notice another interesting phenomenon in the general multiunit auction case. For all the auctions we have considered in this chapter so far, it has always been computationally straightforward to identify the winners. In this setting, however, the problem of finding the social-welfare-maximizing allocation is computationally hard. Specifically, finding a subset of bids to satisfy that maximizes the sum of bidders' valuations for them is equivalent to a weighted knapsack problem, and hence is NP-complete.

winner determination problem, general multiunit auction

Definition 11.2.2 (Winner determination problem (WDP)) *The* winner deter-mination problem *(WDP) for a general multiunit auction, where m denotes the* *total number of units available and $\hat{v}_i(k)$ denotes bidder i's declared valuation for being awarded k units, is to find the social-welfare-maximizing allocation of goods to agents. This problem can be expressed as the following integer program.*

$$\text{maximize} \quad \sum_{i \in N} \sum_{1 \le k \le m} \hat{v}_i(k) x_{k,i} \tag{11.11}$$

$$\text{subject to} \quad \sum_{i \in N} \sum_{1 \le k \le m} k \cdot x_{k,i} \le m \tag{11.12}$$

$$\sum_{1 \le k \le m} x_{k,i} \le 1 \qquad \forall i \in N \tag{11.13}$$

$$x_{k,i} = \{0, 1\} \qquad \forall 1 \le k \le m, i \in N \tag{11.14}$$

This integer program uses a variable $x_{k,i}$ to indicate whether bidder i is allocated exactly k units, and then seeks to maximize the sum of agents' valuations for the chosen allocation in the objective function (11.11). Constraint (11.13) ensures that no more than one of these indicator variables is nonzero for any bidder, and constraint (11.12) ensures that the total number of units allocated does not exceed the number of units available. Constraint (11.14) requires that the indicator variables are integral; it is this constraint that makes the problem computationally hard.

Representing multiunit valuations

We have assumed that agents can communicate their complete valuations to the auctioneer. When a large number of units are available in an auction, this means that bidders must specify a valuation for every number of units. In practice, it is common that bidders would be provided with some *bidding language* that would allow them to convey this same information more compactly.

bidding language

Of course, the usefulness of a bidding language depends on the sorts of underlying valuations that bidders will commonly want to express. A few common symmetric valuations are the following.

- *Additive valuation:* The bidder's valuation of a set is directly proportional to the number of goods in the set, so that $v_i(S) = c|S|$ for some constant c.
- *Single item valuation:* The bidder desires any single item, and only a single item, so that $v_i(S) = c$ for some constant c for all $S \ne \emptyset$.
- *Fixed budget valuation:* Similar to the additive valuation, but the bidder has a maximum budget of B, so that $v_i(S) = \min(c|S|, B)$.
- *Majority valuation:* The bidder values equally any majority of the goods, so that

$$v_i(S) = \begin{cases} 1 & \text{if } |S| \ge m/2; \\ 0 & \text{otherwise.} \end{cases}$$

We can generalize all of these valuations to a general symmetric valuation.

- *General symmetric valuation:* Let $p_1, p_2, \ldots, p_m$ be arbitrary nonnegative prices, so that p_j specifies how much the bidder is willing to pay of the jth item won. Then

$$v_i(S) = \sum_{j=1}^{|S|} p_j.$$

- *Downward sloping valuation:* A downward sloping valuation is a symmetric valuation in which $p_1 \geq p_2 \geq \cdots \geq p_m$.

11.2.4 *Unlimited supply: random sampling auctions*

Earlier, we suggested that MP3 downloads serve as a good example of a multiunit good. However, they differ from the other examples we gave, such as new cars, in an important way. This difference is that a seller of MP3 downloads can produce additional units of the good at zero marginal cost, and hence has an effectively unlimited supply of the good. This does not mean that the units have no value or that the seller should give them away—after all, the *first* unit may be very expensive to produce, requiring the seller to amortize this cost across the sale of multiple units. What it does mean is that the seller will not face any supply restrictions other than those she imposes herself.

We thus face the following multiunit auction problem: how should a seller choose a multiunit auction mechanism for use in an unlimited supply setting if she cares about maximizing her revenue? The goal will be finding an auction mechanism that chooses among bids in a way that achieves good revenue without artificially picking a specific number of goods to sell in advance, and also without relying on distributional information about buyers' valuations. We also want the mechanism to be dominant-strategy truthful, individually rational, and weakly budget balanced. Clearly, it will be necessary to artificially restrict supply (and thus cause allocative inefficiency), because otherwise bidders would be able to win units of the good in exchange for arbitrarily small payments. Although this assumption can be relaxed, to simplify the presentation we will return to our previous assumption that bidders are interested in buying at most one unit of the good.

The main insight that allows us to construct a mechanism for this case is that, if we *knew* bidders' valuations but had to offer the goods at the same price to all bidders, it would be easy to compute the optimal single price.

optimal single price **Definition 11.2.3 (Optimal single price)** *The* optimal single price *is calculated as follows.*

1. *Order the bidders in descending order of valuation; let v_i denote the ith-highest valuation.*
2. *Calculate $opt \in \arg\max_{i \in \{1,\ldots,n\}} i \cdot v_i$.*
3. *The optimal single price is v_{opt}.*

Simply offering the good to the agents at the optimal single price is not a dominant-strategy truthful mechanism: bidders would have incentive to misstate their valuations. However, this procedure can be used as a building block to construct a simple and powerful dominant-strategy truthful mechanism.

random sampling optimal price auction

Definition 11.2.4 (Random sampling optimal price auction) *The* random sampling optimal price auction *is defined as follows.*

1. *Randomly partition the set of bidders N into two sets, N_1 and N_2 (i.e., $N = N_1 \cup N_2$; $N_1 \cap N_2 = \emptyset$; each bidder has probability 0.5 of being assigned to each set).*
2. *Using the procedure above find p_1 and p_2, where p_i is the optimal single price to charge the set of bidders N_i.*
3. *Then set the allocation and payment rules as follows:*

   - *For each bidder $i \in N_1$, award a unit of the good if and only if $b_i \geq p_2$, and charge the bidder p_2;*
   - *For each bidder $j \in N_2$, award a unit of the good if and only if $b_j \geq p_1$, and charge the bidder p_1.*

Observe that this mechanism follows the Wilson doctrine: it works even in the absence of distributional information. Random sampling optimal price auctions also have a number of other desirable properties.

Theorem 11.2.5 *Random sampling optimal price auctions are dominant-strategy truthful, weakly budget balanced and* ex post *individually rational.*

The proof of this theorem is left as an exercise to the reader. The proof of truthfulness is essentially the same as the proof of Theorem 11.1.1: bidders' declared valuations are used only to determine whether or not they win, but beyond serving as a maximum price offer do not affect the price that a bidder pays. Of course the random sampling auction is not efficient, as it sometimes refuses to sell units to bidders who value them. The random sampling auction's most interesting property concerns revenue.

Theorem 11.2.6 *The random sampling optimal price auction always yields expected revenue that is at least a $(\frac{1}{4.68})$ constant fraction of the revenue that would be achieved by charging bidders the optimal single price, subject to the constraint that at least two units of the good must be sold.*

A host of other auctions have been proposed in the same vein, for example covering additional settings such as goods for which there is a limited supply of units. (Here the trick is essentially to throw away low bidders so that the number of remaining bidders is the same as the number of goods, and then to proceed as before with the additional constraint that the highest rejected bid must not exceed the single price charged to any winning bidder.) In this limited supply case, both the random sampling optimal price auction and its more sophisticated counterparts can achieve revenues much higher than VCG, and hence also higher than all the other auctions discussed in Section 11.2.2. Other work considers the

online auction *online auction* case, where bidders arrive one at a time and the auction must decide whether each bidder wins or loses before seeing the next bid.

11.2.5 *Position auctions*

The last auction type we consider in this section goes somewhat beyond the multiunit auctions we have defined previously. Like multiunit auctions in the

position auction case of single-unit demand, *position auctions* never allocate more than one item per bidder and ask bidders to specify their preferences using a single real number. The wrinkle is that these auctions sell a set of goods among which bidders are not indifferent: one of a set of ordered positions. The motivating example for these auctions is the sale of ranked advertisements on a page of search results. (For this reason, these auctions are also called *sponsored search auctions*.) Since the goods are not identical, we cannot consider them to be multiple units of the same good. In this sense position auctions can be understood as combinatorial auctions, the topic of the Section 11.3. Nevertheless, we choose to present them here because they have been called multiunit auctions in the literature, and because their bidding and allocation rules have a multiunit flavor.

Regardless of how we choose to classify them, position auctions are very important. From a theoretical point of view they are interesting and have good properties both in terms of incentives and computation. Practically speaking, major search engines use them to sell many billions of dollars worth of advertising space annually, and indeed did so even before much was known about the auctions' theoretical properties. In these auctions, search engines offer a set of keyword-specific "slots"—usually a list on the right-hand side of a page of search results—for sale to interested advertisers. Slots are considered to be more valuable the closer they are to the top of the page, because this affects their likelihood of being clicked by a user. Advertisers place bids on keywords of interest, which are retained by the system. Every time a user searches for a keyword on which advertisers have bid, an auction is held. The outcome of this auction is a decision about which ads will appear on the search results page and in which order. Advertisers are required to pay only if a user clicks on their ad. Because sponsored search is the dominant application of position auctions, we will use it as our motivating example here.

How should position auctions be modeled? The setting can be understood as inducing an infinitely repeated Bayesian game, because a new auction is held every time a user searches for a given keyword. However, researchers have argued that it makes sense to study an unrepeated, perfect-information model of the setting. The single-shot assumption is considered reasonable because advertisers tend to value clicks additively (i.e., the value derived from a given user clicking on an ad is independent of how many other users clicked earlier), at least when advertisers do not face budget constraints. The perfect-information assumption makes sense because search engines allow bidders either to observe other bids or to figure them out by probing the mechanism.

We now give a formal model. As before, let N be the set of bidders (advertisers), and let v_i be i's (commonly known) valuation for getting a click. Let $b_i \in \mathbb{R}_+$ denote i's bid, and let $b_{(j)}$ denote the jth-highest bid, or 0 if there are fewer than j bids. Let $G = \{1, \ldots, m\}$ denote the set of goods (slots), and let α_j denote the expected number of clicks (the *click-through rate*) that an ad will receive if it is listed in the ith slot. Observe that we assume that α does not depend on the bidder's identity.

click-through
rate

The generalized first-price auction was the first position auction to be used by search engine companies.

generalized
first-price
auction (GFP)

Definition 11.2.7 (Generalized first-price auction) *The* generalized first-price auction *(GFP) awards the bidder with the jth-highest bid the jth slot. If bidder i's ad receives a click, he pays the auctioneer b_i.*

Unfortunately, these auctions do not always have pure-strategy equilibria, even in the unrepeated, perfect-information case. For example, consider three bidders 1, 2, and 3 who value clicks at \$10, \$4, and \$2 respectively, participating in an auction for two slots, where the probability of a click for the two slots is $\alpha_1 = 0.5$ and $\alpha_2 = 0.25$, respectively. Bidder 2 needs to bid at least \$2 to get a slot; suppose he bids \$2.01. Then bidder 1 can win the top slot for a bid of \$2.02. But bidder 2 could get the top slot for \$2.03, increasing his expected utility. If the agents bid by best responding to each other—as has indeed been observed in practice—their bids will increase all the way up to bidder 2's valuation, at which point bidder 2 will drop out, bidder 1 will reduce his bid to bidder 3's valuation, and the cycle will begin again.

The instability of bidding behavior under the GFP led to the introduction of the generalized second-price auction, which is currently the dominant mechanism in practice.

generalized
second-price
auction (GSP)

Definition 11.2.8 (Generalized second-price auction) *The* generalized second-price auction *(GSP) awards the bidder with the jth-highest bid the jth slot. If bidder i's ad is ranked in slot j and receives a click, he pays the auctioneer $b_{(j+1)}$.*

The GSP is more stable than the GFP. Continuing the example from above, if all bidders bid truthfully, then bidder 1 would pay \$4 per click for the first slot, bidder 2 would pay \$2 per click for the second slot, and bidder 3 would lose. Bidder 1's expected utility would be $0.5(\$10 - \$4) = \$3$; if he bid less than \$4 but more than \$2 he would pay \$2 per click for the second slot and achieve expected utility of $0.25(\$10 - \$2) = \$2$, and if he bid even less then his expected utility would be zero. Thus bidder 1 prefers to bid truthfully in this example. If bidder 2 bid more than \$10 then he would win the top slot for \$10, and would achieve negative utility; thus in this example bidder 2 also prefers honest bidding.

This example suggests a connection between the GSP and the VCG mechanism. However, these two mechanisms are actually quite different, as becomes clear when we apply the VCG formula to the position auction setting.

Definition 11.2.9 (VCG) *In the position auction setting, the VCG mechanism awards the bidder with the jth-highest bid the jth slot. If bidder i's ad is ranked in slot j and receives a click, he pays the auctioneer $\frac{1}{\alpha_j} \sum_{k=j+1}^{m+1} b_{(k)}(\alpha_{k-1} - \alpha_k)$.*

Intuitively, the key difference between the GSP and VCG is that the former does not charge an agent his social cost, which depends on the differences between click-through rates that other agents would receive with and without his presence. Indeed, truthful bidding is not always a good idea under the GSP. Consider the same bidders as in our running example, but change the click-through rate of slot 2 to $\alpha_2 = 0.4$. When all bidders bid truthfully we have already shown that bidder 1 would achieve expected utility of $3 (this argument did not depend on α_2). However, if bidder 1 changed his bid to $3, he would be awarded the second slot and would achieve expected utility of $0.4(\$10 - \$2) = \$3.2$. Thus the GSP is not even truthful in equilibrium, let alone in dominant strategies.

What *can* be said about the equilibria of the GSP? Briefly, it can be shown that in the perfect-information setting the GSP has many equilibria. The dynamic nature of the setting suggests that the most stable configurations will be *locally envy free*: no bidder will wish that he could switch places with the bidder who won the slot directly above his. There exists a locally envy-free equilibrium of the GSP that achieves exactly the VCG allocations and payments. Furthermore, all other locally envy-free equilibria lead to higher revenues for the seller, and hence are worse for the bidders.

What about relaxing the perfect information assumption? Here, it is possible to construct a generalized *English* auction that corresponds to the GSP, and to show that this English auction has a unique equilibrium with various desirable properties. In particular, the payoffs under this equilibrium are again the same as the VCG payoffs, and the equilibrium is *ex post* (see Section 6.3.4), meaning that it is independent of the underlying valuation distribution.

11.3 Combinatorial auctions

We now consider an even broader auction setting, in which a whole variety of different goods are available in the same market. This differs from the multiunit setting because we no longer assume that goods are interchangeable. Switching to a multi*good* auction model is important when bidders' valuations depend strongly on which set of goods they receive. Some widely studied practical examples include governmental auctions for the electromagnetic spectrum, energy auctions, corporate procurement auctions, and auctions for paths (e.g., shipping rights; bandwidth) in a network.

More formally, let us consider a setting with a set of bidders $N = \{1, \dots, n\}$ (as before) and a set of goods $G = \{1, \dots, m\}$. Let $v = (v_1, \dots, v_n)$ denote the true *valuation functions* of the different bidders, where for each $i \in N$, $v_i : 2^G \mapsto \mathbb{R}$. There is a substantive assumption buried inside this definition: that there are *no externalities*. (Indeed, we have been making this assumption almost continuously since introducing quasilinear utilities in the previous chapter; however, this is a good time to remind the reader of it.) Specifically, we have asserted that a bidder's valuation depends only on the set of goods he wins. This assumption is quite standard; however, it does not allow us to model a bidder who also cares about the allocations and payments of the other agents.

nonadditive valuation functions

We will usually be interested in settings where bidders have *nonadditive valuation functions*, for example valuing bundles of goods more than the sum of the values for single goods. We identify two important kinds of nonadditivity.

partial substitutes

First, when two items are *partial substitutes* for each other (e.g., a Sony TV and a Toshiba TV, or, more partially, a CD player and an MP3 player), their combined value is less than the sum of their individual values. Strengthening this condition,

strict substitutes

when two items are *strict substitutes* their combined value is the same as the value for either one of the goods. For example, consider two nontransferable tickets for seats on the same plane. Sets of strictly substitutable goods can also be seen as multiple units of a single good.

substitutability

Definition 11.3.1 (Substitutability) *Bidder i's valuation v_i exhibits substitutability if there exist two sets of goods $G_1, G_2 \subseteq G$, such that $G_1 \cap G_2 = \emptyset$ and $v(G_1 \cup G_2) < v(G_1) + v(G_2)$. When this condition holds, we say that the valuation function v_i is subadditive.*

The second form of nonadditivity we will consider is *complementarity*. This condition is effectively the opposite of substitutability: the combined value of goods is greater than the sum of their individual values. For example, consider a left shoe and a right shoe, or two adjacent pieces of real estate.

complementarity

Definition 11.3.2 (Complementarity) *Bidder i's valuation v_i exhibits complementarity if there exist two sets of goods $G_1, G_2 \subseteq G$, such that $G_1 \cap G_2 = \emptyset$ and $v(G_1 \cup G_2) > v(G_1) + v(G_2)$. When this condition holds, we say that the valuation function v_i is superadditive.*

How should an auctioneer sell goods when faced with such bidders? One approach is simply to sell the goods individually, ignoring the bidders' valuations. This is easy for the seller, but it makes things difficult for the bidders. In

exposure problem

particular, it presents them with what is called the *exposure problem*: a bidder might bid aggressively for a set of goods in the hopes of winning a bundle, but succeed in winning only a subset of the goods and therefore pay too much. This problem is especially likely to arise in settings where bidders' valuations exhibit strong complementarities, because in these cases bidders might be willing to pay substantially more for bundles of goods than they would pay if the goods were sold separately.

The next-simplest method is to run essentially separate auctions for the different goods, but to connect them in certain ways. For example, one could hold a multiround (e.g., Japanese) auction, but synchronize the rounds in the different auctions so that as a bidder bids in one auction he has a reasonably good indication of what is transpiring in the other auctions of interest. This approach can be made more effective through the establishment of constraints on bidding that span all the auctions (so-called activity rules). For example, bidders might be allowed to increase their aggregate bid amount by only a certain percentage from one round to the next, thus providing a disincentive for bidders to fail to participate in early rounds of the auction and thus improving the information transfer between auctions. Bidders might also be subject to other constraints: for example a budget constraint could require that a bidder not exceed a certain total commitment across all auctions. Both of these ideas can be seen in some government auctions for electromagnetic spectrum (where the so-called *simulta-*

simultaneous *neous ascending auction* was used) as well as in some energy auctions. Despite
ascending some successes in practice, however, this approach has the drawback that it only
auction mitigates the exposure problem rather than eliminating it entirely.

A third approach ties goods together in a more straightforward way: the auctioneer sells all goods in a single auction, and allows bidders to bid directly
combinatorial on bundles of goods. Such mechanisms are called *combinatorial auctions*. This
auction approach eliminates the exposure problem because bidders are guaranteed that their bids will be satisfied "all or nothing." For example a bidder may be permitted to offer $100 for the pair (TV, DVD player), or to make a disjunctive offer "either $100 for TV1 or $90 for TV2, but not both." However, we will see that while combinatorial auctions resolve the exposure problem they raise many other questions. Indeed, these auctions have been the subject of considerable recent study in both economics and computer science, some of which we will describe in the remainder of this section.

11.3.1 *Simple combinatorial auction mechanisms*

The simplest reasonable combinatorial auction mechanism is probably the one in which the auctioneer computes the allocation that maximizes the social welfare of the declared valuations (i.e., $\chi = \max_{x \in X} \sum_{i \in N} \hat{v}_i(x)$), and charges the winners their bids (i.e., for all $i \in N$, $\wp_i = \hat{v}_i$). This is a direct generalization of the first-price sealed-bid auction, and like it this naive auction is not incentive compatible. Consider the following simple valuations in a combinatorial auction setting.

Bidder 1	Bidder 2	Bidder 3
$v_1(x, y) = 100$	$v_2(x) = 75$	$v_3(y) = 40$
$v_1(x) = v_1(y) = 0$	$v_2(x, y) = v_2(y) = 0$	$v_3(x, y) = v_3(x) = 0$

This example makes it easy to show that the auction is not incentive compatible: for example, if agents 1 and 2 bid truthfully, agent 3 is better off declaring, for example, $v_3(y) = 26$. Unfortunately, it is not apparent how to characterize the equilibria of this auction using the techniques that worked in the single-good case: we do not have a simple analytic expression that describes when a bidder wins the auction, and we also lack a revenue equivalence theorem.

An obvious alternative is the method we applied most broadly in the multigood case: VCG. In the example above, VCG would award x to 2 and y to 3. Bidder 2 would pay 60; without him in the auction bidder 1 would have gotten both goods, gaining 100 in value, while with bidder 2 in the auction the other bids only net a total value of 40 (from good x assigned to 3). Similarly, bidder 3 would pay 25; the difference between 100 and 75. The reader can verify that no bidder can gain by unilaterally deviating from truthful bidding, and that bidders strictly lose from some deviations.

As in the multiunit case, VCG has some attractive properties when applied to combinatorial auctions. Specifically, it is dominant-strategy truthful, efficient, *ex post* individual rational and weakly budget balanced (the latter by Theorems 10.4.8 and 10.4.10). The VCG combinatorial auction mechanism is not without shortcomings, however, as we already discussed in Section 10.4.5. (Indeed, though we did not discuss them above, most of these shortcomings also affect the use of VCG in the multiunit case, and some even impact second-price single-good auctions.) For example, a bidder who declares his valuation truthfully has two main reasons to worry—one is that the seller will examine his bid before the auction clears and submit a fake bid just below, thus increasing the amount

shill bid that the agent would have to pay if he wins. (This is a so-called *shill bid*.) Another possibility is both his competitors and the seller will learn his true valuation and will be able to exploit this information in a future transaction. Indeed, these two reasons are often cited as reasons why VCG auctions are rarely seen in practice. Other issues include the fact that VCG is vulnerable to collusion among bidders,

pseudonymous
bidding and, conversely, to one bidder masquerading as several different ones (so-called *pseudonymous bidding* or *false-name bidding*). Perhaps the biggest potential hurdle, however, is computational, and it is not specific to VCG. This is the subject

false-name
bidding of the next section.

11.3.2 *The winner determination problem*

Both our naive first-price combinatorial auction and the more sophisticated VCG version share an element in common: given the agents' individual declarations $\hat{v}$, they must determine the allocation of goods to agents that maximizes social welfare. That is, we must compute $\max_{x \in X} \sum_{i \in N} \hat{v}_i(x)$. In single-good and single-unit demand multiunit auctions this was simple—we just had to satisfy the agent(s) with the highest valuation(s). In combinatorial auctions, as in the general multiunit auctions we considered in Section 11.2.3, determining the winners is a more challenging computational problem.

winner
determination
problem,
combinatorial
auction **Definition 11.3.3 (Winner determination problem (WDP))** *The winner determination problem (WDP) for a combinatorial auction, given the agents' declared valuations $\hat{v}$, is to find the social-welfare-maximizing allocation of goods to agents. This problem can be expressed as the following integer program.*

$$\text{maximize} \quad \sum_{i \in N} \sum_{S \subseteq G} \hat{v}_i(S) x_{S,i} \tag{11.15}$$

$$\text{subject to} \quad \sum_{S \ni j} \sum_{i \in N} x_{S,i} \leq 1 \qquad \forall j \in G \tag{11.16}$$

$$\sum_{S \subseteq G} x_{S,i} \leq 1 \qquad \forall i \in N \tag{11.17}$$

$$x_{S,i} = \{0, 1\} \qquad \forall S \subseteq G, i \in N \tag{11.18}$$

In this integer programming formulation, the valuations $\hat{v}_i(S)$ are constants and the variables are $x_{S,i}$. These variables are boolean, indicating whether bundle S is allocated to agent i. The objective function (11.15) states that we want to maximize the sum of the agents' declared valuations for the goods they are allocated. Constraint (11.16) ensures that no overlapping bundles of goods are allocated, and constraint (11.17) ensures that no agent receives more than one bundle. (This makes sense since bidders explicitly assign a valuation to *every* subset of the goods.) Finally, constraint (11.18) is what makes this an *integer program*[11] rather than a linear program: no subset can be partially assigned to an agent.

set packing
problem

The fact that the WDP is an integer program rather than a linear program is bad news, since only the latter are known to admit a polynomial-time solution. Indeed, a reader familiar with algorithms and complexity may recognize the combinatorial auction allocation problem as a *set packing problem* (SPP). Unfortunately, it is well known that the SPP is NP-complete. This means that it is not likely that a polynomial-time algorithm exists for the problem. Worse, it so happens this problem cannot even be approximated uniformly, meaning that there does not exist a polynomial-time algorithm and a fixed constant $k > 0$ such that for all inputs the algorithm returns a solution that is at least $\frac{1}{k}s^*$, where s^* is the value of the optimal solution for the given input.

relaxation
method

There are two primary approaches to getting around the computational problem. First, we can restrict ourselves to a special class of problems for which there is guaranteed to exist a polynomial-time solution. Second, we can resort to heuristic methods that give up the guarantee of polynomial running time, optimality of solution, or both. In both cases, *relaxation methods* are a common approach. One instance of the first approach is to relax the integrality constraint, thereby transforming the problem into a linear program, which is solvable by known methods in polynomial time. In general the solution results in "fractional" allocations, in which fractions of goods are allocated to different bidders. If we are lucky, however, our solution to the LP will just happen to be integral.

Polynomial methods

total
unimodularity

There are several sets of conditions under which such luck is assured. The most common of these is called *total unimodularity* (TU). In general terms, a constraint matrix A (see Appendix B) is TU if the determinant of every square submatrix is 0, 1, or -1. In the case of the combinatorial auction WDP, this condition (via constraint (11.16)) amounts to a restriction on the subsets that bidders are permitted to bid on. How do we find out if a particular matrix is TU? There are many ways. First, there exists a polynomial-time algorithm to decide whether an arbitrary matrix is TU. Second, we can characterize important subclasses of TU matrices. While many of these subclasses defy an intuitive interpretation, in the following discussion we will present a few special cases that are relevant to combinatorial auctions.

---

11. Integer programs and linear programs are defined in Appendix B.

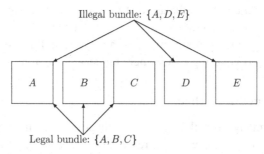

Figure 11.2 Example of legal and illegal bundles for contiguous pieces of lands when we demand the *contiguous ones property*.

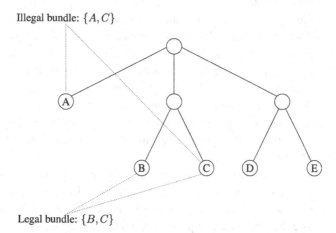

Figure 11.3 Example of a tree-structured bid.

One important subclass of TU matrices is the class of 0–1 matrices with the *consecutive ones property*. In this subclass, all nonzero entries in each column must appear consecutively. This corresponds roughly to contiguous single-dimensional goods, such as time intervals or parcels of land along a shoreline (as shown in Figure 11.2), where bids can only be made on bundles of contiguous goods.

Another subclass of auction problems that have integral polyhedra, and thus can be easily solved using linear programming, corresponds to the set of *balanced matrices*. A 0–1 matrix is balanced if it has no square submatrix of odd order with exactly two 1's in each row and column. One class of auction problems that is known to have a balanced matrix are those that allow only *tree-structured bids*, as illustrated in Figure 11.3. Consider that the set of goods for sale are the vertices of a tree, connected by some set of edges. All bids must be on bundles of the form (j, r), which represents the set of vertices that are within distance r of item j. The constraint matrix for this set of possible bundles is indeed balanced, and so the corresponding polyhedron is integral, and the solution can be found using linear programming.

consecutive ones property

balanced matrix

tree-structured bid

Yet another subclass that can be solved efficiently restricts the bids to be on bundles of no more than two items. The technique here uses dynamic programming, and the algorithm runs in cubic time. Finally, there are a number of positive results covering cases where bidders' valuation functions are subadditive.

Heuristic methods

In many cases the solutions to the associated linear program will not be integral. In these cases we must resort to using *heuristic methods* to find solutions to the auction problem.

Heuristic methods come in two broad varieties. The first is *complete* heuristic methods, which are guaranteed to find an optimal solution if one exists. Despite their discouraging worst-case guarantees, in practice complete heuristic methods are able to solve many interesting problems within reasonable amounts of time. This makes such algorithms the tool of choice for many practical combinatorial auctions. One drawback is that it can be difficult to anticipate how long such algorithms will require to solve novel problem instances, as in the end their performance depends on a problem instance's combinatorial structure rather than on easily-measured parameters like the number of goods or bids. Complete heuristic algorithms tend to perform tree search (to guarantee completeness) along with some sort of pruning technique to reduce the amount of search required (the heuristic).

The second flavor of heuristic algorithm is *incomplete* methods, which are not guaranteed to find optimal solutions. Indeed, as was mentioned earlier, in general there does not even exist a tractable algorithm that can guarantee that you will reach an approximate solution that is within a *fixed fraction* of the optimal solution, no matter how small the fraction. However, methods do exist that can guarantee a solution that is within $1/\sqrt{k}$ of the optimal solution, where k is the number of goods. More importantly, like their complete cousins, incomplete heuristic algorithms often perform very well in practice despite these theoretical caveats. One example of an incomplete heuristic method is a greedy algorithm, a technique that builds up an allocation by adding one bid at a time, and never reconsidering a bid once it has been allocated. Another example is a local search algorithm, in which states in the search space are complete—but possibly infeasible—allocations, and in which the search progresses by modifying these allocations either randomly or greedily.

11.3.3 *Expressing a bid: bidding languages*

We have so far assumed that bidders will specify a valuation for every subset of the goods at auction. Since there are an exponential number of such subsets, this will quickly become impossible as the number of goods grows. If we are to have any hope of finding tractable mechanisms for general combinatorial auctions, we must first find a way for bidders to express their bids in a more succinct manner. In this section we present a number of bidding languages that have been proposed for encoding bids.

As we will see, these languages differ in the ways that they express different classes of bids. We can state some desirable properties that we might like to have in a bidding language. First, we want our language to be *expressive* enough to represent all possible valuation functions. Second, we want our language to be *concise*, so that expressing commonly-used bids does not take space that is exponential in the number of goods. Third, we want our language to be *natural* for humans to both understand and create; thus the structure of the bids should reflect the way in which we think about them. Finally, we want our language to be *tractable* for the auctioneer's algorithms to process when computing an allocation.

In the discussion that follows, for convenience we will often speak about bids as valuation functions. Indeed, in the most general case a bid will contain a valuation for every possible combination of goods. However, be aware that the bid valuations may or may not reflect the players' true underlying valuations. We also limit the scope of our discussion to valuation functions in which the following properties hold:

- *Free disposal:* Goods have nonnegative value, so that if $S \subseteq T$ then $v_i(S) \leq v_i(T)$.
- *Nothing-for-nothing:* $v_i(\emptyset) = 0$ (In other words, a bidder who gets no goods also gets no utility.)

We already discussed multiunit valuations in Section 11.2.3 (additive; single item; fixed budget; majority; general). Combinatorial auctions are different, however, because bidders are expected to value the different goods asymmetrically. For example, there may be different classes of goods, and valuations for sets of goods may be a function of the classes of goods in the set. Imagine that our set G consists of two classes of goods: some red items and some green items, and the bidder requires only items of the same color. Alternatively, it could be the case that the bidder wants exactly one item from each class.

Atomic bids

atomic bid   Let us begin to build up some languages for expressing such bids. Perhaps the most basic bid requests just one particular subset of the goods. We call such a bid an *atomic bid*. An atomic bid is a pair (S, p) that indicates that the agent is willing to pay a price of p for the subset of goods S; we denote this value by $v(S) = p$. Note that an atomic bid implicitly represents an AND operator between the different goods in the bundle. We stated an atomic bid above when we wanted to bid on the TV *and* the DVD player for \$100.

OR bids

OR bid   Of course, many simple bids cannot be expressed as an atomic bid; for example, it is easy to verify that an atomic bid cannot represent even the additive valuation defined earlier. In order to represent this valuation, we will need to be able to bid on disjunctions of atomic valuations. An *OR bid* is a disjunction of atomic bids $(S_1, p_1) \vee (S_2, p_2) \vee \cdots \vee (S_k, p_k)$ that indicates that the agent is willing to pay a price of p_1 for the subset of goods S_1, or a price of p_2 for the subset of goods S_2, etc.

To define the semantics of an OR bid precisely, we interpret OR as an operator for combining valuation functions. Let V be the space of possible valuation functions, and $v_1, v_2 \in V$ be arbitrary valuation functions. Then we have that

$$(v_1 \vee v_2)(S) = \max_{R,T \subseteq S, R \cap T = \emptyset} (v_1(R) + v_2(T)).$$

It is easy to verify that an OR bid can express the additive valuation. As the following result shows, its power is still quite limited; for example, it cannot express the single item valuation described earlier.

Theorem 11.3.4 *OR bids can express all valuation functions that exhibit no substitutability, and only these.*

For example, in the consumer auction setting described earlier, we may have wanted to bid on either the TV and the DVD player for $100, or the TV and the satellite dish for $150, but not both. It is not possible for us to express this using OR bids.

XOR bids

XOR bid *XOR bids* do not have this limitation. An XOR bid is an exclusive OR of atomic bids $(S_1, p_1) \oplus (S_2, p_2) \oplus \cdots \oplus (S_k, p_k)$ that indicates that the agent is willing to accept one but no more than one of the atomic bids.

Once again, the XOR operator is actually defined on the space of valuation functions. We can define its semantics precisely as follows. Let V be the space of possible valuation functions, and $v_1, v_2 \in V$ be arbitrary valuation functions. Then we have that

$$(v_1 \oplus v_2)(S) = \max(v_1(S), v_2(S)).$$

We can use XOR bids to express our example from above:

$$(\{TV, DVD\}, 100) \oplus (\{TV, Dish\}, 150).$$

It is easy to see that XOR bids have unlimited representational power, since it is possible to construct a bid for an arbitrary valuation using an XOR of the atomic valuations for every possible subset $S \subseteq G$.

Theorem 11.3.5 *XOR bids can represent all possible valuation functions.*

However, this does not imply that XOR bids represent every valuation function efficiently. In fact, as the following result states, there are simple valuations that can be represented by short OR bids but that require XOR bids of exponential size.

Theorem 11.3.6 *Additive valuations can be represented by OR bids in linear space, but require exponential space if represented by XOR bids.*

Note that for the purposes of the present discussion, we consider the size of a bid to be the number of atomic formulas that it contains.

Combining the OR and XOR operators

We can also create bidding languages by combining the OR and XOR operators
on valuation functions. Consider a language that allows bids that are of the form
of an OR of XOR of atomic bids. We call these bids *OR-of-XOR bids*. An *OR-
of-XOR bid* is a set of XOR bids, as defined above, such that the bidder is willing
to obtain any number of these bids.

OR-of-XOR bid

Like XOR bids, OR-of-XOR bids have unlimited representational power.
However, unlike XOR bids, they can specialize to plain OR bids, which affords
greater simplicity of expression, as we have seen above. As a specific example,
OR-of-XOR bids can express any downward sloping symmetric valuation on m
items in size of only m^2. However, this language's compactness is still limited. For
example, even simple asymmetric valuations can require size of at least $2^{m/2+1}$
to express in the OR-of-XOR language.

It is also possible to define a language of XOR-of-OR bids, and even a language
allowing arbitrary nesting of OR and XOR statements here (we refer to the latter
as generalized OR/XOR bids). These languages vary in their compactness.

The OR* bidding language

Now we turn to a slightly different sort of bidding language that is powerful
enough to simulate all of the preceding languages with a relatively succinct
representation. This language results from the insight that it is possible to simulate
the effect of an XOR by allowing bids to include *dummy* (or *phantom*) items.
The only difference between an OR and an XOR is that the latter is exclusive;
we can enforce this exclusivity in the OR by ensuring that all of the sets in the
disjunction share a common item. We call this language *OR**.

Definition 11.3.7 (OR* bid) *Given a set of dummy items G_i for each agent $i \in$*
OR* bid   *N, an* OR* *bid is a disjunction of atomic bids $(S_1, p_1) \vee (S_2, p_2) \vee \cdots \vee (S_k, p_k)$,*
where for each $l = 1, \ldots, k$, the agent is willing to pay a price of p_l for the set
of items $S_l \subseteq G \cup G_i$.

An example will help make this clearer. If we wanted to express our TV bid
from above using dummy items, we would create a single dummy item D, and
express the bid as follows.

$$(\{\text{TV, DVD, D}\}, 100) \vee (\{\text{TV, Dish, D}\}, 150)$$

Any auction procedure that does not award one good to two people will select
at most one of these disjuncts. The following results show us that the OR*
language is surprisingly expressive and simple.

Theorem 11.3.8 *Any valuation that can be represented by OR-of-XOR bids of*
size s can also be represented by OR* bids of size s, using at most s dummy items.

Theorem 11.3.9 *Any valuation that can be represented by OR/XOR bids of size*
s can also be represented by OR* bids of size s, using at most s^2 dummy items.

By the definition of OR/XOR bids, we have the following corollary.

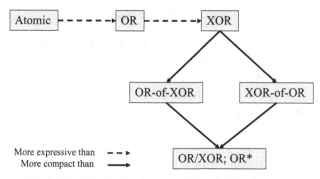

Figure 11.4 Relationships between different bidding languages.

Corollary 11.3.10 *Any valuation that can be represented by XOR-of-OR bids of size s can also be represented by OR* bids of size s, using at most s^2 dummy items.*

Let us briefly review the properties of the languages we have discussed. The XOR, OR-of-XORs, XOR-of-OR and OR* languages are all powerful enough to express all valuations. Second, the efficiencies of the OR-of-XOR and XOR-of-OR languages are incomparable: there are bids that can be expressed succinctly in one but not the other, and vice-versa. Third, the OR* language is strictly more compact than both the OR-of-XOR and XOR-of-OR languages: it can efficiently simulate both languages, and can succinctly express some valuations that require exponential size in each of them. These properties are summarized in Figure 11.4.

Interpretation and verification complexity

Recall that in the auction setting these languages are used for communicating bids to the auctioneer. It is the auctioneer's job to first interpret these bids, and then calculate an allocation of goods to agents. Thus it is natural to be concerned about the computational complexity of a given bidding language. In particular, we may want to know how difficult it is to take an arbitrary bid in some language and compute the valuation of some arbitrary subset of goods according to that bid. We call this the *interpretation complexity*. The interpretation complexity of a bidding language is the minimum time required to compute the valuation $v(S)$, given input of an arbitrary subset $S \subseteq G$ and arbitrary bid v in the language.

interpretation complexity

Not surprisingly, the atomic bidding language has interpretation complexity that is polynomial in the size of the bid. To compute the valuation of some arbitrary subset S, one need only check whether all members of S are in the atomic bid. If they are, the valuation of S is just that given in the bid (because of free disposal); and if they are not, then the valuation of S is 0. The XOR bidding language also has interpretation complexity that is polynomial in the size of the bid; just perform the above procedure for each of the atomic bids in turn. However, all of the other bidding languages mentioned above have interpretation complexity that is exponential in the size of the bid. For example, given the OR bid $(S_1, p_1) \vee (S_2, p_2) \vee \cdots \vee (S_k, p_k)$, computing the valuation of S requires checking all possible combinations of the atomic bids, and there are 2^k such possible combinations.

One might ask why we even consider bidding languages that have exponential interpretation complexity. Simply stated, the answer is that languages with only polynomial interpretation complexity are either not expressive enough or not compact enough. This brings us to a more relaxed criterion. It may be sufficient to require that a given claim about a bid's valuation for a set is *verifiable* in polynomial time. We define the *verification complexity* of a bidding language as the minimum time required to verify the valuation $v(S)$, given input of an arbitrary subset $S \subseteq G$, an arbitrary bid v in the language, and a proof of the proposed valuation $v(S)$. All of the languages we have presented in this section have polynomial verification complexity.

verification complexity

11.3.4   *Iterative mechanisms*

We have argued that in a combinatorial auction setting, agents' valuations can be so complex that they cannot be tractably communicated. The idea behind bidding languages (Section 11.3.3) is that this communication limitation can be overcome if an agent's valuation can be succinctly represented. In this section we consider another, somewhat independent idea: replacing the sealed-bid mechanism we have discussed so far with an indirect mechanism that probes agents' valuations only as necessary.

Intuitively, the use of indirect mechanisms in combinatorial auctions offers the possibility of several benefits. Most fundamentally, allowing the mechanism to query bidders selectively can reduce communication. For example, if the mechanism arrived at the desired outcome (say the VCG allocation and payments) after a small number of queries, other agents could realize that they were unable to make a bid that would improve the allocation, and could thus quit the auction without communicating any more information to the auctioneer. This sort of reduction in communication is still useful even if the auction is small enough that agents' full valuations *could* be tractably conveyed. First, reducing communication can benefit bidders who want to reveal as little as possible about their valuations to their competitors and/or to the auctioneer. Second, iterative mechanisms can help in cases where it is difficult for bidders to determine their own valuations. For example, in a logistics domain a bidder might have to solve a traveling salesman problem to determine his valuation for a given bundle and thus could face a computational barrier to determining his whole valuation function.

Indirect mechanisms also have benefits that go beyond reducing communication. First, they can be easier for bidders to understand than complex direct mechanisms like VCG, and so can be seen as more transparent. This matters, for example, in government auctions of public assets like radio spectrum, where taxpayers want to be assured that the auction was conducted fairly. Finally, while no general result shows this formally, experience with single-good auctions in the common and affiliated values case (Section 11.1.10) suggests that allowing bidders to iteratively exchange partial information about their valuations may lead to improvements in both revenue and efficiency.

Of course, considering iterative mechanisms also invites new challenges. Most of all, such mechanisms are tremendously complicated, and hence can require extensive effort to design. Furthermore, small flaws in this design can lead to huge problems. For example, iterative mechanisms can give rise to considerably richer strategy spaces than direct mechanisms do: an agent can condition his actions on everything he has learned about the actions taken previously by other agents. Beyond potentially making things complicated for the agents, this strategic flexibility can also facilitate undesirable behavior. For example, agents can bid insincerely in order to signal each other (e.g., "do not bid on my bundle and I will not bid on yours"), and thus collude against the seller. Another problem is that agents can often gain by waiting for others to reveal information, especially in settings where determining one's own valuation is costly. The auction must therefore be designed in a way that gives agents some reason to bid early. One approach is to establish activity rules that restrict an agent's future participation in the auction if he does not remain sufficiently active. This idea was already discussed at the beginning of Section 11.3 in the context of decoupled auctions for combinatorial settings such as the simultaneous ascending auction.

Because iterative mechanisms can become quite complex, we will not formally describe any of them here. (As always, interested readers should consult the references cited at the end of the chapter.) However, we will note some of the main questions addressed in this literature, and some of the general trends in the answers to these questions. The first question is what social choice functions to implement, and thus what payments to impose. Here a popular choice is to design mechanisms that converge to an efficient allocation, and that elicit enough information to guarantee that agents pay the same amounts that they would under VCG. Even if an indirect mechanism mimics VCG, it does not automatically inherit its equilibrium properties—the revelation principle only covers the transformation of indirect mechanisms *into* direct mechanisms. Indeed, under indirect mechanisms that mimic VCG, answering queries honestly is no longer a dominant strategy. For example, if agent i knows that agent j will overbid, agent i may also want to do so, as dishonest declarations by j can affect the queries that i will receive in the future. Nevertheless, it can be shown that it is an *ex post* equilibrium (see Section 6.3.4) for agents to cooperate with any indirect mechanism that achieves the same allocation and payment as VCG when all bidders but some bidder i bid truthfully and i bids arbitrarily. In some cases mechanism designers have considered mechanisms that do *not* always converge to VCG payments. Here they usually also assume that agents will bid "straightforwardly"—that is, that they will answer queries truthfully even if it is not in their interest to do so. This assumption is typically justified by demonstrations that agents can gain very little by deviations even in the worst case (i.e., straightforward bidding is an ϵ-equilibrium for small ϵ), claims that determining a profitable deviation would be computationally intractable, and/or an appeal to a complete-information Nash analysis. As long as bidders do behave in a way consistent with this assumption, these iterative mechanisms are able to avoid some of the undesirable properties of VCG discussed in Section 10.4.5.

value query

demand query

A second question is what sorts of queries the indirect mechanisms should ask. The most popular are probably *value queries* and *demand queries*. A mechanism asks a value query when it suggests a bundle and asks how much it is worth to a bidder. Demand queries are in some sense the reverse: the mechanism asks which bundle the bidder would prefer at given prices. Demand queries come in various different forms: the simplest have prices only on single goods and offer the same prices to all bidders, while the most complicated attach bidder-specific prices to bundles. When only demand queries are used and prices (of whatever form) are

ascending combinatorial auction

guaranteed to rise as the auction progresses, we call the auction an *ascending combinatorial auction*. (Observe that a Japanese auction is a single-good special case of such an auction.) Of course, all sorts of other query families are also

order query

bounding query

possible. For example, bidders can be asked *order queries* (state which of two bundles is preferred) and *bounding queries* (answer whether a given bundle is worth more or less than a given amount). Some mechanisms even allow *push-pull queries*: bidders can answer questions they weren't asked, and can decline to answer questions they were asked.

The final general lesson to convey from the literature on iterative combinatorial auctions is that in the worst case, any mechanism that achieves an efficient or approximately efficient[12] allocation in equilibrium must receive an amount of information equal in size to a single agent's complete valuation. Since the size of an agent's valuation is exponential in the number of goods, this is discouraging. However, this result speaks only about the worst case and only about bidders with unrestricted valuations. Researchers have managed to show theoretically that worst-case communication requirements are polynomial under some restricted classes of valuations or when the query complexity is parameterized by the minimal representation size in a given bidding language, and to demonstrate empirically that iterative mechanisms can terminate after reasonable amounts of communication when used with real bidders.

11.3.5   *A tractable mechanism*

tractability

Recall that a *tractable* mechanism is one that can determine the allocation and payments using only polynomial-time computation (Definition 10.3.10). We have seen that such a mechanism can easily be achieved by restricting bidders to expressing valuations from a set that makes the winner determination problem tractable (as discussed in Section 11.3.2), and then using VCG. Here we look beyond such bidding restrictions, seeking more general mechanisms that nevertheless remain computationally feasible. The idea here is to build a mechanism around an optimization algorithm that is guaranteed to run in polynomial time regardless of its inputs.

We will give one example of such a mechanism: a dominant-strategy truthful mechanism for combinatorial auctions that is built around a greedy algorithm.

---

12. Formally, this result holds for any auction that always finds allocations that achieve more than a $\frac{1}{n}$-fraction of the optimal social welfare.

This mechanism only works for bidders with a restricted class of valuations, called *single minded*.

single-minded **Definition 11.3.11 (Single-minded bidder)** *A bidder is* single-minded *if he has the valuation function:*

$$\forall s \in 2^G, \qquad v_i(s) = \begin{cases} v_i > 0 & \textit{if } s \supseteq b_i; \\ 0 & \textit{otherwise.} \end{cases}$$

Intuitively, a bidder is single-minded if he is only interested in a single bundle; he values this bundle and all supersets of it[13] at the same amount, v_i, and values all other bundles at zero. Although this is a severe restriction on agents' valuations, it does not make the winner determination problem any easier. Intuitively, this is because agents remain free to choose *any* bundle s from the set of possible bundles 2^G, and so the auctioneer can receive bids on any set of bundles.

In a direct mechanism for single-minded bidders, every bidder i places a bid $b_i = (\widehat{s_i}, \widehat{v_i})$ indicating a bundle of interest $\widehat{s_i}$ and an amount $\widehat{v_i}$. (Observe that we have assumed that the auctioneer does not know bidder i's bundle of interest s_i; this is why i's bid must have two parts.) Let $apg_i = \widehat{v_i}/|\widehat{s_i}|$ denote bidder i's declared amount per good, and as before let n be the number of bidders. Now consider the following greedy algorithm.

greedy allocation scheme **Definition 11.3.12 (Greedy allocation scheme)** *The* greedy allocation scheme *is defined as follows.*

> store the bidders' bids in a list L, sorted in decreasing order of apg
> let $L(j)$ denote the jth element of L
> $a \leftarrow \emptyset$
> $j \leftarrow 1$
> **while** $j \le n$ **do**
> > **if** $a \cap \widehat{s_j} = \emptyset$ **then**
> > > bid b_j wins
> > > $a \leftarrow a \cup \widehat{s_j}$
>
> **foreach** *winning bidder* i **do**
> > look for a bidder $inext$, the first bidder whose bid follows i's in L, whose bid does not win, and whose bid does win if the greedy allocation scheme is run again with i's bid omitted
> > **if** *a bidder* $inext$ *exists* **then**
> > > bidder i's payment is $p_i \leftarrow \frac{|\widehat{s_i}| \cdot \widehat{v_{inext}}}{|\widehat{s_{inext}}|}$
> > **else**
> > > $p_i \leftarrow 0$
>
> **foreach** *losing bidder* i **do**
> > $p_i =\leftarrow 0$

---

13. Implicitly, this amounts to an assumption of free disposal, as defined in Section 11.3.3.

Intuitively, the greedy allocation scheme ranks all bids in decreasing order of apg, and then greedily allocates bids starting from the top of L. The payment of a winning bidder i is the apg of the highest-ranked bidder that would have won but for i's participation, multiplied by the number of goods allocated to i. Bidder i pays zero if he does not win or if there is no bidder $inext$.

Theorem 11.3.13 *When bidders are single minded, the greedy allocation scheme is dominant-strategy truthful.*

We leave the proof of this theorem as an exercise; however, it is not hard. Observe that, because it is based on a greedy algorithm, this mechanism does not select an efficient allocation. It is natural to wonder whether this mechanism can come close. It turns out that the best that can be achieved comes from modifying the algorithm, replacing apg_i with the ratio $\widehat{v}_i/\sqrt{|\widehat{s}_i|}$. This can be shown to preserve dominant-strategy truthfulness, and to achieve the $1/\sqrt{k}$-bound on efficiency discussed above.

11.4 Exchanges

So far, we have surveyed single-good, multiunit, and combinatorial auctions. Despite the wide variety within these families, we have not yet exhausted the space of auctions. We now briefly discuss one last, important category of auctions.

exchange   These are *exchanges*: auctions in which agents are able to act as both buyers and sellers. We discuss two varieties, which differ more in their purposes than in their mechanics. The first is intended to allocate goods; the second is designed to aggregate information.

11.4.1 *Two-sided auctions*

two-sided auction   In *two-sided auctions*, otherwise known as *double auctions*, there are many buyers and sellers. A typical example is the stock market, where many people are interested in buying or selling any given stock. It is important to distinguish

double auction   this setting from certain marketplaces (such as popular consumer auction sites) in which there are multiple separate single-sided auctions. We will not have much to say about double auctions, in part because the relative dearth of theoretical results about them. However, let us mention two primary models of single-dimensional double markets, that is, markets in which there are many potential buyers and sellers of many units of the same good (e.g., the shares of a given

continuous double auction (CDA)   company). We distinguish here between two kinds of markets, the *continuous double auction* (CDA) and the *periodic double auction* (otherwise known as the *call market*).

periodic double auction   In both the CDA and the call market agents bid at their own pace and as many times as they want. Each bid consists of a price and quantity, where the quantity is either positive (signifying a "buy" order) or negative (signifying a "sell" order).

call market   There are no constraints on what the price or quantity might be. Also in both

order book   cases, the bids received are put in a central repository, the *order book*. Where the

before	Sell:	5@\$1	3@\$2	6@\$4	2@\$6	4@\$9	
	Buy:	6@\$9	4@\$5	6@\$4	3@\$3	5@\$2	2@\$1

$$\Downarrow$$

after	Sell:	2@\$6	4@\$9		
	Buy:	2@\$4	3@\$3	5@\$2	2@\$1

Figure 11.5 A call-market order book, before and after market clears.

CDA and call market diverge is on the question of when a trade occurs. In the CDA, as soon as the bid is received, an attempt is made to match it with one or more bids on the order book; for example, a new sell order for 10 units may be matched with one existing buy bid for 4 units and another buy bid for 6 units, so long as both the buy-bid prices are higher than the sell price. In cases of partial matches, the remaining units (either of the new bid or of one of order-book bids) is put back on the order book. For example, if the new sell order is for 13 units and the only buy bids on the order book with a higher price are the ones described (one buy bid for 4 units and another buy bid for 6 units), two trades are arranged—one for 4 units, and one for 6—and the remaining 3 units of the new bid are put on the order book as a sell order. (We have not mentioned the price of the trades arranged; obviously, they must lie in the interval between the price in the buy bid and the price in the sell bid—the so called bid-ask spread—but are unconstrained otherwise. Indeed, the amount paid to the seller could be less than the amount charged to the buyer, allowing a commission for the exchange or broker.)

In contrast, when a bid arrives in the call market, it is simply placed in the order book. No trade is attempted. Then, at some predetermined time, an attempt is made to arrange the maximal amount of trade possible (called clearing the market). In this case this is done simply by ranking the sell bids in ascending order, the buy bids in descending order, and finding the point at which supply meets demand. Figure 11.5 depicts a typical call market. In this example 14 units are traded when the market clears, after the remaining bids are left on the order book awaiting the next market clear.

11.4.2 *Prediction markets*

prediction market

information market

A *prediction market*, also called an *information market*, is a double-sided auction that is used to aggregate agents' beliefs rather than allocating goods. For example, such a market could be used to assess different candidates' chances in an upcoming presidential election. To set up a prediction market, the market designer establishes contracts $(c_1, \ldots, c_k)$, where each contract c_i is a commitment to pay the bearer \$1 if candidate i wins the election, and \$0 otherwise. However, the market designer does not actually sell such contracts himself. Instead, he simply opens up his market, and allows interested agents to both buy and sell contracts with each other. The reason that such a market is interesting is that a risk-neutral bidder should value a contract c_i at exactly the probability that i will win the election. If a bidder believes that he has information about the election that other bidders do not have, and consequently believes that the probability of i winning is greater than the current asking price for contract c_i then i will want to buy contracts. Conversely, if a bidder believes the true probability of i winning is less than the current price, he will want to sell contracts. In equilibrium, prices reflect bidders' aggregate beliefs. This is similar to the idea that price of a company's stock in a stock market will reflect the market's belief about the company's future profitability. Indeed, an even tighter collection is to futures markets, in which traders can purchase delivery of a commodity (e.g., oil) at a future date; prices in these markets are commonly treated as forecasts about future commodity prices. The key distinction is that a prediction market is designed primarily to elicit such information, rather than doing so as a side effect of solving an allocation problem. On the other hand, futures markets are used primarily for hedging risk (e.g., an airline company might buy oil futures in order to defend itself against the risk that oil prices will rise, causing it to lose money on tickets it has already sold).

Prediction markets have been used in practice for a wide variety of belief aggregation tasks, including political polling, forecasting the outcomes of sporting events, and predicting the box office returns of new movies. Of course, there are other methods that can be used for all of these tasks. In particular, opinion polls and surveys of experts are common approaches. However, there is mounting empirical evidence that prediction markets can outperform both of these methods. This is interesting because prediction markets do not need to work with unbiased samples of the population: bidders are not asked what *they* think, but what they think *the whole population* thinks, and they are given an economic incentive to answer correctly. Similarly, prediction markets do not need to explicitly identify or score experts: participants are self-selecting, and weight their input themselves by the size of the position they take in the market. Finally, prediction markets are able to update their forecasts in real time, for example, immediately updating prices to reflect the consequences of a political scandal on an election.

A variety of different mechanisms can be used to implement prediction markets. One straightforward approach is to use continuous double auctions or call markets, as described in the previous section. These mechanisms have the drawback that they can suffer from a lack of liquidity: trades can occur only when agents want contracts on both sides of the market at the same time. Liquidity can be introduced by a market maker. However, such a market maker therefore assumes financial risk, which means the mechanism is not budget balanced. Another approach is a parimutuel market: bidders place money on different outcomes, and when the true outcome is observed, the winning bidders split the pot in proportion to the amount each one gambled. These markets are budget balanced and do not suffer from illiquidity; however, agents have no incentive to place bets before the last moment, and so the market loses the property of aggregating beliefs in real time. Yet more complicated mechanisms, such as dynamic parimutuel markets and market scoring rules, are able to achieve real-time belief aggregation, bound the market-maker's worst-case loss, and still provide substantial liquidity. These markets are too involved to describe briefly here; the reader is referred to the references for details.

11.5   History and references

Krishna [2002] is an excellent book that provides a formal introduction to auction theory. Klemperer [1999b] is a large edited collection of many of the most important papers on the theory of auctions, preceded by a thorough survey by the editor; this survey is reproduced in Klemperer [1999a]. Earlier surveys include Cassady [1967], Wilson [1987a], and McAfee and MacMillan [1987]. These texts cover most of the canonical single-good and multi-unit auction types we discuss in the chapter. (One exception is the elimination auction, which we gave as an example of the diversity of auction types; it was introduced by Fujishima et al. [1999b].)

Vickrey's seminal contribution [Vickrey, 1961] is still recommended reading for anyone interested in auctions. In it Vickrey introduced the second-price auction and argued that bidders in such an auction do best when they bid sincerely. He also provided the analysis of the first-price auction under the independent private value model with the uniform distribution described in the chapter. He even proved an early version of the revenue-equivalence theorem (Theorem 11.1.4), namely that in the independent private value case, the English, Dutch, first-price, and second-price auctions all produce the same expected revenue for the seller. For his work, Vickrey received a Nobel Prize in 1996.

The more general form of the revenue-equivalence theorem, Theorem 11.1.4, is due to Myerson [1981] and Riley and Samuelson [1981], who also investigated optimal (i.e., revenue-maximizing) auctions. Our proof of the theorem follows Klemperer [1999a]. The "auctions as structured negotiations" point of view was advanced by Wurman et al. [2001]. McAfee and McMillan [1987] introduced the notion of auctions with an uncertain number of bidders, and Harstad et al. [1990] analyzed its equilibrium. The so-called Wilson doctrine was articulated in Wilson [1987b]. The result that one additional bidder yields more revenue than an optimal reserve price is due to Bulow and Klemperer [1996]. The most influential theoretical studies of collusion were by Graham and Marshall [1987] for second-price auctions and McAfee and McMillan [1992] for first-price auctions. Early important results on the common-value (CV) model include Wilson [1969] and Milgrom [1981]. The former showed that when bidders are uncertain about their values their bids are not truthful, but rather are lower that their assessment of that value. Milgrom [1981] analyzed the symmetric equilibrium for second-price auctions under common values. The affiliated value model was introduced by Milgrom and Weber [1982].

The equilibrium analysis of sequential auctions is due to Milgrom and Weber [2000], from a seminal paper written in 1982 but only published recently. The random sampling optimal price auction and the first proof that this auction achieves a constant fraction of optimal fixed-price revenue is due to Goldberg et al. [2006]; the $\frac{1}{4.68}$ bound on revenue is due to Saeed et al. [2008]. Our discussion of position auctions generally follows Edelman et al. [2007]; see also Varian [2007].

Combinatorial auctions are covered in depth in the edited collection Cramton et al. [2006], which probably provides the best single-source overview of the area. Several chapters of this book are especially worthy of note here. The computational complexity of the WDP is discussed in a chapter by Lehmann et al. Algorithms for the WDP have an involved history, and are reprised in chapters by Müller and Sandholm. A detailed discussion of bidding languages can be found in a chapter by Nisan; the OR* language is due to Fujishima et al. [1999a]. Iterative mechanisms are covered in chapters by Parkes, Ausubel and Milgrom, and Sandholm and Boutilier; the worst-case communication complexity analysis appears in a chapter by Segal. Finally, the tractable greedy mechanism is due to Lehmann et al. [2002].

There is a great deal of empirical evidence that prediction markets can effectively aggregate beliefs; two prominent examples of this literature consider election results [Berg et al., 2001] and movie box office revenues [Spann and Skiera, 2003]. Dynamic parimutuel markets are due to Pennock [2004], and the market scoring rule is described by Hanson [2003].

12

Teams of Selfish Agents: An Introduction
to Coalitional Game Theory

In Chapters 1 and 2 we looked at how teams of cooperative agents can accomplish
more together than they can achieve in isolation. Then, in Chapter 3 and many
of the chapters that followed, we looked at how self-interested agents make
individual choices. In this chapter we interpolate between these two extremes,
asking how self-interested agents can combine to form effective teams. As the
title of the chapter suggests, this chapter is essentially a crash course in *coalitional
game theory*, also known as *cooperative game theory*. As was mentioned at the
beginning of Chapter 3, when we introduced noncooperative game theory, the
term "cooperative" can be misleading. It does not mean that, as in Chapters 1
and 2, each agent is agreeable and will follow arbitrary instructions. Rather,
it means that the basic modeling unit is the group rather than the individual
agent. More precisely, in coalitional game theory we still model the individual
preference of agents, but not their possible actions. Instead, we have a coarser
model of the capabilities of different groups.

We proceed as follows. First, we define the most widely studied model of coalitional games, give examples of situations that can be modeled in this way, and
discuss a series of refinements to the model. Then we consider how such games
can be analyzed. The main solution concepts we discuss here are the *Shapley
value*, the *core*, and the *nucleolus*. Finally, we consider compact representations
of coalitional games and their computational implications. We conclude by sur-
veying further directions that have been explored in the literature.

12.1 Coalitional games with transferable utility

In coalitional game theory our focus is on what groups of agents, rather than
individual agents, can achieve. Given a set of agents, a coalitional game defines
how well each group (or *coalition*) of agents can do for itself. We are not concerned with how the agents make individual choices within a coalition, how they
coordinate, or any other such detail; we simply take the payoff[1] to a coalition as
given.

---

1. Alternatively, one might assign *costs* instead of payoffs to coalitions. Throughout this chapter, we will
focus on the case of payoffs; the concepts defined herein can be extended analogously to the case of costs.

12.1.1 *Definition*

For most of this chapter we will make the *transferable utility assumption*—that the payoffs to a coalition may be freely redistributed among its members. This assumption is satisfied whenever there is a universal *currency* that is used for exchange in the system. When this assumption holds, each coalition can be assigned a single value as its payoff.

coalitional game
with transferable
utility

Definition 12.1.1 (Coalitional game with transferable utility) *A* coalitional game with transferable utility *is a pair* (N, v), *where:*

characteristic
function

- N *is a finite*[2] *set of players, indexed by* i; *and*
- $v : 2^N \mapsto \mathbb{R}$ *associates with each coalition* $S \subseteq N$ *a real-valued payoff* $v(S)$ *that the coalition's members can distribute among themselves. The function* v *is also called the* characteristic function, *and a coalition's payoff is also called its* worth. *We assume that* $v(\emptyset) = 0$.

Most of the time, coalitional game theory is used to answer two fundamental questions:

1. Which coalition will form?
2. How should that coalition divide its payoff among its members?

It turns out that the answer to (1) is often "the grand coalition"—the name given to the coalition of all the agents in N—though this answer can depend on having made the right choice about (2). Before we go any further in answering these questions, however, we provide several examples of coalitional games to help motivate the model.

12.1.2 *Examples*

Coalitional games can be used to describe problems arising in a wide variety of different contexts. To emphasize the relevance of coalitional game theory to other topics covered in this book, we give examples motivated by problems from social choice (Chapter 9), mechanism design (Chapter 10), and auctions (Chapter 11). We will also use these examples to highlight some important classes of coalitional games in Section 12.1.3. We note that here we do *not* describe how payments could or should be divided among the agents; we defer such discussion to Section 12.2. Our first example draws on social choice, in the vein of the discussion in Section 9.3.

Example 12.1.2 (Voting game) *A parliament is made up of four political parties, A, B, C, and D, which have 45, 25, 15, and 15 representatives, respectively. They are to vote on whether to pass a $100 million spending bill and how much of this amount should be controlled by each of the parties. A majority vote, that is, a minimum of 51 votes, is required in order to pass any legislation, and if the bill does not pass then every party gets zero to spend.*

---

2. Observe that we consider only finite coalitional games. The infinite case is also considered in the literature; many but not all of the results from this chapter also hold in this case.

More generally, in a voting game, there is a set of agents N and a set of coalitions $\mathcal{W} \subseteq 2^N$ that are winning *coalitions, that is, coalitions that are sufficient for the passage of the bill if all its members choose to do so. To each coalition $S \in \mathcal{W}$, we assign $v(S) = 1$, and to the others we assign $v(S) = 0$.*

Many voting games that arise in practice can be represented succinctly as *weighted majority* or *weighted voting* games. We discuss these representations in Section 12.3.1.

Our next example concerns sharing the cost of a public good, along the lines of the road-building referendum example given in Section 10.4.

Example 12.1.3 (Airport game) *A number of cities need airport capacity. If a new regional airport is built the cities will have to share its cost, which will depend on the largest aircraft that the runway can accommodate. Otherwise each city will have to build its own airport.*

This situation can be modeled as a coalitional game (N, v), where N is the set of cities, and $v(S)$ is the sum of the costs of building runways for each city in S minus the cost of the largest runway required by any city in S.

Next, we consider a situation in which agents need to get *connected* to the public good in order to enjoy its benefit. One such setting is the problem of multicast cost sharing that we previously examined in Section 10.6.3.

Example 12.1.4 (Minimum spanning tree game) *A group of customers must be connected to a critical service provided by some central facility, such as a power plant or an emergency switchboard. In order to be served, a customer must either be directly connected to the facility or be connected to some other connected customer. Let us model the customers and the facility as nodes on a graph, and the possible connections as edges with associated costs.*

This situation can be modeled as a coalitional game (N, v). N is the set of customers, and $v(S)$ is the cost of connecting all customers in S directly to the facility minus the cost of the minimum spanning tree that spans both the customers in S and the facility.

Finally, we consider a coalitional game in an auction setting.

Example 12.1.5 (Auction game) *Consider an auction mechanism in which the allocation rule is efficient (i.e., social welfare maximizing). Our analysis in Chapter 11 treated the set of participating agents as given. We might instead want to determine whether the seller would prefer to exclude some interested agents to obtain higher payments. (Indeed, it turns out that this can occur; see Section 12.2.2.) To find out, we can model the auction as a coalitional game.*

Let N_B be the set of bidders, and let 0 be the seller. The agents in the coalitional game are $N = N_B \cup \{0\}$. Choosing a coalition means running the auction with the appropriate set of agents. The value of a coalition S is the sum of agents' utilities for the efficient allocation when the set of participating agents is restricted to S.[3] A coalition that does not include the seller has value 0, because in this case a trade cannot occur.

---

3. The value of a coalition can be understood as the sum of the agents' utilities for the auction's outcome (their valuations for bundles received minus payments) plus the seller's utility (the sum of payments

12.1.3 *Classes of coalitional games*

In this section we will define a few important classes of coalitional games, which have interesting applications as well as useful formal properties. We start with the notion of superadditivity, a property often assumed for coalitional games.

superadditive
game

Definition 12.1.6 (Superadditive game) *A game* $G = (N, v)$ *is* superadditive *if for all* $S, T \subset N$, *if* $S \cap T = \emptyset$, *then* $v(S \cup T) \geq v(S) + v(T)$.

Superadditivity is justified when coalitions can always work without interfering with one another; hence, the value of two coalitions will be no less than the sum of their individual values. Note that superadditivity implies that the value of the entire set of players (the "grand coalition") is no less than the sum of the value of any nonoverlapping set of coalitions. In other words, the grand coalition has the highest payoff among all coalitional structures. All of the examples we gave earlier describe superadditive games.

Taking noninterference across coalitions to the extreme, when coalitions can never affect one another, either positively or negatively, then we have *additive* (or *inessential*) games.

additive game

Definition 12.1.7 (Additive game) *A game* $G = (N, v)$ *is* additive *(or* inessential*) if for all* $S, T \subset N$, *if* $S \cap T = \emptyset$, *then* $v(S \cup T) = v(S) + v(T)$.

A related class of games is that of constant-sum games.

constant-sum
game

Definition 12.1.8 (Constant-sum game) *A game* $G = (N, v)$ *is* constant sum *if for all* $S \subset N$, $v(S) + v(N \setminus S) = v(N)$.

Note that every additive game is necessarily constant sum, but not vice versa. As in noncooperative game theory, the most commonly studied constant-sum games zero-sum game   are *zero-sum games*.

An important subclass of superadditive games are convex games.

convex game

Definition 12.1.9 (Convex game) *A game* $G = (N, v)$ *is* convex *if for all* $S, T \subset N$, $v(S \cup T) \geq v(S) + v(T) - v(S \cap T)$.

Clearly, convexity is a stronger condition than superadditivity. While convex games may therefore appear to be a very specialized class of coalitional games, these games are actually not so rare in practice. For example, the Airport game as described in Example 12.1.3 is convex. Convex games have a number of useful properties, as we will discuss in the next section.

Finally, we present a class of coalitional games with restrictions on the values that payoffs are allowed to take.

simple game

Definition 12.1.10 (Simple game) *A game* $G = (N, v)$ *is* simple *if for all* $S \subset N$, $v(S) \in \{0, 1\}$.

---

received). Note that because payments are transfers between members of the coalition they cancel out and do not affect the coalition's value.

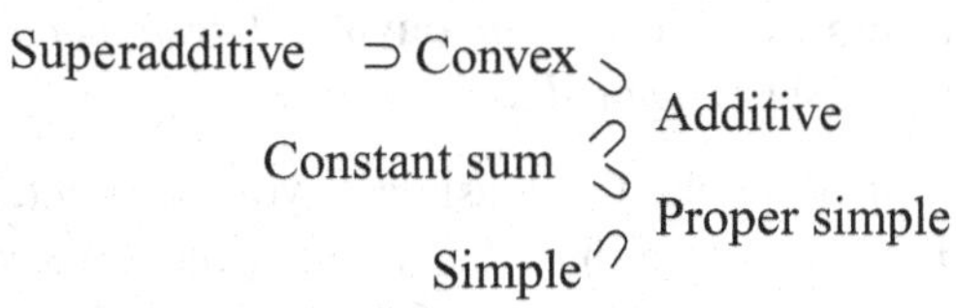

Superaddifive ⊃ Convex
Additive
Constant sum
Proper simple
Simple

Figure 12.1 A hierarchy of coalitional game classes; $X \supset Y$ means that class X is a superclass of class Y.

Simple games are useful for modeling voting situations, such as those described in Example 12.1.2. In simple games we often add the requirement that if a coalition wins, then all larger sets are also winning coalitions (i.e., if $v(S) = 1$, then for all $T \supset S$, $v(T) = 1$). This condition might seem to imply superadditivity, but it does not quite. For example, the condition is met by a voting game in which only 50% of the votes are required to pass a bill, but such a game is not superadditive. Consider two disjoint winning coalitions S and T; when they join to form the coalition $S \cup T$ they do not achieve at least the sum of the values that they achieve separately as superadditivity requires.

proper simple game   When simple games are also constant sum, they are called *proper simple games*. In this case, if S is a winning coalition, then $N \setminus S$ is a losing coalition.

Figure 12.1 graphically depicts the relationship between the different classes of games that we have discussed in this section.

12.2 Analyzing coalitional games

The central question in coalitional game theory is the division of the payoff to the grand coalition among the agents. This focus on the grand coalition is justified in two ways. First, since many of the most widely studied games are superadditive, the grand coalition will be the coalition that achieves the highest payoff over all coalitional structures, and hence we can expect it to form. Second, there may be no choice for the agents but to form the grand coalition; for example, public projects are often legally bound to include all participants.

If it is easy to decide to concentrate on the grand coalition, however, it is less easy to decide how this coalition should divide its payoffs. In this section we explore a variety of solution concepts that propose different ways of performing this division.

Before presenting the solution concepts, it is helpful to introduce some terminology. First, let $\psi : \mathbb{N} \times \mathbb{R}^{2^{|N|}} \mapsto \mathbb{R}^{|N|}$ denote a mapping from a coalitional game (that is, a set of agents N and a value function v) to a vector of $|N|$ real values, and let $\psi_i(N, v)$ denote the ith such real value. Denote such a vector of $|N|$ real values as $x \in \mathbb{R}^{|N|}$. Each x_i denotes the share of the grand coalition's payoff that agent $i \in N$ receives. When the coalitional game (N, v) is understood from context, we write x as a shorthand for $\psi(N, v)$.

Now we can give some basic definitions about payoff division. Each has an analogue in the properties we required of quasilinear mechanisms in Section 10.3.2, which we name as we come to each definition.

feasible payoff

Definition 12.2.1 (Feasible payoff) *Given a coalitional game* (N, v), *the* feasible payoff set *is defined as* $\{x \in \mathbb{R}^N \mid \sum_{i \in N} x_i \leq v(N)\}$.

In other words, the feasible payoff set contains all payoff vectors that do not distribute more than the worth of the grand coalition. We can view this as requiring the payoffs to be *weakly budget balanced*.

pre-imputation

Definition 12.2.2 (Pre-imputation) *Given a coalitional game* (N, v), *the* pre-imputation set, *denoted* $\mathcal{P}$, *is defined as* $\{x \in \mathbb{R}^N \mid \sum_{i \in N} x_i = v(N)\}$.

We can view the pre-imputation set as the set of feasible payoffs that are *efficient*, that is, they distribute the entire worth of the grand coalition. Looked at another way, the pre-imputation set is the set of feasible payoffs that are *strictly budget balanced*. (In this setting these two concepts are equivalent; do you see why?)

imputation

Definition 12.2.3 (Imputation) *Given a coalitional game* (N, v), *the* imputation set, $\mathcal{I}$, *is defined as* $\{x \in \mathcal{P} \mid \forall i \in N, x_i \geq v(i)\}$.

Imputations are payoff vectors that are not only efficient but *individually rational*. That is, each agent is guaranteed a payoff of at least the amount that he could achieve by forming a singleton coalition.

Now we are ready to delve into different solution concepts for coalitional games.

12.2.1   *The Shapley value*

Perhaps the most straightforward answer to the question of how payoffs should be divided is that the division should be *fair*. As we did in Section 9.4.1, let us begin by laying down axioms that describe what fairness means in our context.

interchangeable
agents

First, say that agents i and j are *interchangeable* if they always contribute the same amount to every coalition of the other agents. That is, for all S that contains neither i nor j, $v(S \cup \{i\}) = v(S \cup \{j\})$. The *symmetry* axiom states that such agents should receive the same payments.

Axiom 12.2.4 (Symmetry) *For any* v, *if* i *and* j *are interchangeable then* $\psi_i(N, v) = \psi_j(N, v)$.

dummy player

Second, say that an agent i is a *dummy player* if the amount that i contributes to any coalition is exactly the amount that i is able to achieve alone. That is, for all S such that $i \notin S$, $v(S \cup \{i\}) - v(S) = v(\{i\})$. The *dummy player* axiom states that dummy players should receive a payment equal to exactly the amount that they achieve on their own.

Axiom 12.2.5 (Dummy player) *For any* v, *if* i *is a dummy player then* $\psi_i(N, v) = v(\{i\})$.

Finally, consider two different coalitional game theory problems, defined by two different characteristic functions v_1 and v_2, involving the same set of agents. The *additivity* axiom states that if we re-model the setting as a single game in which each coalition S achieves a payoff of $v_1(S) + v_2(S)$, the agents' payments in each coalition should be the sum of the payments they would have achieved for that coalition under the two separate games.

Axiom 12.2.6 (Additivity) *For any two v_1 and v_2, we have for any player i that $\psi_i(N, v_1 + v_2) = \psi_i(N, v_1) + \psi_i(N, v_2)$, where the game $(N, v_1 + v_2)$ is defined by $(v_1 + v_2)(S) = v_1(S) + v_2(S)$ for every coalition S.*

If we accept these three axioms, we are led to a strong result: there is always exactly one pre-imputation that satisfies them.

Theorem 12.2.7 *Given a coalitional game (N, v), there is a unique pre-imputation $\phi(N, v) = \phi(N, v)$ that satisfies the Symmetry, Dummy player, Additivity axioms.*

Note that our requirement that $\phi(N, v)$ be a pre-imputation implies that the payoff division be feasible and efficient (or strictly budget balanced). Because we do not insist on an imputation, individual rationality is not required to hold, though of course it still may.

What is this unique payoff division $\phi(N, v)$? It is called the *Shapley value*, and it is defined as follows.

Shapley value

Definition 12.2.8 (Shapley value) *Given a coalitional game (N, v), the* Shapley value *of player i is given by*

$$\phi_i(N, v) = \frac{1}{|N|!} \sum_{S \subseteq N \setminus \{i\}} |S|!(|N| - |S| - 1)! \Big[v(S \cup \{i\}) - v(S) \Big].$$

This expression can be viewed as capturing the "average marginal contribution" of agent i, where we average over all the different sequences according to which the grand coalition could be built up from the empty coalition. More specifically, imagine that the coalition is assembled by starting with the empty set and adding one agent at a time, with the agent to be added chosen uniformly at random. Within any such sequence of additions, look at agent i's marginal contribution at the time he is added. If he is added to the set S, his contribution is $[v(S \cup \{i\}) - v(S)]$. Now multiply this quantity by the $|S|!$ different ways the set S could have been formed prior to agent i's addition and by the $(|N| - |S| - 1)!$ different ways the remaining agents could be added afterward. Finally, sum over all possible sets S and obtain an average by dividing by $|N|!$, the number of possible orderings of all the agents.

For a concrete example of the Shapley value in action, consider the voting game given in Example 12.1.2. Recall that the four political parties A, B, C, and D have 45, 25, 15, and 15 representatives, respectively, and a simple majority (51 votes) is required to pass the $100 million spending bill. If we want to analyze how much money it is fair for each party to demand, we can calculate the Shapley values of the game. Note that every coalition with 51 or more members has a value of $100 million,[4] and others have $0. In this game, therefore, the parties

---

4. Notice that for these calculations we scale the value function to 100 for winning coalitions and 0 for losing coalitions in order to make it align more tightly with our example.

B, C, and D are interchangeable, since they add the same value to any coalition. (They add \$100 million to the coalitions $\{B, C\}$, $\{C, D\}$, $\{B, D\}$ that do not include them already and to $\{A\}$; they add \$0 to all other coalitions.) The Shapley value of A is given by:

$$\phi_A = \frac{1}{4!}[3!0!(100 - 100) + 3 \cdot 2!1!(100 - 0) + 3 \cdot 1!2!(100 - 0)$$
$$+ 0!3!(0 - 0)]$$
$$= \frac{1}{24}[0 + 600 + 600 + 0] = \$50 \text{ million.}$$

The Shapley value for B (and, by symmetry, also for C and D) is given by:

$$\phi_B = \frac{1}{4!}[3!0!(100 - 100) + 2 \cdot 2!1!(100 - 100) + 2!1!(100 - 0)$$
$$+ 1!2!(100 - 0) + 2 \cdot 1!2!(0 - 0) + 0!3!(0 - 0)]$$
$$= \frac{1}{24}[0 + 0 + 200 + 200 + 0 + 0] = \$16.67 \text{ million.}$$

Thus the Shapley values are $(50, 16.67, 16.67, 16.67)$, which add up to the entire \$100 million.

To continue with an example mentioned earlier, in Section 10.6.3 we discussed the *Shapley mechanism* for sharing the cost of multicast transmissions. Now that we have learned about the Shapley value—what is the connection? It turns out to depend on a probabilistic interpretation of the Shapley value. Suppose that the agents to be served arrive in a random order in the fixed multicast tree, and that each agent is responsible for the cost of the remaining edges needed to be built for him to get connected. The Shapley mechanism charges the agents their expected connection costs in such a model, averaging over all orders chosen uniformly at random.

12.2.2   *The core*

The Shapley value defined a fair way of dividing the grand coalition's payment among its members. However, this analysis ignored questions of stability. We can also ask: would the agents be *willing* to form the grand coalition given the way it will divide payments, or would some of them prefer to form smaller coalitions? Unfortunately, sometimes smaller coalitions can be more attractive for subsets of the agents, even if they lead to lower value overall. Considering the majority voting example, while A does not have a unilateral motivation to vote for a different split, A and B have incentive to defect and divide the \$100 million between themselves (e.g., dividing it $(75, 25)$).

This leads to the question of what payment divisions would make the agents want to form the grand coalition. The answer is that they would want to do so if and only if the payment profile is drawn from a set called the *core*, defined as follows.

core   **Definition 12.2.9 (Core)** *A payoff vector x is in the* core *of a coalitional game* (N, v) *if and only if*

$$\forall S \subseteq N, \sum_{i \in S} x_i \geq v(S).$$

Thus, a payoff is in the core if and only if no sub-coalition has an incentive to break away from the grand coalition and share the payoff it is able to obtain independently. That is, it requires that the sum of payoffs to any group of agents $S \subseteq N$ must be at least as large as the amount that these agents could share among themselves if they formed a coalition on their own. Notice that Definition 12.2.9 implies that payoff vectors in the core must always be imputations: that is, they must always be strictly budget balanced and individually rational.

Since the core provides a concept of stability for coalitional games we can see it as an analog of the Nash equilibrium from noncooperative games. However, it is actually a stronger notion: Nash equilibrium describes stability only with respect to deviation by a single agent. Instead, the core is an analog of the concept of *strong equilibrium* (discussed in Section 10.7.3), which requires stability with respect to deviations by arbitrary coalitions of agents.

How can the core be computed? The answer is conceptually straightforward and is given by the linear feasibility problem[5] that follows.

$$\sum_{i \in S} x_i \geq v(S) \qquad \forall S \subseteq N \tag{12.1}$$

As a notion of stability for coalitional games, the core is appealing. However, the alert reader might have two lingering doubts, arising due to its implicit definition through inequalities:

1. Is the core always nonempty?
2. Is the core always unique?

Unfortunately, the answer to both questions is no. Let us consider again the Parliament example with the four political parties. The set of minimal coalitions that meet the required 51 votes is $\{A, B\}$, $\{A, C\}$, $\{A, D\}$, and $\{B, C, D\}$. We can see that if the sum of the payoffs to parties B, C, and D is less than $100 million, then this set of agents has incentive to deviate. On the other hand, if B, C, and D get the entire payoff of $100 million, then A will receive $0 and will have incentive to form a coalition with whichever of B, C, and D obtained the smallest payoff. Thus, the core is empty for this game.

On the other hand, when the core is nonempty it may not define a unique payoff vector either. Consider changing our example so that instead of a simple majority, an 80% majority is required for the bill to pass. The minimal winning coalitions are now $\{A, B, C\}$ and $\{A, B, D\}$. Any complete distribution of the $100 million among parties A and B now belongs to the core, since all winning coalitions must have both the support of these two parties.

---

5. Linear feasibility problems are linear programs with only constraints but no objective function. Linear programs are defined in Appendix B.

These examples call into question the universality of the core as a solution concept for coalitional games. We already saw in the context of noncooperative game theory that solution concepts—notably, the Nash equilibrium—do not yield unique solutions in general. Here we are in an arguably worse situation, in that the solution concept may yield no solution at all.

Can we characterize when a coalitional game has a nonempty core? Fortunately, that at least is possible. To do so, we first need to define a concept known as *balancedness*.

Definition 12.2.10 (Balanced weights) *A set of nonnegative weights (over 2^N),*

balanced set λ, *is balanced if*

$$\forall i \in N, \sum_{S:i \in S} \lambda(S) = 1.$$

Intuitively, the weights on the coalitions involving any given agent i can be interpreted as the conditional probabilities that these coalitions will form, given that i will belong to a coalition.

Theorem 12.2.11 (Bondereva–Shapley) *A coalitional game (N, v) has a nonempty core if and only if for all balanced sets of weights λ,*

$$v(N) \geq \sum_{S \subseteq N} \lambda(S)v(S). \tag{12.2}$$

Proof. Consider the linear feasibility problem used to compute the core, given in Equation (12.1). We can construct the following linear program:

$$\text{minimize} \quad \sum_{i \in N} x_i$$

$$\text{subject to} \quad \sum_{i \in S} x_i \geq v(S) \qquad \forall S \subseteq N$$

Note that when the value of this program is no bigger than $v(N)$, then the payoff vector x is feasible, and belongs to the core. Indeed, the value of the program is equal to $v(N)$ if and only if the core is nonempty. Now consider this linear program's dual.

$$\text{maximize} \quad \sum_{S \subseteq N} \lambda(S)v(S)$$

$$\text{subject to} \quad \sum_{S \subseteq N} \lambda(S) = 1 \qquad \forall i \in N$$

$$\lambda(S) \geq 0 \qquad \forall S \subseteq N$$

Note that the linear constraints in the dual ensure that λ is balanced. By weak duality, the optimal value of the dual is at most the optimal value of the primal, and hence the core is nonempty if and only if the optimal value of the dual is no greater than $v(N)$. ∎

While the Bondereva–Shapley theorem completely characterizes when a coalitional game has a nonempty core, it is not always easy or feasible to check that a game satisfies Equation (12.2) for all balanced sets of weights. Luckily, there exist several results that allow us to predict the emptiness or nonemptiness of the core based on a coalitional game's membership in one of the classes we defined in Section 12.1.3.

Theorem 12.2.12 *Every constant-sum game that is not additive has an empty core.*

veto player   We say that a player i is a *veto player* if $v(N \setminus \{i\}) = 0$.

Theorem 12.2.13 *In a simple game the core is empty iff there is no veto player. If there are veto players, the core consists of all payoff vectors in which the nonveto players get zero.*

Theorem 12.2.14 *Every convex game has a nonempty core.*

A final question we consider regards the relationship between the core and the Shapley value. We know that the core may be empty, but if it is not, is the Shapley value guaranteed to lie in the core? The answer in general is no, but the following theorem gives us a sufficient condition for this property to hold. We already know from Theorem 12.2.14 that the core of convex games is nonempty. The following theorem further tells us that for such games the Shapley value belongs to that set.

Theorem 12.2.15 *In every convex game, the Shapley value is in the core.*

We now consider an application of the core to our Auction game (Example 12.1.5). Earlier we asked whether any coalition (consisting of bidders and the seller) could do better than the payoffs they receive when everyone participates in the mechanism. Now that we have defined the core, we can see that the question can be rephrased as asking whether the seller's and the agents' payoffs from the auction are in the core.

First, let us consider the case of single-item, second-price auctions. If the bidders follow their weakly dominant strategy of truthful reporting, it turns out that the payoffs are always in the core. This is because the seller receives revenue equal to or greater than the valuations of all the losing bidders (specifically, equal to the second-highest valuation), and hence cannot entice any of the losing bidders to pay him more.

Now let us consider the case of the VCG mechanism applied to combinatorial auctions. Interestingly, though this mechanism generalizes the second-price auction discussed earlier, it does *not* guarantee payoffs from the core. For example, consider an auction with three bidders and two goods x and y, with the following valuations.

Bidder 1	Bidder 2	Bidder 3
$v_1(x, y) = 90$	$v_2(x) = v_2(x, y) = 100$	$v_3(y) = v_3(x, y) = 100$
$v_1(x) = v_1(y) = 0$	$v_2(y) = 0$	$v_3(x) = 0$

The efficient allocation awards x to bidder 2 and y to bidder 3. Neither bidder is pivotal, so both pay 0. However, both bidder 1 and the seller would benefit from forming a coalition in which bidder 1 wins the bundle x, y and pays any amount $0 < p_1 < 90$. Thus in a combinatorial auction the VCG payoffs are not guaranteed to belong to the core.

12.2.3   *Refining the core: ϵ-core, least core, and nucleolus*

We now consider some refinements of the core that address its possible nonexistence and nonuniqueness.

We first define a concept analogous to ϵ-equilibrium (defined in Section 3.4.7).

ϵ-core   **Definition 12.2.16 (ϵ-core)** *A payoff vector x is in the ϵ-core of a coalitional game (N, v) if and only if*

$$\forall S \subset N, \ \sum_{i \in S} x_i \geq v(S) - \epsilon. \tag{12.3}$$

One interpretation of the ϵ in Equation (12.3) is that there is an ϵ cost for deviating from the grand coalition. As a result, even if the payoffs to agents in coalition S are less than their worth, $v(S)$, as long as the difference is less than ϵ, the payoff vector is still stable.

Mathematically speaking, there is no requirement that ϵ has to be nonnegative; when ϵ is negative, $-\epsilon$ can be seen as a "bonus" for forming a new coalition. Thus, when ϵ is negative, a payoff vector that is in the ϵ-core is *more* stable than a vector that is only in the core. Note that in Equation (12.3), constraints are quantified only over coalitions that are strict subsets of N. Since payoff vectors are efficient, $\sum_{i \in N} x_i = v(N)$ is always true. Thus, adding a constraint for N is unnecessary: it would be trivially satisfied when ϵ is nonnegative and violated when ϵ is negative, and would shed no light on whether the grand coalition would form.

Note that just like the core, for a given ϵ, the ϵ-core of a game may be empty. On the other hand, it is easy to see that given a game, there always exists some ϵ that is sufficiently large to ensure that the ϵ-core of the game is nonempty. A natural problem, therefore, is to find the smallest ϵ for which the ϵ-core of a game is nonempty. This leads to a solution concept called the *least core*.

least core   **Definition 12.2.17 (Least core)** *A payoff vector x is in the* least core *of a coalitional game (N, v) if and only if x is the solution to the following linear program.*

$$\text{minimize} \ \ \epsilon$$
$$\text{subject to} \ \ \sum_{i \in S} x_i \geq v(S) - \epsilon \qquad\qquad \forall S \subset N$$

The objective function in the linear program given in Definition 12.2.17 is nonpositive if and only if the core of the game is nonempty. As explained, for sufficiently large ϵ, the constraints in the linear program can always be satisfied; hence, the least core of a game is never empty. Thus the least core can be considered a generalization of the core. On the other hand, the least core does not contain every payoff vector in the core when the core is nonempty; rather, it consists only of payoff vectors that will give all coalitions as little incentive to deviate as possible. In this sense, the least core is also a refinement of the core.

Although it refines the core, the least core does not uniquely determine a payoff vector to a game: it can still return a set of payoff vectors. The intuition is that beyond those coalitions for which the constraints in the linear program are tight (i.e., are realized as equality), there are extra degrees of freedom available for distributing the payoffs to agents in the other coalitions. Based on this intuition, it is easy to construct counterexamples to the uniqueness of the least core. (Can you find one?)

It seems that we could further strengthen the least core by requiring that coalitions whose constraints are slack (i.e., not tight) in the linear program must be as stable as possible. To formalize this idea, let ϵ_1 be the objective value of the linear program, and let $\mathcal{S}_1$ be the set of coalitions corresponding to the set of constraints that are tight in the optimal solution. We now optimize.

$$
\begin{aligned}
\text{minimize} \quad & \epsilon \\
\text{subject to} \quad & \sum_{i \in S} x_i = v(S) - \epsilon_1 && \forall S \in \mathcal{S}_1 \\
& \sum_{i \in S} x_i \geq v(S) - \epsilon && \forall S \in 2^N \setminus \mathcal{S}_1
\end{aligned}
$$

But what if some constraints remain slack even in the solution to the new optimization problem? We can simply solve yet another optimization problem to make the remaining coalitions as stable as possible. Repeating this procedure, the payoff vector gets progressively tightened. Since at each step, at least one more constraint will be made tight, this process must terminate. Indeed, a careful argument that counts the number of dimensions shows that in fact the process must terminate after at most $|N|$ steps. At the end of this process, we reach a unique payoff vector, known as the *nucleolus*.

nucleolus **Definition 12.2.18 (Nucleolus)** *A payoff vector x is in the nucleolus of a coalitional game (N, v) if it is the solution to the series of optimization programs $O_1, O_2, \ldots, O_{|N|}$, where these programs are defined recursively as follows.*

$$
(O_1) \quad \left\{
\begin{aligned}
\text{minimize} \quad & \epsilon \\
\text{subject to} \quad & \sum_{i \in S} x_i \geq v(S) - \epsilon \quad \forall S \subset N
\end{aligned}
\right.
$$

$$
(O_i) \quad \left\{
\begin{aligned}
\text{minimize} \quad & \epsilon \\
\text{subject to} \quad & \sum_{i \in S} x_i = v(S) - \epsilon_0 && \forall S \in \mathcal{S}_1 \\
& \quad\quad \vdots \\
& \sum_{i \in S} x_i = v(S) - \epsilon_{i-1} && \forall S \in \mathcal{S}_{i-1} \setminus \mathcal{S}_{i-2} \\
& \sum_{i \in S} x_i \geq v(S) - \epsilon && \forall S \in 2^N \setminus \mathcal{S}_{i-1}
\end{aligned}
\right.
$$

ϵ_{i-1} is the optimal objective value to program O_{i-1} and S_{i-1} is the set of coalitions for which in the optimal solution to O_{i-1}, the constraints are realized as equalities.[6]

Unlike the core, the ϵ-core, and the least core, the nucleolus possesses the desirable property that it is unique, regardless of the game.

Theorem 12.2.19 *For any coalitional game (N, v), the nucleolus of the game always exists and is unique.*

Proof Sketch. For existence of the nucleolus, one can solve the series of optimization programs as defined in Definition 12.2.18 and end up with some assignment of values to the variables x that corresponds to some payoff vector. These programs are linear programs with finitely many constraints; hence, they can always be solved.

For uniqueness, first observe that the earlier optimization problems can only influence the later ones through the values. Therefore, the set of programs will always yield the same set of solutions $\{\epsilon_1, \ldots, \epsilon_{|N|}\}$. After $|N|$ optimization, we are left with a system of $2^{|N|}$ equations over $|N|$ variables. This system is of rank at most $|N|$, and therefore if a solution exists, it must be unique. ∎

The nucleolus has an alternate definition in terms of *excess*, which we define next. This definition is worth understanding because it provides additional intuition about the meaning of the nucleolus.

excess of a coalition

Definition 12.2.20 (Excess of a coalition) *The* excess of a coalition S in game (N, v) with respect to a payoff vector x, $e(S, x, v)$, is defined as $v(S) - \sum_{i \in S} x_i$, that is, the amount a coalition gain by deviating, as compared to that coalition's payoff as part of the grand coalition.

Given a coalitional game (N, v) and a payoff vector x, compute the excesses of all coalitions except coalition N and $\emptyset$; we call this $(2^{|N|} - 2)$-dimensional vector the *raw excess vector*. When this vector is sorted in decreasing order of excess, we call it the *sorted excess vector* and denote it as $ev(x, v)$.

Given two payoff vectors x and y, we say the excesses due to x are lexicographically smaller than those due to y, written $x \prec_{e(v)} y$, if for the smallest index such that $ev(x, v)$ and $ev(y, v)$ differ, $ev(x, v) < ev(y, v)$. The nucleolus can then be defined as the payoff vector that is smallest according to the $\prec_{e(v)}$ relation.

Definition 12.2.21 (Nucleolus, alternate definition) *Given a coalitional game (N, v), the* nucleolus *is the payoff vector x such that for all other payoff vectors y, $y \succ_{e(v)} x$, that is, x lexicographically minimizes the excesses of all coalitions except N and $\emptyset$.*

---

6. We have to be careful and terminate before the $|N|$-th program if $S_i = 2^N$ for some $i < |N|$.

12.3   Compact representations of coalitional games

Our focus so far has been on analyzing coalitional games. Now that we have some solution concepts under our belts, a natural question is whether and how we can efficiently compute these solution concepts. However, we immediately run into the problem that a straightforward representation of coalitional games by enumeration requires space exponential in the number of agents. This has the odd side-effect that simple brute-force approaches appear to have "good" (i.e., low-order polynomial) complexity, since complexity measures express the amount of time it will take to solve a problem as a function of the input size. Nevertheless, applying these algorithms to the straightforward representation will only allow us to compute solution concepts for problems with very few agents.

In order to ask more interesting questions we must first find a game representation that is more compact. In general one cannot compress the representation, but many natural settings contain inherent symmetries and independencies that do give rise to more compact representations. Conceptually, this section mirrors the discussion in Section 6.5, which considered compact representations of noncooperative games.

In the following, we discuss a number of compactly-represented coalitional games. These games can be roughly categorized into two types. First, we look at games that arise from specific applications, such as weighted voting games and weighted graph games. We then look at "languages" designed to represent coalitional games. This includes the synergy representation for superadditive games and the multi-issue and marginal contribution nets representations for general games.

12.3.1   *Weighted majority games and weighted voting games*

The weighted majority game representation is a compact way of encoding voting situations. Its definition is straightforward.

weighted
majority game

Definition 12.3.1 (Weighted majority game) *A* weighted majority game *is defined by weights w_i assigned to each player $i \in N$. Let W be $\sum_{i \in N} w(i)$. The value of a coalition is 1 if $\sum_{i \in S} w(i) > \frac{W}{2}$ and 0 otherwise.*

Since this game is simple (in the sense of Definition 12.1.10), testing the nonemptiness of the core is equivalent to testing the existence of a veto player, which can be done quickly. However, it is not so easy to compute the Shapley value.

Theorem 12.3.2 *Computing the Shapley value in weighted majority games is #P-complete.*[7]

This can be proved by a reduction from the counting version of KNAPSACK.

weighted voting
games

Weighted voting games are natural generalization of weighted majority games.

---

7. Recall that #P consists of the counting versions of the deterministic polynomial-time decision problems; for example, not simply deciding whether a Boolean formula is satisfiable, but rather counting the number of truth assignments that satisfy the formula.

Instead of stipulating that all coalitions with more than half the votes win, an explicit minimum number of votes, known as the *threshold*, is specified. This representation can be used to represent voting situations in which the number of votes required for the selection of a candidate is not a simple majority.

12.3.2 *Weighted graph games*

A weighted graph game (WGG) is a coalitional game defined by an undirected weighted graph (i.e., a set of nodes and a real-valued weight associated with every unordered pair of nodes). Intuitively, the nodes in the graph represent the players, and the value of a coalition is obtained by summing the weights of the edges that connect pairs of vertices corresponding to members of the coalition. WGGs thus explicitly model pairwise synergies among players and assume that all such synergies increase the coalition's value additively. In exchange for this reduced expressiveness we gain a much more compact representation: a game with n agents is represented by only $\frac{n(n-1)}{2}$ weights.

Definition 12.3.3 (Weighted graph game) *Let (V, W) denote an undirected weighted graph, where V is the set of vertices and $W \in \mathbb{R}^{V \times V}$ is the set of edge weights; denote the weight of the edge between the vertices i and j as*
weighted graph   *$w(i, j)$. This graph defines a* weighted graph game *(WGG), where the coalitional*
game   *game is constructed as follows:*

- $N = V$;
- $v(S) = \sum_{i,j \in S} w(i, j)$.

An example that WGGs model well is the Revenue Sharing game.

Example 12.3.4 (Revenue Sharing game) *Consider the problem of dividing the revenues from toll highways between the cities that the highways connect. The pair of cities connected by a highway get to share in the revenues only when they form an agreement on revenue splitting; otherwise, the tolls go to the state. This problem can be represented as a weighted graph game (V, W), where the nodes V represent the cities, each edge represents a highway between a pair of cities, and the weight $w(i, j)$ of a given edge indicates that highway's toll revenues.*

The following is a direct consequence of the definitions.

Proposition 12.3.5 *If all the weights are nonnegative then the game is convex.*

Thus we know that in this case WGGs have a nonempty core and, furthermore, that the core contains the Shapley value. But the core may contain additional payoff vectors, and it is natural to ask whether testing membership is easy or hard. The answer is given by the following proposition.

Proposition 12.3.6 *If all the weights are nonnegative then membership of a payoff vector in the core can be tested in polynomial time.*

The proof is achieved by providing a maxflow-type algorithm.

The Shapley value is also easy to compute, even when we lift our restriction that the weights must be nonnegative.

Theorem 12.3.7 *The Shapley value of the coalitional game (N, v) induced by a weighted graph game (V, W) is*

$$\phi_i(N, v) = \frac{1}{2} \sum_{j \neq i} w(i, j).$$

Proof. Consider the contribution of the edge (i, j) to $\phi_i(N, v)$. For every subset S containing it, it contributes $\frac{(n-|S|)!(|S|-1)!}{n!} w(i, j)$. There are $\binom{n-2}{k-2}$ subsets of size k that contain both i and j. So all subsets of size k contribute $\binom{n-2}{k-2} \frac{(n-|S|)!(|S|-1)!}{n!} w(i, j) = \frac{(k-1)}{n(n-1)} w(i, j)$. And by summing over $k = 2, \ldots, n$ we obtain the result.   ∎

It follows that we can compute the Shapley value in $O(n^2)$ time.

Answering questions regarding the core of WGGs is more complex. Recall that a cut in a graph is a set of edges that divide the nodes into two disjoint sets, the weight of a cut is the sum of its weights, and a negative cut is a cut whose weight is negative. We begin by noting the following proposition.

Proposition 12.3.8 *The Shapley value is in the core of a weighted graph game if and only if there is no negative cut in the weighted graph.*

Proof. Note that while the value of a coalition S is the sum of the weights within S, the Shapley values in the same coalition is the same sum plus half the total weights of edges between S and $N \setminus S$. But the edges between S and $N \setminus S$ form a cut. Clearly, if that cut is negative, the Shapley value cannot be in the core, and since this holds for all sets S, the converse is also true.   ∎

And so we get as a consequence that if the weighted graph contains no negative cuts, the core cannot be empty. The next theorem turns this into a necessary and sufficient condition.

Theorem 12.3.9 *The core of a weighted graph game is nonempty if and only if there is no negative cut in the weighted graph.*

Proof. The if part follows from the preceding proposition.

For the only-if part, suppose we have a negative cut in the graph between S and $N \setminus S$. By virtue of being a cut, we have

$$\sum_{i \in S} \phi_i(N, v) - v(S) = \sum_{i \in (N \setminus S)} \phi_i(N, v) - v(N \setminus S) = \frac{\sum_{i \in S, j \in (N \setminus S)} w(i, j)}{2} < 0.$$

For any payoff vector x, we have

$$v(N) = \sum_{i \in N} x_i = \sum_{i \in S} x_i + \sum_{i \in (N \setminus S)} x_i$$

$$= \sum_{i \in N} \phi_i(N, v) = \sum_{i \in S} \phi_i(N, v) + \sum_{i \in (N \setminus S)} \phi_i(N, v).$$

Combining the two, and summing up,

$$\left(\sum_{i \in S} x_i - v(S)\right) + \left(\sum_{i \in (N \setminus S)} x_i - v(N \setminus S)\right) < 0.$$

Hence, either the first or the second term (possibly both) has to be negative. The payoff vecor x is not in the core. Since x is an arbitrary payoff vector, the core is empty.   ■

These theorems suggest that one test for nonemptiness of the core is to check whether the Shapley solution lies in the core. However, despite all these promising indications, testing membership in WGGs remains elusive in general.

Theorem 12.3.10 *Testing the nonemptiness of the core of a general WGG is NP-complete.*

The proof is based on a reduction from MAXCUT, a well-known NP-complete problem.

12.3.3   *Capturing synergies: a representation for superadditive games*

So far, we have looked at compact representations of coalitional games that can express only very restricted classes of games, but that are extremely compact for those classes. We now switch gears to consider compact representations that are designed with the intent to represent more diverse families of coalitional games.

The first one we will examine can be used to represent any superadditive game. As mentioned earlier in the chapter, superadditivity is sometimes assumed for coalitional games and is justified when coalitions do not exert negative externalities on each other.

synergy
representation

Definition 12.3.11 (Synergy representation) *The* synergy representation *of superadditive games is given by the pair (N, s), where N is the set of agents and s is a set function that maps each coalition to a value interpreted as the* synergy *the coalition generates when the agents of the coalition work together. Only coalitions with strictly positive synergies will be included in the specification of the game.*

The underlying coalitional game (N, v) under representation (N, s) is given by

$$v(S) = \left(\max_{\{S_1, S_2, \dots, S_k\} \in \pi(S)} \sum_{i=1}^{k} v(S_i)\right) + s(S),$$

where $\pi(S)$ denotes the set of all partitions of S.

Note that for some superadditive games, the representation may still require space exponential in the number of agents. This is unavoidable as the space of coalitional games of n agents, when treated as a vector space, is of $2^n - 1$ dimensions. However, for many games, the space required is much less.

We can evaluate the usefulness of a representation based on a number of criteria. One criterion is whether the representation exposes the structure of the underlying game to facilitate efficient computation. For example, is it easy to find out the value of a given coalition? Unfortunately, the answer is negative for the synergy representation.

Proposition 12.3.12 *It is NP-complete to determine the value of some coalitions for a coalitional game specified with the synergy representation. In particular, it is NP-complete to determine the value of the grand coalition.*

Intuitively, the reason why it is hard to determine the value of a coalition under the synergy representation is that we need to pick the best partitioning for a coalition, for which there is a number of choices exponential in the size of the coalition. As a result, it is (co)NP-hard to determine whether a given payoff vector is in the core, and it is NP-hard to compute the Shapley value of the game, as solution to either problem can be used to compute the value of the grand coalition. It is also (co)NP-hard to determine whether the core is empty or not, which follows from a reduction from Exact Cover by 3 Sets.

Interestingly, if the value of the grand coalition is given as part of the input of the problem, then the emptiness of the core can be determined efficiently.

Theorem 12.3.13 *Given a superadditive coalitional game specified with the synergy representation and the value of the grand coalition, we can determine in polynomial time whether the core of the game is empty or not.*

The proof is achieved by showing that a payoff in the core can be found by solving a linear program.

12.3.4   *A decomposition approach: multi-issue representation*

The central idea behind the multi-issue representation, a representation based on game decomposition, is that of *addition* of games (in the sense of Axiom 12.2.6 (Additivity) in the axiomatization of the Shapley value). Formally, the multi-issue representation is given as follows.

multi-issue representation **Definition 12.3.14 (Multi-issue representation)** *A multi-issue representation is composed of a collection of coalitional games, each known as an issue, (N_1, v_1), $(N_2, v_2), \ldots, (N_k, v_k)$, which together constitute the coalitional game (N, v), where:*

- *$N = N_1 \cup N_2 \cup \cdots \cup N_k$; and*
- *for each coalition $S \subseteq N$, $v(S) = \sum_{i=1}^{k} v_i(S \cap N_i)$.*

Intuitively, each issue of the game involves some set of agents, which may be partially or completely overlapping with the set of agents for another issue. The value of a coalition in the game is the sum of the values achieved by the coalition in each issue. For example, consider a robotics domain where there are certain tasks to be performed, each of which can be performed by a certain subset of a group of robots. We can then treat each of these tasks as an issue and model the system as a coalitional game where the value of any group of robots is the sum of the values of the tasks the group can perform.

Clearly, the multi-issue representation can be used to represent any coalitional game, as we can always choose to treat the coalitional game as a single big issue.

From the computational standpoint, due to its close relationship to the additivity axiom, it is perhaps not surprising that the Shapley value of a coalitional game specified in the multi-issue representation can be found efficiently.

Proposition 12.3.15 *The Shapley value of a coalitional game specified with the multi-issue representation can be computed in time linear in the size of the input.*

This is not hard to see. First, note that the Shapley value of a game can be computed in linear time when the input is given by the enumeration of the value function. This is because the direct approach of computing the Shapley value requires summing over each coalition once, and so the total number of operations is linear in the size of the enumeration. Observe that the factorials can be computed quickly to any desired accuracy using the Stirling approximation. Then, to prove the proposition, we must simply use the fact that the Shapley value satisfies the additivity axiom.

On the other hand, the multi-issue representation does not help with computational questions about the core. For example, it is coNP-hard to determine if a given payoff vector belongs to the core when the game is specified with the multi-issue representation.

12.3.5   *A logical approach: marginal contribution nets*

Marginal contribution nets (MC-nets) contstitute a representation scheme for coalitional games that attempts to harness the power of boolean logic to reduce the space required for specifying a coalitional game. The basic idea is to treat each agent in a game as a boolean variable and to treat the (binary) characteristic vector of a coalition as a truth assignment. This truth assignment can be used to evaluate whether a boolean formula is satisfied, which can in turn be used to determine the value of a coalition.

marginal
contribution net
(MC-net)

Definition 12.3.16 (MC-net representation) *An* MC-net *consists of a set of rules. Each rule has the syntactic form* (Pattern, weight), *where the pattern is given by a boolean formula and the weight is a real value.*

The MC-net $(p_1, w_1), (p_2, w_2), \ldots, (p_k, w_k)$ *specifies a coalitional game* (N, v), *where* N *is the set of propositions that appear in the patterns and the value function is given by*

$$v(S) = \sum_{i=1}^{k} p_i(e^S) w_i,$$

where $p_i(e^S)$ *evaluates to 1 if the boolean formula* p_i *evaluates to true for the truth assignment* e^S *and 0 otherwise.*

As an example, consider an MC-net with two rules: $(a \wedge b, 5), (b, 2)$. The coalitional game represented has two agents, a and b, and the following value function.

$$v(\emptyset) = 0 \qquad\qquad v(\{a\}) = 0$$
$$v(\{b\}) = 2 \qquad\qquad v(\{a, b\}) = 5 + 2 = 7$$

An alternative interpretation of MC-nets is a graphical representation. We can treat the agents as nodes on a graph, and for each pattern, a clique is drawn on the graph for the agents that appear in the same pattern. The weight of a rule is the weight associated with the corresponding clique.

A natural question for a representation language is whether there is a limit on the class of objects it can represent.

Proposition 12.3.17 *MC-nets can represent any game when negative literals are allowed in the patterns or when the weights can be negative. When the patterns are limited to conjunctive formula over positive literals and the weights are nonnegative, MC-nets can represent all and only convex games.*

Intuitively, when negative literals are allowed, we can specify the value of each coalition S directly by having a boolean formula that can be satisfied if and only if the truth assignment corresponds to the characteristic vector of S, and hence arbitrary games can be represented.

Another question is how a language relates to other representation languages. We can show that MC-nets generalize two of the previously discussed representations: WGGs and the multi-issue representation. First, MC-nets can be viewed as a generalization of WGGs that assigns weights to hyper-edges rather than to simple edges. A pattern in this case specifies the agents that share the same edge. However, since MC-nets can represent any coalitional games, MC-nets constitute a strict generalization of WGGs. Second, MC-nets generalize the multi-issue representation. Each issue is represented by patterns that only involve agents relevant to the issue. Comparing the two representations, MC-nets require at most $O(n)$ more space than the multi-issue representation; however, there exist coalitional games for which MC-nets are exponentially more succinct (in the number of agents) than the multi-issue representation.

From a computational standpoint, when only limited to conjunctions in the Boolean formula, MC-nets still make the Shapley value just as easy to compute as did the multi-issue representation.

Theorem 12.3.18 *Given a coalitional game specified with an MC-net limited to conjunctive patterns, the Shapley value can be computed in time linear in the size of the input.*

However, when other logical connectives are allowed, there is no known algorithm for finding the Shapley value efficiently. Essentially, finding the Shapley value involves summing factors of hypergeometric distributed variables, a problem for which there is no known closed-form solution.

Since MC-nets generalize WGGs, the problems of determining whether the core is empty and whether a payoff vector belongs to the core are both coNP-hard. However, there exists an algorithm that can solve both problems in time exponential only in the tree-width of the graphical representation of the MC-net.

12.4 Further directions

Before we conclude the chapter, we briefly survey some more advanced topics in coalitional game theory.

12.4.1 *Alternative coalitional game models*

Let us first revisit the transferable utility assumption that we made at the beginning of the chapter. In some situations, this assumption is not reasonable, for example, due to legal reasons (agents cannot engage in side payments), or because the agents do not have access to a common currency. Such settings are described as nontransferable utility (NTU) games.

coalitional game
with
nontransferable
utility

Definition 12.4.1 (Coalitional game with nontransferable utility) *A coalitional game (with nontransferable utility) is a pair (N, v), where:*

- *N is a finite set of players, indexed by i; and*
- *$v : 2^N \mapsto 2^{\mathbb{R}^{|S|}}$ associates each coalition $S \subseteq N$ with a set of value vectors $v(S) \subseteq \mathbb{R}^{|S|}$, which can be interpreted as the different sets of payoffs that S is able to achieve for each of its members.*

Note that the function v returns *sets* of value vectors rather than single real numbers, as in the case of coalitional games with transferable utility. Thus, rather than giving the total amount of utility and allowing agents to divide it arbitrarily among themselves, coalitional games with nontransferable utility explicitly list all the divisions that are possible and prohibit the rest.

It might seem that there is no problem left to be solved in the case of NTU games—after all, earlier we largely focused on payoff division, and in these games payoffs cannot be divided at all. However, in these games it is interesting to study which coalitions form.

For example, consider the matching problem we introduced in Section 10.6.4, pairing graduate students with advisors. This setting can be modeled as a coalitional game with nontransferable utility as follows. Let $N = A \cup S$ be the set of players in the game. Let Λ be the set of all possible matchings, and let $\mu(i)$ denote the agent matched with agent i, where $\mu \in \Lambda$. For each coalition T, the payoffs achievable by its members, $A' \cup S'$, are the preferences induced by all possible matchings among the group's members. As a concrete example, suppose each student in S achieves a payoff of $|A|$ by being matched to his most preferred advisor, $|A| - 1$ for being matched to his second most preferred advisor, and so on, with a payoff of 0 for not being matched. In general, write $u_s(\mu)$ to denote student s's payoff for matching μ. Similarly, let there be such a payoff function over students for each advisor. For each coalition $T \subseteq N$, each different matching within the group gives rise to different payoff vectors for its members.

A matching μ is in the core if there is no coalition of advisors and students where all members of the coalition are weakly better off matching only among themselves, and at least one member is strictly better off. Mathematically, μ is in the core if and only if there is no matching μ' and set $S \subseteq N$ such that

1. $\forall i \in S, \mu'(i) \in S$;
2. $\forall i \in S, u_i(\mu') \geq u_i(\mu)$; and
3. $\exists i \in S, u_i(\mu') > u_i(\mu)$.

As it turns out, the core of the matching game is always nonempty.

Another commonly-made assumption is that the value (or in the case of NTU games, the set of achievable payoff vectors) of a coalition is independent of the other coalitions. This is at best an approximation for many situations. If the value of a coalition can depend meaningfully on what other coalitions form, one has to take into account the whole *coalitional structure* when assigning payoffs to the coalitions.

coalitional game **Definition 12.4.2 (Coalitional game in partition form)** *A coalitional game in*
in partition form *partition form is a pair* (N, p), *where:*

- *N is a finite set of players, indexed by i; and*
coalitional - *p associates each partition π of N (also known as a* coalitional structure*)*
structure *and a coalition $S \in \pi$ with $p(\pi, S) \subseteq \mathbb{R}^S$, to be interpreted as the set of payoffs that S can achieve for its members under partition π.*

Yet another important direction is the incorporation of uncertainties into coalitional game models. Unlike incomplete-information games in noncooperative game theory, where a well-developed theory exists, a universally-accepted theory of coalitional games with uncertainty has yet to emerge. Such a theory would be useful. In many situations, it is natural to suppose that the values to coalitions are not known with certainty. It is also reasonable to assume that the payoffs to a coalition could depend on private information held by the agents, about which other agents have only probabilistic information. Some efforts toward modeling coalitional games under uncertainty have been made in the work cited at the end of the chapter; yet more work is still needed to create a fully complete theory.

12.4.2   *Advanced solution concepts*

There are a number of other solution concepts that we have not discussed in this chapter. One interesting phenomenon in the analysis of coalitional games is that researchers seem to have quite diverse opinions as to which solution concepts make the most sense. This is perhaps natural since different conflicts call for different notions of stability and fairness. Some other important solution concepts in the literature include *stable sets*, the *bargaining set*, and the *kernel*. Some of these, for example, attempt to capture the intuition that should a coalition deviate due to the violation of some stability property, that it should deviate to a stable coalition itself.

However, interestingly, most solution concepts to date focus on dividing the payoff of the grand coalition. While this appears quite reasonable when the core of the game is nonempty, when the core is empty, one might expect coalitions other than the grand coalition to form. This problem is worse when the game is not superadditive; in this case, it is possible that some partitioning of the coalitions could achieve strictly higher total payoffs than the grand coalition. It is therefore important to consider solution concepts that allow payoffs to depend on the coalitional structure.

Finally, computing the agents' payoffs is often only part of the problem. It can also be important to find out what coalitions would and should form, and how agents should coordinate their actions. This has been an area traditionally ignored in the literature, perhaps due to a focus on the abstract properties of coalitional games. For certain applications, the coalitional formation process cannot be ignored. Indeed, much work in artificial intelligence has been devoted to analyzing the process of coalition formation. By applying coalitional game theory to analyze such process, it may be possible to learn more about the strategic properties of different coordination mechanisms in the presence of selfish agents.

12.5   History and references

In the early days of game theory research, coalitional game theory was a major focus, particularly of economists. This is partly because the theory is closely related to equilibrium analysis and seemingly bridges a gap between game theory and economics. Von Neumann and Morgenstern, for example, devoted more than half of their classic text, *Theory of Games and Economic Behavior* [von Neumann and Morgenstern, 1944], to an analysis of coalitional games. A large body of theoretical work on coalitional game theory has focused on the development of solution concepts, possibly in an attempt to explain the behavior of large systems such as markets. Solid explanations of the many solution concepts and their properties are given by Osborne and Rubinstein [1994] and Peleg and Sudhölter [2003].

Some examples used in this chapter have appeared in other contexts. The connection between matching and coalitional game theory has been explored by a number of economists and surveyed in the works of Al Roth (see e.g., Roth and Sotomayor [1990]). The airport game and the minimum spanning tree game appeared in Peleg and Sudhölter [2003]. The connection between auctions and core was explored, for example, by Day and Milgrom [2008].

The first systematic investigation of the computational complexity of solution concepts in coalitional game theory were carried out in Deng and Papadimitriou [1994]. This paper defined weighted graph games and studied the complexity of computing the core and the Shapley value, as well as a few of the other solution concepts we mentioned. For weighted voting games, a systematic study of the computational complexity of the various solution concepts have appeared in Elkind et al. [2007]. Languages for succinct representation of coalitional games have been developed mostly in the AI community. The superadditive representation was developed by Conitzer and Sandholm [2003a]; it also naturally extends to representing superadditive games with nontransferable utilities. The multi-issue representation was also developed by Conitzer and Sandholm [2004]. The marginal contribution nets representation was first proposed in Ieong and Shoham [2005], and later generalized in Ieong and Shoham [2006].

Work on coalitional games under uncertainty includes Suijs et al. [1999], Chalkiadakis and Boutilier [2004], Myerson [2007], and Ieong and Shoham [2008]; as mentioned earlier, many open problems remain.

13

Logics of Knowledge and Belief

In this chapter we look at how one might represent statements such as "John knows that it is raining," "John believes that it will rain tomorrow," "Mary knows that John believes that it will rain tomorrow" and "It is common knowledge between Mary and John that it is raining."

13.1 The partition model of knowledge

Consider a distributed system, in which multiple processors autonomously performing some joint computation. Of course, the joint nature of the computation means that the processors need to communicate with one another. One set of problems comes about when the communication is error prone. In this case the system analyst may find himself saying something like the following: "Processor A sent the message to processor B. The message may not arrive, and processor A knows this. Furthermore, this is common knowledge, so processor A knows that processor B knows that it (A) knows that if a message was sent it may not arrive." The topic of this chapter is how to make such reasoning precise.

13.1.1 *Muddy children and warring generals*

Often the modeling is done in the context of some stylized problem, with an associated entertaining story. Thus, for example, when we return to the distributed computing application in Section 13.4, rather than speak about computer processors, we will tell the story of two generals who attempt to coordinate among themselves to gang up on a third. For now, however, consider the following less violent story.

> A group of n children enters their house after having played in the mud outside. They are greeted in the hallway by their father, who notices that k of the children have mud on their foreheads. He makes the following announcement, "At least one of you has mud on his forehead." The children can all see each other's foreheads, but not their own. The father then says, "Do any of you know that you have mud on your forehead? If you do, raise

your hand now." No one raises his hand. The father repeats the question, and again no one moves. The father does not give up and keeps repeating the question. After exactly k rounds, all the children with muddy foreheads raise their hands simultaneously.

How can this be? On the face of it only the father's initial statement conveyed new information to the children, and his subsequent questions add nothing. If a child did not have information at the beginning, how could he later on?

Here is an informal argument. Let us start with the simple case in which $k = 1$. In this case the single muddy child knows that all the others are clean, and when the father announces that at least one child is dirty he can conclude that he himself is that child. Note that none of the *other* children know at this point whether or not they are muddy. (After the muddy child raises his hand, however, they do; see next.) Now consider $k = 2$. Imagine that you are one of the two muddy children. After the father's first announcement, you look around the room and see that there is a muddy child other than you. Thus after the father's announcement you do not have enough information to know whether you are muddy (you might be, but it could also be that the other child is the only muddy one). But you note that after the father's first question the other muddy child does not raise his hand. You then realize that you yourself must be muddy as well, or else—based on the reasoning in the $k = 1$ case—that child would have raised his hand. So you raise your hand. Of course, so does the other muddy child.

You could extend this argument to $k = 3, 4, \ldots$, showing in each case that all of the k muddy children raise their hands together after the kth time that the father asks the question. But of course, you would rather have a general theorem that applies to all k. In particular, you might want to prove by induction that after rounds $1, 2, \ldots, k - 1$, none of the children know whether they are dirty, but after the next round exactly the muddy children do. However, for this we will need a formal model of "know" that applies in this example.

13.1.2   *Formalizing intuitions about the partition model*

partition model **Definition 13.1.1 (Partition model)** *An (n-agent) partition model over a language* Σ *is a tuple* $A = (W, \pi, I_1, \ldots, I_n)$, *where:*

- *W is a set of possible worlds;*
- *$\pi : \Sigma \mapsto 2^W$ is an interpretation function that determines which sentences in the languages are true in which worlds; and*
- *each I_i denotes a set of possible worlds that are equivalent from the point of view of agent i. Formally, I_i is a partition of W; that is, $I_i = (W_{i_1}, \ldots, W_{i_r})$ such that $W_{i_j} \cap W_{i_k} = \emptyset$ for all $j \neq k$, and $\cup_{1 \leq j \leq r} W_{i_j} = W$. We also use the following notation: $I_i(w) = \{w' \mid w \in W_{i_j} \text{ and } w' \in W_{i_j}\}$; that is, $I_i(w)$ includes all the worlds in the partition of world w, according to agent i.*

Thus each possible world completely specifies the concrete state of affairs, at least insofar as the language Σ can describe it. For example, in the context of the Muddy Children Puzzle, each possible world will specify precisely which of the children are muddy. The choice of Σ is not critical for current purposes, but for concreteness we will take it to be the languages of propositional logic over some set of primitive propositional symbols.[1] For example, in the context of the muddy children we will assume the primitive propositional symbols muddy1, muddy2, ... , muddy$n$.

We will use the notation $K_i(\varphi)$ (or simply $K_i\varphi$, when no confusion arises) as "agent i knows that φ."[2] The following defines when a statement is true in a partition model.

Definition 13.1.2 (Logical entailment for partition models) *Let* $A = (W, \pi,$ $I_1, \ldots, I_n)$ *be a partition model over* Σ, *and* $w \in W$. *Then we define the* $\models$ *(logical entailment) relation as follows:*

- *For any* $\varphi \in \Sigma$, *we say that* $A, w \models \varphi$ *if and only if* $w \in \pi(\varphi)$.
- $A, w \models K_i\varphi$ *if and only if for all worlds* w', *if* $w' \in I_i(w)$, *then* $A, w' \models \varphi$.

The first part of the definition gives the intended meaning to the interpretation function π from Definition 13.1.1. The second part states that we can only conclude that agent i knows φ when φ is true in all possible worlds that i considers indistinguishable from the true world.

Let us apply this modeling to the Muddy Children story. Consider the following instance of $n = k = 2$ (i.e., the instance with two children, both muddy). There are four possible worlds, corresponding to each of the children being muddy or not. There are two equivalence relations I_1 and I_2, which allow us to express each of the children's perspectives about which possible worlds can be distinguished. There are two primitive propositional symbols—muddy1 and muddy2. At the outset, before the children see or hear anything, all four worlds form one big equivalence class for each child.

After the children see each other (and can tell apart worlds in which *other* children's state of cleanliness is different) but before the father speaks, the state of knowledge is as illustrated in Figure 13.1. The ovals illustrate the four possible worlds, with the dark oval indicating the true state of the world. The solid boxes indicate the equivalence classes in I_1, and the dashed boxes indicate the equivalence classes in I_2.

Note that in this state of knowledge, in the real world both the sentences K_1muddy2 and K_2muddy1 are true (along with, for example, $K_1\neg K_2$muddy2). However, neither K_1muddy1 nor K_2muddy2 is true in the real world.

---

1. See Appendix D for a review of propositional logic.
2. The reader familiar with modal logic will note that a partition model is nothing but a special case of a propositional Kripke model with n modalities, in which each of the n accessibility relations is an equivalence relation (i.e., a binary relation that is reflexive, transitive, and symmetric). What is remarkable is that the modalities defined by these accessibility relations correspond well to the notion of knowledge that we have been discussing. This is why we use the modal operator K_i rather than the generic necessity operator $\Box_i$. (The reader unfamiliar with modal logic should ignore this remark; we review modal logic in the next section.)

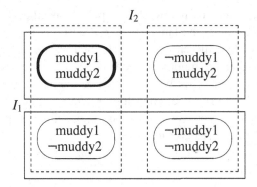

Figure 13.1 Partition model after the children see each other.

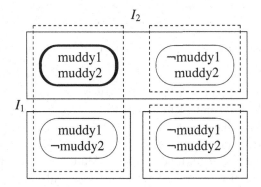

Figure 13.2 Partition model after the father's announcement.

Once the father announces publicly that at least one child is dirty, the world in which neither child is muddy is ruled out. This world then becomes its own partition, leaving the state of knowledge as shown in Figure 13.2.

Thus, were it the case that only one of the children were dirty, at this point he would be able to uniquely identify the real world (and in particular the fact that he was dirty). However, in the real world (where both children are dirty) it is still the case that in this world neither K_1muddy1 nor K_2muddy2 holds. However, once the children each observe that the other child does not know his state of cleanliness, the state of knowledge becomes as shown in Figure 13.3. And now indeed both K_1muddy1 and K_2muddy2 hold.

Thus, we can reason about knowledge rigorously in terms of partition models. This is a big advance over the previous informal reasoning. But it is also quite cumbersome, especially as these models get larger. Fortunately, we can sometimes reason about such models more concisely using an axiomatic system, which provides a complementary perspective on the same notion of knowledge.

13.2   A detour to modal logic

In order to reason about the partition model of knowledge and later about models of belief and other concepts, we must briefly discuss modal logic. This discussion

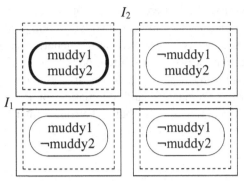

Figure 13.3 Final partition model.

presupposes familiarity with classical logic. For the most part we will consider propositional logic, but we will also make some comments about first-order logic. Both are reviewed in Appendix D.

From the syntactic point of view, modal logic augments classical logic with one or more (usually, unary) *modal operators*, and a modal logic is a logic that includes one or more modal operators. The classical notation for a modal operator is $\Box$, often pronounced "necessarily" (and thus $\Box\varphi$ is read as "φ is necessarily true"). The dual modal operator is $\Diamond$, often pronounced "possibly," and is typically related to the necessity operator by the formula $\Diamond\varphi \equiv \neg\Box\neg\varphi$.

What does a modal operator represent, and how is it different from a classical connective such as negation? In general, a modality represents a particular type of judgment regarding a sentence. The default type of judgment, captured in classical logic and not requiring an explicit modal operator, is whether the sentence is true or false. But one might want to capture other sorts of judgments. The original motivation within philosophy for introducing the modal operator is to distinguish between different "strengths of truth." In particular, the wish was to distinguish between accidental truths such as "it is sunny in Palo Alto" (represented, say, by the propositional symbol p), necessary truths such as "either it is sunny in Palo Alto or it is not" ($\Box(p \lor \neg p)$), and possible truths as captured by "it may be sunny in Palo Alto" ($\Diamond p$). A natural hierarchy exists among these three attitudes, with necessary truth implying accidental truth and both implying possible truth. However, the formal machinery has since been used for a variety of other purposes. For example, some logics interpret the modality as quantifying over certain contexts. A case in point are tense logics, or modal temporal logics, in which the context is time. In particular, in some tense logics $\Box\varphi$ is read as "φ is true now and will always be true in the future." We will encounter a similar temporal operator later in the chapter when we discuss robot motion planning. Logics of knowledge and belief read $\Box$ yet differently, as "φ is known" or "φ is believed." These inherently relate the sentence to an *agent*, who is doing the knowing or believing. Indeed, in these logics that interpret the modality in a rather specific way, the $\Box$ notation is usually replaced by other notation that is more indicative of the intended interpretation, but in this section we stick to the generic notation.

modal operator

Of course, the different interpretations of $\Box$ suggest that there are different modal logics, each endowing $\Box$ with different properties. This indeed is the case. In this section we briefly present the generic framework of modal logic, and in later sections we specialize it to model knowledge, belief, and related notions.

As with any logic, in order to discuss modal logic we need to discuss in turn syntax, semantics, and axiomatics (or proof theory). We concentrate on propositional modal logic, but make some remarks about first-order modal logic at the end.

13.2.1 *Syntax*

The set of sentences in the modal logic with propositional symbols P is the smallest set $\mathcal{L}$ containing P such that if $\varphi, \psi \in \mathcal{L}$ then also $\neg\varphi \in \mathcal{L}, \varphi \wedge \psi \in \mathcal{L}$, and $\Box\varphi \in \mathcal{L}$. As usual, other connectives such as $\vee, \rightarrow$ and $\equiv$ can be defined in terms of $\wedge$ and $\neg$. In addition, it is common to define the *dual operator* to $\Box$, often denoted $\Diamond$, and pronounced "possibly." It is defined by $\Diamond\varphi \equiv \neg\Box\neg\varphi$, which can be read as "the statement that φ is possibly true is equivalent to the statement that not φ is not necessarily true."

dual operator

13.2.2 *Semantics*

The semantics are defined in terms of *possible-worlds structures*, also called *Kripke structures*. A (single-modality) Kripke structure is a pair (W, R), where W is a collection of (not necessarily distinct) classical propositional models (i.e., models that give a truth value to all sentences that do not contain $\Box$), and R is binary relation on these models. Each $w \in W$ is called a *possible world*. R is called the *accessibility relation*, and sometimes also the *reachability relation* or *alternativeness relation*. It is convenient to think of Kripke structures as directed graphs, with the nodes being the classical models and the arcs representing accessibility.

Kripke structure

possible world

accessibility relation

reachability relation

alternativeness relation

This is where the discussion can start to be related back to the partition model of knowledge. The partition is of course nothing but a binary relation, albeit one with special properties. We will return to these special properties in the next section, but for now let us continue with the generic treatment of modal logic, one that allows for arbitrary accessibility relations.

A truth of a modal sentence is evaluated relative to a particular possible world w in a particular Kripke structure (W, R). (The pair is called a *Kripke model*.) The satisfaction relation is defined recursively as follows:

Kripke model

- $M, w \models p$ if p is true in w, for any primitive proposition p;
- $M, w \models \varphi \wedge \psi$ iff $M, w \models \varphi$ and $M, w \models \psi$;
- $M, w \models \neg\varphi$ iff it is not the case that $M, w \models \varphi$;
- $M, w \models \Box\varphi$ iff, for any $w' \in W$ such that $R(w, w')$, it is the case that $M, w' \models \varphi$.

validity As in classical logic, we overload the $\models$ symbol. In addition to denoting the satisfaction relation, it is used to denote *validity*. $\models \varphi$ means that φ is true in all Kripke models, and, given a class of Kripke models M, $\models_M \varphi$ means that φ is true in all Kripke models within M.

13.2.3 *Axiomatics*

Now that we have a well-defined notion of validity, we can ask whether there exists an axiom system that allows us to derive precisely all the valid sentences, or the valid sentences within a given class. Here we discuss the first question of capturing the sentences that are valid in all Kripke structures; in future sections we discuss specific classes of models of particular interest.

Consider the following axiom system, called the axiom system **K**.

Axiom 13.2.1 (Classical) *All propositional tautologies are valid.*

Axiom 13.2.2 (K) $(\Box\varphi \wedge \Box(\varphi \to \psi)) \to \Box\psi$ *is valid.*

Rule 13.2.3 (Modus Ponens) *If both φ and $\varphi \to \psi$ are valid, infer the validity of ψ.*

Rule 13.2.4 (Necessitation) *From the validity of φ infer the validity of $\Box\varphi$.*

It is not hard to see that this axiom system is *sound*; all the sentences pronounced valid by these axioms and inference rules are indeed true in every Kripke model. What is less obvious, but nonetheless true, is that this is also a *complete* system for the class of all Kripke models; there do not exist additional sentences representation theorem that are true in all Kripke models. Thus we have the following *representation theorem*:

Theorem 13.2.5 *The system* **K** *is sound and complete for the class of all Kripke models.*

13.2.4 *Modal logics with multiple modal operators*

We have so far discussed a single modal operator and a single accessibility relation corresponding to it. But it is easy to generalize the formulation to include multiple of each. Rather than have a single modal operator $\Box$ we have operators $\Box_1, \Box_2, \ldots, \Box_n$. The set of possible worlds is unchanged, but now we have n accessibility relations: $R_1, R_2, \ldots, R_n$. The last semantic truth condition is changed to:

- For any $i = 1, \ldots, n$, $M, w \models \Box_i\varphi$ iff, for any $w' \in W$ such that $R_i(w, w')$, it is the case that $M, w' \models \varphi$.

Finally, by making similar substitutions in the axiom system **K**, we get a sound and complete axiomatization for modal logic with n modalities. We sometimes denote the system $\mathbf{K_n}$ to emphasize the fact that it contains n modalities. Again, we emphasize that this is soundness and completeness for the class of all n-modality Kripke models. The system remains sound if we restrict the class of

Kripke models under consideration. In general, it is no longer complete, but often we can add axioms and/or inference rules and recapture completeness.

13.2.5 *Remarks about first-order modal logic*

We have so far discussed propositional model logic, that is, modal logic in which the underlying classical logic is propositional. But we can also look at the case in which the underlying logic is a richer first-order language. In such a first-order modal logic we can express sentences such as $\Box \forall x \exists y Father(y, x)$ (read "necessarily, everyone has a father"). We will not discuss first-order in detail, since it will not play a role in what follows. And in fact the technical development is for the most part unremarkable, and it simply mirrors the additional richness of first-order logic as compared to the propositional calculus.

There are however some interesting subtleties, which we point out briefly here.
Barcan formula The first has to do with the question of whether the so-called *Barcan formula* is valid.

$$\forall x \Box R(x) \rightarrow \Box \forall x R(x)$$

For example, when we interpret $\Box$ as "know," we might ask whether, if for every person it is known individually that the person is honest, it follows that it is known that all people are honest. One can imagine some settings in which the intuitive answer is yes, and others in which it is no. From the technical point of view, the answer depends on the domains of the various models. In first-order modal logic, possible worlds are first-order models. This means in particular that each possible world has a domain, or a set of individuals to which terms in the language are mapped. It is not hard to see that the Barcan formula is valid in the class of first-order Kripke models in which all possible worlds have the same domain, but not in general.

A similar problem has to do with the interaction between modality and equality. If $a = b$, then is it always the case that $\Box(a = b)$? Again, for intuition, consider the knowledge interpretation of $\Box$ and the following famous philosophical example. Suppose there are two objects, the "morning star" and the "evening star." These are two stars that are observed regularly, one in the evening and one in the morning. It so happens these are the same object, namely, the planet Venus. Does it follow that one knows that they are the same object? Intuitively, the answer is no. From the technical point of view, this is because, even if the domains of all possible worlds are the same, the interpretation function which maps language terms to objects might be different in the different worlds. For similar reasons, in general one cannot infer that the world's greatest jazz soprano saxophonist was born in 1897 based on the fact that Sidney Bechet was born that year, since one may not know that Sidney Bechet was indeed the best soprano sax player that ever graced this planet.

13.3 S5: An axiomatic theory of the partition model

We have seen that the axiom system **K** precisely captures the sentences that are valid in the class of all Kripke models. We now return to the partition model of knowledge, and search for an axiom system that is sound and complete for this more restricted class of Kripke models. Since it is a smaller class, it reasonable to expect that it will be captured by an expanded set of axioms. This is indeed the case.

We start with system **K**, but now use the K_i to denote the modal operators rather than $\Box_i$, representing the interpretation we have in mind.[3]

Axiom 13.3.1 (Classical) *All propositional tautologies are valid.*

Axiom 13.3.2 (K) $(K_i\varphi \wedge K_i(\varphi \rightarrow \psi)) \rightarrow K_i\psi$

Rule 13.3.3 (Modus Ponens) *From φ and $\varphi \rightarrow \psi$ infer ψ.*

Rule 13.3.4 (Necessitation) *From φ infer $K_i\varphi$.*

Note that the generic axiom **K** takes on a special meaning under this interpretation of the modal operator—it states that an agent knows all of the tautological consequences of that knowledge. Thus, if you know that it is after midnight, and if you know that the president is always asleep after midnight, then you know that the president is asleep. Less plausibly, if you know the Peano axioms, you know all the theorems of arithmetic. For this reason, this property is sometimes called *logical omniscience*. One could argue that this property is too strong, that it is too much to expect of an agent. After all, humans do not seem to know the full consequences of all of their knowledge. Similar criticisms can be levied also against the other axioms we discuss below. This is an important topic, and we returned to it later. However, it is a fact that the idealized notion of "knowledge" defined by the partition model has the property of logical omniscience. So to the extent that the partition model is useful (and it is, as we discuss later in the chapter), one must learn to live with some of the strong properties it induces on "knowledge."

logical omniscience

Other properties of knowledge do not hold for general Kripke structures, and are thus not derivable in the **K** axiom system. One of these properties is *consistency*, which states that an agent cannot know a contradiction. The following axiom, called axiom **D** for historical reasons, captures this property:

consistency

Axiom 13.3.5 (D) $\neg K_i(p \wedge \neg p)$.

Note that this axiom cannot be inferred from the axiom system **K** and is thus not valid in the class of all n-Kripke structures. It is, however, valid in the more restricted class of *serial models*, in which each accessibility is serial. (A binary relation X over domain Y is serial if and only if $\forall y \in Y \ \exists y' \in Y$ such that $(y, y') \in X$.) Indeed, as we shall see, it is not only sound but also complete. The axiom system obtained by adding axiom **D** to the axiom system **K** is called axiom system **KD**.

serial accessibility relation

Another property that holds for our current notion of knowledge is that of *veridity*; it is impossible for an agent to know something that is not actually true.

veridity

---

3. We stick to this notation for historical accuracy, but the reader should not be confused by it. The axiomatic system **K** and the axiom **K** that gives rise to the name have nothing to do with the syntactic symbol K_i we use to describe the knowledge of an agent.

Indeed, this is often taken to be the property that distinguishes knowledge from other informational attitudes, such as belief. We can express this property with the so-called axiom **T**:

Axiom 13.3.6 (T) $K_i\varphi \to \varphi$.

This axiom also cannot be inferred from the axiom system **KD**. Again, the class of Kripke structures for which axiom **T** is sound and complete can be defined succinctly—it consists of the Kripke structures in which each accessibility relation is *reflexive*. (A binary relation X over domain Y is reflexive if and only if $\forall y \in Y, (y, y) \in X$. Note that $y = y'$ is allowed.) By adding axiom **T** to the system **KD** we get the axiom system **KDT**. However, in this system the axioms are no longer independent; the axiom system **KDT** is equivalent to the axiom system **KT**, since the axiom **D** can be derived from the remaining axioms and the inference rules. Indeed, this follows easily from the completeness properties of the individual axioms; every reflexive relation is trivially also serial.

There are two additional properties of knowledge induced by the partition model which are not captured by the axioms discussed thus far. They both have to do with the introspective capabilities of a given agent, or the nesting of the knowledge operation. We first consider *positive introspection*, the property that, when an agent knows something, he knows that he knows it. It is expressed by the following axiom, historically called axiom **4**:

Axiom 13.3.7 (4) $K_i\varphi \to K_iK_i\varphi$.

Again, it does not follow from the other axioms discussed so far. The class of Kripke structures for which axiom **4** is sound and complete consists of the structures in which each accessibility relation is transitive. (A binary relation X is transitive if and only if for all $y, y', y'' \in Y$ it is the case that if $(y, y') \in X$ and $(y', y'') \in X$ then $(y, y'') \in X$.)

The last property of knowledge we will consider is *negative introspection*. This is quite similar to positive introspection, but here we are concerned that if an agent does not know something, then he knows that he does not know it. We express this with the following axiom, called axiom **5**:

Axiom 13.3.8 (5) $\neg K_i\varphi \to K_i\neg K_i\varphi$.

Again, it does not follow from the other axioms. Consider now the class of Kripke structures in which axiom **5** is sound and complete. Informally speaking, we want to ensure that if two worlds are accessible from the current world, then they are also accessible from each other. Formally, we say that the accessibility relation must be *Euclidean*. (A binary relation X over domain Y is Euclidean if and only if for all $y, y', y'' \in Y$ it is the case that if $(y, y') \in X$ and $(y, y'') \in X$ then $(y', y'') \in X$.)

At this point the reader may feel confused. After all, we started with a very simple class of Kripke structures—partition models—in which each accessibility relation is a simple equivalence relation. Then, in our pursuit of axioms that capture the notion of knowledge defined in partition models, we have looked at increasingly baroque properties of accessibility relations and associated axioms, as summarized in Table 13.1.

reflexive accessibility relation

positive introspection

negative introspection

Euclidean accessibility relation

Name	Axiom	Accessibility Relation
Axiom **K**	$(K_i(\varphi) \wedge K_i(\varphi \rightarrow \psi)) \rightarrow K_i(\psi)$	NA
Axiom **D**	$\neg K_i(p \wedge \neg p)$	Serial
Axiom **T**	$K_i\varphi \rightarrow \varphi$	Reflexive
Axiom **4**	$K_i\varphi \rightarrow K_i K_i\varphi$	Transitive
Axiom **5**	$\neg K_i\varphi \rightarrow K_i\neg K_i\varphi$	Euclidean

Table 13.1 Axioms and corresponding constraints on the accessibility relation.

What do these complicated properties have to do with the simple partition models that started this discussion? The answer lies in the following observation.

Proposition 13.3.9 *A binary relation is an equivalence relation if and only if it is reflexive, transitive, and Euclidean.*

Indeed, the system **KT45** (which results from adding to the axiom system **K** all the axioms **T**, **4**, and **5**), exactly captures the properties of knowledge as defined by the partition model. System **KT45** is also known by another, more common name—the **S5** axiom system. **S5** is both sound and complete for the class of all partition models. However, we are able to state an even more general theorem, which will serve us well later when we discuss moving from knowledge to belief.

Theorem 13.3.10 *Let **X** be a subset of $\{\mathbf{D}, \mathbf{T}, \mathbf{4}, \mathbf{5}\}$ and let $\mathcal{X}$ be the corresponding subset of $\{serial, reflexive, transitive, Euclidean\}$. Then $\mathbf{K} \cup \mathbf{X}$ (which is the basic axiom system **K** with the appropriate subset of axioms added) is a sound and complete axiomatization of K_i for the class of Kripke structures whose accessibility relations satisfy $\mathcal{X}$.*

13.4   Common knowledge, and an application to distributed systems

Earlier we discussed the domain of distributed systems. The following example illustrates the sort of reasoning one would like to perform in that context. (In case this problem does not sound like distributed computing to you, imagine that the generals are computer processes, which are trying to communicate reliably over a faulty communication line.)

> Two generals standing on opposing hilltops are trying to communicate in order to coordinate an attack on a third general, whose army sits in the valley between them. The two generals are communicating via messengers who must travel across enemy lines to deliver their messages. Any messenger carries the risk of being caught, in which case the message is lost. (Alas, the fate of the messenger is of no concern in this story.) Each of the two generals wants to attack, but only if the other does; if they both attack they will win, but either one will lose if he attacks alone. Given this context,

what protocol can the generals establish that will ensure that they attack simultaneously?

You might imagine the following naive communication protocol. The protocol for the first general, S, is to send an "attack tomorrow" message to general R and to keep sending this message repeatedly until he receives an acknowledgment that the message was received. The protocol for the second general, R, is to do nothing until he receives the message from S, and then send a single acknowledgment message back to general S. The question is whether the agents can have a plan of attack based on this protocol, which always guarantees—or, less ambitiously, can sometimes guarantee—that they attack simultaneously. And if not, can a different protocol achieve this guarantee?

Clearly, it would be useful to reason about this scenario rigorously. It seems intuitive that the formal notion of "knowledge" should apply here, but the question is precisely how. In particular, what is the knowledge condition that must be attained in order to ensure a coordinated attack?

To apply the partition model of knowledge we need to first define the possible worlds. We will do this by first defining the *local state* of each agent (i.e., each general); together the two local states will form a *global state*. To reason about the naive protocol, we will have the local state for S represent two binary pieces of information: whether or not an "attack" message was sent and whether or not an acknowledgment was received. The local state for R will also represent two binary pieces of information: whether or not an "attack" message was received and whether or not an acknowledgment message was sent. Thus we end up with the four possible local states for each general—$(0, 0)$, $(0, 1)$, $(1, 0)$, and $(1, 1)$—and thus sixteen possible global states.

We are now ready to define the possible worlds of our model. The initial global state is well defined—by convention, we call it $\langle(0, 0), (0, 0)\rangle$. It will then evolve based on two factors—the dictates of the protocol, and the nondeterministic effect of nature (which decides whether a message is received or not). Thus, among all the possible sequences of global states, only some are legal according to the protocol. We will call any finite prefix of such a legal sequence of global states a

history   *history*. These histories will be the possible worlds in our model.

For example, consider the following possible sequence of events, given the naive protocol: S sends an "attack" message to R, R receives the message and sends an acknowledgment to S, and S receives the acknowledgment. The history corresponding to this scenario is

$$\langle(0, 0), (0, 0)\rangle, \langle(1, 0), (1, 0)\rangle, \langle(1, 1), (1, 1)\rangle.$$

The structure of our possible worlds suggests a natural definition of the partition associated with each of the agents. We will say that two histories are in the same equivalence class of agent i ($i \in \{S, R\}$) if their respective final global states have identical local state for agent i. That is, history

$$\langle(0, 0), (0, 0)\rangle, \langle x_{S,1}, x_{R,1}\rangle, \ldots, \langle x_{S,k}, x_{R,k}\rangle$$

is indistinguishable in the eyes of agent i from history

$$\langle (0,0), (0,0) \rangle, \langle y_{S,1}, y_{R,1} \rangle, \ldots, \langle y_{S,l}, y_{R,l} \rangle$$

if and only if $x_{i,k} = y_{i,l}$. Note a very satisfying aspect of this definition; the accessibility relation, which in general is thought of as an abstract notion, is given here a concrete interpretation in terms of local and global states.

We can now reason formally about the naive protocol. Consider again the possible history mentioned above:

$$\langle (0,0), (0,0) \rangle, \langle (1,0), (1,0) \rangle, \langle (1,1), (1,1) \rangle.$$

The reader can verify that in this possible world the following sentences are true: $K_S\text{attack}, K_R\text{attack}, K_S K_R\text{attack}$. However, it is also the case that this world satisfies $\neg K_R K_S K_R\text{attack}$. Intuitively speaking, R does not know that S knows that R knows that S intends to attack, since for all R knows its acknowledgment could have been lost. Thus R cannot proceed to attack; he reasons that if indeed the acknowledgement was lost then S will not dare attack.

This of course suggests a fix—have S acknowledge R's acknowledgement. To accommodate this we would need to augment each of the local states with another binary variable, representing for S whether the second acknowledgement was sent and for R whether it was received. However it is not hard to see that this also has a flaw; assuming S's second acknowledgement indeed goes through, in the resulting history we will have $K_R K_S K_R\text{attack}$ hold, but not $K_S K_R K_S K_R\text{attack}$. Can this be fixed, and what is the general knowledge condition we must aim for?

common
knowledge It turns out that the required condition is what is known as *common knowledge*. This is a very intuitive notion, and we define it in two steps. We first define what it means that *everybody knows* a particular sentence to be true. To represent this we use the symbol E_G, where G is a particular group of agents. The "everybody knows" operator has the same syntactic rules as the knowledge operator. As one might expect, the semantics can be defined easily in terms of the basic knowledge operator. We define the semantics as follows.

Definition 13.4.1 ("Everyone knows") *Let M be a Kripke structure, w be a possible world in M, G be a group of agents, and φ be a sentence of modal logic. Then $M, w \models E_G\varphi$ if and only if $M, w' \models \varphi$ for all $w' \in \cup_{i \in G} I_i(w)$. (Equivalently, we can require that $\forall i \in G$ it is the case that $M, w \models K_i\varphi$.)*

In other words, everybody knows a sentence when the sentence is true in all of the worlds that are considered possible in the current world by any agent in the group.

Using this concept we can define the notion of *common knowledge*, or, as it is sometimes called lightheartedly, "what any fool knows." If "any fool" knows something, then we can assume that everybody knows it, and everybody knows that everybody knows it, and so on. An example from the real world might be the assumption we use when driving a car that all of the other drivers on the road also know the rules of the road, and that they know that we know the rules of the road, and that they know that we know that they know them, and so on. We require an infinite series of "everybody knows" in order to capture this intuition. For this reason we use the following recursive definition.

Definition 13.4.2 (Common knowledge) *Let M be a Kripke structure, w be a possible world in M, G be a group of agents, and φ be a sentence of modal logic. Then $M, w \models C_G \varphi$ if and only if $M, w \models E_G(\varphi \wedge C_G \varphi)$.*

fixed-point axiom

In other words, a sentence is common knowledge if everybody knows it and knows that it is common knowledge. This formula is called the *fixed-point axiom*, since $C_G \varphi$ can be viewed as the fixed-point solution of the equation $f(x) = E_G(\varphi \wedge f(x))$. Fixed-point definitions are notoriously hard to understand intuitively. Fortunately, we can give alternative characterizations of C_G. First, we can give a direct semantic definition.

Theorem 13.4.3 *Let M be a Kripke structure, w be a possible world in M, G be a group of agents, and φ be a sentence of modal logic. Then $M, w \models C_G \varphi$ if and only if $M, w' \models \varphi$ for every sequence of possible worlds $(w = w_0, w_1, \ldots, w_n) = w'$ for which the following holds: for every $0 \leq i < n$ there exists an agent $j \in G$ such that $w_{i+1} \in I_j(w_i)$.*

Second, it is worth noting that our **S5** axiomatic system can also be enriched to provide a sound and complete axiomatization of C_G. It turns out that what are needed are two additional axioms and one new inference rule.

Axiom 13.4.4 (A3) $E_G \varphi \leftrightarrow \bigwedge_{i \in G} K_i \varphi$

Axiom 13.4.5 (A4) $C_G \varphi \rightarrow E_G(\varphi \wedge C_G \varphi)$

Rule 13.4.6 (R3) *From $\varphi \rightarrow E_G(\psi \wedge \varphi)$ infer $\varphi \rightarrow C_G \psi$*

This last inference rule is a form of an *induction rule*.

Armed with this notion of common knowledge, we return to our warring generals. It turns out that, in a precise sense, whenever any communication protocol guarantees a coordinated attack in a particular history, in that history it must achieve common knowledge between the two generals that an attack is about to happen. It is not hard to see that no finite exchange of acknowledgments will ever lead to such common knowledge. And thus it follows that there is *no* communication protocol that solves the Coordinated Attack problem, at least not as the problem is stated here.

13.5 Doing time, and an application to robotics

We now move to another application of reasoning about knowledge, in robotics. The domain of robotics is characterized by *uncertainty*; a robot receives uncertain readings from its input devices, and its motor controls produce imprecise motion. Such uncertainty is not necessarily the end of the world, so long as its magnitude is not too large for the task at hand, and that the robot can reason about it effectively. We will explicitly consider the task of robot motion planning under uncertainty, and in particular the question of how a robot knows to stop despite

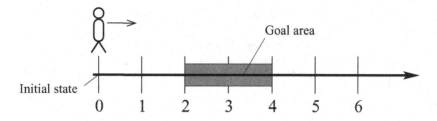

Figure 13.4 A one-dimensional robot motion problem.

having an imprecise sensor and perhaps also motion controller. We first discuss the single-robot case, where we see the power of the knowledge abstraction. We then move to the multiagent case, where we show the importance of each agent being able to model the other agents in the system.

13.5.1 *Termination conditions for motion planning*

Imagine a point robot moving in one dimension along the positive reals from the origin to the right, with the intention of reaching the interval [2, 4], the goal region (see Figure 13.4). The robot is moving at a fixed finite velocity, so there is no question about whether it will reach the destination; the question is whether the robot will know when to stop. Assume the robot has a position sensor that is inaccurate; if the true position is L, the sensor will return any value in the interval $[L - 1, L + 1]$. We assume that time is continuous and that the sensor provides readings continuously, but that the reading values are not necessarily continuous. We are looking for a termination condition—a predicate on the readings such that as soon as it evaluates to "true" the robot will stop. What is a reasonable termination condition?

Consider the termination condition "$R = 3$" where R is the current reading. This is a "safe" condition, in the sense that if that is the reading then the robot is guaranteed to be in the goal region. The problem is that, because of the discontinuity of reading values, the robot may never get that reading. In other words, this termination condition is sound but not complete. Conversely, the condition "$2 \leq R \leq 4$" is complete but not sound. Is there a sound and complete termination condition?

On reflection it is not hard to see that "$R \geq 3$" is such a condition. Although there are locations outside the goal region that can give rise to readings that satisfy the predicate (e.g., in location 10 the reading necessarily does), what matters is the first time the robot encounters that reading. Given its starting location and its motion trajectory, the first time can be no earlier than 2 and no later than 4.

Let us now turn to a slightly more complex, two-dimensional robotic motion planning problem, depicted in Figure 13.5. A point robot is given a command to move in a certain direction, and its goal is to arrive at the depicted rectangular goal region. As in the previous example, its sensor is error prone; it returns a reading that is within ρ from the true location (i.e., the uncertainty is captured by a disk of radius ρ). Unlike the previous example, the robot's motion is also subject to

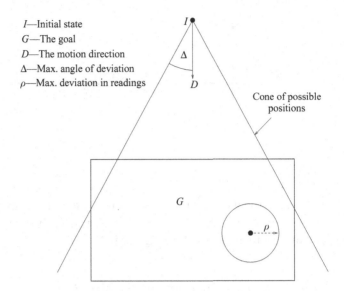

Figure 13.5 A two-dimensional robot motion problem.

error; given a command to move in a certain direction, the robot's actual motion can deviate up to Δ degrees from the prescribed direction. It is again assumed that when the robot is given the command "Go in direction D" it moves at some finite velocity, always moving within Δ degrees from that direction, but that the deviation is not consistent and can change arbitrarily within the tolerance Δ. It is not hard to see that the space of locations reachable by the robot is described by a cone whose angle is 2Δ, as depicted in Figure 13.5. Again we ask, what is a good termination condition for the robot? One candidate is a rectangle inside the goal region whose sides are Δ away from the corresponding side of the goal rectangle. Clearly this is a sound condition, but not a complete one. Figure 13.6 depicts another candidate, termed the "naive termination condition"—all the sensor readings in the region with the dashed boundary. Although it is now less obvious, this termination condition is also sound—but, again, not complete. In contrast, Figure 13.7 shows a termination condition that is both sound and complete.

Consider now the two sound and complete termination conditions identified for the two problems, the one-dimensional one and the two-dimensional. Geometrically, they seem to bear no similarity, and yet one's intuition might be that they embody the same principle. Can one articulate this principle, and thus perhaps later on apply it to yet other problems? To do so we abstract to what is sometimes called the knowledge level. Rather than speak about the particular geometry of the situation, let us reason more abstractly about the knowledge available to the robot. Not surprisingly, we choose to do it using possible-worlds semantics and specifically the S5 model of knowledge. Define a history of the robot as a mapping from time to both location and sensor reading. That is, a history tells us where the robot has been at any point in time and its sensor reading at that time. Clearly, every motion planning problem—including both the ones presented here—defines a set of legal histories. For every motion planning

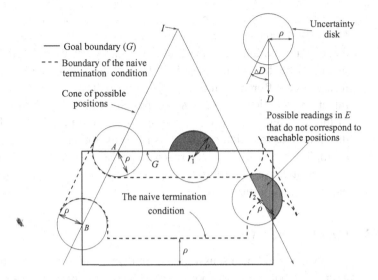

Figure 13.6 A sound but incomplete termination condition for the two-dimensional robot motion problem.

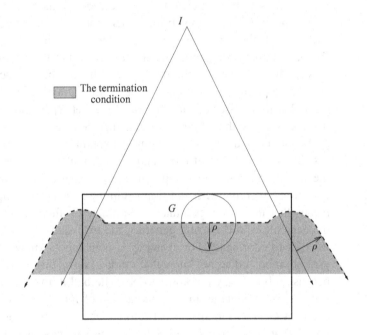

Figure 13.7 A sound and complete termination condition for the two-dimensional robot motion problem.

problem, our set of possible worlds will consist of pairs (h, t), where h is a legal history and t is a time point. We will say that (h, t) is accessible from (h', t') if the sensor reading in h at t is identical to the sensor reading in h' at t'. Clearly, this defines an equivalence relation. We now need a language to speak about such possible-worlds structures. As usual, we will use the K operator to denote knowledge, that is, truth in all accessible worlds. We will also use a temporal operator $\diamond$ whose meaning will be "sometime in the past."

Specifically, $h, t \models \Diamond \varphi$ holds at h, t just in case there exists a $t' \le t$ such that $h, t' \models \varphi$.[4] Armed with this, we are ready to provide a knowledge–level condition that captures both the specific geometric ones given earlier. It is deceptively simple. Let g be the formula meaning "I am in the goal region" (which is of course instantiated differently in the two examples). Then the sound and complete termination condition is simply $K \Diamond g$, reading informally "I know that either I am in the goal region now, or else sometime in the past I was." It turns out that this simple termination condition is not only sound and complete for these two examples, but is so also in a much larger set of examples, and is furthermore optimal: it causes the robot to stop no later than any other sound and complete termination condition.

13.5.2   *Coordinating robots*

We now extend this discussion to a multiagent setting. Add to the first example a second robot that must coordinate with the first robot (number the robots 1 and 2). Specifically, robot 2 must take an action at the exact time (and never before) robot 1 stops in the goal area. For concreteness, think of the robots as teammates in the annual robotic competition, where robotic soccer teams compete against each other. Robot 2 needs to pass the ball to robot 1 in front of the opposing goal. Robot 1 must be stopped to receive the pass, but the pass must be completed as soon it stops, or else the defense will have time to react. The robots must coordinate on the pass without the help of communication. Instead, we provide robot 2 with a sensor of robot 1's position. Let R_1 be the sensor reading of the first robot, R_2 be that of the second, and p the true position of robot 1.

The sound and complete termination condition that we seek in this setting is actually a conjunction of termination conditions for the two robots. Soundness means that robot 1 never stops outside the goal area and robot 2 never takes its action when robot 1 is not stopped in the goal area. Completeness obviously means that the robots eventually coordinate on their respective actions. As in the example of the warring generals, the two robots need common knowledge in order to coordinate. For example, if both robots know that robot 1 is in the goal but robot 1 does not know that robot 2 knows this, then robot 1 cannot stop, because robot 2 may not know to pass the ball at that point in time. Thus, the sound and complete termination condition of $K(\Diamond g)$ in the single robot setting becomes $C_{1,2}(\Diamond g)$ in the current setting, where g now means "Robot 1 is in the goal." That is, when this fact becomes common knowledge for the first time, the robots take their respective actions.

So far, we have left the sensor of robot 2 unspecified. In the first instance we analyze, let robot 2 be equipped with a sensor identical to robot 1's. That is, the sensor returns the position of robot 1 within the interval $[L - 1, L + 1]$, and the noise of the sensor is deterministic so that this value is exactly equal to the value observed by robot 1. Further, assume that this setting is common knowledge between the two robots. We can formalize the setting as a partition

---

4. This operator is taken from the realm of temporal logic, where it usually appears in conjunction of other modalities; however, those are not required for our example.

model in which a possible world (which was (h, t) above) is a tuple $\langle p, R_1, R_2 \rangle$. The partitions of the two robots define equivalence classes based on the sensor of that agent. That is, $P_i(\langle p, R_1, R_2 \rangle) = \{\langle p', R_1', R_2' \rangle | (R_i = R_i')\}$, for $i = 1, 2$. The common knowledge of the properties of the sensors reduces the space of possible worlds to all $\langle p, R_1, R_2 \rangle$ such that $|p - R_1| \leq 1$ and $R_1 = R_2$. The latter property implies that the partition structures for the two agents are identical.

We can now analyze this partition model to find our answer. From the single robot case, we know that $R_1 \geq 3 \rightarrow K_1(\Diamond g)$. This statement can be verified easily using the partition structure. Recall the semantic definition of common knowledge: the proposition must be true in the last world of every sequence of possible worlds that starts at the current world and only transitions to a world that is in the partition of an agent. Because the partition structures are identical, starting from a world in which $R_1 \geq 3$, we cannot transition out of the identical partition we are in for both agents, and thus $C_{1,2}(\Diamond g)$ holds. Therefore, the robots can coordinate by using the same rule as the previous section: robot 1 stops when $R_1 \geq 3$ is first true, and robot 2 takes its action when $R_2 \geq 3$ is first true. Since any lower sensor reading does not permit either robot to know $\Diamond g$, it is also the case that this rule is optimal.

One year later, we are preparing for the next robotic competition. We have received enough funding to purchase a perfect sensor (one that always returns the true position of robot 1). We replace the sensor for robot 2 with this sensor, but we do not have enough money left over to buy either a new sensor for robot 1 or a means of communication between the two robots. Still, we are optimistic that our improved equipment will allow us to fare better this year by improving the termination condition so that the pass can be completed earlier. Since we have different common knowledge about the sensors, we have a different set of possible worlds. In this instance, each possible world $\langle p, R_1, R_2 \rangle$ must satisfy $p = R_2$ instead of $R_1 = R_2$ while also satisfying $|p - R_1| \leq 1$ as before. This change causes the robots to no longer have identical partition structures. Analyzing the new structure, we quickly realize that not only can we not improve the termination condition, but also that our old rule no longer works. As soon as $R_2 \geq 3$ becomes true, we have both $K_1(\Diamond g)$ and $K_2(\Diamond g)$. However, in the partition for robot 2 in which $R_2 = 3$, we find possible worlds in which $R_1 < 3$. In these worlds, robot 1 does not know that it is in the goal. Thus, we have $\neg K_2(K_1(\Diamond g))$, which means that robot 2 cannot take the action because robot 1 might not stop to receive the pass.

Suppose we try to salvage the situation by implementing a later termination condition. This idea obviously fails, because if robot 2 instead waits for R_2 to be a value greater than 3, then robot 1 may have already stopped. However, our problems run deeper. Even if we only require that the robots coordinate at some time after robot 1 enters the goal area (implicitly extending the goal to the range $[2, \infty]$), we can never find a sound and complete termination condition. Common knowledge of $\Diamond g$ is impossible to achieve, because, from any state of the world $\langle p, R_1, R_2 \rangle$, we can follow a path of possible worlds (alternating between the robot whose partition we transition in) until we get to a world in which robot 1 is not in the goal. For example, consider the world $\langle 5, 6, 5 \rangle$. Robot 2 considers it possible that robot 1 observes a value of 4 (in world $\langle 5, 4, 5 \rangle$). In that world, robot 1 would consider it possible that the true world is 3 (in world $\langle 3, 4, 3 \rangle$).

We can then make two more transitions to $\langle 3, 2, 3 \rangle$ and then to $\langle 1, 2, 1 \rangle$, in which robot 1 is not in the goal. Thus, we have $\neg K_2(K_1(K_2(K_1(\Diamond g))))$. Since we can obviously make a similar argument for any world with a finite value for p, there does not exist a world that implies $C_{1,2}(\Diamond g)$.

This example illustrates how, when coordination is necessary, knowledge of the knowledge of the other agents can be more important than objective knowledge of state of the world. In fact, in many cases it takes common knowledge to achieve common knowledge. When robot 2 had the flawed sensor, we needed common knowledge of the fact that the sensors are identically flawed (which was encoded in the space of possible worlds) in order for the robots to achieve common knowledge of $\Diamond g$. Also, the fact that the agents have to coordinate is a key restriction in this setting. If instead, we only needed robot 1 to stop in the goal and robot 2 to take its action at some point after robot 1 stopped, then all we need is $K_1(\Diamond g)$ and $K_2(K_1(\Diamond g))$. This is achieved by the termination conditions of $R_1 \geq 3$ for robot 1 and $R_2 \geq 4$ for robot 2.

13.6   From knowledge to belief

We have so far discussed a particular informational attitude, namely "knowledge," but there are others. In this section we discuss "belief," and in the next section we will mention a third—"certainty."

Like knowledge, belief is a mental attitude, and concerns an agent's view of different state of affairs. Indeed, we will model belief using Kripke structures (i.e., possible worlds with binary relations defined on them). However, intuition tells us that belief has different properties from those of knowledge, which suggests that these Kripke structures should be different from the partition models capturing knowledge.

We will return to the semantic model of belief shortly, but it is perhaps easiest to make the transition from knowledge to belief via the axiomatic system. Recall that in the case of knowledge we had the veridity axiom, **T**: $K_i \varphi \rightarrow \varphi$. Suppose we take the **S5** system and simply drop this axiom—what do we get? Hidden in this simple question is a subtlety, it turns out. Recall that **S5** was shorthand for the system **KDT45**. Recall also that **KDT45** was logically equivalent to **KT45**. However, **KD45** is *not* logically equivalent to **K45**; axiom **D** is not derivable without axiom **T** (why?). It turns out that both **KD45** and **K45** have been put forward as a logic of idealized belief, and in fact both have been called *weak* **S5**. There are good reasons, however, to stick to **KD45**, which we will do henceforth.

The standard logic of belief **KD45** therefore consists of the following axioms; note that we change the notation K_i to B_i to reflect the fact that we are modeling belief rather than knowledge.

Axiom 13.6.1 (K) $(B_i \varphi \wedge B_i(\varphi \rightarrow \psi)) \rightarrow B_i \psi$

Axiom 13.6.2 (D) $\neg B_i(p \wedge \neg p)$

Axiom 13.6.3 (4) $B_i \varphi \rightarrow B_i B_i \varphi$

Axiom 13.6.4 (5) $\neg B_i \varphi \rightarrow B_i \neg B_i \varphi$

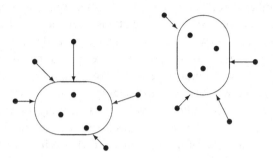

Figure 13.8 Graphical depiction of a quasi-partition structure.

These axioms—logical omniscience, consistency, positive and negative introspection—play the same roles as in the logic of knowledge, as do the two inference rules, *Modus Ponens* and *Necessitation*.

The next natural question is whether we can complement this axiomatic theory of belief with a semantic one. That is, is there a class of Kripke structures for which this axiomatization is sound and complete, just as **S5** is sound and complete for the class of partition models? The theorem in Section 13.3 has already provided us the answer; it is the class of Kripke structures in which the accessibility relations are serial, transitive, and Euclidean. Just as in the case of knowledge, there is a relatively simple way to understand this class, although it is not as simple as the class of partitions. As shown in Figure 13.8, we can envision such a structure as composed of several clusters of worlds, each of which is completely connected internally. In addition there are possibly some singleton worlds, each of which is connected only to all the worlds within exactly one cluster. We call such a model a *quasi-partition*.

We saw that it was useful to define the notion of common knowledge. Can we similarly define *common belief*, and would it be a useful one? The answers are yes, and maybe. We can certainly define common belief by mirroring the definitions for knowledge. However, the resulting notion will necessarily be weaker. In particular, it is not hard to verify the validity of the following sentence:

$$K_i C_G \varphi \equiv C_G \varphi.$$

This equivalence breaks down in the case of belief and common belief.

13.7   Combining knowledge and belief (and revisiting knowledge)

Up until this point we have discussed how to model knowledge and belief separately from each other. But of course we may well want to combine the two, so that we can formalize sentences such as "if Bob knows that Alice believes it is raining, then Alice knows that Bob knows it." Indeed, it is easy enough to just merge the languages of knowledge and belief, so as to allow the sentence:

$$K_B B_A \texttt{rain} \rightarrow K_A K_B B_A \texttt{rain}.$$

Margin notes: Modus Ponens · Necessitation · quasi-partition · common belief

Furthermore, there is no difficulty in merging the semantic structures and having two sets of accessibility relations over the possible worlds—partition models representing the knowledge of the agents and quasi-partition models representing their beliefs. Finally, we can merge the axiom systems of knowledge and belief and obtain a sound and complete axiomatization of merged structures.

However, doing just these merges, while preserving the individual properties of knowledge and belief, will not capture any interaction between them. And we do have some strong intuitions about such interactions. For example, according to the intuition of most people, knowledge implies belief. That is, the following sentence ought to be valid:

$$K_i \varphi \to B_i \varphi.$$

This sentence is not valid in the class of all merged Kripke structures defined earlier. This means that we must introduce further restrictions on these models in order to capture this property, and any other property about the interaction between knowledge and belief that we care about.

We will introduce a particular way of tying the two notions together, which has several conceptual and technical advantages. It will force us, however, to reconsider some of the assumptions we have made about knowledge and belief thus far, and to enter into philosophical discussion more deeply than we have heretofore.

To begin, we should distinguish two types of belief. Both of them are dis-
defeasible   tinguished from knowledge in that they are *defeasible*; the agent may believe something false. However, in one version of belief the believing agent is aware of this defeasibility, while in the other it is not. We will call the first type of belief "mere belief" (or sometimes "belief" for short) and the second type of belief "certainty." An agent who is certain of a fact will not admit that he might be wrong; to him, his beliefs look like knowledge. It is only *another* agent who might label his beliefs as only that and deny them the status of knowledge. Imagine that John is certain that the bank is open, but in fact the bank is closed. If you were to ask John, he would tell you that the bank is open. If you pressed him with "Are you sure?" he would answer "What do you mean 'am I sure,' I *know* that it is open." But of course this cannot be knowledge, since it is false. In contrast, you can imagine that John is pretty sure that the bank is open, sufficiently so to make substantial plans based on this belief. In this case if you were to ask John "Is the bank open?" he might answer "I believe so, but I am not sure." In this case John would be the first to admit that this is a mere belief on his part, not knowledge.[5]

---

5. A reader seeped in the Bayesian methodology would at this point be strongly tempted to protest that if John is not sure he should quantify his belief probabilistically, rather than make a qualitative statement. We will indeed discuss probabilistic beliefs in Section 14.1. But one should not discard qualitative beliefs casually. Not only can psychological and commonsense arguments be made on behalf of this notion, but in fact the notion plays an important role in game theory, which is for the most part Bayesian in nature. We return to this later in this chapter, when we discuss *belief revision* in Section 14.2.1.

From a technical point of view, we can split the B_i operator into two versions, B_i^m for "mere belief" and B_i^c for certainty. Even before entering into a formal account of such operators, we expect that the sentence $B_i^c \varphi \rightarrow B_i^c K_i \varphi$ will be valid, but the sentence $B_i^m \varphi \rightarrow B_i^m K_i \varphi$ will not.

This distinction between certainty and mere belief also calls into question some of our assumptions about knowledge, in particular the negative introspection property. Consider the following informal argument. Suppose John is certain that the bank in open (B_j^copen). According to our interpretation of certainty, we have that John is certain that he knows that the bank is open ($B_j^c K_J$open). Suppose that the bank is in fact closed ($\neg$open). This means that John does not know that the bank is closed, because of the veridity property ($\neg K_J$open). Because of the negative introspection property, we can conclude that John knows that he does not know that the bank is open ($K_J \neg K_J$open). If we reasonably assume that knowledge implies certainty, we also get that John is certain that he does not know that the bank is open ($B_j^c \neg K_J$open). Thus we get that John is certain of both the sentence $K_J$open and its negation, which contradicts the consistency property of certainty. So something has to give—and the prime candidate is the negative introspection axiom for knowledge. And indeed, even absent a discussion of belief, this axiom has attracted criticism early on as being counter-intuitive.

This is a good point at which to introduce a caveat. Commonsense can serve at best as a crude guideline for selecting formal models. Human language and thinking are infinitely flexible, and we must accept at the outset a certain discrepancy between the meaning of a commonsense term and any formalization of it. That said, some discrepancies are more harmful than others, and negative introspection for knowledge may be one of the biggest offenders. This criticism does not diminish the importance of the partition model, which as we have seen is quite useful for modeling a variety of information settings. What it does mean is that perhaps the partition model is better thought of as capturing something other than knowledge; perhaps "possesses the implicit information that" would be a more apt descriptor of the **S5** modal operator. It also means then that we need to look for an alternative model that better captures our notion of knowledge, albeit still in a highly idealized fashion.

Here is one such model. Before we present it, one final philosophical musing. Philosophers have offered two informal slogans to explain the connection between knowledge and belief. The first is "knowledge is justified, true belief." The intuition is that knowledge is a belief that not only is true, but is held with proper justification. What "proper" means is open to debate, and we do not offer a formal account of this slogan. The second slogan is "knowledge is belief that is stable with respect to the truth." The intuition is that, when presented with evidence to the contrary, an agent would discard any of his false belief. It is only correct beliefs that cannot be contradicted by correct evidence, and are thus stable in the face of such evidence. Our formal model can be thought of a formal version of the second informal slogan; we will return to this intuition when we discuss *belief revision* later in Section 14.2.1.

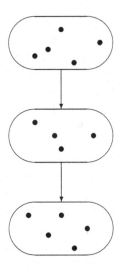

Figure 13.9 A KB structure.

In developing our combined formal model of knowledge and belief, we need to be clear about which sense of belief we intend. In the following we restrict the attention to the certainty kind, namely, B_i^c. Since there is no ambiguity, we will use the simpler notation B_i.

We have so far encountered two special classes of Kripke models—partitions and quasi-partitions. We will now introduce a third special class, the class of *total preorders* (a total preorder $\leq$ over domain Y is a reflexive and transitive binary relation, such that for all $y, y' \in Y$ it is the case that either $y \leq y'$, or $y' \leq y$, or both).

KB-structure **Definition 13.7.1 (KB- structure)** *An (n-agent) KB-structure is a tuple $(W, \leq_1, \ldots, \leq_n)$, where:*

- *W is a set of possible worlds; and*
- *each $\leq_i$ is a finite total preorder over W.*

KB-model *An (n-agent) KB-model over a language Σ is a tuple $A = (W, \pi, \leq_1, \ldots, \leq_n)$, where:*

- *W and $\leq_i$ are as earlier; and*
- *$\pi : \Sigma \mapsto 2^W$ is an interpretation function that determines which sentences in the languages are true in which worlds.*

Although there is merit in considering more general cases, we will confine our attention to *well-founded* total preorders; that is, we will assume that for no preorder $\leq_i$ does there exist an infinite sequence of worlds $w_1, w_2, \ldots$ such that $w_j <_i w_{j+1}$ for all $j > 0$ (where $<$ is the anti-reflexive closure of $\leq$).

A graphical illustration of a KB structure for a single agent is given in Figure 13.9. In this structure there are three clusters of pairwise connected worlds, and each world in a given cluster is connected to all worlds in lower clusters.

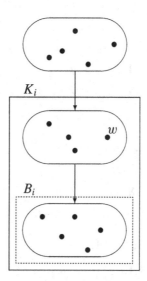

Figure 13.10 Knowledge and belief in a KB model.

We use KB structures to define both knowledge and certainty-type belief. First, it is useful to attach an intuitive interpretation to the accessibility relation. Think of it as describing the cognitive bias of an agent; $w' \leq w$ iff w' is at least as easy for the agent to imagine as w. The agent's beliefs are defined as truth in the most easily imagined worlds; intuitively, those are the only worlds the agent considers, given his current evidence. Since we are considering only finite hierarchies, these consist of the "bottom-most" cluster. However, there are other worlds that are consistent with the agent's evidence; as implausible as they are, the agent's knowledge requires truth in them as well. (Beside being relevant to defining knowledge, these worlds are relevant in the context of belief revision, discussed in Section 14.2.1.) Figure 13.10 depicts the definitions graphically; the full definitions follow.

Definition 13.7.2 (Logical entailment for KB models) *Let* $A = (W, \pi, \leq_1, \ldots, \leq_n)$ *be a KB- model over* Σ, *and* $w \in W$. *Then we define the* $\models$ *relation as:*

- *If* $\varphi \in \Sigma$, *then* $A, w \models \varphi$ *if and only if* $w \in \pi(\varphi)$.
- $A, w \models K_i\varphi$ *if and only if for all worlds* w', *if* $w' \leq_i w$ *then* $A, w' \models \varphi$.
- $A, w \models B_i\varphi$ *if and only if* φ *is true in all worlds minimal in* $\leq_i$.

We can now ask whether there exists an axiom system that captures the properties of knowledge and belief, as defined by KB models. The answer is yes; the system that does it, while somewhat more complex than **S5** or **KD45**, is nonetheless illuminating. If we only wanted to capture the notion of knowledge, we would need the so-called **S4.3** axiomatic system, a well-known system of modal logic capturing total preorders. But we want to capture knowledge as well as belief. This is done by the following axioms (to which one must add the two usual inference rules of modal logic).

Knowledge:

Axiom 13.7.3 $K_i(\varphi \rightarrow \psi) \rightarrow (K_i\varphi \rightarrow K_i\psi)$

Axiom 13.7.4 $K_i\varphi \rightarrow \varphi$

Axiom 13.7.5 $K_i\varphi \rightarrow K_iK_i\varphi$

Belief:

Axiom 13.7.6 $B_i(\varphi \rightarrow \psi) \rightarrow (B_i\varphi \rightarrow B_i\psi)$

Axiom 13.7.7 $B_i\varphi \rightarrow \neg B_i\neg\varphi$

Axiom 13.7.8 $B_i\varphi \rightarrow B_iB_i\varphi$

Axiom 13.7.9 $\neg B_i\varphi \rightarrow B_i\neg B_i\varphi$

Knowledge and belief:

Axiom 13.7.10 $K_i\varphi \rightarrow B_i\varphi$

Axiom 13.7.11 $B_i\varphi \rightarrow B_iK_i\varphi$

Axiom 13.7.12 $B_i\varphi \rightarrow K_iB_i\varphi$

Axiom 13.7.13 $\neg B_i\varphi \rightarrow K_i\neg B_i\varphi$

The properties of belief are precisely the properties of a **KD45** system, the standard logic of belief. The properties of knowledge as listed consist of the **KD4** system, also known as **S4**. However, by virtue of the relationship between knowledge and belief, one obtains additional properties of knowledge. In particular, one obtains the property of the so-called **S4.2** system, a weak form of introspection:

$$\neg K_i\neg K_i\varphi \rightarrow K_i\neg K_i\neg K_i\varphi.$$

This property is unintuitive, until one notes another surprising connection that can be derived between knowledge and belief in our system:

$$\neg K_i\neg K_i\varphi \leftrightarrow B_i\varphi.$$

While one can debate the merits of this property, it is certainly less opaque than the previous one; and, if this equivalence is substituted into the previous formula, then we simply get one of the introspection properties of knowledge and belief!

13.8   History and references

The most comprehensive one-stop shop for logics of knowledge and belief is Fagin et al. [1995]. It is written by computer scientists, but covers well the perspectives of philosophy and game theory. Readers who would like to go directly to the sources should start with Hintikka [1962] in philosophy, Aumann [1976] in game theory (who introduced the partition model), Moore [1985] in computer science for an artificial intelligence perspective, and Halpern and Moses [1990] in computer science for a distributed systems perspective. The Muddy Children puzzle has its origin in the Cheating Wives puzzle in Gamow and Stern

[1958], and the Coordinated Attack problem in Gray [1978]. Another good reference is provided by the proceedings of Theoretical Aspects of Rationality and Knowledge—or TARK—which can be found online at www.tark.org. (The acronym originally stood for Theoretical Aspects of Reasoning about Knowledge.) There are many books on modal logic in general, not only when applied to reasoning about knowledge and belief; from the classic Chellas [1980] to more modern ones such as Fitting and Mendelsohn [1999] or Blackburn et al. [2002]. The application of logics of knowledge in robotics is based on Brafman et al. [1997] and Brafman et al. [1998].

14

Beyond Belief: Probability, Dynamics,
and Intention

In this chapter we go beyond the model of knowledge and belief introduced in
the previous chapter. Here we look at how one might represent statements such
as "Mary believes that it will rain tomorrow with probability $> .7$," and even
"Bill knows that John believes with probability .9 that Mary believes with probability $> .7$ that it will rain tomorrow." We will also look at rules that determine
how these knowledge and belief statements can change over time, more broadly
at the connection between logic and games, and consider how to formalize the
notion of intention.

14.1 Knowledge and probability

In a Kripke structure, each possible world is either possible or not possible for
a given agent, and an agent knows (or believes) a sentence when the sentence
is true in all of the worlds that are accessible for that agent. As a consequence,
in this framework both knowledge and belief are binary notions in that agents
can only believe or not believe a sentence (and similarly for knowledge). We
would now like to add a quantitative component to the picture. In our quantitative
setting we will keep the notion of knowledge as is, but will be able to make
statements about the degree of an agent's belief in a particular proposition. This
will allow us to express not only statements of the form "the agent believes with
probability .3 that it will rain" but also statements of the form "agent i believes
with probability .3 that agent j believes with probability .9 that it will rain."
These sort of statements can become tricky if we are not careful—for example,
what happens if $i = j$ in the last sentence?[1]

There are several ways of formalizing a probabilistic model of belief in a
multiagent setting, which vary in their generality. We will define a relatively
restrictive class of models, but will then mention a few ways in which these can
be generalized.

---

1. Indeed, some years back the *New Yorker* magazine published a cartoon with the following caption:
*There is now 60% chance of rain tomorrow, but there is 70% chance that later this evening the chance of
rain tomorrow will be 80%.*

Our technical device will be to simply take our partition model and overlay
a commonly known probability distribution (called the *common prior*) over the
possible worlds.

common prior

Definition 14.1.1 (Multiagent probability structure) *Given a set X, let* $\Pi(X)$
be the class of all probability distributions over X. Then we define a (common-
prior) *multiagent probability structure*

common prior

M over a nonempty set Φ *of primitive propositions as the tuple* $(W, \pi, I_1,$
$\ldots, I_n, \mathcal{P})$, *where:*

- *W is a nonempty set of* possible worlds;
- $\pi : \Phi \mapsto 2^W$ *is an* interpretation *that associates with each primitive propo-*

interpretation

  sition $p \in \Phi$ *the set of possible worlds* $w \in W$ *in which p is true;*
- *each* I_i *is a partition relation, just as in the original partition model (see*
  Definition 13.1.1); and
- $\mathcal{P} \in \Pi(W)$ *is the* common prior probability.

Adding the probability distribution does not change the set of worlds that an
agent considers possible from a world w, but it does allow us to quantify how
likely an agent considers each of these possible worlds. In world w, agent i can
condition on the fact that it is in the partition $I(w)$ to determine the probability
that it is in each $w' \in I(w)$. For all $w' \notin I_i(w)$ it is the case that $P_i(w' \mid w) = 0$,
and for all $w' \in I_i(w)$ we have:

$$P_i(w'|w) = \frac{\mathcal{P}(w')}{\sum_{v|v \in I(w)} \mathcal{P}(v)}. \tag{14.1}$$

Note that this means that if w' and w'' lie in the same partition for agent i,
then $P_i(w|w') = P_i(w|w'')$. This is one of the restrictions of our formulation to
which we return later.

We will often drop the designations "common prior" and "multiagent" when
they are understood from the context and simply use the term "probability structure." Next we discuss the syntax of a language to reason about probability
structures and its semantics. We define the syntax of this new language formally
as follows.

Definition 14.1.2 (Well-formed probabilistic sentences) *Given a set* Φ *of*
primitive propositions, the set of well-formed probabilistic sentences $\mathcal{L}_P$ *is de-*
fined by:

- $\Phi \subseteq \mathcal{L}_P$.
- *If* $\varphi, \psi \in \mathcal{L}_P$, *then* $\varphi \wedge \psi \in \mathcal{L}_P$ *and* $\neg\varphi \in \mathcal{L}_P$.
- *If* $\varphi \in \mathcal{L}_P$, *then* $P_i(\varphi) \geq a \in \mathcal{L}_P$, *for* $i = 1, \ldots, n$ *and* $a \in [0, 1]$.

The intuitive meaning of these statements is clear; in particular, $P_i(\varphi) \geq a$
states that the probability of φ is at least a. For convenience, we can use several
other comparison operators on probabilities, including $P_i(\varphi) \leq a \equiv P_i(\neg\varphi) \geq a$,
$P_i(\varphi) = a \equiv P_i(\varphi) \geq a \wedge P_i(\varphi) \leq a$, and $P_i(\varphi) > a \equiv P_i(\varphi) \geq a \wedge \neg P_i(\varphi) = a$.

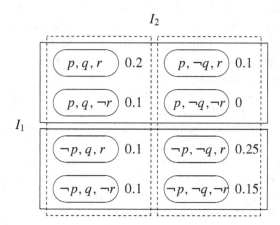

Figure 14.1 A KP structure with a common prior.

Now note that we can use this language and its abbreviations to express formally our sample higher-order sentences given earlier. To express the sentence "agent 1 believes with probability 0.3 that agent 2 believes with probability 0.7 that q," we would write $P_1(P_2(q) = 0.7) = 0.3$. Similarly, to express the sentence "agent 1 believes with probability 1 that she herself believes with probability at least 0.5 that p," we would write $P_1(P_1(p) \geq 0.5) = 1$.

Now we define $\models$, the satisfaction relation, which links our syntax and semantics.

Definition 14.1.3 ($\models$ relation) *Let $p \in \Phi$ be a primitive proposition and φ and ψ be sentences of modal logic. We define the $\models$ relation as:*

- $M, w \models p$ *if and only if $w \in \pi(p)$.*
- $M, w \models \neg\varphi$ *if and only if $M, w \not\models \varphi$.*
- $M, w \models \varphi \wedge \psi$ *if and only if $M, w \models \varphi$ and $M, w \models \psi$.*
- $M, w \models P_i(\varphi) \geq a$ *if and only if $\dfrac{\sum_{v \mid (v \in I(w)) \wedge (M, v \models \varphi)} P(v)}{\sum_{v \mid v \in I(w)} P(v)} \geq a$.*

The following example illustrates the definitions. Consider Figure 14.1. The interpretation of this structure is that each agent knows the truth of exactly one proposition: p for agent 1, and q for agent 2. If the real world is $w = (p, q, r)$, then we have $M, w \models (P_1(q) = 0.75)$, because conditioning on the partition $I_1(w)$ yields: $\frac{0.2+0.1}{0.2+0.1+0.1+0} = 0.75$. We now go a level deeper to consider agent 2 modeling agent 1. In the partition for agent 2 containing the true world, $I_2(w)$, the probability that agent 2 assigns to being in the partition $I_1(w)$ is $\frac{0.2+0.1}{0.2+0.1+0.1+0.1} = 0.6$. Since the other partition for agent 1 yields a different $P_1(q)$, we have the sentence $M, w \models P_2(P_1(q) = 0.75) = 0.6$.

This is a good point at which to discuss to some of the constraints of the theory as we have presented it. To begin with, we have implicitly assumed that the partition structure of each agent is common knowledge among all agents. Second, we assumed that the beliefs of the agents are based on a common prior and are obtained by conditioning on the worlds in the partition. Both are substantive assumptions and have ramifications. For example, we have the fact that the beliefs of an agent are the same within all worlds of any given partition. Also note that we have a strong property of discreteness in the higher-order beliefs. Specifically, the number of statements of the form "$M, w \models P_i(P_j(\varphi) = a) = b$" in which $b > 0$ is equal to at most the number of partitions for agent j, and the sum of all b's from these statements is equal to 1. Thus, you do not need intervals to account for all of the probability mass, as you would with a continuous distribution. In fact, it is the case that for any depth of recursive modeling, you can always decompose a probability over a range into a finite number of probabilities of individual points. One could imagine an alternative formulation in which agent i did not know the prior of agent j, but instead had a continuous distribution over agent j's possible priors. In that case, you could have $M, w \models P_i(P_j(\varphi) \geq b) > 0$ without there being any specific $a \geq b$ such that $M, w \models P_i(P_j(\varphi) = a) > 0$.

We could in fact endow agents with probabilistic beliefs without assuming a common prior and with a much more limited use of the partition structure. For example, we could replace, in each world and for each agent, the set of accessible worlds by a probability distribution over a set of worlds $\mathcal{P}_{i,w}$. In this case the semantic condition for belief would change to:

- $M, w \models P_i(\varphi) \geq a$ if and only if $\sum_{v;M,v\models\varphi} \mathcal{P}_{i,w}(v) \geq a$.

If we want to retain the partition structure so we can speak about knowledge as well as (probabilistic) belief, we could add the requirement that $\mathcal{P}_{i,w}(w') = 0$ for all $w' \notin I(w)$.

However, such interesting extensions bring up a variety of complexities, including, in particular, the axiomatization of such a system, and throughout this book we stay within the confines of the theory as presented.

As we have discussed, every probability space gives rise to an infinite hierarchy of beliefs for each agent: beliefs about the values of the primitive propositions, about other agents' beliefs, beliefs about other agents' beliefs about other agents' beliefs, and so on. This collection of beliefs is called the "epistemic type space" in game theory and plays an important role in the study of games of incomplete information, or Bayesian games, discussed in Section 6.3. (There, each possible world is a different game, and the partition of each agent represents the set of games that the agent cannot distinguish between, given the signal it receives from nature.) One can ask whether the converse holds: Is it the case that every type space is given rise to by some multiagent probability structure? The perhaps surprising answer is yes so long as the type space is *self coherent*. Self coherence, which plays the role analogous to a combination of the positive and negative introspection properties of knowledge and (qualitative) belief, is defined as follows.

self-coherent
belief

Definition 14.1.4 (Self-coherent belief) *A collection of higher-order beliefs is self-coherent iff, for any agent i, any proposition φ, and any $a \in [0, 1]$, if $P_i(\varphi) \geq a$ then $P_i(P_i(\varphi) \geq a) = 1$.*

common
probabilistic
belief

Recall that in modal logics of knowledge and belief we augmented the single-agent modalities with group ones. In particular, we defined the notion of *common knowledge*. We can now do the same for probabilistic beliefs and define the notion of *common (probabilistic) belief*.

Definition 14.1.5 (Common belief) *A sentence* φ *is* commonly believed *among two agents a and b, written* $C^p_{a,b}(\varphi)$, *if* $P_a(\varphi) = 1$, $P_b(\varphi) = 1$, $P_a(P_b(\varphi) = 1) = 1$, $P_b(P_a(\varphi) = 1) = 1$, $P_a(P_b(P_a(\varphi) = 1) = 1) = 1$, *and so on. The definition extends naturally to common (probabilistic) belief among an arbitrary set G of agents, denoted* C^p_G.

Now that we have the syntax and semantics of the language, we may ask whether there exists an axiomatic system for this language that is sound and complete for the class of common-prior probability spaces. The answer is that one exists, and it is the following.

Definition 14.1.6 (Axiom system AX$_P$) *The axiom system* **AX**$_P$ *consists of the following axioms and inference rules (in the schemas that follow,* φ *and* ψ *range over all sentences, and* $n_i, n_j \in [0, 1]$*).*

Axiom 14.1.7 (A1) *All of the tautological schema of propositional logic*

Axiom 14.1.8 (P1: Nonnegativity) $P_i(\varphi) \geq 0$

Axiom 14.1.9 (P2: Additivity) $P_i(\varphi \wedge \psi) + P_i(\varphi \wedge \neg\psi) = P_i(\varphi)$

Axiom 14.1.10 (P3: Syntax independence) $P_i(\varphi) = P_i(\psi)$ *if* $\varphi \Leftrightarrow \psi$ *is a propositional tautology*

Axiom 14.1.11 (P4: Positive introspection) $(P_i(\varphi) \geq a) \rightarrow P_i(P_i(\varphi) \geq a) = 1$

Axiom 14.1.12 (P5: Negative introspection) $(\neg P_i(\varphi) \geq a) \rightarrow P_i(\neg P_i(\varphi) \geq a) = 1$

Axiom 14.1.13 (P6: Common-prior assumption) $C^p_{i,j}(P_i(\varphi) = n_i \wedge P_j(\varphi) = n_j) \rightarrow n_i = n_j$

Axiom 14.1.14 (R1) *From* φ *and* $\varphi \rightarrow \psi$ *infer* ψ

Axiom 14.1.15 (R2) *From* φ *infer* $P_i(\varphi) = 1$

Note that axioms P4 and P5 are the direct analogs of the positive and negative introspection axioms in the modal logics of knowledge and belief. Axiom P6 is novel and captures the common-prior assumption.

Now we are ready to give the soundness and completeness result.

Theorem 14.1.16 *The axiom system* **AX**$_P$ *is sound and complete with respect to the class of all common-prior multiagent probability structures.*

14.2 Dynamics of knowledge and belief

We have so far discussed how to represent "snapshots" of knowledge and belief. We did speak a little about how, for example, knowledge changes over time, for example in the context of the Muddy Children problem. But the theories presented were all static ones. For example, in the Muddy Children problem we used the partition model to represent the knowledge state of the children after each time the father asks his question, but did not give a formal account for how the system transitions from one state to the other. This section is devoted to discussing such dynamics.

We will first consider the problem of *belief revision*, which is the process of revising an existing state of belief on the basis of newly learned information. We will consider the revision of both qualitative and quantitative (i.e., probabilistic) beliefs. Then we will briefly look at dynamic operations on beliefs that are different from revision, such as *update* and *fusion*.

14.2.1   *Belief revision*

belief revision

One way in which the knowledge and belief of an agent change is when the agent learns new facts, whether by observing the world or by being informed by another agent. Whether the beliefs are categorical (as in the logical setting) or quantitative (as in the probabilistic setting), we call this process *belief revision*.

When the new information is consistent with the old beliefs, the process is straightforward. In the case of categorical beliefs, one simply adds the new beliefs to the old ones and takes the logical closure of the union. That is, consider a knowledge base (or belief base—here it does not matter) K and new information φ such that $K \not\models \neg\varphi$. The result of revising K by φ, written $K * \varphi$, is simply $Cn(K, \varphi)$, where Cn denotes the logical closure operator. Or thought of semantically, the models of $K * \varphi$ consist of the intersection of the models of K and the models of φ.

The situation is equally straightforward in the probabilistic case. Consider a prior belief in the form of a probability distribution $P(\cdot)$ and new information φ such that $P(\varphi) > 0$. $P * \varphi$ is then simply the posterior distribution $P(\cdot \mid \varphi)$.

Note that in both the logical and probabilistic settings, the assumption that the new information is consistent with the prior belief (or knowledge) is critical. Otherwise, in the logical setting the result of revision yields the empty set of models, and in the probabilistic case the result is undefined. Thus the bulk of the work in belief revision lies in trying to capture how beliefs are revised by information that is inconsistent with the prior beliefs (and, given their defeasibility, these are really beliefs, as opposed to knowledge).

One might be tempted to argue that this is a waste of time. If an agent is misguided enough to hold false beliefs, let him suffer the consequences. But this is not a tenable position. First, it is not only the agent who suffers the consequences, but also other agents—and we, the modelers—who must reason about him. But beyond that, there are good reasons why agents might hold firm beliefs and later retract them. In the logical case, if an agent were to wait for foolproof evidence

default reasoning

before adopting any belief, the agent would never believe anything but tautologies. Indeed, the notion of belief is intimately tied to that of *default reasoning* and

nonmonotonic reasoning

nonmonotonic reasoning, which are motivated by just this observation.

In the probabilistic case too there are times at which it is unnatural to assign an event a probability other than zero. We encounter this, in particular, in the context of noncooperative game theory. There are situations in which it is a *strictly dominant strategy* (see Section 3.4.3) for an agent to take a certain action, as in the following example.

Two foes, Lance and Lot, about to enter into a duel. Each of them must choose a weapon and then decide on a fighting tactic. Lance can choose among two swords—an old, blunt sword, and a new, sharp one. The new one is much better than the old, regardless of the weapon selected by Lot and the tactics selected by either foe. So in selecting his fighting tactic, Lot is justified in assuming that Lance will have selected the new sword with probability one. But what should Lot do if he sees that Lance selected the old sword after all?

If this example seems a bit unnatural, the reader might refer to the discussion of *backward induction* in Chapter 3.

backward
induction

Indeed, a great deal of attention has been paid to belief revision by information inconsistent with the initial beliefs. We first describe the account of belief revision in the logical setting; we then show how the account in the probabilistic setting is essentially the same.

Logical belief revision: The AGM model

We start our discussion of belief revision semantically and with a familiar structure—the KB-models of Section 13.7. In that section, KB-models were used to distinguish knowledge from (a certain kind of) belief. Here we use them for an additional purpose, namely, to reason about conditional beliefs. Technically, in Section 13.7, KB-models were used to give meaning to the two modal operators K_i and B_i. Given a particular world, K_i was defined as truth in all worlds "downward" from it and B_i was defined (loosely speaking) as truth in all worlds in the "bottom-most" cluster. But the KB-model can be mined for further information, and here is the intuition. Imagine a KB-model, and a piece of evidence φ. Now erase from that model all the worlds that do not satisfy φ and all the links to and from those worlds; the result is a new, reduced KB-model. The new beliefs of the agent, after taking φ into account, are the beliefs in this reduced KB-model. The following definition makes this precise.

Definition 14.2.1 (Belief revision in a KB-model) *Given a KB-model* $A = (W, \pi, \leq_1, \ldots, \leq_n)$ *over* Σ *and any* $\varphi \in \Sigma$, *let* $W(\varphi) = \{w \in W \mid A, w \models \varphi\}$, *let* $\leq_i (\varphi) = \{(w_1, w_2) \in \leq_i | A, w_1 \models \varphi, A, w_2 \models \varphi\}$, *and let* $A(\varphi) = (W(\varphi), \pi, \leq_1 (\varphi), \ldots, \leq_n (\varphi))$. *Then the beliefs of agent* i *after receiving evidence* φ, *denoted* B_i^φ, *are defined as*

$$A, w \models B_i^\varphi(\psi) \text{ iff } A(\varphi), w \models B\psi.$$

Figure 14.2 illustrates this definition. The checkered areas contain all worlds that do no satisfy φ.

Note some nice properties of this definition. First, note that you have that $B_i\psi \leftrightarrow B_i^{true}\psi$, where *true* is any tautology. Second, recall the philosophical slogan regarding knowledge as belief that is stable with respect to the truth. We now see, in a precise sense, why our definitions of knowledge and belief can be viewed as embodying this slogan. Consider ψ such that $A, w \models K_i\psi$. Obviously, $A, w \models B_i\psi$, but it is also the case that $A, w \models B_i^\varphi \psi$ for any φ such

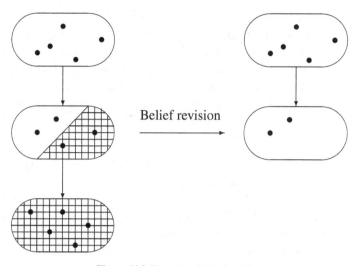

Figure 14.2 Example of belief revision.

that $A, w \models \varphi$. So we get one direction of the slogan—if a proposition is known, then indeed it is a belief that will be retained in the face of any correct evidence. While the converse—that any believed proposition that is not known can be falsified by some evidence—is less straightforward, we point that it would hold if the language Σ were rich enough to capture all propositions.

As usual, we ask whether this semantic definition can be characterized through alternative means. The answer is yes, and in fact we will discuss two ways of doing
AGM postulate   so. The first is via the so-called *AGM postulates*, so named after Alchourrón, Gärdenfors, and Makinson. These set-theoretic postulates take arbitrary belief revision operator $*$ and the initial belief set K and ask, for any evidence φ, what
prioritization   the properties should be of $K * \varphi$, the revision K by φ. For example, one property
rule   is the following: $\varphi \in K * \varphi$. This is the *prioritization rule*, or the *gullibility rule*—
gullibility rule   new information is always accepted and is thus given priority over old one. (As we see below, other rules require that the theory be consistent, which can cause old information to be discarded as a result of the revision.)

The AGM postulates for revising a theory K are as follows. (Recall that $Cn(T)$ denotes the tautological consequence of the theory T.)

($*$1) $K * \varphi = Cn(K * \varphi)$.

($*$2) $\varphi \in K * \varphi$.

($*$3) $K * \varphi \subseteq Cn(K, \varphi)$.

($*$4) If $\neg\varphi \notin K$ then $Cn(K, \varphi) \subseteq K * \varphi$.

($*$5) If φ is consistent then $K * \varphi$ is consistent.

($*$6) If $\models \varphi \leftrightarrow \psi$ then $K * \varphi = K * \psi$.

($*$7) $K * (\varphi \wedge \psi) \subseteq Cn(K * \varphi, \psi)$

($*$8) If $\neg\psi \notin K * \varphi$ then $Cn(K * \varphi, \psi) \subseteq K * (\varphi \wedge \psi)$.

In a sense that we will make precise later, these exactly characterize belief revision with KB-models. But before making this precise, let us discuss an alternative axiomatic characterization of belief revision operators. This is an axiomatic theory of consequence relations. By definition a meta-theory, this axiomatic theory consists of rules of the form "if derivation x is valid then so is derivation y."

It turns out that this meta-theoretic characterization lends particularly deep insight into belief dynamics. For example, in the classical logic setting, the *monotonicity rule* is valid.

monotonicity rule

(Monotonicity) $\frac{\alpha \vdash \beta}{\alpha, \gamma \vdash \beta}$

The analogous rule for belief revision would be "if $\beta \in K * \alpha$ then $\beta \in K * (\alpha \wedge \gamma)$." This does not hold for belief revision, and indeed for this reason belief revision is often called *nonmonotonic reasoning*. There are however other meta-rules that hold for belief revision. Consider the following rules; in these we use the symbol $\vdash\!\!\sim$ to represent the fact that we are reasoning about a nonmonotonic consequence relation, and not the classical $\vdash$.

nonmonotonic reasoning

(Left logical equivalence) $\frac{\models \alpha \leftrightarrow \beta, \alpha \vdash\!\!\sim \gamma}{\beta \vdash\!\!\sim \gamma}$

(Right weakening) $\frac{\models \alpha \to \beta, \gamma \vdash\!\!\sim \alpha}{\gamma \vdash\!\!\sim \beta}$

(Reflexivity) $\alpha \vdash\!\!\sim \alpha$

(And) $\frac{\alpha \vdash\!\!\sim \beta, \alpha \vdash\!\!\sim \gamma}{\alpha \vdash\!\!\sim \beta \wedge \gamma}$

(Or) $\frac{\alpha \vdash\!\!\sim \gamma, \beta \vdash\!\!\sim \gamma}{\alpha \vee \beta \vdash\!\!\sim \gamma}$

(Cautious monotonicity) $\frac{\alpha \vdash\!\!\sim \beta, \alpha \vdash\!\!\sim \gamma}{\alpha \wedge \beta \vdash\!\!\sim \gamma}$

(Rational monotonicity) $\frac{\alpha \wedge \beta \not\vdash\!\!\sim \gamma, \alpha \not\vdash\!\!\sim \neg \beta}{\alpha \not\vdash\!\!\sim \gamma}$

A consequence relation that satisfies all of these properties is called a *rational consequence relation*.

rational consequence relation

The following theorem ties together the semantic notion, the axiomatic one, and the meta-axiomatic one.

Theorem 14.2.2 *Consider propositional language L with a finite alphabet, a revision operator ∗, and a theory $K \subseteq L$. Then the following are equivalent:*

1. *∗ is defined by a finite total preorder: There is a single-agent KB-model A and a world w such that $A, w \models B\rho$ for each $\rho \in K$, and for each $\varphi, \psi \in L$ it is the case that $\psi \in K * \varphi$ iff $A, w \models B^\varphi \psi$;*
2. *∗ satisfies the AGM postulates;*
3. *∗ is a rational consequence relation.*

Probabilistic belief revision

As we have discussed, the crux of the problem in probabilistic belief revision is the inability, in the traditional Bayesian view, to condition beliefs on evidence whose prior probability is zero (also known as measure-zero events). Specifically, the definition of conditional probability,

$$P(A \mid B) = \frac{P(A \cap B)}{P(B)},$$

leaves the conditional belief $P(A \mid B)$ undefined when $P(B) = 0$.

Popper function

nonstandard
probability

There exist various extensions of traditional probability theory to deal with this problem. In one of them, based on so-called *Popper functions*, one takes the expression $P(A \mid B)$ as primitive, rather than defined. In another, called the theory of *nonstandard probabilities*, one allows as probabilities not only the standard real numbers but also *nonstandard reals*. These two theories, as well as several others, turn out to be essentially the same, except for some rather fine mathematical subtleties. We discuss explicitly one of these theories, the theory of lexicographic probability systems (LPSs).

lexicographic
probability
system (LPS)

Definition 14.2.3 (Lexicographic probability system) *A (finite) lexicographic probability system (LPS) is a sequence $p = p_1, p_2, \ldots, p_n$ of probability distributions. Given such an LPS, we say that event $p(A \mid B) = c$ if there is an index $1 \leq i \leq n$ such that for all $1 \leq j < i$, it is the case that $p_j(B) = 0$, $p_i(B) \neq 0$, and (now using the classical notion of conditional probability) $p_i(A|B) = c$. Similarly, we say that event A has a higher probability than event B $(p(A) > p(B))$ if there is an index $1 \leq i \leq n$ such that for all $1 \leq j < i$, it is the case that $p_j(A) = p_j(B)$, and $p_i(A) > p_i(B)$.*

In other words, to determine the probability of an event in an LPS, one uses the first probability distribution if it is well defined for this event. If it is not, one tries the second distribution, and so on until the probability is well defined. A standard example of this involves throwing a die. It may land on one of its six faces, with equal probability. There is also a possibility that it will land on one of its twelve edges; minuscule as this probability is, it is nonetheless a possible event. Finally, even more improbably, the die may land and stay perched on one of its eight corners. In a classical probabilistic setting one usually accords each of the first six events—corresponding to the die landing on one of the faces—a probability of 1/6, and the other 20 events a probability of zero. In this more general setting, however, one can define an LPS (p_1, p_2, p_3) as follows. p_1 would be the classical distribution just described. p_2 would give a probability of 1/12 to the die landing on each of the edges, and 0 to all other events. Finally, the p_3 would give a probability of 1/8 to the die landing on each of the corners, and 0 to all other events.

LPSs are closely related to AGM-style belief revision.

Theorem 14.2.4 *Consider propositional language L with a finite alphabet, a revision operator $*$, and a theory $K \subseteq L$. Then the following are equivalent.*

1. *$*$ satisfies the AGM postulates for revising K.*
2. *There exists an LPS $p = p_1, \ldots, p_n$ such that $p_1(K) = 1$, and such that for every φ and ψ it is the case that $\psi \in K * \varphi$ iff $p(\psi \mid \varphi) = 1$.*

Thus, we have extended the three-way equivalence of Theorem 14.2.2 to a four-way one.

14.2.2 *Beyond AGM: update, arbitration, fusion, and friends*

In discussing the dynamics of belief we have so far limited ourselves to belief revision, and a specific type of revision at that (namely, AGM-style belief revision). But there are other forms of belief dynamics, and this section discusses some of them. We do so more briefly than with belief revision; at the end of the chapter we provide references to further readings on these topics.

Expansion and contraction

belief expansion

To begin with, there are two simple operations closely linked to revision that are worth noting. *Expansion* is simply the addition of a belief, regardless of whether it leads to a contradiction. The expansion of a theory K by a formula φ, written $K + \varphi$, is defined by

$$K + \varphi = Cn(K \cup \{\varphi\}).$$

belief contraction

Harper identity

Contraction is the operation of removing just enough from a theory to make it consistent with new evidence. The contraction of a theory K by a formula φ, written $K - \varphi$, is reduced to the revision operator via the *Harper identity*,

$$K - \varphi = K \cap (K * \neg\varphi).$$

Levi identity

The *Levi identity* relates the three operations of revision, expansion, and contraction:

$$K * \varphi = (K - \neg\varphi) + \varphi.$$

Update

belief update

Belief *update* is another interesting operation. Similar to revision, it incorporates new evidence into an existing belief state, ensuring that consistency is maintained. But the intuition behind update is subtly different from the intuition underlying revision. In revision, the second argument represents new evidence of facts that were true all along. In update, the second argument represents facts that have possibly become true only after the original beliefs were formed. Thus if an agent believes that it is not raining and suddenly feels raindrops, in the case of revision the assumption is that he was wrong to have those original beliefs, but in the case of update the assumption is that he was right, but that subsequently it started raining.

The different intuitions underlying revision and update translate to different conclusions that they can yield. Consider the following initial belief: "Either the room is white, or else the independence day of Micronesia is November 2 (or both)." Now consider two scenarios. In the first one, you look in the room and see that it is green; the white hypothesis is ruled out and you infer that the independence day of Micronesia is November 2. This is an instance of belief revision. But now consider a different scenario, in which a painter emerges from the room and informs you that he has just painted the room green. You now have no business inferring anything about Micronesia. This is an instance of belief update.[2]

---

2. And good thing too, as the Federated States of Micronesia, which have been independent since 1986, celebrate independence day on November 3.

Since revision and update are different, it is to be expected that the set of postulates governing update will differ from the AGM postulates. The update of a theory K by a formula φ is written $K \diamond \varphi$. The so-called KM postulates for updating a (consistent) theory K (so named after Katsuno and Mendelzon) read as follows.

($\diamond 1$) $K \diamond \varphi = Cn(K \diamond \varphi)$.

($\diamond 2$) $\varphi \in K \diamond \varphi$.

($\diamond 3$) If $\varphi \in K$ then $K \diamond \varphi = K$.

($\diamond 4$) $K \diamond \varphi$ is inconsistent iff φ is inconsistent.

($\diamond 5$) If $\models \varphi \equiv \psi$ then $K \diamond \varphi = K \diamond \psi$.

($\diamond 6$) $K \diamond (\varphi \wedge \psi) \subseteq (K \diamond \varphi) + \psi$.

($\diamond 7$) If $\psi \in K \diamond \varphi$ and $\varphi \in K \diamond \psi$ then $K \diamond \varphi = K \diamond \psi$.

($\diamond 8$) If K is complete[3] then $K \diamond (\varphi \wedge \psi) \subseteq K \diamond \varphi \cap K \diamond \psi$.

($\diamond 9$) $K \diamond \varphi = \cap_{M \in Comp(K)} M \diamond \varphi$, where $Comp(K)$ denotes the set of all complete theories that entail K.

As in the case of revision, the model theory of update is also well understood, and is as follows (this discussion is briefer than that for revision, and we point the reader to further reading at the end of the chapter).

Revision and update can be related by the identity

$$K \diamond \varphi = \cap_{M \in Comp(K)} M * \varphi.$$

Comparing this property to the last postulate for update, one can begin to gain intuition for the model theoretic characterization of the operator. In particular, it is the case that there is no distinction between revision and update when dealing with complete theories. More generally, the following completely characterizes the class of update operators that obey the KM postulates.

Theorem 14.2.5 *The following two statements about an update operator $\diamond$ are equivalent:*

1. *$\diamond$ obeys the KM postulates;*
2. *There exists a function that maps each interpretation M to a partial preorder $\leq_M$ such that $Mod(K \diamond \varphi) = \cup_{M \in Comp(K)} Min(Mod(M), \leq_M)$.*

Note two important contrasts with the model-theoretic characterization of belief revision. First, here the preorders need not be total. Second, and more critically, in revision there is one global preorder on worlds; here each interpretation induces a different preorder.

Arbitration

Returning to AGM-style revision, we note two aspects of asymmetry. The first is blatant—new evidence is given priority over existing beliefs. The second is more subtle—the first argument to any AGM revision operator is a "richer" type

---

3. A theory K is complete if for each sentence φ, either $\varphi \in K$ or $\neg\varphi \in K$.

of object than the second. In this subsection we discuss the first asymmetry, and in the next subsection the second.

There are certainly situations in which new evidence is not necessarily given precedence over existing beliefs. Indeed, from the multiagent perspective, each revision can be seen as involving at least two agents—the believer and the informer. When synthesizing a new belief, the believer may wish to accord himself higher priority, to not favor one of them over the other, or to give priority to one over the other depending on the subject matter. Various theories exist to capture this intuition. One of them is the theory of belief *arbitration*, which takes an egalitarian approach: it does not favor either of the two sides over the other. Technically speaking, it does so by jettisoning the second AGM postulate $\varphi \in K * \varphi$, and replaces it with a "fairness" axiom:

belief arbitration

$$\text{if } K \cup \{\varphi\} \models \bot, \text{ then } K \not\subseteq K * \varphi \text{ and } \varphi \notin K * \varphi.$$

The intuition is that if the new evidence is inconsistent with the initial beliefs, then each must "give up something."

Fusion

The other source of asymmetry in AGM revision is more subtle. The AGM postulates obscure the asymmetry between the two arguments. In $K * \varphi$, K is a *belief set*, or a set of sentences (which happen to be tautologically closed). True, usually we think of φ as a single sentence, but for most purposes it would matter little if we have it represent a set of sentences. However, it must be remembered that the AGM postulates do not define the operator $*$, but only constrain it. As is seen from Theorem 14.2.2, any particular $*$ is defined with respect to a complete *belief state* or total preorder on possible worlds. This belief state defines the initial belief set, but it defines much more, namely, all the beliefs conditional on new evidence. Thus every specific AGM operator takes as its first input a belief state and as its second a mere belief set. We now consider what happens when the second argument is also a belief state.

belief set

belief state

The first question to ask is what does it mean intuitively to take a second belief state as an argument. Here again the multiagent perspective is useful. We think of a belief set as describing "current" beliefs and a belief state as describing "conditional" (i.e., "current" as well as "hypothetical") beliefs. According to AGM revision, then, the believer has access to his full conditional beliefs, but the informer reveals only (some of his) current beliefs. However, one could imagine situations in which the informer engages in "full disclosure," revealing not only all his current beliefs but also all his hypothetical ones. When this is the case, the believer is faced with the task of merging two belief states; this process is called belief *fusion*.

Belief fusion calls for resolving conflicts among the two belief states. From the technical standpoint, the basic unit of conflict here is no longer the inconsistency of two beliefs, but rather the inconsistency of the two orderings on possible worlds. Here is how one theory resolves such conflicts; we present it in some detail since it very explicitly adopts a multiagent perspective.

To develop intuition for the following definitions, imagine a set of information sources and a set of agents. The sources can be thought of as primitive agents with fixed belief states. Each source informs some of the agents of its belief state; in effect, each source offers the opinion that certain worlds are more likely than others and remains neutral about other pairs.

An agent's belief state is simply the amalgamation of all these opinions, each annotated by its origin (or "pedigree"). Of course, these opinions in general conflict with one another. To resolve these conflicts, the agent places a strict "credibility" ranking on the sources and accepts the highest-ranked opinion offered on every pair of worlds.

We define the *pedigreed belief state* as follows.

pedigreed belief
state

Definition 14.2.6 (Pedigreed belief state) *Given a finite set of belief states S (over W), the* pedigreed belief state *(over W) induced by S is a function $\Psi : W \times W \mapsto 2^{S \cup \{s_0\}}$ such that $\Psi(w_1, w_2) = \{(W, \leq) \in S : w_2 \not\leq w_1\} \cup \{s_0\}$.*

In words, $\Psi(w_1, w_2)$ is the set of all agents who do not believe that world w_2 is at least as likely as w_1. The agent s_0 has no beliefs and is thus always present.

We will use S to denote the set of all of sources over W, and throughout this section we will consider pedigreed belief states that are induced by subsets of S. Note that both $\{\}$ and s_0 induce the same pedigreed belief state; by slight abuse of notation we will denote it too by s_0.

Next we define a particular policy for resolving conflicts within a pedigreed belief state, since for any two worlds w_1 and w_2, we have the competing camps of $\Psi(w_1, w_2)$ and $\Psi(w_2, w_1)$. We assume a strict ranking $\sqsubset$ on S (and thus also on the sources that induce any particular Ψ). We interpret $s_1 \sqsubset s_2$ as "s_2 is more credible than s_1." As usual, we define $\sqsubseteq$, read "as credible as," as the reflexive closure of $\sqsubset$.

We also assume that s_0 is the least credible source, which may merit some explanation. It might be asked why equate the most agnostic source with the least credible one. In fact we do not have to, but since in the definitions that follow, agnosticism is overridden by any opinion regardless of credibility ranking, we might as well assume that all agnosticism originates from the least credible source, which will permit simpler definitions.

Intuitively, given a pedigreed belief state Ψ, $\Psi_\sqsubset$ will retain from Ψ the highest-ranked opinion about the relative likelihood between any two worlds.

dominating
belief state

Definition 14.2.7 (Dominating belief state) *Given W, S, Ψ and $\sqsubset$ as defined earlier, the* dominating belief state *of Ψ is the function $\Psi_\sqsubset : W \times W \mapsto S$ such that $\forall w_1, w_2 \in W$ the following holds: If $\max(\Psi(w_2, w_1)) \sqsubset \max(\Psi(w_1, w_2))$ then $\Psi_\sqsubset(w_1, w_2) = \max(\Psi(w_1, w_2))$. Otherwise, $\Psi_\sqsubset(w_1, w_2) = s_0$.*[4]

Clearly, for any $w_1, w_2 \in W$ either $\Psi_\sqsubset(w_1, w_2) = s_0$ or $\Psi_\sqsubset(w_2, w_1) = s_0$ or both.

It is not hard to see that $\Psi_\sqsubset$ induces a standard (anonymous) belief state.

---

4. Note the use of the restrictions. Finiteness assures that a maximal source exists; we could readily replace it by weaker requirements on the infinite set. The absence of ties in the ranking $\sqsubset$ ensures that the maximal source is unique; removing this restriction is not straightforward.

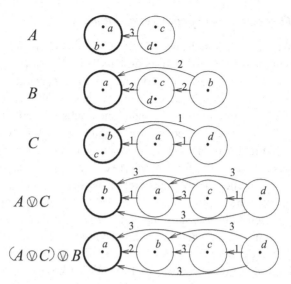

Figure 14.3 Example of a fusion operator. Only the dominating belief states are shown.

Definition 14.2.8 (Ordering induced by $\Psi_\sqsubset$) *The ordering induced by $\Psi_\sqsubset$ is the relation $\preceq$, a binary relation on $\mathcal{W}$, such that $w_1 \preceq w_2$ iff $\Psi_\sqsubset(w_2, w_1) = s_0$.*

Clearly, $\preceq$ is a total preorder on $\mathcal{W}$. Thus a dominating belief state is a generalization of the standard notion of belief state.

We are now ready to define the fusion operator.

belief fusion

Definition 14.2.9 (Belief fusion) *Given a set of sources S and $\sqsubset$ as previously, $S_1, S_2 \subset S$, the pedigreed belief state Ψ_1 induced by S_1, and pedigreed belief state Ψ_2 induced by S_2, the fusion of Ψ_1 and Ψ_2, denoted $\Psi_1 \,\varobslash\, \Psi_2$, is the pedigreed belief state induced by $S_1 \cup S_2$.*

Figure 14.3 illustrates the operation of the fusion operator. Of the three agents, A has the highest priority at 3, followed by B and C with priorities of 2 and 1, respectively. The first three lines describe the beliefs of the agents over the four worlds a,b,c, and d in the form of a dominating belief state. An arrow from one circle to another means that all worlds in the second circle are considered at least as likely as all worlds in the first. Each arrow is labeled with the priority of the agent who holds those beliefs. The final two lines show examples of fusion, with the corresponding diagrams showing the resulting dominating belief state. When fusing the beliefs of A and C, we see that all of A's beliefs are retained because it has a higher priority. Since A has no opinion on the pairs (a, b) and (c, d), we take C's beliefs. Of C's two remaining beliefs, $c \leq_C a$ is overruled by $a \leq_A c$, while $b \leq_C d$ is consistent with A's belief and thus does not show up in the output. When we fuse this dominating belief state with that of B, the only possible changes that can be made are on the pairs (a, b) and (c, d), since A has a belief on all other pairs and has a higher priority than B. Agent B has no opinion on (c, d) but disagrees with C on (a, b), causing a reversal of the arrow, which is now labeled with B's priority.

14.2.3 *Theories of belief change: a summary*

We have discussed a number of ways in which beliefs change over time. Starting with AGM revision and its probabilistic counterpart, we went on to discuss the operations of expansion, contraction, update, arbitration, and fusion.

It is important to emphasize that these examples are not exhaustive. Because it is so important, we devote this short subsection to making just this point. There are other specific theories, for example, theories accounting for the *iteration* of belief revision; most of the theories we discussed do not say what happens if one wishes to conduct, for example, two revisions in sequence. Other theories are more abstract, and provide a general framework for belief change, of which revision, update, and other operators are special instances. Among them, importantly, are theories with *information change operators*. These are couched in propositional dynamic logic, which we will encounter in Section 14.4, but in which the modal action operators are indexed by a logical proposition, denoting, for example, the learning of that proposition. In Section 14.5 we provide some references to this literature.

iterated belief revision

information change operators

14.3 Logic, games, and coalition logic

So far in this chapter we looked at the use of logic to reason about purely "informational" aspects of agents. Now we broaden the discussion to include also "motivational" aspects. In this section we (briefly) look at the interaction between modal logic and game theory.

The connection between logic and games is multifaceted. Even before the advent of modern-day game theory, in the context of classical logic, it was proposed that a logical system be viewed as a game between a prover (the "asserter") and a disprover (the "respondent"), with a sentence being valid if the prover had a winning strategy. In this case, games are used to reason about logic. But much recent work has been aimed in the opposite direction, with logic being used to reason about games. Here games are meant in the formal sense of modern game theory, and the logics are modal rather than classical.

We have so far seen how modal logic can be used to model and reason about agents' knowledge and the beliefs and how those change in time. But modal logic can be used to reason also about their actions, preferences, and hence also about games and (certain) solution concepts. Much of the literature here focuses on extensive-form games of both perfect and imperfect (though not yet incomplete) information. There exists a rapidly expanding literature on the topic which is somewhat complex. We will just mention here that these logics allow us to reason about certain solution concepts, particularly those involving only pure strategies. And so one can recapture in logic the notion of (pure strategy) dominant strategy, iterated elimination of dominated strategies, rationalizability, and pure-strategy Nash equilibria.

In lieu of full discussion of this somewhat complex material, to give a feel for what can be expressed in such logics we briefly look at one particular coalition logic   exemplar—so-called *coalition logic* (CL). In the language of CL we do not model the actions of the agents, but rather the capabilities of groups. (In this respect, CL is similar to coalitional game theory.) For any given set of agents C, the modal operator $[C]$ is meant to capture the capability of the group. $[C]\varphi$ means that the group can bring about φ (or, equivalently, can ensure that φ is the case), regardless of the actions of agents outside the set C.

The formal syntax of CL is defined as follows. Given a (finite, nonempty) set of agents N and a set of primitive propositions Φ_0, the set Φ of well-formed formulas is the smallest set satisfying the following:

$\Phi_0 \subset \Phi$;

If $\varphi_1, \varphi_2 \in \Phi$, then $\neg\varphi_1 \in \Phi$ and $\varphi_1 \vee \varphi_2 \in \Phi$;

If $\varphi \in \Phi$, $C \subset N$, and φ is $[C']$-free for all C', then $[C]\varphi \in \Phi$.

$\top$ is shorthand for $\neg\bot$, and $\wedge$, $\rightarrow$, and $\leftrightarrow$ are defined as usual. $\bot$ can be viewed as shorthand for $p \wedge \neg p$, and $[i]\varphi$ is shorthand for $[\{i\}]\varphi$ for any $i \in N$.[5]

The formal semantics of CL are as follows. A CL model is a triple (S, E, V), such that:

S is a set of states or worlds;

$V : \Phi_0 \mapsto 2^S$ is the valuation function, specifying the worlds in which primitive propositions hold;

$E : 2^N \mapsto 2^{2^S}$ such that:

- $E(\{\}) = \{S\}$,
- if $C \subset C'$, then $E(C')$ is a refinement of $E(C)$.

The satisfaction relation is defined as follows:

$(S, E, V) \not\models \bot$;

for $p \in \Phi_0$, $(S, E, V) \models p$ iff $s \in V(p)$;

$(S, E, V) \models \neg\varphi$ iff $(S, E, V) \not\models \varphi$;

$(S, E, V) \models \varphi_1 \vee \varphi_2$ iff $(S, E, V) \models \varphi_1$ or $(S, E, V) \models \varphi_2$;

$(S, E, V) \models [C]\varphi$ iff there exists $S' \in E(C)$ such that for all $s \in S'$ it is the case that $s \models \varphi$ (here $\models$ is used in the classical sense).

What can be said about the sentences that are valid in this logic? For example, clearly, both $\neg[C]\bot$ and $[C]\top$ are valid (no coalition can force a contradiction, and tautologies are true in any model, and thus in any set forced by a coalition). Equally intuitively, $[C](\varphi_1 \wedge \varphi_2) \rightarrow [C]\varphi_2$ is also valid (after all, if a coalition can enforce an outcome to lie within a given set of worlds, it can also enforce it to lie within a superset). Perhaps more insightful is the following valid sentence: $([C_1]\varphi_1 \wedge [C_2]\varphi_2) \rightarrow [C_1 \cup C_2](\varphi_1 \wedge \varphi_2)$, for any $C_1 \cap C_2 = \emptyset$ (if one coalition

---

5. This is in fact a simplified version of CL, which might be called *flat CL*. The full definition allows for the nesting of $[C]$ operators, but that requires semantics that are too involved to include here.

can force some set of worlds, and a disjoint coalition can force another set of worlds, then together they can enforce their intersection). The discussion at the end of the chapter points the reader to a more complete discussion of this and related logics.

14.4   Towards a logic of "intention"

intention

In this section we look, briefly again, at modal logics that have explicit "motivational" modal operators, ones that capture the motivation of agents and their reason for action. The specific notion we will try to capture is that of *intention* first as it is attributed to a single agent and then as it is attributed to set of agents functioning as a group. As we shall see, in service of "intention" we will need to define several auxiliary notions. These extended theories are sometimes called

BDI theories

jokingly *BDI (pronounced 'beady eye') theories*, because, beside the notion of belief (B), they include the notions of desire (D) and intention (I), as well as several others.

It should be said at the outset that these extended theories are considerably more complex and messy than the theories of knowledge and belief that were covered in previous sections, and mathematically less well developed than those on games and information dynamics. We will therefore not treat this subject matter with the same amount of detail as we did the earlier topics. Instead we will do the following. We will start with an informal discussion of notions such as intention and some requirements on a formal theory of them. We will then outline only the syntax of one such theory, starting with a single-agent theory and then adding a multiagent component.

14.4.1   *Some preformal intuitions*

Let us start by considering what we are trying to achieve. Recall that our theories of "knowledge" and "belief" only crudely approximated the everyday meanings of those terms, but were nonetheless useful for certain applications. We would like to strike a similar balance in a theory of "intention." Such theories could be use to reason about cooperative or adversarial planning agents, to create intelligent dialog systems, or to create useful personal assistants, to mention a few applications.

What are some of the requirements on such theories? Different applications will give rise to different answers. For motivation behind a particular theory we will outline, consider the following scenario.

> Phil is having trouble with his new household robot, Hector. He says, "Hector, bring me a beer." The robot replies, "OK, boss." Twenty minutes later, Phil yells, "Hector, why didn't you bring that beer?" It answers, "Well, I had intended to get you the beer, but I decided to do something else." Miffed, Phil sends the wise guy back to the manufacturer, complaining about a lack of commitment. After retrofitting, Hector is returned, marked "Model C: The Committed Assistant." Again, Phil asks Hector to bring a beer. Again,

it accedes, replying "Sure thing." Then Phil asks, "What kind do we have?" It answers, "Anchor Steam." Phil says, "Never mind." One minute later, Hector trundles over with an Anchor Steam in its gripper. This time, Phil angrily return Hector for overcommitment. After still more tinkering, the manufacturer sends Hector back, promising no more problems with its commitments. So, being a somewhat trusting consumer, Phil accepts the rascal back into his household, but as a test, he asks Hector to bring him the last beer remaining in the fridge. Hector again accedes, saying, "Yes, sir." The robot gets the beer and starts toward Phil. As it approaches, it lifts its arm, wheels around, deliberately smashes the bottle, and trundles off. Back at the plant, when interrogated by customer service as to why it had abandoned its commitments, the robot replies that according to its specifications, it could not have a commitment that it believed to be unachievable. Once the last remaining bottle was smashed, the commitment became unachievable. Despite the impeccable logic, and the correct implementation, Hector is dismantled.

intention

This example suggests that in order to capture the notions of *intention* we must also tackle those of *commitment* and *capability*. In addition, we will consider the notions of *desire*, *goal*, and even *agency*. This scenario also suggests some constraints. Here are some of the intuitions that will underlie the formal development.

commitment

capability

desire

goal

agency

1. Desires are unconstrained; they need not be achievable, and not even consistent (one can desire smoking and good health simultaneously). Goals in contrast must be consistent with each other, and furthermore must be believed to be achievable, or at least not believed to be unachievable. The same is true of intentions, which have the additional property of being persistent in time. Loosely speaking, these three notions form a hierarchy; goals imply desires, and intentions imply goals. These mutual constraints among the difference notions are sometimes called *rational balance*. In the formulation in the next section we will consider goals and intentions, but not desires.

rational balance

2. Intentions come in two varieties—intentions to achieve a particular state (such as being in San Francisco) and intentions to take a particular action (such as boarding a particular train to San Francisco). In particular, intentions are future directed. The same is true of goals.

3. Plans consist of a set of intentions and goals. Plans are in general *partial*; that is, they have some goals or intentions that are not directly achievable by the agent. For example, an agent may intend to go to San Francisco, even though he may not yet know whether he wants to drive there or take the train.

4. Plans give rise to further goals and intentions. For example, an intention to go to San Francisco requires that the agent further specify some means for getting there (e.g., driving or taking the train).

5. At the same time, since the plan of an agent must be internally consistent, a given plan constrains the addition of new goals and intentions. For example, if the agent's plan already contains the intention of leaving his car at home

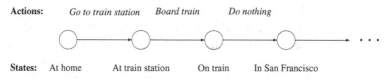

Actions: *Go to train station*   *Board train*   *Do nothing*

States: At home   At train station   On train   In San Francisco

Figure 14.4 An event sequence: actions lead from one state to another.

for his wife to use, then he cannot adopt the intention of driving it to San Francisco as the means of getting there.

6. Intentions are persistent, but not irrevocable. For example, an intention to go to San Francisco may be part of a larger plan serving a higher-level goal, such as landing a new job. Our agent may find another means to achieve the same goal and thus drop his intention to go to San Francisco. Or, he may obtain new information that makes his trip to San Francisco infeasible, in which case too that intention must be dropped.

7. Agents need not intend the anticipated side effects of their intentions. Driving to San Francisco may necessarily increase one's risk of a traffic accident, but an agent intending the former does not generally intend the latter. Similarly, an agent intending to go to the dentist does not typically have the intention of experiencing great pain even if he may fully expect such pain.

Philosophers have written at considerably greater length on these topics, but let us stop here and look at what a formal theory might look like in light of some of these considerations.

14.4.2 *The road to hell: elements of a formal theory of intention*

We will sketch a theory aimed at capturing the notion of intention. Our sketch will be incomplete, and will serve primarily to highlight some of the surprising elements that a complete theory requires. Certainly we will only consider a propositional theory; there is no special difficulty in extending the treatment to the first-order case. More critically, we will discuss only the axiomatic theory, and will have little to say about its formal semantics.

Since intentions are future directed, we must start by representing the passage of time. Although we have just said that we will not discuss semantics, it is useful to keep in mind the picture of a sequence of events, leading from one state to another. Figure 14.4 shows an example.

Such event sequences will form the basis for reasoning about intentions and other notions. The language we use to speak about such events is based on *dynamic logic*. Dynamic logic takes the basic events (such as going to the train station) as the primitive objects. These basic events can be combined in a number of ways to make complex events; see later. By associating an agent with an event (where basic or complex) we get an *action*. Since we will not consider agent-less events here, we will use the term "event" and "action" interchangeably.

Thus one component of dynamic logic is a language to speak about actions. Another component is a language to speak about what is true and false in states, the end points of actions. In this language we will have two primitive modal operators—B (belief) and G (goal)—and define intention in terms of those. We will not define desire in the language since our notion of intention will not depend on it.

Interestingly, the language of actions and the language of states are mutually dependent. They are defined as follows.

Definition 14.4.1 (Action and state expressions) *Given a set of agents N, a set E of primitive actions, a set V of variables (which will range over actions), and a set P of primitive propositions (describing what is true and false in states), the set of action expressions A and the set of state expressions $\mathcal{L}_{BG}$ are the smallest sets satisfying the following constraints:*

- *$E \subseteq A$.*
- *If $a, b \in A$, then $a; b \in A$.*
- *If $a, b \in A$, then $a|b \in A$.*
- *If $a \in A$, then $a^* \in A$.*
- *If $\varphi \in \mathcal{L}_{BG}$, then $\varphi? \in A$.*

Here is an intuitive explanation of these complex actions. $a; b$ denotes composition: doing a followed by b. $a|b$ denotes nondeterministic choice: doing a or b, nondeterministically. a^* denotes iteration: repeating a zero or more times, nondeterministically. $\varphi?$ is perhaps the most unusual action. It is the test action; it does not change the state, but is successful only in states that satisfy φ. Specifically, if the action $\varphi?$ is taken in a state, then φ must be true in that state.

- *$P \subseteq \mathcal{L}_{BG}$.*
- *If $\varphi, \psi \in \mathcal{L}_{BG}$, then $\varphi \wedge \psi, \neg\varphi \in \mathcal{L}_{BG}$.*
- *If $a \in A$, then $JustHappened(a) \in \mathcal{L}_{BG}$.*
- *If $a \in A$, then $AboutToHappen(a) \in \mathcal{L}_{BG}$.*
- *If $i \in N$ and $a \in A$, then $Agent(a) = i \in \mathcal{L}_{BG}$.*
- *If $v \in V$ and $\varphi \in \mathcal{L}_{BG}$, then $\forall v\varphi \in \mathcal{L}_{BG}$.*
- *If $a, b \in A \cup V$, then $a < b \in \mathcal{L}_{BG}$.*
- *If $i \in N$ and $\varphi \in \mathcal{L}_{BG}$, then $B_i\varphi \in \mathcal{L}_{BG}$.*
- *If $i \in N$ and $\varphi \in \mathcal{L}_{BG}$, then $G_i\varphi \in \mathcal{L}_{BG}$.*

Again, an explanation is in order. JustHappened(a) captures the fact that action a ended in the current state, and AboutToHappen(a) that it is about to start in the current state. Agent(a) = i identifies i as the agent of action a. $a < b$ means that action a took place before action b. Finally, B_i and G_i of course denote the belief and goal operators, respectively. However, especially for G_i, it is important to be precise about the reading; $G_i\varphi$ means that φ is true in all the states that satisfy the goals of agent i.

While we do not discuss formal semantics here, let us briefly discuss the belief and goal modal operators. They are both intended to be interpreted via possible-worlds semantics. So the question arises as to the individual properties of these operators and the interaction between them. B is a standard (KD45) belief operator. We place no special restriction on the G operator, other than it that must be serial (to ensure that goals are consistent). However, since goals must also be consistent with beliefs, we require that the goal accessibility relation be a subset of the belief accessibility relation. That is, if a possible world is ruled out by the agent's beliefs, it cannot be a goal. This assumption means that you cannot intend that be in San Francisco on Saturday if you believe that you will be in Europe at the same time. (Of course, you can intend to be in San Francisco on Saturday even though you are in Europe today.)[6]

Before proceeding further, it will be useful to define several auxiliary operators.

Definition 14.4.2 $Always(\varphi) \stackrel{\text{def}}{=} \forall a(AboutToHappen(a) \rightarrow AboutToHappen(a; \varphi?))$.

$Eventually(\varphi) \stackrel{\text{def}}{=} \neg Always(\neg\varphi)$.

$Later(\varphi) \stackrel{\text{def}}{=} (\neg\varphi) \wedge (Eventually(\varphi))$.

$Before(\varphi, \psi) \stackrel{\text{def}}{=} \forall c(AboutToHappen(c; \psi?)) \rightarrow \exists a((a \leq c)\wedge AboutToHappen(a; \varphi?))$.

$JustDid_i(a) \stackrel{\text{def}}{=} JustHappened(a) \wedge Agent(a) = i$.

$AboutToDo_i(a) \stackrel{\text{def}}{=} AboutToHappen(a) \wedge Agent(a) = i$.

With these in place, we proceed to define the notion of intention as follows. We first strengthen the notion of having a goal. The G_i operator defines a weak notion; in particular, it includes goals that are already satisfied and thus provides achievement goal no impetus for action. An *achievement goal*, or *AGoal*, focuses on the goals that are yet to be achieved.

Definition 14.4.3 (Achievement goal)

$$AGoal_i\varphi \stackrel{\text{def}}{=} G_i(Later(\varphi)) \wedge B_i(\neg\varphi)$$

An achievement goal is useful, but it is still not an intention. This is because an achievement goal has no model of commitment. An agent might form an achievement goal, only to drop it moments later, for no apparent reason. In order to model commitment, we should require that the agent not drop the goal until he reaches it. We call such a goal a persistent goal, defined as follows.

Definition 14.4.4 (Persistent goal)

$$PGoal_i\varphi \stackrel{\text{def}}{=} AGoal_i\varphi \wedge$$
$$Before(B_i(\varphi) \vee B_i(Always(\neg\varphi)), \neg G_i(Later(\varphi)))$$

In other words, a persistent goal is an achievement goal that the agent will not give up until he believes that it is true or will never be true.

The notion of a persistent goal brings us closer to a reasonable model of intention, but it still misses an essential element of intentionality. In particular, it

---

6. Note that this constraint means that the formula $B_i\varphi \rightarrow G_i\varphi$ is valid in our models; this is where the careful reading of the G_i operator is required.

does not capture the requirement that the agent himself do something to fulfill the goal, let alone do so knowingly. The following definitions attempt to add these ingredients.

Recall that an agent may intend either to do an action or to achieve a state. For this reason, we give two definitions of intention. The following is a definition of intending an action.

Definition 14.4.5 (Intending an action)

$$IntendsA_i a \overset{\text{def}}{\equiv} PGoal_i(JustDid_i(B_i(AboutToDo_i(a))?; a))$$

In other words, if an agent intends an action, he must have a persistent goal to first believe that he is about to take that action and then to actually take it.

The second definition captures the intention to bring about a state with certain properties.

Definition 14.4.6

$$IntendS_i(\varphi) \overset{\text{def}}{\equiv} PGoal_i \exists e JustDid_i(B_i(\exists e' AboutToDo_i(e'; \varphi?)) \wedge \\ \neg G_i(\neg AboutToDo_i(e; \varphi?))?; e; \varphi?))$$

We explain this definition in a number of steps. Notice that to intend a state in which φ holds, an agent is committed to taking a number of actions e himself, after which φ holds. However, in order to avoid allowing him to intend φ by doing something accidentally, we require that he believe he is about to do some series of events e' that brings about φ. In other words, we require that he has a plan e', which he believes that he is executing, which will achieve his goal φ.

As was mentioned at the beginning of this section, the logic of intention is considerably more complex, messy, and controversial than that of knowledge and belief. We will not discuss the pros and cons of this line of definitions further, nor alternative definitions; the notes at the end of the chapter provide references to these. Let us just note that these definitions do have some desirable formal properties. For example, early intentions preempt later potential intentions (namely those that undermine the objectives of the earlier intentions); agents cannot intend to achieve something if they believe that no sequence of actions on their part will bring it about; and agents do not necessarily intend all the side effects of their intentions.

14.4.3   Group intentions

The theory of intention outlined in the previous section considers a single agent. We have seen that the extension of the logic of (e.g.,) knowledge to the multi-agent setting is interesting; in particular, it gives rise to the notion of common knowledge, which is so central to reasoning about coordination. Is there similar motivation for considering intention in a multiagent setting?

Consider a set of agents acting as a group. Consider specifically the notion of convoy. Consisting of multiple agents, each making local decisions, a convoy nonetheless moves purposefully as a single body. It is tempting to attribute beliefs and intentions to this body, but how do these group notions relate to the beliefs and intentions of the individual agents?

It certainly will not do to equate a group intention simply with all the individual agents having that intention; it is not even sufficient to have those individual intentions be common knowledge among the agents. Consider the following scenario.

> Two ancient Inca kings, Euqsevel and Nehoc, aim to ride their horses in a convoy to the Temple of Otnorot. Specifically, since King Euqsevel knows the way, he intends to simply ride there and King Nehoc intends to follow him. They set out, and since King Euqsevel's horse is faster, he quickly loses King Nehoc, who never makes it to Otnorot.
>
>   Centuries later, in old Wales, two noblemen—Sir Ffegroeg and Sir Hgnis—set out on another journey. Here again only one of them, Sir Ffegroeg, knows the way. However, having learned from the Incan mishap, they do not simply adopt their individual intentions. Instead, Sir Ffegroeg agrees to go ahead on his faster horse, but to wait at road junctions in order to show Sir Hgnis the way. They proceed in this fashion until at some point Sir Ffegroeg discovers that snow is covering the mountain pass and that it is impossible to get to their destination. So he correctly jettisons his intention to get there and therefore also the dependent intention to show Sir Hgnis the way. Instead he makes his way to a comfortable hotel he knows in a nearby town; Sir Hgnis is left to wander the unfamiliar land to his last day.

Clearly, the interaction between the intentions of the individual agents and the collective mental state is involved.

One way to approach the problem is to mimic the development in the single-agent case. Specifically, one can first define the notion of *joint persistent goal* and then use it to define the notion of a *joint intention*. We give an informal outline of such definitions.

weak achievement goal — **Definition 14.4.7 (Weak achievement goal; informal)** *An agent has a* weak achievement goal *with respect to a team of agents iff one of the following is true:*

1. *The agent has a standard achievement goal (AGoal) to bring about φ;*
2. *The agent believes that φ is true or will never be true, but has an achievement goal that the status of φ be common belief within the team.*

joint persistent goal — **Definition 14.4.8 (Joint persistent goal; informal)** *A team of agents has a joint persistent goal to achieve φ iff the following are all true:*

1. *They have common belief that φ is currently false;*
2. *They have common (and true) belief that they each have the goal that φ hold eventually;*
3. *They have common (and true) belief that, until they come to have common belief that φ is either true or never will be true, they will each continue to have φ as a weak achievement goal.*

joint intention   **Definition 14.4.9 (Joint intention; informal)** *A team of agents has a* joint intention *to take a set of actions A (each action by one of the agents) iff the agents have a joint persistent goal of (a) having done A and (b) having common belief throughout the execution of A that they are doing A.*

In addition to avoiding the pitfalls demonstrated in the earlier stories, these definitions have several potentially attractive properties, including the following.

- The joint persistent goals of a team consisting of a single agent coincide with the intentions of that agent.
- If a team has φ as a joint persistent goal then so does every individual agent.
- If a team has a joint intention to take a set of actions A, and action $a \in A$ belongs to agent i, then agent i intends to take action a.

However, these definitions are nothing if not complex, which is why we omit their formal details here. Furthermore, as was said, there is not yet universal agreement on the best definitions. The references point to the reader to further reading on this fascinating, yet incomplete, body of work.

14.5   History and references

The combination of knowledge and probability as discussed in this chapter is based on Fagin and Halpern [1994], and is covered also in Halpern [2005], which remains a good technical introduction to the topic.

Theories of belief dynamics—revision, update, and beyond—are covered in Pappas [2007], as well as in the older but still excellent Gärdenfors and Rott [1995]. Early seminal work includes that of Gärdenfors [1988], Alchourron, Gärdenfors and Makinson [1985] (after whom "AGM revision" is named), and Katsuno and Mendelzon [1991], who introduced the distinction between belief revision and belief update. The material in the chapter on knowledge, certainty, and belief is based on Boutilier [1992] and Lamarre and Shoham [1994]. The material in the section on belief fusion is based on Maynard-Reid and Shoham [2001] and Maynard-Zhang and Lehmann [2003]. A broader introduction to logical theories of belief fusion can be found in Grégoire and Konieczny [2006].

nonmonotonic   Belief dynamics are closely related to the topic of *nonmonotonic logics*, a rich
logic   source of material. An early comprehensive coverage of nonmonotonic logics can be found in Ginsberg [1987], and a more recent survey is provided in Brewka et al. [2007].

The recent trend toward modal logics of information dynamics is heralded in van Benthem [1997]. The early seminal work with a game-like perspective on logic is due to Peirce [1965]. A recent review of modal logic for games and information change appears in van der Hoek and Pauly [2006]. Our discussion of coalition logic is covered there in detail, and originally appeared in Pauly [2002].

In general, belief dynamics still constitute an active area of research, and the interested reader will undoubtedly want to study the recent literature, primarily in artificial intelligence and philosophical logic.

The literature on formal theories of "motivational" attitudes, such as desires, goals, and intentions, is much sparser than the literature on the "informational" attitudes discussed earlier. There are fewer publications on these topics, and the results are still preliminary (which is reflected in the style and length of the book section). The material presented is based largely on work in artificial intelligence by Cohen and Levesque [Cohen and Levesque, 1990, 1991], which in turn was inspired by philosophical work such as that of Bratman [1987]. The Cohen-Levesque formulation has attracted criticism and alternative formulations, some of which can be found in [Rao and Georgeff, 1991, 1998], Singh [1992], Meyer et al. [1999], and van der Hoek and Wooldridge [2003]. Much of this literature is surveyed in Wooldridge [2000]. This area presents many opportunities for further research.

Appendices: Technical Background

A

Probability Theory

Probability theory provides a formal framework for the discussion of chance or uncertainty. This appendix reviews some key concepts of the theory and establishes notation. However, it glosses over some details (e.g., pertaining to measure theory). Therefore, the interested reader is encouraged to consult a textbook on the topic for a more comprehensive picture.

A.1 Probabilistic models

A probabilistic model is defined as a tuple $(\Omega, \mathcal{F}, P)$, where:

sample space

event space

- Ω is the *sample space*, also called the *event space*;
- $\mathcal{F}$ is a σ-algebra over Ω; that is, $\mathcal{F} \subseteq 2^{\Omega}$ and is closed under intersection and countable union; and

probability
density function

- $P : \mathcal{F} \mapsto [0, 1]$ is the *probability density function* (PDF).

Intuitively, the sample space is a set of things that can happen in the world according to our model. For example, in a model of a six-sided die, we might have $\Omega = \{1, 2, 3, 4, 5, 6\}$. The σ-field $\mathcal{F}$ is a collection of measurable events. $\mathcal{F}$ is required because some outcomes in Ω may not be measurable; thus, we must define our probability density function P over $\mathcal{F}$ rather than over Ω. However, in many cases, such as the six-sided die example, all outcomes *are* measurable. In those cases we can equate $\mathcal{F}$ with 2^{Ω} and view the probability space as the pair (Ω, P) and P as $P : 2^{\Omega} \mapsto [0, 1]$. We assume this in the following.

A.2 Axioms of probability theory

The probability density function P must satisfy the following axioms.

1. For any $A \subseteq \Omega$, $P(\emptyset) = 0 \le P(A) \le P(\Omega) = 1$.
2. For any pair of disjoint sets $A, A' \subset \Omega$, $P(A \cup A') = P(A) + P(A')$.

That is, all probabilities must be bounded by 0 and 1; 0 is the probability of the empty set and 1 the probability of the whole sample space. Second, when sets of outcomes from the sample space are nonoverlapping, the probability of achieving an outcome from either of the sets is the sum of the probabilities of achieving an outcome from each of the sets. We can infer from these rules that if two sets $A, A' \subseteq \Omega$ are not disjoint, $P(A \cup A') = P(A) + P(A') - P(A \cap A')$.

A.3 Marginal probabilities

We are often concerned with sample spaces Ω that are defined as the Cartesian product of a set of random variables $X_1, \ldots, X_n$ with domains $\mathcal{X}_1, \ldots, \mathcal{X}_n$ respectively. Thus, in this setting, $\Omega = \prod_{i=1}^{n} \mathcal{X}_i$. Our density function P is thus defined over full assignments of values to our variables, such as $P(X_1 = x_1, \ldots, X_n = x_n)$. However, sometimes we want to ask about *marginal probabilities*: the probability that a single variable X_i takes some value $x_i \in \mathcal{X}_i$. We define

marginal probability

$$P(X_i = x_i) = \sum_{x_1 \in \mathcal{X}_1} \cdots \sum_{x_{i-1} \in \mathcal{X}_{i-1}} \sum_{x_{i+1} \in \mathcal{X}_{i+1}} \cdots \sum_{x_n \in \mathcal{X}_n} P(X_1 = x_1, \ldots, X_n = x_n).$$

From this definition and from the axioms given earlier we can also infer that, for example,

$$P(X_i = x_i) = \sum_{x_j \in \mathcal{X}_j} P(X_i = x_i \text{ and } X_j = x_j).$$

A.4 Conditional probabilities

We say that two random variables X_i and X_j are *independent* when $P(X_i = x_i$ and $X_j = x_j) = P(X_i = x_i) \cdot P(X_j = x_j)$ for all values $x_i \in \mathcal{X}_i, x_j \in \mathcal{X}_j$.

Often, random variables are not independent. When this is the case, it can be important to know the probability that X_i will take some value x_i given that $X_j = x_j$ has already been observed. We define this probability as

$$P(X_i = x_i | X_j = x_j) = \frac{P(X_i = x_i \text{ and } X_j = x_j)}{P(X_j = x_j)}.$$

conditional probability

joint probability

We call $P(X_i = x_i | X_j = x_j)$ a *conditional probability*; $P(X_i = x_i$ and $X_j = x_j)$ is called a *joint probability* and $P(X_j = x_j)$ a marginal probability, already discussed previously.

Bayes' rule

Finally, *Bayes' rule* is an important identity that allows us to reverse conditional probabilities. Specifically,

$$P(X_i = x_i | X_j = x_j) = \frac{P(X_j = x_j | X_i = x_i) P(X_i = x_i)}{P(X_j = x_j)}.$$

B

Linear and Integer Programming

Linear programs and integer programs are optimization problems with linear objective functions and linear constraints. These problems are very general: linear programs can be used to express a wide range of problems such as bipartite matching and network flow, while integer programs can express an even larger range of problems, including all of those in NP. We define each of these formalisms in turn.

B.1 Linear programs

Defining linear programs

linear program   A *linear program* is defined by:

- a set of real-valued variables;
- a linear objective function (i.e., a weighted sum of the variables); and
- a set of linear constraints (i.e., the requirement that a weighted sum of the variables must be less than or equal to some constant).

Let the set of variables be $\{x_1, \ldots, x_n\}$, with each $x_i \in \mathbb{R}$. The objective function of a linear program, given a set of constants $w_1, \ldots, w_n$, is

$$\text{maximize} \sum_{i=1}^{n} w_i x_i.$$

Linear programs can also express minimization problems: these are just maximization problems with all weights in the objective function negated.

Constraints express the requirement that a weighted sum of the variables must be greater than or equal to some constant. Specifically, given a set of constants $a_{1j}, \ldots, a_{nj}$ and a constant b_j, a constraint is an expression

$$\sum_{i=1}^{n} a_{ij} x_i \leq b_j.$$

This form actually allows us to express a broader range of constraints than might immediately be apparent. By negating all constants, we can express greater-than-or-equal constraints. By providing both less-than-or-equal and greater-than-or-equal constraints with the same constants, we can express equality constraints.

By setting some constants to zero, we can express constraints that do not involve all of the variables. Furthermore, even problems with piecewise-linear constraints (e.g., involving functions like a max of linear terms) can sometimes be expressed as linear programs by adding both new constraints and new variables. Observe that we *cannot* always write strict inequality constraints, though sometimes such constraints can be enforced through changes to the objective function. (For example, see the linear program given in Equations (4.42)–(4.44) on p. 108, which enforces the strict inequality constraints given in Equations (4.39)–(4.41).)

Bringing it all together, if we have m different constraints, we can write a linear program as follows.

$$\text{maximize} \quad \sum_{i=1}^{n} w_i x_i$$

$$\text{subject to} \quad \sum_{i=1}^{n} a_{ij} x_i \leq b_j \qquad \forall j = 1 \ldots m$$

$$x_i \geq 0 \qquad \forall i = 1 \ldots n$$

Observe that the requirement that each x_i must be nonnegative is not restrictive: problems involving negative variables can always be reformulated into equivalent problems that satisfy the constraint.

A linear program can also be written in matrix form. Let $\mathbf{w}$ be an $n \times 1$ vector containing the weights w_i, let $\mathbf{x}$ be an $n \times 1$ vector containing the variables x_i, let $\mathbf{A}$ be an $m \times n$ matrix of constants a_{ij}, and let $\mathbf{b}$ be an $m \times 1$ vector of constants b_j. We can then write a linear program in matrix form as follows.

$$\text{maximize} \quad \mathbf{w}^T \mathbf{x}$$

$$\text{subject to} \quad \mathbf{A}\mathbf{x} \leq \mathbf{b}$$

$$\mathbf{x} \geq \mathbf{0}$$

In some cases we care to satisfy a given set of constraints, but do not have an associated objective function; any solution will do. In this case the LP reduces to a constraint satisfaction or *feasibility* problem, but we will sometimes still refer to one as an LP with an empty objective function (or, equivalently, the trivial one).

primal problem   Finally, every linear program (a so-called *primal problem*) has a corresponding *dual problem* which shares the same optimal solution. For the linear program

dual problem   given earlier, the dual program is as follows.

$$\text{minimize} \quad \mathbf{b}^T \mathbf{y}$$

$$\text{subject to} \quad \mathbf{A}^T \mathbf{y} \geq \mathbf{w}$$

$$\mathbf{y} \geq \mathbf{0}$$

In this linear program our variables are $\mathbf{y}$. Variables and constraints effectively trade places: there is one variable $y \in \mathbf{y}$ in the dual problem for every constraint from the primal problem and one constraint in the dual problem for every variable $x \in \mathbf{x}$ from the primal problem.

Solving linear programs

In order to solve linear programs, it is useful to observe that the set of feasible solutions to a linear program corresponds to a convex polyhedron in n-dimensional space. This is true because all of the constraints are linear: they correspond to hyperplanes in this space, and so the set of feasible solutions is the region bounded by all of the hyperplanes. The fact that the objective function is also linear allows us to conclude two useful things: any local optimum in the feasible region will be a global optimum, and at least one optimal solution will exist at a vertex of the polyhedron. (More than one optimal solution may exist if an edge or even a whole face of the polyhedron is a local maximum.)

simplex
algorithm
    The most popular algorithm for solving linear programs is the *simplex algorithm*. This algorithm works by identifying one vertex of the polyhedron and then taking uphill (i.e., objective-function-improving) steps to neighboring vertices until an optimum is found. This algorithm requires an exponential number of steps in the worst case, but is usually very efficient in practice.

interior-point
method
    Although the simplex algorithm is not polynomial, it can be shown that other algorithms called *interior-point methods* solve linear programs in worst-case polynomial time. These algorithms get their name from the fact that they move through the interior region of the polyhedron rather than jump from one vertex to another. Surprisingly, although these algorithms dominate the simplex method in the worst case, they can be much slower in practice.

B.2 Integer programs

Defining integer programs

integer program
Integer programs are linear programs in which one additional constraint holds: the variables are required to take integral (rather than real) values. This makes it possible to express combinatorial optimization problems such as satisfiability or set packing as integer programs. A useful subclass of integer programs are 0–1 *integer programs*, in which each variable is constrained to take either the value 0 or the value 1. These programs are sufficient to express any problem in NP. The form of a 0–1 integer program is as follows.

$$\text{maximize} \quad \sum_{i=1}^{n} w_i x_i$$

$$\text{subject to} \quad \sum_{i=1}^{n} a_{ij} x_i \leq b_j \qquad\qquad \forall j = 1 \ldots m$$

$$x_i \in \{0, 1\} \qquad\qquad \forall i = 1 \ldots n$$

mixed-integer
program
    Another useful class of integer programs is *mixed-integer programs*, which involve a combination of integer and real-valued variables.

    Finally, as in the LP case, both integer and mixed-integer programs can come without an associated objective function, in which case they reduce to constraint-satisfaction problems.

Solving integer programs

The introduction of an integrality constraint to linear programs leads to a much harder computational problem: NP-hard even when variables are restricted to two discrete values. Thus, it should not be surprising that there is no efficient procedure for solving integer programs.

branch-and-bound search   The most commonly used technique is *branch-and-bound search*. The space of variable assignments is explored depth-first: first one variable is assigned a value, then the next, and so on; when a constraint is violated or a complete variable assignment is achieved, the search backtracks and tries other assignments. The best feasible solution found so far is recorded as a lower bound on the value linear program relaxation of the optimal solution. At each search node the *linear program relaxation* of the integer program is solved: this is the linear program where the remaining variables are allowed to take real rather than integral values between the minimum and maximum values in their domains. It is easy to see that the value of a linear program relaxation of an integer program is an upper bound on the value of that integer program, since it involves a loosening of the latter problem's constraints. Branch-and-bound search differs from standard depth-first search because it sometimes prunes the tree. Specifically, branch-and-bound backtracks whenever the upper bound at a search node is less than or equal to the lower bound. In this way it can skip over large parts of the search tree while still guaranteeing that it will find the optimal solution.

Other, more complex techniques for solving integer programs include branch-and-cut and branch-and-price search. These methods offer no advantage over branch-and-bound search in the worst case, but often outperform it in practice.

Although they are computationally intractable in the worst case, sometimes integer programs are provably easy. This occurs when it can be shown that the solution to the linear programming relaxation is integral, meaning that the integer program can be solved in polynomial time. One important example is when the total unimodularity constraint matrix is *totally unimodular* and the vector **b** is integral. A unimodular matrix is a square matrix whose determinant is either -1 or 1; a totally unimodular matrix is one for which every square submatrix is unimodular. This definition implies that the entries in a totally unimodular matrix can only be -1, 0, and 1.

C

Markov Decision Problems (MDPs)

We briefly review the main ingredients of Markov Decision Problems or MDPs, which, as we discuss in Chapter 6, can be viewed as single-agent stochastic games. The literature on MDPs is rich, and the reader is referred to the many textbooks on the subject for further reading.

C.1 The model

An MDP is a model for decision making in an uncertain, dynamic world. The (single) agent starts out in some state, takes an action, and receives some immediate rewards. The state then transitions probabilistically to some other state and the process repeats. Formally speaking, an MDP is a tuple (S, A, p, r). S is a set of states and A is a set of actions. The function $p : S \times A \times S \mapsto \mathbb{R}$ specifies the transition probability among states: $p(s, a, s')$ is the probability of ending in state s' when taking action a in state s. Finally, the function $r : S \times A \mapsto \mathbb{R}$ returns the reward for each state-action pair.

The individual rewards are aggregated in one of two ways.

The limit-average reward: $\lim_{T=1}^{\infty} \frac{\sum_{t=1}^{T} r^{(t)}}{T}$, where $r^{(t)}$ is the reward at time t.

The future-discounted reward: $\sum_{t=1}^{\infty} \beta r^{(t)}$, where $0 < \beta < 1$ is the discount factor.

(The reader will notice that both of these definitions must be refined to account for cases in which (in the first case) the limit is ill defined, and in which (in the second case) the sum is infinite. We do not discuss these subtleties here.)

A (stationary, deterministic) policy $\Pi : S \mapsto A$ maps each state to an action.

For concreteness, in the next section we focus on the future-discounted reward case.

C.2 Solving known MDPs via value iteration

Every policy yields a reward under either of the reward-aggregation schemes. A policy that maximizes the total reward is called an *optimal policy*. The primary computational challenge associated with MDPs is to find an optimal policy for a given MDP.

optimal policy

It is possible to use linear programming techniques (see Appendix B) to calculate an optimal policy for a known MDP in polynomial time. However, we *value iteration* will focus on an older, dynamic-programming-style method called *value iteration*. We do so for two reasons. First, in typical real-world cases of interest, the LP-formulation of the MDP is too large to solve. Value iteration provides the basis for a number of more practical solutions, such as those providing approximate solutions to very large MDPs. Second, value iteration is relevant to the discussion of learning in MDPs, discussed in Chapter 7.

Value iteration defines a value function $V^\pi : S \mapsto \mathbb{R}$, which specifies the value of following policy π starting in state s. Similarly, we can define a state-action value function $Q^\pi : S \times A \mapsto \mathbb{R}$ as a function that captures the value of starting in state s, taking action a, and then continuing according to policy π. These two functions are related to each other by the following pair of equations.

$$Q^\pi(s, a) = r(s, a) + \beta \sum_{\hat{s}} p(s, a, \hat{s}) V^\pi(\hat{s})$$

$$V^\pi(s) = Q^\pi(s, \pi(s))$$

For the optimal policy π^*, the second equation becomes $V^{\pi^*}(s) = \max_a$
Bellman $Q^{\pi^*}(s, a)$ and the set of equations is referred to as the *Bellman equations*. Note
equations that the optimal policy is easily recovered from the solution to the Bellman equations, specifically from the Q function; the optimal action in state s is $\arg\max_a Q^{\pi^*}(s, a)$.

The Bellman equations are interesting not only because they characterize the optimal policy, but also—indeed, primarily—because they give rise to a procedure for calculating the Q and V values of the optimal policy, and hence the optimal policy itself. Consider the following two assignment versions of the Bellman equations.

$$Q_{t+1}(s, a) \leftarrow r(s, a) + \beta \sum_{\hat{s}} p(s, a, \hat{s}) V_t(\hat{s})$$

$$V_t(s) \leftarrow \max_a Q_t(s, a)$$

Given an MDP, and starting with arbitrary initial Q values, we can repeatedly iterate these two sets of assignment operators ("sets," since each choice of s and a produces a different instance). It is well known that any "fair" order of iteration (by which we mean that each instance of the rules is updated after a finite amount of time) converges on the Q and V values of an optimal policy.

D

Classical Logic

The following is not intended as an introduction to classical logic, but rather as a review of the concepts and a setting of notation. We start with propositional calculus and then move to first-order logic. (We do the latter for completeness, but in fact first-order logic plays almost no role in this book.)

D.1 Propositional calculus

Syntax

Given a set P of propositional symbols, the set of sentences in the propositional calculus is the smallest set $\mathcal{L}$ containing P such that if $\varphi, \psi \in \mathcal{L}$ then also $\neg\varphi \in \mathcal{L}$ and $\varphi \wedge \psi \in \mathcal{L}$. Other connectives such as $\vee$, $\rightarrow$, and $\equiv$ can be defined in terms of $\wedge$ and $\neg$.

Semantics

interpretation

model

A propositional *interpretation* (or a *model*) is a set $M \subset P$, the subset of true primitive propositions. The satisfaction relation $\models$ between models and sentences is defined recursively as follows.

- For any $p \in P$, $M \models p$ iff $p \in M$.
- $M \models \varphi \wedge \psi$ iff $M \models \varphi$ and $M \models \psi$.
- $M \models \neg\varphi$ iff it is not the case that $M \models \varphi$.

validity

entailment

We overload the $\models$ symbol. First, it is used to denote *validity*; $\models \varphi$ means that φ is true in all propositional models. Second, it is used to denote *entailment*; $\varphi \models \psi$ means that any model that satisfies φ also satisfies ψ.

Axiomatics

The following axiom system is sound and complete for the class of all propositional models:

A1. $A \rightarrow (B \rightarrow A)$

A2. $(A \rightarrow (B \rightarrow C)) \rightarrow ((A \rightarrow B) \rightarrow (A \rightarrow C))$

A3. $(\neg A \rightarrow \neg B) \rightarrow (B \rightarrow A)$

R1 (Modus Ponens). $A, A \rightarrow B \vdash B$.

D.2   First-order logic

This book makes very little reference to first-order constructs, but we include the basic material on the first-order logic for completeness.

Syntax

Given a set C of constant symbols, V of variables, F of function symbols each of a given arity, and R of relation (or predicate) symbols each of a given arity. The set of terms is the smallest set T such that $C \cup V \subset T$, and if $f \in F$ is an n-ary functions symbol and $t_1, \ldots, t_n \in T$ then also $f(t_1, \ldots, t_n) \in T$. The set of sentences is the smallest set $\mathcal{L}$ satisfying the following conditions.

> If r is an n-ary relation symbol and $t_1, \ldots, t_n \in T$ then $r(t_1, \ldots, t_n) \in \mathcal{L}$. These are the atomic sentences.
>
> If $\varphi, \psi \in \mathcal{L}$ then also $\neg\varphi \in \mathcal{L}$ and $\varphi \wedge \psi \in \mathcal{L}$.
>
> If $\varphi \in \mathcal{L}$ and $v \in V$ then $\forall v \varphi \in L$.

Semantics

interpretation

model

A first-order *interpretation* (or a *model*) is a tuple $M = (D, G, S, \mu)$. D is the domain of M, a set. G is a set of functions from the domain onto itself, each of a given arity. S is a set of relations over the domain, each of a given arity. μ is an (overloaded) interpretation function: $\mu : C \cup V \mapsto D$, $\mu : F \mapsto G$, $\mu : R \mapsto S$ (we assume that μ respects the arities of the function and relations). We can lift μ to apply to any term by the recursive definition $\mu(f(t_1, \ldots, t_n)) = \mu(f)(\mu(t_1), \ldots, \mu(t_n))$.

The satisfaction relation $\models$ between models and sentences is defined recursively.

- For any atomic sentence $\varphi = r(t_1, \ldots, t_n)$, $M \models \varphi$ iff $(\mu(t_1), \ldots, \mu(t_n)) \in \mu(r)$.
- $M \models \varphi \wedge \psi$ iff $M \models \varphi$ and $M \models \psi$.
- $M \models \neg\varphi$ iff it is not the case that $M \models \varphi$.
- $M \models \forall v \varphi$ iff $M \models \varphi[v/t]$ for all terms t, where $\varphi[v/t]$ denotes φ with all free instances of v replaced by t.

We overload the $\models$ symbol as before.

Axiomatics

We omit the axiomatics of first-order logic since they play no role at all in this book.

Bibliography

Alchourron, C. E., Gärdenfors, P., and Makinson, D. (1985). On the logic of theory change: Partial meet contraction and revision functions. *Journal of Symbolic Logic*, *50*(2), 510–530.

Allen, J. F., Shubert, L. K., Ferguson, G., Heeman, P., Hwang, C. H., Kato, T., Light, M., Martin, N. G., Miller, B. W., Poesio, M., and Traum, D. R. (1995). The TRAINS project: A case study in building a conversational planning agent. *Journal of Experimental and Theoretical AI*, *7*, 7–48.

Altman, A., and Tennenholtz, M. (2005). Ranking systems: The PageRank axioms. *EC: Proceedings of the ACM Conference on Electronic Commerce* (pp. 1–8).

Altman, A., and Tennenholtz, M. (2007a). An axiomatic approach to personalized ranking systems. *Technion Israel Institute of Technology Technical Report*.

Altman, A., and Tennenholtz, M. (2008). Axiomatic foundations for ranking systems. *JAIR: Journal of Artificial Intelligence Research*, *31*, 473–495.

Archer, A., and Tardos, E. (2002). Frugal path mechanisms. *DA: Proceedings of the ACM-SIAM Symposium on Discrete Algorithms* (pp. 991–999).

Arrow, K. (1977). *The property rights doctrine and demand revelation under incomplete information*. Stanford, CA: Institute for Mathematical Studies in the Social Sciences, Stanford University.

Arrow, K. J. (1970). *Social choice and individual values*. New Haven, CT: Yale University Press.

Aumann, R. (1959). Acceptable points in general cooperative *n*-person games. *Contributions to the Theory of Games*, *4*, 287–324.

Aumann, R. (1974). Subjectivity and correlation in randomized strategies. *Journal of Mathematical Economics*, *1*, 67–96.

Aumann, R. (1995). Backward induction and common knowledge of rationality. *GEB: Games and Economic Behavior*, *8*(1), 6–19.

Aumann, R. (1996). Reply to Binmore. *GEB: Games and Economic Behavior*, *17*(1), 138–146.

Aumann, R. J. (1976). Agreeing to disagree. *Annals of Statistics*, *4*, 1236–1239.

Austin, J. L. (1962). *How to do things with words: The William James lectures delivered at Harvard University in 1955*. Oxford: Clarendon.

Austin, J. L. (2006). *How to do things with words: The William James lectures delivered at Harvard University in 1955, 2nd edition*. Harvard University Press.

Ausubel, L. M., and Milgrom, P. (2006). Ascending proxy auctions. In [Cramton et al., 2006], chapter 3, 79–98.

Axelrod, R. (1984). *The evolution of cooperation*. New York: Basic Books.

Beckmann, M. J., McGuire, C. B., and Winsten, C. B. (1956). *Studies in the economics of transportation*. New Haven, CT: Yale University Press.

Bell, D. E. (1982). Regret in decision making under uncertainty. *Operations Research*, *30*, 961–981.

Bellman, R. (1957). *Dynamic programming*. Princeton, NJ: Princeton University Press.

Ben-Porath, E. (1990). The complexity of computing a best response automaton in repeated games with mixed strategies. *GEB: Games and Economic Behavior*, *2*(1), 1–12.

Berg, J., Forsythe, R., Nelson, F., and Rietz, T. (2001). Results from a dozen years of election futures markets research. *Handbook of Experimental Economics Results*.

Berger, U. (2005). Fictitious play in 2 × n games. *Journal of Economic Theory, 120*(2), 139–154.

Bernheim, B. D. (1984). Rationalizable strategic behavior. *Econometrica, 52*, 1007–1028.

Bertsekas, D. (1982). Distributed dynamic programming. *IEEE Transactions on Automatic Control, 27*(3), 610–616.

Bertsekas, D. P. (1991). *Linear network optimization: Algorithms and codes.* Cambridge, MA: MIT Press.

Bertsekas, D. P. (1992). Auction algorithms for network flow problems: A tutorial introduction. *Computational Optimization and Applications, 1*(1), 7–66.

Bhat, N. A. R., and Leyton-Brown, K. (2004). Computing Nash equilibria of action-graph games. *UAI: Proceedings of the Conference on Uncertainty in Artificial Intelligence* (pp. 35–42).

Binmore, K. (1996). A note on backward induction. *GEB: Games and Economic Behavior, 17*(1), 135–137.

Blackburn, P., de Rijke, M., and Venema, Y. (2002). *Modal logic.* Cambridge, UK: Cambridge University Press.

Blackwell, D. (1956). Controlled random walks. *Proceedings of the International Congress of Mathematicians* (pp. 336–338). North-Holland.

Blum, B., Shelton, C., and Koller, D. (2006). A continuation method for Nash equilibria in structured games. *Journal of Artificial Intelligence Resarch, 25*, 457–502.

Border, K. C. (1985). *Fixed point theorems with applications to economics and game theory.* Cambridge University Press.

Borodin, A., Roberts, G., Rosenthal, J., and Tsaparas, P. (2005). Link analysis ranking: Algorithms, theory, and experiments. *ACM Transactions on Internet Technology, 5*(1), 231–297.

Boutilier, C. (1992). *Conditional logics for default reasoning and belief revision.* Doctoral dissertation, University of Toronto, Toronto, Canada.

Bowling, M., and Veloso, M. (2001). Rational and convergent learning in stochastic games. *IJCAI: Proceedings of the International Joint Conference on Artificial Intelligence* (pp. 1021–1026).

Braess, D. (1968). Über ein Paradoxon aus der Verkehrsplanung. *Unternehmensforschung, 12*, 258–268.

Brafman, R., Halpern, J., and Shoham, Y. (1998). On the knowledge requirements of tasks. *Journal of Artificial Intelligence, 98*(1–2), 317–350.

Brafman, R., Latombe, J. C., Moses, Y., and Shoham, Y. (1997). Applications of a logic of knowledge to motion planning under uncertainty. *Journal of the ACM, 44*(5), 633–668.

Brafman, R., and Tennenholtz, M. (2002). R-max, a general polynomial time algorithm for near-optimal reinforcement learning. *Journal of Machine Learning Research, 3*, 213–231.

Bratman, M. E. (1987). *Intention, plans, and practical reason.* Stanford, CA: CSLI Publications, Stanford University.

Brewka, G., Niemela, I., and Truszczynski, M. (2007). Nonmonotonic Reasoning. In F. van Harmelen, V. Lifschitz, B. Porter (Eds.), *Handbook of Knowledge Representation.* St. Louis, MO: Elsevier.

Brouwer, L. E. J. (1912). Über Abbildung von Mannigfaltigkeiten. *Mathematische Annalen, 71*, 97–115.

Brown, G. (1951). Iterative solution of games by fictitious play. In *Activity analysis of production and allocation.* New York: John Wiley and Sons.

Bulow, J., and Klemperer, P. (1996). Auctions versus negotiations. *The American Economic Review, 86*(1), 180–194.

Cassady, R. (1967). *Auctions and auctioneering.* University of California Press.

Cavallo, R. (2006). Optimal decision-making with minimal waste: Strategyproof redistribution of VCG payments. *AAMAS: Proceedings of the International Joint Conference on Autonomous Agents and Multiagent Systems* (pp. 882–889).

Chalkiadakis, G., and Boutilier, C. (2004). Bayesian reinforcement learning for coalition formation under uncertainty. *AAMAS: Proceedings of the International Joint Conference on Autonomous Agents and Multiagent Systems* (pp. 1090–1097).

Chellas, B. F. (1980). *Modal logic: An introduction.* Cambridge, UK: Cambridge University Press.

Chen, X., and Deng, X. (2006). Settling the complexity of 2-player Nash-equilibrium. *FOCS: Proceedings of the Annual IEEE Symposium on Foundations of Computer Science.*

Clarke, E. H. (1971). Multipart pricing of public goods. *Public Choice, 11,* 17–33.

Claus, C., and Boutilier, C. (1998). The dynamics of reinforcement learning in cooperative multiagent systems. *AAAI: Proceedings of the AAAI Conference on Artificial Intelligence* (pp. 746–752).

Cohen, P. R., and Levesque, H. J. (1990). Intention is choice with commitment. *Artificial Intelligence, 42*(2–3), 213–261.

Cohen, P. R., and Levesque, H. J. (1991). Teamwork. *Noûs, 25*(4), 487–512.

Conitzer, V. (2006). *Computational aspects of preference aggregation.* Doctoral dissertation, Carnegie Mellon University.

Conitzer, V., and Sandholm, T. (2003a). Complexity of determining nonemptiness of the core. *IJCAI: Proceedings of the International Joint Conference on Artificial Intelligence.*

Conitzer, V., and Sandholm, T. (2003b). Complexity results about Nash equilibria. *IJCAI: Proceedings of the International Joint Conference on Artificial Intelligence* (pp. 761–771).

Conitzer, V., and Sandholm, T. (2004). Computing Shapley values, manipulating value division schemes, and checking core membership in multi-issue domains. *AAAI: Proceedings of the AAAI Conference on Artificial Intelligence.*

Conitzer, V., and Sandholm, T. (2005). Complexity of (iterated) dominance. *EC: Proceedings of the ACM Conference on Electronic Commerce* (pp. 88–97).

Correa, J. R., Schulz, A. S., and Stier-Moses, N. E. (2005). On the inefficiency of equilibria in nonatomic congestion games. *IPCO: Proceedings of the Conference on Integer Programming and Combinatorial Optimization* (pp. 167–181).

Cramton, P., Shoham, Y., and Steinberg, R. (Eds.). (2006). *Combinatorial auctions.* Cambridge, MA: MIT Press.

Crawford, V. P., and Sobel, J. (1982). Strategic information transmission. *Econometrica, 50*(6), 1431–1451.

Daskalakis, C., Fabrikant, A., and Papadimitriou, C. (2006a). The game world is flat: The complexity of Nash equilibria in succinct games. *ICALP: Proceedings of the International Colloquium on Automata, Languages and Programming.*

Daskalakis, C., Goldberg, P. W., and Papadimitriou, C. H. (2006b). The complexity of computing a Nash equilibrium. *STOC: Proceedings of the Annual ACM Symposium on Theory of Computing.*

Daskalakis, C., and Papadimitriou, C. (2006). Computing pure Nash equilibria via Markov random fields. *EC: Proceedings of the ACM Conference on Electronic Commerce* (pp. 91–99).

d'Aspremont, C., and Gérard-Varet, L. (1979). Incentives and incomplete information. *Journal of Public Economics, 11*(1), 25–45.

Davis, R., and Smith, R. G. (1983). Negotiation as a metaphor for distributed problem solving. *Artificial Intelligence, 20,* 63–109.

Day, R., and Milgrom, P. (2008). Core-selecting package auctions. *International Journal of Game Theory, 36*(3), 393–407.

de Borda, J.-C. C. (1781). Mémoire sur les élections au scrutin. *Histoire de l'Académie Royale des Sciences.*

de Condorcet, M. J. A. N. C. (1784). Essay on the application of analysis to the probability of majority decisions.

Deng, X., and Papadimitriou, C. H. (1994). On the complexity of cooperative solution concepts. *Mathematics of Operations Research, 19,* 257.

Dubins, L., and Freedman, D. (1981). Machiavelli and the Gale-Shapley algorithm. *The American Mathematical Monthly, 88*(7), 485–494.

Edelman, B., Schwarz, M., and Ostrovsky, M. (2007). Internet advertising and the generalized second price auction: Selling billions of dollars worth of keywords. *American Economic Review, 97*(1), 242–259.

Elkind, E., Goldberg, L. A., Goldberg, P. W., and Wooldridge, M. (2007). Computational complexity of weighted threshold games. *AAAI: Proceedings of the AAAI Conference on Artificial Intelligence*.

Elkind, E., Sahai, A., and Steiglitz, K. (2004). Frugality in path auctions. *DA: Proceedings of the ACM-SIAM Symposium on Discrete Algorithms* (pp. 701–709).

Fabrikant, A., Papadimitriou, C., and Talwar, K. (2004). The complexity of pure Nash equilibria. *STOC: Proceedings of the Annual ACM Symposium on Theory of Computing*.

Fagin, R., and Halpern, J. Y. (1994). Reasoning about knowledge and probability. *Journal of the ACM, 41*(2), 340–367.

Fagin, R., Halpern, J. Y., Moses, Y., and Vardi, M. Y. (1995). *Reasoning about knowledge.* Cambridge, MA: MIT Press.

Faltings, B. (2006). Distributed Constraint Programming. In F. Rossi, P. van Beck, and T. Walsh (Eds.), *Handbook of Constraint Programming*, 699–729. Foundations of Artificial Intelligence. Elsevier.

Farrell, J. (1987). Cheap talk, coordination, and entry. *RAND Journal of Economics, 18*(1), 34–39.

Farrell, J. (1993). Meaning and credibility in cheap-talk games. *GEB: Games and Economic Behavior, 5*(4), 514–531.

Farrell, J. (1995). Talk is cheap. *American Economic Review, 85*(2), 186–190. Papers and Proceedings of the Meeting of the American Economic Association.

Farrell, J., and Rabin, M. (1996). Cheap talk. *Journal of Economic Perspectives, 10*(3), 103–118.

Feigenbaum, J., Krishnamurthy, A., Sami, R., and Shenker, S. (2003). Hardness results for multicast cost sharing. *Theoretical Computer Science, 304*(1-3), 215–236.

Feigenbaum, J., Papadimitriou, C. H., and Shenker, S. (2001). Sharing the cost of multicast transmissions. *Journal of Computer System Sciences, 63*(1), 21–41.

Feigenbaum, J., Schapira, M., and Shenker, S. (2007). Distributed algorithmic mechanism design. In [Nisan et al., 2007], chapter 14, 363–384.

Feldman, A. M., and Serrano, R. (2006). *Welfare economics and social choice theory.* Kluwer Academic Publishers.

Ferguson, G., and Allen, J. F. (1998). TRIPS: An intelligent integrated problem-solving assistant. *AAAI: Proceedings of the AAAI Conference on Artificial Intelligence* (pp. 567–573).

Filar, J., and Vrieze, K. (1997). *Competitive Markov decision processes.* Springer-Verlag.

Finin, T., Labrou, Y., and Mayfield, J. (1997). KQML as an agent communication language. In J. Bradshaw (Ed.), *Software agents.* Cambridge, MA: MIT Press.

Fitting, M., and Mendelsohn, R. L. (1999). *First-order modal logic.* New York: Springer.

Flores, F., Graves, M., Hartfield, B., and Winograd, T. (1988). Computer systems and the design of organizational interaction. *ACM Transactions on Office Information Systems, 6*(2), 153–172.

Foster, D., and Vohra, R. (1999). Regret in the on-line decision problem. *GEB: Games and Economic Behavior, 29*, 7–36.

Fudenberg, D., and Levine, D. (1995). Universal consistency and cautious fictitious play. *Journal of Economic Dynamics and Control, 19*, 1065–1089.

Fudenberg, D., and Levine, D. (1999). Conditional universal consistency. *GEB: Games and Economic Behavior, 29*, 104–130.

Fudenberg, D., and Levine, D. K. (1998). *The theory of learning in games.* Cambridge, MA: MIT Press.

Fudenberg, D., and Tirole, J. (1991). *Game theory.* Cambridge, MA: MIT Press.

Fujishima, Y., Leyton-Brown, K., and Shoham, Y. (1999a). Taming the computational complexity of combinatorial auctions: Optimal and approximate approaches. *IJCAI: Proceedings of the International Joint Conference on Artificial Intelligence* (pp. 548–553).

Fujishima, Y., McAdams, D., and Shoham, Y. (1999b). Speeding up ascending-bid auctions. *IJCAI: Proceedings of the International Joint Conference on Artificial Intelligence* (pp. 554–563).

Gaertner, W. (2006). *A primer in social choice theory.* New York: Oxford University Press.

Gale, D., and Shapley, L. S. (1962). College admissions and the stability of marriage. *The American Mathematical Monthly, 69*(1), 9–15.

Gamow, G., and Stern, M. (1958). *Puzzle math.* New York: Viking Press.

García, C. B., and Zangwill, W. I. (1981). *Pathways to solutions, fixed points and equilibria.* Englewood Cliffs, NJ: Prentice Hall.

Gärdenfors, P. (1988). *Knowledge in flux: Modeling the dynamics of epistemic states.* Cambridge, MA: Bradford Book, MIT Press.

Gärdenfors, P., and Rott, H. (1995). Belief revision. In *Handbook of logic in artificial intelligence and logic programming*, vol. 4, 35–132. Oxford, UK: Oxford University Press.

Geanakoplos, J. (2005). Three brief proofs of Arrow's impossibility theorem. *Economic Theory, 26*(1), 211–215.

Gibbard, A. (1973). Manipulation of voting schemes: A general result. *Econometrica, 41,* 587–601.

Gilboa, I. (1988). The complexity of computing best-response automata in repeated games. *Journal of Economic Theory, 45,* 342–352.

Gilboa, I., Kalai, E., and Zemel, E. (1989). *The complexity of eliminating dominated strategies* (Technical Report). Northwestern University, Center for Mathematical Studies in Economics and Management Science. Available at http://ideas.repec.org/p/nwu/cmsems/853.html.

Gilboa, I., and Zemel, E. (1989). Nash and correlated equilibria: Some complexity considerations. *GEB: Games and Economic Behavior, 1,* 80–93.

Ginsberg, M. L. (Ed.). (1987). *Readings in nonmonotonic reasoning.* San Francisco, CA: Morgan Kaufmann Publishers Inc.

Goldberg, A., Hartline, J., Karlin, A., Saks, M., and Wright, A. (2006). Competitive auctions. *GEB: Games and Economic Behavior, 55*(2), 242–269.

Goldberg, P. W., and Papadimitriou, C. H. (2006). Reducibility among equilibrium problems. *STOC: Proceedings of the Annual ACM Symposium on Theory of Computing* (pp. 61–70).

Gottlob, G., Greco, G., and Scarcello, F. (2003). Pure Nash equilibria: Hard and easy games. *TARK: Proceedings of the ACM Conference on Theoretical Aspects of Rationality and Knowledge.*

Govindan, S., and Wilson, R. (2003). A global Newton method to compute Nash equilibria. *Journal of Economic Theory, 110,* 65–86.

Govindan, S., and Wilson, R. (2005a). Essential equilibria. *Proceedings of the National Academy of Sciences USA, 102,* 15706–15711.

Govindan, S., and Wilson, R. (2005b). Refinements of Nash equilibrium. In S. Durlauf, L. Blume (Eds.), *The new Palgrave dictionary of economics*, vol. II. New York: Macmillan.

Graham, D., and Marshall, R. (1987). Collusive bidder behavior at single-object second-price and English auctions. *Journal of Political Economy, 95,* 579–599.

Gray, J. (1978). Notes on database operating systems. In R. Bayers, R. M. Graham, G. Seegmuller (Eds.), *Operating systems: An advanced course*, vol. 66 of *Lecture Notes in Computer Science.* New York: Springer-Verlag.

Green, J., and Laffont, J. (1977). Characterization of satisfactory mechanisms for the revelation of preferences for public goods. *Econometrica, 45*(2), 427–438.

Grégoire, E., and Konieczny, S. (2006). Special issue on logic-based approaches to information fusion. *Information Fusion, 7*(1), 4–18.

Grice, H. P. (1969). Utterer's meaning and intention. *The Philosophical Review, 78,* 147–177.

Grice, H. P. (1989). *Studies in the way of words.* Harvard University Press.

Groves, T. (1973). Incentives in teams. *Econometrica, 41*(4), 617–31.

Guestrin, C. E. (2003). *Planning under uncertainty in complex structured environments.* Doctoral dissertation, Department of Computer Science, Stanford University.

Guo, M., and Conitzer, V. (2007). Worst-case optimal redistribution of VCG payments. *EC: Proceedings of the ACM Conference on Electronic Commerce.*

Halpern, J. Y. (2005). *Reasoning about uncertainty.* Cambridge, MA: MIT Press.

Halpern, J. Y., and Moses, Y. (1990). Knowledge and common knowledge in a distributed environment. *Journal of the ACM (JACM)*, *37*(3), 549–587.

Hannan, J. F. (1957). Approximation to Bayes risk in repeated plays. *Contributions to the Theory of Games*, *3*, 97–139.

Hanson, R. (2003). Combinatorial Information Market Design. *Information Systems Frontiers*, *5*(1), 107–119.

Harsanyi, J. (1967–1968). Games with incomplete information played by "Bayesian" players, parts I, II and III. *Management Science*, *14*, 159–182, 320–334, 486–502.

Harstad, R., Kagel, J., and Levin, D. (1990). Equilibrium bid functions for auctions with an uncertain number of bidders. *Economics Letters*, *33*(1), 35–40.

Hart, S., and Mas-Colell, A. (2000). A simple adaptive procedure leading to correlated equilibrium. *Econometrica*, *68*, 1127–1150.

Hillas, J., and Kohlberg, E. (2002). Foundations of strategic equilibrium. In R. Aumann, S. Hart (Eds.), *Handbook of game theory*, vol. III, chapter 42, 1597–1663. Amsterdam: Elsevier.

Hintikka, J. (1962). *Knowledge and belief*. Ithaca, NY: Cornell University Press.

Hu, J., and Wellman, P. (1998). Multiagent reinforcement learning: Theoretical framework and an algorithm. *Proceedings of the Fifteenth International Conference on Machine Learning* (pp. 242–250).

Hurwicz, L. (1960). Optimality and informational efficiency in resource allocation processes. In K. Arrow, S. Karlin, P. Suppes (Eds.), *Mathematical methods in the social sciences*, 27–46. Stanford, CA: Stanford University Press.

Hurwicz, L. (1972). On informationally decentralized systems. In C. McGuire, R. Radner (Eds.), *Decision and organization*, 297–336. London: North-Holland.

Hurwicz, L. (1975). On the existence of allocation systems whose manipulative Nash equilibria are Pareto optimal. Unpublished.

Hyafil, N., and Boutilier, C. (2004). Regret minimizing equilibria and mechanisms for games with strict type uncertainty. *UAI: Proceedings of the Conference on Uncertainty in Artificial Intelligence* (pp. 268–277).

Ieong, S., and Shoham, Y. (2005). Marginal contribution nets: A compact representation scheme for coalitional games. *EC: Proceedings of the ACM Conference on Electronic Commerce*.

Ieong, S., and Shoham, Y. (2006). Multi-attribute coalitional games. *EC: Proceedings of the ACM Conference on Electronic Commerce* (pp. 170–179).

Ieong, S., and Shoham, Y. (2008). Bayesian coalitional games. *AAAI: Proceedings of the AAAI Conference on Artificial Intelligence*.

Jiang, A. X., and Leyton-Brown, K. (2006). A polynomial-time algorithm for action-graph games. *AAAI: Proceedings of the AAAI Conference on Artificial Intelligence* (pp. 679–684).

Johari, R. (2007). The price of anarchy and the design of scalable resource allocation mechanisms. In [Nisan et al., 2007], chapter 21, 543–568.

Johari, R., and Tsitsiklis, J. N. (2004). Efficiency loss in a network resource allocation game. *Mathematics of Operations Research*, *29*(3), 407–435.

Johari, R., and Tsitsiklis, J. N. (2005). Communication requirements of VCG-like mechanisms in convex environments. *Allerton Conference on Communication, Control, and Computing*.

Kaelbling, L. P., Littman, M. L., and Moore, A. P. (1996). Reinforcement learning: A survey. *JAIR: Journal of Artificial Intelligence Research*, *4*, 237–285.

Kalai, E., and Lehrer, E. (1993). Rational learning leads to Nash equilibrium. *Econometrica*, *61*(5), 1019–1045.

Katsuno, H., and Mendelzon, A. (1991). On the difference between updating a knowledge base and revising it. *KR: Proceedings of the International Conference on Knowledge Representation and Reasoning* (pp. 387–394).

Kearns, M., Littman, M., and Singh, S. (2001). Graphical models for game theory. *UAI: Proceedings of the Conference on Uncertainty in Artificial Intelligence*.

Kearns, M., and Singh, S. (1998). Near-optimal reinforcement learning in polynomial time. *ICML: Proceedings of the International Conference on Machine Learning* (pp. 260–268).

Kelly, F. P. (1997). Charging and rate control for elastic traffic. *European Transactions on Telecommunications, 8*, 33–37.

Kleinberg, J. (1999). Authoritative sources in a hyperlinked environment. *Journal of the ACM (JACM), 46*(5), 604–632.

Klemperer, P. (1999a). Auction theory: A guide to the literature. *Journal of Economic Surveys, 13*(3), 227–286.

Klemperer, P. (Ed.). (1999b). *The economic theory of auctions*. Edward Elgar.

Knuth, D. E. (1976). *Marriages stables*. Montreal: Les Presses de I'Universite de Montreal.

Kohlberg, E., and Mertens, J.-F. (1986). On the strategic stability of equilibria. *Econometrica, 54*, 1003–1038.

Koller, D., and Megiddo, N. (1992). The complexity of two-person zero-sum games in extensive form. *GEB: Games and Economic Behavior, 4*, 528–552.

Koller, D., Megiddo, N., and von Stengel, B. (1996). Efficient computation of equilibria for extensive two-person games. *GEB: Games and Economic Behavior, 14*, 247–259.

Koller, D., and Milch, B. (2003). Multi-agent influence diagrams for representing and solving games. *GEB: Games and Economic Behavior, 45*(1), 181–221.

Koller, D., and Pfeffer, A. (1995). Generating and solving imperfect information games. *IJCAI: Proceedings of the International Joint Conference on Artificial Intelligence* (pp. 1185–1193).

Korf, R. E. (1990). Real-time heuristic search. *Artificial Intelligence, 42*(2-3), 189–211.

Koutsoupias, E., and Papadimitriou, C. H. (1999). Worst-case equilibria. *Proceedings of the 16th annual symposium on Theoretical Aspects of Computer Science (STACS)* (pp. 404–413).

Kreps, D., and Wilson, R. (1982). Sequential equilibria. *Econometrica, 50*, 863–894.

Kreps, D. M. (1988). *Notes on the theory of choice*. Boulder, CO: Westview Press.

Krishna, V. (2002). *Auction theory*. New York: Elsevier Science.

Krishna, V., and Perry, M. (1998). *Efficient Mechanism Design* (Technical Report). Pennsylvania State University.

Kuhn, H. (1953). Extensive games and the problem of information. *Contributions to the Theory of Games* (pp. 193–216). Princeton, NJ: Princeton University Press. Reprinted in H. Kuhn (Ed.), *Classics in Game Theory*, Princeton, NJ: Princeton University Press, 1997.

La Mura, P. (2000). Game networks. *UAI: Proceedings of the Conference on Uncertainty in Artificial Intelligence* (pp. 335–342).

Lamarre, P., and Shoham, Y. (1994). Knowledge, certainty, belief, and conditionalisation. *KR: Proceedings of the International Conference on Knowledge Representation and Reasoning* (pp. 415–424).

Lehmann, D., Müller, R., and Sandholm, T. (2006). The winner determination problem. In [Cramton et al., 2006], chapter 12, 297–318.

Lehmann, D., O'Callaghan, L., and Shoham, Y. (2002). Truth revelation in approximately efficient combinatorial auctions. *JACM: Journal of the ACM, 49*(5), 577–602.

Lemke, C. (1978). Some pivot schemes for the linear complementarity problem. *Mathematical Programming Study, 7*, 15–35.

Lemke, C., and Howson, J. (1964). Equilibrium points of bimatrix games. *Society for Industrial and Applied Mathematics Journal of Applied Mathematics, 12*, 413–423.

Littman, M., Ravi, N., Talwar, A., and Zinkevich, M. (2006). An efficient optimal equilibrium algorithm for two-player game trees. *UAI: Proceedings of the Conference on Uncertainty in Artificial Intelligence*.

Littman, M. L. (1994). Markov games as a framework for multi-agent reinforcement learning. *Proceedings of the 11th International Conference on Machine Learning* (pp. 157–163).

Littman, M. L. (2001). Friend-or-foe Q-learning in general-sum games. *ICML: Proceedings of the International Conference on Machine Learning*.

Littman, M. L., and Szepesvari, C. (1996). A generalized reinforcement-learning model: Convergence and applications. *Proceedings of the 13th International Conference on Machine Learning* (pp. 310–318).

Loomes, G., and Sugden, R. (1982). Regret theory: An alternative theory of rational choice under uncertainty. *Economic Journal*, *92*, 805–824.

Luce, R., and Raiffa, H. (1957a). *Games and decisions*. New York: John Wiley and Sons.

Luce, R. D., and Raiffa, H. (1957b). *Games and decisions: Introduction and critical survey*. New York: John Wiley and Sons.

Mas-Colell, A., Whinston, M. D., and Green, J. R. (1995). *Microeconomic theory*. Oxford: Oxford University Press.

Maynard-Reid, P., and Shoham, Y. (2001). Belief fusion: Aggregating pedigreed belief states. *Journal of Logic, Language and Information*, *10*(2), 183–209.

Maynard Smith, J. (1982). *Evolution and the theory of games*. Cambridge University Press.

Maynard Smith, J., and Price, G. R. (1973). The logic of animal conflict. *Nature*, *246*, 15–18.

Maynard-Zhang, P., and Lehmann, D. (2003). Representing and aggregating conflicting beliefs. *JAIR: Journal of Artificial Intelligence Research*, *19*, 155–203.

McAfee, R., and MacMillan, J. (1987). Auctions and bidding. *Journal of Economic Literature*, *25*(3), 699–738.

McAfee, R., and McMillan, J. (1987). Auctions with a stochastic number of bidders. *Journal of Economic Theory*, *43*, 1–19.

McAfee, R., and McMillan, J. (1992). Bidding rings. *American Economic Review*, *82*, 579–599.

McCarthy, J. (1994). A programming language based on speech acts. Stanford University, `http://www-formal.stanford.edu/jmc/elephant/elephant.html`.

McGrew, R., and Shoham, Y. (2004). Using contracts to influence the outcome of a game. *AAAI: Proceedings of the AAAI Conference on Artificial Intelligence* (pp. 238–244).

McKelvey, R., and McLennan, A. (1996). Computation of equilibria in finite games. *Handbook of Computational Economics*, *1*, 87–142.

McKelvey, R. D., McLennan, A. M., and Turocy, T. L. (2006). Gambit: Software tools for game theory. `http://econweb.tamu.edu/gambit`.

Megiddo, N. and Papadimitriou, C.H. (1991). A note on total functions, existence theorems, and computational complexity. *Theoretical Computer Science*, *81*(1), 317–324.

Meisels, A. (2008). *Distributed search by constrained agents*. Springer-Verlag.

Meyer, J.-J. C., van der Hoek, W., and van Linder, B. (1999). A logical approach to the dynamics of commitments. *Artificial Intelligence*, *113*(1-2), 1–40.

Milgrom, P., and Weber, R. (1982). A theory of auctions and competitive bidding. *Econometrica*, *50*(5), 1089–1122.

Milgrom, P., and Weber, R. (2000). A theory of auctions and competitive bidding, II. In P. Klemperer (Ed.), *The economic theory of auctions*. Edward Elgar.

Milgrom, P. R. (1981). Rational expectations, information acquisition, and competitive bidding. *Econometrica*, *49*, 921–943.

Monderer, D., and Shapley, L. (1996a). Potential games. *GEB: Games and Economic Behavior*, *14*, 124–143.

Monderer, D., and Shapley, L. S. (1996b). Fictitious play property for games with identical interests. *Journal of Economic Theory*, *1*, 258–265.

Monderer, D., and Tennenholtz, M. (2003). *k*-implementation. *EC: Proceedings of the ACM Conference on Electronic Commerce* (pp. 19–28). San Diego, CA: ACM Press.

Monderer, D., and Tennenholtz, M. (2006). Strong mediated equilibrium. *AAAI: Proceedings of the AAAI Conference on Artificial Intelligence*.

Moore, R. C. (1985). A formal theory of knowledge and action. In *Formal theories of the commonsense world*. Ablex Publishing Corporation.

Moulin, H. (1994). Social choice. In R. Aumann, S. Hart (Eds.), *Handbook of game theory with economic applications*, vol. II. New York: Elsevier.

Muller, E., and Satterthwaite, M. (1977). The equivalence of strong positive association and strategy-proofness. *Journal of Economic Theory*, *14*, 412–418.

Müller, R. (2006). Tractable cases of the winner determination problem. In [Cramton et al., 2006], chapter 13, 319–336.

Myerson, R. (1978). Refinements of the Nash equilibrium concept. *International Journal of Game Theory*, *7*, 73–80.

Myerson, R. (1979). Incentive compatibility and the bargaining problem. *Econometrica*, *47*(1), 61–73.

Myerson, R. (1981). Optimal auction design. *Mathematics of Operations Research*, *6*(1), 58–73.

Myerson, R. (1982). Optimal coordination mechanisms in generalized principal-agent models. *Journal of Mathematical Economics*, *11*, 67–81.

Myerson, R. (1986). Multistage games with communication. *Econometrica*, *54*(2), 323–358.

Myerson, R. (1991). *Game theory: Analysis of conflict*. Harvard Press.

Myerson, R., and Satterthwaite, M. (1983). Efficient mechanisms for bilateral trading. *Journal of Economic Theory*, *29*(2), 265–281.

Myerson, R. B. (2007). Virtual utility and the core for games with incomplete information. *Journal of Economic Theory*, *136*(1), 260–285.

Nachbar, J. (1990). Evolutionary selection dynamics in games: Convergence and limit properties. *International Journal of Game Theory*, *19*, 59–89.

Nachbar, J. H., and Zame, W. R. (1996). Non-computable strategies and discounted repeated games. *Journal of Economic Theory*, *8*(1), 103–122.

Nanson, E. J. (1882). Methods of election. *Transactions and Proceedings of the Royal Society of Victoria*, *18*, 197–240.

Nash, J. (1950). Equilibrium points in n-person games. *Proceedings of the National Academy of Sciences USA*, *36*, 48–49. Reprinted in H. Kuhn (Ed.), *Classics in Game Theory*, Princeton, NJ: Princeton University Press, 1997.

Nash, J. (1951). Non-cooperative games. *Annals of Mathematics*, *54*, 286–295.

Neyman, A. (1985). Bounded complexity justifies cooperation in finitely repeated prisoner's dilemma. *Economic Letters*, 227–229.

Neyman, A., and Sorin, S. (2003). *Stochastic games and applications*. Kluwer Academic Press.

Nisan, N. (2006). Bidding languages for combinatorial auctions. In [Cramton et al., 2006], chapter 9, 215–232.

Nisan, N. (2007). Introduction to mechanism design (for computer scientists). In [Nisan et al., 2007], chapter 9, 209–242.

Nisan, N., and Ronen, A. (2001). Algorithmic mechanism design. *GEB: Games and Economic Behavior*, *35*(1-2), 166–196.

Nisan, N., and Ronen, A. (2007). Computationally feasible VCG mechanisms. *JAIR: Journal of Artificial Intelligence Research*, *29*, 19–47.

Nisan, N., Roughgarden, T., Tardos, E., and Vazirani, V. (Eds.). (2007). *Algorithmic game theory*. Cambridge, UK: Cambridge University Press.

Nudelman, E., Wortman, J., Shoham, Y., and Leyton-Brown, K. (2004). Run the GAMUT: A comprehensive approach to evaluating game-theoretic algorithms. *AAMAS: Proceedings of the International Joint Conference on Autonomous Agents and Multiagent Systems* (pp. 880–887).

Osborne, M. J., and Rubinstein, A. (1994). *A course in game theory*. Cambridge, MA: MIT Press.

Page, L., Brin, S., Motwani, R., and Winograd, T. (1998). *The PageRank citation ranking: Bringing order to the web* (Technical Report). Stanford Digital Library Technologies Project.

Papadimitriou, C. (2005). Computing correlated equilibria in multiplayer games. *STOC: Proceedings of the Annual ACM Symposium on Theory of Computing*.

Papadimitriou, C., and Yannakakis, M. (1994). On bounded rationality and computational complexity. *STOC: Proceedings of the Annual ACM Symposium on Theory of Computing* (pp. 726–733).

Papadimitriou, C. H. (1992). On players with a bounded number of states. *GEB: Games and Economic Behavior*, *4*(1), 122–131.

Papadimitriou, C. H. (1994). On the complexity of the parity argument and other inefficient proofs of existence. *Journal of Computer and System Sciences, 48*(3), 498–532.

Pappas, P. (2007). Belief revision. In F. van Harmelen, V. Lifschitz, B. Porter (Eds.), *Handbook of knowledge representation*. St. Louis, MO: Elsevier.

Parikh, P. (2001). *The use of language*. Stanford, CA: CSLI Publications.

Parkes, D. (2001). *Iterative combinatorial auctions: Achieving economic and computational efficiency*. Doctoral dissertation, University of Pennsylvania.

Parkes, D. C. (2006). Iterative combinatorial auctions. In [Cramton et al., 2006], chapter 2, 41–78.

Parkes, D. C., and Ungar, L. H. (2000). Iterative combinatorial auctions: Theory and practice. *AAAI: Proceedings of the AAAI Conference on Artificial Intelligence* (pp. 74–81).

Pauly, M. (2002). A modal logic for coalitional power in games. *Logic and Computation, 12*(1), 149–166.

Pearce, D. (1984). Rationalizable strategic behavior and the problem of perfection. *Econometrica, 52*, 1029–1050.

Peirce, C., Hartshorne, C., and Weiss, P. (1965). *Collected papers of Charles Sanders Peirce*. Cambridge, MA: Harvard University Press.

Peleg, B., and Sudhölter, P. (2003). *Introduction to the theory of cooperative games*. Kluwer Academic Publishers.

Pennock, D. (2004). A dynamic pari-mutuel market for hedging, wagering, and information aggregation. *EC: Proceedings of the ACM Conference on Electronic Commerce*.

Pigou, A. C. (1920). *The economics of welfare*. Macmillan.

Poole, D., Mackworth, A., and Goebel, R. (1997). *Computational intelligence: A logical approach*. Oxford, UK: Oxford University Press.

Porter, R., Nudelman, E., and Shoham, Y. (2004a). Simple search methods for finding a Nash equilibrium. *AAAI: Proceedings of the AAAI Conference on Artificial Intelligence* (pp. 664–669).

Porter, R., Shoham, Y., and Tennenholtz, M. (2004b). Fair imposition. *Journal of Economic Theory, 118*(2), 209–228.

Powers, R., and Shoham, Y. (2005a). Learning against opponents with bounded memory. *IJCAI: Proceedings of the International Joint Conference on Artificial Intelligence*.

Powers, R., and Shoham, Y. (2005b). New criteria and a new algorithm for learning in multi-agent systems. In *Advances in Neural Information Processing Systems 17*. MIT Press.

Rabin, M. (1990). Communication between rational agents. *Journal of Economic Theory, 51*(1), 144–170.

Rao, A. S., and Georgeff, M. P. (1991). Modeling rational agents within a BDI-architecture. *KR: Proceedings of the International Conference on Knowledge Representation and Reasoning* (pp. 473–484).

Rao, A. S., and Georgeff, M. P. (1998). Decision procedures for BDI logics. *Logic and Computation, 8*(3), 293–342.

Rastegari, B., Condon, A., and Leyton-Brown, K. (2007). Revenue monotonicity in combinatorial auctions. *AAAI: Proceedings of the AAAI Conference on Artificial Intelligence* (pp. 122–127).

Riley, J., and Samuelson, W. (1981). Optimal auctions. *The American Economic Review, 71*(3), 381–392.

Roberts, K. (1979). The characterization of implementable choice rules. *Aggregation and Revelation of Preferences*, 321–348.

Robinson, J. (1951). An iterative method of solving a game. *Annals of Mathematics, 54*, 298–301.

Rosenthal, R. (1981). Games of perfect information, predatory pricing and the chain-store paradox. *Journal of Economic Theory, 25*(1), 92–100.

Rosenthal, R. W. (1973). A class of games possessing pure-strategy Nash equilibria. *International Journal of Game Theory, 2*, 65–67.

Roth, A. E. (1984). The evolution of the labor market for medical interns and residents: A case study in game theory. *Journal of Political Economy, 92*, 991–1016.

Roth, A. E., and Sotomayor, M. A. O. (1990). *Two-sided matching: A study in game-theoretic modeling and analysis.* Cambridge University Press.

Roughgarden, T. (2005). *Selfish routing and the price of anarchy.* Cambridge, MA: MIT Press.

Roughgarden, T., and Tardos, E. (2004). Bounding the inefficiency of equilibria in nonatomic congestion games. *GEB: Games and Economic Behavior, 47*(2), 389–403.

Rubinstein, A. (1998). *Modeling bounded rationality.* Cambridge, MA: MIT Press.

Rubinstein, A. (2000). *Economics and language: Five essays.* Cambridge University Press.

Russell, S., and Norvig, P. (2003). *Artificial intelligence: A modern approach, 2nd edition.* Englewood Cliffs, NJ: Prentice Hall.

Sandholm, T. (2006). Optimal winner determination algorithms. In [Cramton et al., 2006], chapter 14, 337–368.

Sandholm, T., and Boutilier, C. (2006). Preference elicitation in combinatorial auctions. In [Cramton et al., 2006], chapter 10, 233–264.

Sandholm, T. W. (1993). An implementation of the contract net protocol based on marginal cost calculations. *Proceedings of the 12th International Workshop on Distributed Artificial Intelligence* (pp. 295–308).

Sandholm, T. W. (1998). Contract types for satisficing task allocation: I Theoretical results. *AAAI: Proceedings of the AAAI Conference on Artificial Intelligence.*

Satterthwaite, M. (1975). Strategy-proofness and Arrow's conditions: Existence and correspondence theorems for voting procedures and social welfare functions. *Journal of Economic Theory, 10*, 187–217.

Savage, L. J. (1954). *The foundations of statistics.* New York: John Wiley and Sons. (2nd edition, Mineola, NY: Dover Press, 1972).

Savani, R., and von Stengel, B. (2004). Exponentially many steps for finding a Nash equilibrium in a bimatrix game. *FOCS: Proceedings of the Annual IEEE Symposium on Foundations of Computer Science* (pp. 258–267).

Scarf, H. (1967). The approximation of fixed points of continuous mappings. *SIAM Journal of Applied Mathematics, 15*, 1328–1343.

Schelling, T. C. (1960). *The strategy of conflict.* Cambridge, MA: Harvard University Press.

Schmeidler, D. (1973). Equilibrium points of nonatomic games. *Journal of Statistical Physics, 7*(4), 295–300.

Schummer, J., and Vohra, R. V. (2007). Mechanism design without money. In [Nisan et al., 2007], chapter 10, 243–265.

Schuster, P., and Sigmund, K. (1982). Replicator dynamics. *Theoretical Biology, 100*, 533–538.

Searl, J. R. (1979). *Expression and meaning: Studies in the theory of speech acts.* London: Cambridge University Press.

Segal, I. (2006). The communication requirements of combinatorial allocation problems. In [Cramton et al., 2006], chapter 11, 265–295.

Selten, R. (1965). Spieltheoretische Behandlung eines Oligopolmodells mit Nachfrageträgheit. *Zeitschrift für die gesamte Staatswissenschaft, 12*, 301–324.

Selten, R. (1975). Reexamination of the perfectness concept for equilibrium points in extensive games. *International Journal of Game Theory, 4*, 25–55.

Shapley, L. (1964). Some topics in two-person games. In M. Drescher, L. Shapley, A. Tucker (Eds.), *Advances in game theory.* Princeton, NJ: Princeton University Press.

Shapley, L. (1974). *A note on the Lemke-Howson algorithm.* RAND.

Shapley, L. S. (1953). Stochastic games. *Proceedings of the National Academy of Sciences, 39*, 1095–1100.

Shoham, Y. (1993). Agent oriented programming. *Artificial Intelligence, 60*(1), 51–92.

Shoham, Y. (1997). *Rational programming* (Technical Report). Stanford University Computer Science Department, Stanford, CA.

Shoham, Y., Powers, W. R., and Grenager, T. (2007). If multiagent learning is the answer, what is the question? *Artificial Intelligence, 171*(1). Special issue on foundations of multiagent learning.

Shoham, Y., and Tennenholtz, M. (1995). On social laws for artificial agent societies: Off-line design. *Artificial Intelligence, 73*(1-2), 231–252.

Shoham, Y., and Tennenholtz, M. (1997). On the emergence of social conventions: Modeling, analysis and simulations. *Journal of Artificial Intelligence, 94*(1-2), 139–166.

Singh, M. P. (1992). A critical examination of the Cohen–Levesque theory of intentions. *ECAI: Proceedings of the European conference on Artificial intelligence* (pp. 364–368).

Smith, J. H. (1973). Aggregation of preferences with variable electorate. *Econometrica, 41,* 1027–1041.

Smith, R. G. (1980). The contract net protocol: High-level communication and control in a distributed problem solver. *IEEE Transactions on Computers, C-29*(12), 1104–1113.

Solan, E., and Vohra, R. (2002). Correlated equilibrium payoffs and public signalling in absorbing games. *International Journal of Game Theory, 31,* 91–121.

Spann, M., and Skiera, B. (2003). Internet-based virtual stock markets for business forecasting. *Management Science, 49*(10), 1310–1326.

Spence, A. M. (1973). Job market signaling. *Quarterly Journal of Economics, 87,* 355–374.

Sperner, E. (1928). Neuer Beweis fur die Invarianz der Dimensionszahl und des Gebietes. *Abhandlungen aus dem Mathematischen Seminar der Hamburgischen Universität, 6,* 265–272.

Stalnaker, J. M. (1953). The matching program for intern replacement: The second year of operation. *Journal of Medical Education, 28,* 13–19.

Suijs, J., Borm, P., Wagenaere, A. D., and Tijs, S. (1999). Cooperative games with stochastic payoffs. *European Journal of Operations Research, 113,* 193–205.

Taylor, P., and Jonker, L. B. (1978). Evolutionarily stable strategies and game dynamics. *Mathematical Biosciences, 40,* 145–156.

van Benthem, J. (1997). *Exploring logical dynamics.* Stanford, CA: Center for the Study of Language and Information.

van der Hoek, W., and Pauly, M. (2006). Modal logic for games and information. In P. Blackburn, van J. Benthem, F. Wolter (Eds.), *Handbook of modal logic, 3.* St. Louis, MO: Elsevier.

van der Hoek, W., and Wooldridge, M. (2003). Towards a logic of rational agency. *Logic Journal of the IGPL, 11*(2), 135–159.

Varian, H. (2007). Position auctions. *International Journal of Industrial Organization, 25*(6), 1163–1178.

Vickrey, W. (1961). Counterspeculation, auctions, and competitive sealed tenders. *The Journal of Finance, 16*(1), 8–37.

Vlassis, N., Elhorst, R., and Kok, J. (2004). Anytime algorithms for multiagent decision making using coordination graphs. *SMC: IEEE Transactions on Systems, Man, and Cybernetics.* The Hague, The Netherlands.

Vohra, R., and Wellman, M. P. (Eds.). (2007). *Special issue on foundations of multiagent learning,* vol. 171. New York: Elsevier.

von Neumann, J. (1928). Zur Theorie der Gesellschaftsspiele. *Mathematische Annalen, 100,* 295–320.

von Neumann, J., and Morgenstern, O. (1944). *Theory of games and economic behavior.* Princeton, NJ: Princeton University Press.

von Neumann, J., and Morgenstern, O. (1947). *Theory of games and economic behavior, 2nd edition.* Princeton, NJ: Princeton University Press.

von Stackelberg, H. F. (1934). *Marktform und Gleichgewicht (market and equilibrium).* Vienna: Julius Springer.

von Stengel, B. (1996). Efficient computation of behavior strategies. *GEB: Games and Economic Behavior, 14,* 220–246.

von Stengel, B. (2002). Computing equilibria for two-person games. In R. Aumann, S. Hart (Eds.), *Handbook of game theory,* vol. III, chapter 45, 1723–1759. Amsterdam: Elsevier.

Vu, T., Powers, R., and Shoham, Y. (2006). Learning against multiple opponents. *AAMAS: Proceedings of the International Joint Conference on Autonomous Agents and Multiagent Systems.*

Wardrop, J. G. (1952). Some theoretical aspects of road traffic research. *Proceedings of the Institute of Civil Engineers, Pt. II* (pp. 325–378).

Watkins, C. J. (1989). *Learning from delayed rewards.* Doctoral dissertation, Cambridge University.

Watkins, C. J. C. H., and Dayan, P. (1992). Technical note: Q-learning. *Machine Learning, 8,* 279–292.

Wellman, M. P. (1993). A market-oriented programming environment and its application to distributed multicommodity flow problems. *JAIR: Journal of Artificial Intelligence Research, 1,* 1–23.

Wellman, M. P., Walsh, W. E., Wurman, P. R., and MacKie-Mason, J. K. (2001). Auction protocols for decentralized scheduling. *GEB: Games and Economic Behavior, 35*(1-2), 271–303.

Williams, S. (1999). A characterization of efficient, Bayesian incentive compatible mechanisms. *Economic Theory, 14*(1), 155–180.

Wilson, R. (1969). Competitive bidding with disparate information. *Management Science, 15*(7), 446–448.

Wilson, R. (1987a). Auction theory. In J. Eatwell, M. Milgate, P. Newman (Eds.), *The new Palgrave dictionary of economics*, vol. I. London: Macmillan.

Wilson, R. (1987b). Game-theoretic approaches to trading processes. *Advances in Economic Theory: Fifth World Congress* (pp. 33–77).

Winograd, T., and Flores, C. F. (1986). *Understanding computers and cognition: A new foundation for design.* Ablex Publishing Corporation.

Wooldridge, M. (2000). *Reasoning about rational agents.* Cambridge, MA: The MIT Press.

Wurman, P., Wellman, M., and Walsh, W. (2001). A parametrization of the auction design space. *GEB: Games and Economic Behavior, 35*(1-2), 304–338.

Yokoo, M. (1994). Weak-commitment search for solving constraint satisfaction problems. *AAAI: Proceedings of the AAAI Conference on Artificial Intelligence* (pp. 313–318).

Yokoo, M. (2001). *Distributed constraint satisfaction: Foundations of cooperation in multi-agent systems.* Springer-Verlag.

Yokoo, M. (2006). Pseudonymous bidding in combinatorial auctions. In [Cramton et al., 2006], chapter 7, 161–188.

Yokoo, M., and Hirayama, K. (2000). Algorithms for distributed constraint satisfaction: A review. *AAMAS: Proceedings of the International Joint Conference on Autonomous Agents and Multiagent Systems* (pp. 185–207).

Yokoo, M., and Ishida, T. (1999). Search algorithms for agents. In G. Weiss (Ed.), *Multiagent systems: A modern approach to distributed artificial intelligence*, 165–200. Cambridge, MA: MIT Press.

Young, H. P. (2004). *Strategic learning and its limits.* Oxford University Press.

Zermelo, E. F. F. (1913). Über eine Anwendung der Mengenlehre auf die Theorie des Schachspiels. *Fifth International Congress of Mathematicians, II,* 501–504.

Index